Extracting Bioactive Compounds for Food Products

Theory and Applications

Contemporary Food Engineering

Series Editor

Professor Da-Wen Sun, Director
Food Refrigeration & Computerized Food Technology
National University of Ireland, Dublin
(University College Dublin)
Dublin, Ireland
http://www.ucd.ie/sun/

Extracting Bioactive Compounds for Food Products: Theory and Applications, *edited by M. Angela A. Meireles* (2009)

Advances in Food Dehydration, *edited by Cristina Ratti* (2009)

Optimization in Food Engineering, *edited by Ferruh Erdoğdu* (2009)

Optical Monitoring of Fresh and Processed Agricultural Crops, *edited by Manuela Zude* (2009)

Food Engineering Aspects of Baking Sweet Goods, *edited by Servet Gülüm Sumnu and Serpil Sahin* (2008)

Computational Fluid Dynamics in Food Processing, *edited by Da-Wen Sun* (2007)

Contemporary Food Engineering Series
Da-Wen Sun, Series Editor

Extracting Bioactive Compounds for Food Products

Theory and Applications

Edited by
M. Angela A. Meireles

CRC Press
Taylor & Francis Group
Boca Raton London New York

CRC Press is an imprint of the
Taylor & Francis Group, an **informa** business

CRC Press
Taylor & Francis Group
6000 Broken Sound Parkway NW, Suite 300
Boca Raton, FL 33487-2742

Printed in the United States of America on acid-free paper
10 9 8 7 6 5 4 3 2 1

International Standard Book Number-13: 978-1-4200-6237-3 (Hardcover)

Library of Congress Cataloging-in-Publication Data

Extracting bioactive compounds for food products : theory and applications / edited by M. Angela A. Meireles.
p. ; cm. -- (Contemporary food engineering ; 6)
Includes bibliographical references and index.
ISBN-13: 978-1-4200-6237-3 (hardcover : alk. paper)
ISBN-10: 1-4200-6237-9 (hardcover : alk. paper)
1. Extraction (Chemistry) 2. Food industry and trade. I. Meireles, M. Angela A. II. Title III. Series.
[DNLM: 1. Plant Extracts--isolation & purification. 2. Food Technology--methods. 3. Food-Processing Industry--methods. 4. Plants, Edible. QV 766 E96 2009]

TP371.E98 2009
664'.022--dc22 2008025560

Visit the Taylor & Francis Web site at
http://www.taylorandfrancis.com

and the CRC Press Web site at
http://www.crcpress.com

To my husband, Ademir, and my sons, Marcelo and Guilherme

Contents

Series Preface ix
Series Editor xi
Preface xiii
Editor xv
Contributors xvii
Acknowledgments xix

Chapter 1 Extraction and Purification of Bioactive Compounds 1

M. Angela A. Meireles

Chapter 2 Steam Distillation Applied to the Food Industry 9

Manuel G. Cerpa, Rafael B. Mato, Maria José Cocero, Roberta Ceriani, Antonio J. A. Meirelles, Juliana M. Prado, Patrícia F. Leal, Thais M. Takeuchi, and M. Angela A. Meireles

Chapter 3 Distillation Applied to the Processing of Spirits and Aromas 75

Antonio J. A. Meirelles, Eduardo A. C. Batista, Helena F. A. Scanavini, Fábio R. M. Batista, Roberta Ceriani, and Luiz F. L. Luz, Jr.

Chapter 4 Low-Pressure Solvent Extraction (Solid–Liquid Extraction, Microwave Assisted, and Ultrasound Assisted) from Condimentary Plants 137

Thais M. Takeuchi, Camila G. Pereira, Mara E. M. Braga, Mário R. Maróstica, Jr., Patrícia F. Leal, and M. Angela A. Meireles

Chapter 5 Liquid–Liquid Extraction Applied to the Processing of Vegetable Oil 219

Eduardo A. C. Batista, Antonio J. A. Meirelles, Christianne E. C. Rodrigues, and Cintia B. Gonçalves

Chapter 6 Supercritical and Pressurized Fluid Extraction Applied to the Food Industry 269

Paulo T. V. Rosa, Juan Carlos Parajó, Herminia Domínguez, Andrés Moure, Beatriz Díaz-Reinoso, Richard L. Smith, Jr., Masaaki Toyomizu, Louw J. Florusse, Cor J. Peters, Motonobu Goto, Susana Lucas, and M. Angela A. Meireles

Chapter 7 Concentration of Bioactive Compounds by Adsorption/Desorption ... 403

Lourdes Calvo and María José Cocero

Index 441

Series Preface

CONTEMPORARY FOOD ENGINEERING

Food engineering is the multidisciplinary field of applied physical sciences combined with the knowledge of product properties. Food engineers provide the technological knowledge transfer essential to the cost-effective production and commercialization of food products and services. In particular, food engineers develop and design processes and equipment in order to convert raw agricultural materials and ingredients into safe, convenient, and nutritious consumer food products. However, food engineering topics are continuously undergoing changes to meet diverse consumer demands, and the subject is being rapidly developed to reflect market needs.

In the development of food engineering, one of the many challenges is to employ modern tools and knowledge, such as computational materials science and nanotechnology, to develop new products and processes. Simultaneously, improving food quality, safety, and security remain critical issues in food engineering study. New packaging materials and techniques are being developed to provide more protection to foods, and novel preservation technologies are emerging to enhance food security and defense. Additionally, process control and automation regularly appear among the top priorities identified in food engineering. Advanced monitoring and control systems are developed to facilitate automation and flexible food manufacturing. Furthermore, energy saving and minimization of environmental problems continue to be important food engineering issues, and significant progress is being made in waste management, efficient utilization of energy, and reduction of effluents and emissions in food production.

Consisting of edited books, the *Contemporary Food Engineering* book series attempts to address some of the recent developments in food engineering. Advances in classical unit operations in engineering applied to food manufacturing are covered as well as such topics as progress in the transport and storage of liquid and solid foods; heating, chilling, and freezing of foods; mass transfer in foods; chemical and biochemical aspects of food engineering and the use of kinetic analysis; dehydration, thermal processing, nonthermal processing, extrusion, liquid food concentration, membrane processes and applications of membranes in food processing; shelf-life, electronic indicators in inventory management, and sustainable technologies in food processing; and packaging, cleaning, and sanitation. The books aim at professional food scientists, academics researching food engineering problems, and graduate-level students.

The editors of the books are leading engineers and scientists from many parts of the world. All the editors were asked to present their books in a manner that will address the market need and pinpoint the cutting-edge technologies in food engineering. Furthermore, all contributions are written by internationally renowned experts who have both academic and professional credentials. All authors have attempted to provide critical, comprehensive, and readily accessible information on the art and science of a relevant topic in each chapter, with reference lists to be used by readers for further information. Therefore, each book can serve as an essential reference source to students and researchers in universities and research institutions.

Da-Wen Sun
Series Editor

Series Editor

Born in Southern China, Professor Da-Wen Sun is a world authority on food engineering research and education. His main research activities include cooling, drying, and refrigeration processes and systems, quality and safety of food products, bioprocess simulation and optimization, and computer vision technology. Especially, his innovative studies on vacuum cooling of cooked meats, pizza quality inspection by computer vision, and edible films for shelf-life extension of fruits and vegetables have been widely reported in national and international media. Results of his work have been published in over 180 peer-reviewed journal papers and more than 200 conference papers.

He received a first class BSc Honours and MSc in mechanical engineering, and a PhD in chemical engineering in China before working in various universities in Europe. He became the first Chinese national to be permanently employed in an Irish university when he was appointed college lecturer at National University of Ireland, Dublin (University College Dublin) in 1995, and was then continuously promoted in the shortest possible time to senior lecturer, associate professor, and full professor. Sun is now professor of Food Biosystems Engineering and director of the Food Refrigeration and Computerized Food Technology Research Group at University College Dublin.

As a leading educator in food engineering, Sun has contributed significantly to the field of food engineering. He has trained many PhD students, who have made their own contributions to the industry and academia. He has also given lectures on advances in food engineering on a regular basis in academic institutions internationally and delivered keynote speeches at international conferences. As a recognized authority in food engineering, he has been conferred adjunct/visiting/consulting professorships from ten top universities in China including Zhejiang University, Shanghai Jiaotong University, Harbin Institute of Technology, China Agricultural University, South China University of Technology, and Jiangnan University. In recognition of his significant contribution to food engineering worldwide and for his outstanding leadership in the field, the International Commission of Agricultural Engineering (CIGR) awarded him the CIGR Merit Award in 2000 and again in 2006. The Institution of Mechanical Engineers (IMechE) based in the United Kingdom named him Food Engineer of the Year 2004. In 2008 he was awarded the CIGR Recognition Award in honor of his distinguished achievements in the top one percent of agricultural engineering scientists in the world.

He is a fellow of the Institution of Agricultural Engineers. He has also received numerous awards for teaching and research excellence, including the President's Research Fellowship, and he twice received the President's Research Award of University College Dublin. He is a member of the CIGR Executive Board and honorary vice-president of CIGR, editor-in-chief of *Food and Bioprocess Technology—An International Journal* (Springer), series editor of the *Contemporary Food Engineering* book series (CRC Press/Taylor & Francis), former editor of *Journal of Food Engineering* (Elsevier), and editorial board member for *Journal of Food Engineering* (Elsevier), *Journal of Food Process Engineering* (Blackwell), *Sensing and Instrumentation for Food Quality and Safety* (Springer), and *Czech Journal of Food Sciences.* He is also a chartered engineer registered in the U.K. Engineering Council.

Preface

Bioactive compounds found in extracts of a variety of vegetable matrices, such as bulbs, flowers, fruits, leaves, seeds, stems, and other botanical fruits, are presently used in a variety of formulations for the food, cosmetic, and pharmaceutical industries. In some cases, the same extract or purified compound is used as food seasoning, for instance, the turmeric oleoresin that is a seasoning agent for the food industry is used to impart color in cosmetic or pharmaceutical formulations. With the growing concern of the population about the benefits of a balanced diet, new product developers are seeking bioactive compounds that can be used for their functional properties. Because of the value society is nowadays imparting to products made from natural resources and using technologically friendly processes (that is, green processes), some classical unit operations, such as steam distillation, require improvement from the process design point of view in order to fulfill the consumers' demands, while simultaneously emerging technologies are also considered as a viable alternative to produce certain extracts/bioactive compounds. For instance, in spite of steam distillation being an ancient process for producing volatile oil, there are innumerable opportunities to improve the process. Additionally, supercritical fluid technology may be the answer for obtaining bioactive ingredients from some solid matrices.

This book was organized for engineers and technologists working with the development of extraction processes for obtaining bioactive mixtures/compounds. The core idea was to have the book cover subjects that are not traditionally covered in the unit operations reference books, such as the application of extraction techniques to obtain bioactive compounds. Therefore, in this volume the reader will get an overview of the fundamentals of heat and mass transfer as well as the thermodynamics of the processes of steam distillation, distillation, low-pressure solvent extraction (solid–liquid) from vegetable matrices, high-pressure extraction from vegetable matrices, and liquid–liquid extraction and adsorption, which are processes used to obtain high-quality bioactive extracts and purified compounds from botanical sources. Each chapter in the book is organized in three major sections: (1) fundamental aspects of transport phenomena and thermodynamics related to the chapter topic, (2) a state of the art mini-review of the literature for the chapter topic, and (3) in one or more sections, examples of novelty (from the industrial point of view) applications that were chosen from case studies of actual or near to industrial applications. These are very specific examples; nonetheless, they will provide enough details so the readers can use them as a guide to develop other applications.

This volume covers the basic and applied aspects of two groups of extraction processes. The first group of processes deals with obtaining extracts from solid matrices such as (1) steam distillation for obtaining volatile oil from aromatic, condimentary, spice, and similar plants; (2) low-pressure solvent extraction or solid–liquid extraction for obtaining pigments, antioxidants, flavonoids, vegetable oils, protein concentrates, and so on; and (3) supercritical fluid extraction (SFE, with and without cosolvent), including extraction with solvents that are gases, such as carbon dioxide, at room conditions as well as pressurized liquid solvents (PSL) at the same conditions, for instance, hot water extraction (HWE). The second group of processes is devoted to processing liquid mixtures and includes processes generally used in sequence for steam distillation, low-pressure solvent extraction, and, more recently, for the removal of cosolvents and liquid solvents in SFE and PSL processes: (1) distillation, a process required for the removing of the solvent from the output of low-pressure solvent extraction as well as high-pressure extraction processes; (2) liquid–liquid extraction, which is generally employed as an intermediate step after low-pressure solvent extraction; and (3) adsorption/desorption, which is also used for the removal of solvent (or cosolvent) from solvent/extract mixtures as well as the removal of impurities from extract or purified compounds. In some cases, there is an overlapping of the applications just mentioned, such as the case of steam distillation, which is broadly used and is denoted as stripping for the removal of impurities at the final stages of vegetable oil production. The operating conditions such as solid matrices preprocessing (for instance, comminution and drying), steam pressure, temperature, solvent-to-feed ratio, pressure, cosolvent, packing type and shape, tray type, and so on will be discussed as applied to each process. The kinetics of the process will be discussed where appropriate.

Because the strength of this book is on engineering design of extraction processes, in spite of the importance of several other separation processes, such as membrane-based separation, they were not included here. Other extraction techniques used as analytical tools, such as microwave- and ultrasound-assisted extraction, are discussed in Chapter 4 along with solid–liquid extraction.

M. Angela A. Meireles
Campinas, Brazil

Editor

M. Angela A. Meireles is a professor of food engineering at UNICAMP (State University of Campinas), which she joined in 1983 as an assistant professor. She holds a PhD in chemical engineering from Iowa State University (1982); she also holds an MSc and a BS in food engineering from UNICAMP (1979 and 1977). Dr. Meireles published 92 research papers in peer-reviewed journals and has more than 340 presentations in scientific meetings. She has supervised 26 PhD dissertations, 20 MSc theses, and about 40 undergraduate research projects. Her research is in the field of production of extracts from aromatic, medicinal, and spice plants by supercritical fluid extraction and conventional techniques such as steam distillation and GRAS (or green) solvent extraction, including process parameters determination, process integration and optimization, extraction fractionation, and technical and economical analysis. She has coordinated scientific exchange projects between UNICAMP and European universities (France, Holland, and Germany). Nationally she coordinated a project called SuperNat that involved five Brazilian institutions (UFPA, UFRN, UEM, UFSC, and IAC) and a German university (TUHH); she coordinated a thematic project financed by FAPESP (State of São Paulo Science Foundation) from 2000 to 2005 (supercritical technology applied to the processing of essentials oils, vegetable oils, pigments, stevia, and other natural products). Presently she is coordinating two technology transfer projects in the area of supercritical fluid extraction from native Brazilian plants. She belongs to the editorial boards of the *Brazilian Journal of Medicinal Plants, Journal of Food Process Engineering* (Blackwell Publishing), *Recent Patents on Engineering* and *Open Chemical Engineering Journal* (Bentham Science Publishers), *Pharmacognosy Reviews* (Pharmacognosy Network), and *Food and Bioprocess Technology* (Springer). She was associate editor of the journals *Ciência e Tecnologia de Alimentos (Food Science and Technology)* and *Boletim da SBCTA* (newsletter from SBCTA, the Brazilian Society of Food Science and Technology) from 1994 to 1998.

Contributors

Eduardo A. C. Batista
School of Food Engineering
State University of Campinas
Campinas, Brazil

Fábio R. M. Batista
School of Food Engineering
State University of Campinas
Campinas, Brazil

Mara E. M. Braga
Department of Chemical Engineering
University of Coimbra
Coimbra, Portugal

Lourdes Calvo
Department of Chemical Engineering
Complutense University of Madrid
Madrid, Spain

Roberta Ceriani
School of Food Engineering
State University of Campinas
Campinas, Brazil

Manuel G. Cerpa
Department of Chemical Engineering and Environmental Technology
University of Valladolid
Valladolid, Spain

Maria José Cocero
Department of Chemical Engineering and Environmental Technology
University of Valladolid
Valladolid, Spain

Beatriz Díaz-Reinoso
Department of Chemical Engineering
University of Vigo
Ourense, Spain

Herminia Domínguez
Department of Chemical Engineering
University of Vigo
Ourense, Spain

Louw J. Florusse
Laboratory of Physical Chemistry and Molecular Thermodynamics
Delft University of Technology
Delft, The Netherlands

Cintia B. Gonçalves
Faculty of Animal Science and Food Engineering
University of São Paulo
Pirassununga, Brazil

Motonobu Goto
Department of Applied Chemistry and Biochemistry
Kumamoto University
Kumamoto, Japan

Patrícia F. Leal
School of Food Engineering
State University of Campinas
Campinas, Brazil

Luiz F. L. Luz, Jr.
Department of Chemical Engineering
Federal University of Paraná
Curitiba, Brazil

Susana Lucas
Department of Chemical Engineering and Environmental Technology
University of Valladolid
Valladolid, Spain

Mário Maróstica, Jr.
Research and Development Department
Centroflora Group
Botucatu, Brazil

Rafael B. Mato
Department of Chemical Engineering and Environmental Technology
University of Valladolid
Valladolid, Spain

M. Angela A. Meireles
School of Food Engineering
State University of Campinas
Campinas, Brazil

Antonio J. A. Meirelles
School of Food Engineering
State University of Campinas
Campinas, Brazil

Andrés Moure
Department of Chemical Engineering
University of Vigo
Ourense, Spain

Juan Carlos Parajó
Department of Chemical Engineering
University of Vigo
Ourense, Spain

Camila G. Pereira
Department of Chemical Engineering
Federal University of Rio Grande do Norte
Natal, Brazil

Cor J. Peters
Laboratory of Physical Chemistry and Molecular Thermodynamics
Delft University of Technology
Delft, The Netherlands

Juliana M. Prado
School of Food Engineering
State University of Campinas
Campinas, Brazil

Christianne E. C. Rodrigues
Faculty of Animal Science and Food Engineering
University of São Paulo
Pirassununga, Brazil

Paulo T. V. Rosa
Department of Physical Chemistry
State University of Campinas
São Paulo, Brazil

Helena F. A. Scanavini
School of Food Engineering
State University of Campinas
Campinas, Brazil

Richard L. Smith, Jr.
Department of Chemical Engineering
Tohoku University
Sendai, Japan

Thais M. Takeuchi
School of Food Engineering
State University of Campinas
Campinas, Brazil

Masaaki Toyomizu
Department of Life Science
Tohoku University
Sendai, Japan

Acknowledgments

I thank all contributors for accepting my invitation to be part of the challenging task given to me by Professor Da-Wen Sun. I also thank the College of Food Engineering, University of Campinas (UNICAMP, Brazil) for allowing me to expend part of my working time on organizing this book. I express my gratitude to my sponsors: FAPESP (The São Paulo State Research Foundation), CNPq (National Council for Scientific and Technological Development, Brazil), and CAPES for supporting the research done in LASEFI/DEA/FEA/UNICAMP, part of which is presented in this book. Finally, I thank Professor Da-Wen Sun for inviting me to edit this book, the reviewers of the book proposal who have positively contributed to enhance the quality of its contents, and the CRC Press team who made it possible.

M. Angela A. Meireles
Campinas, Brazil

1 Extraction and Purification of Bioactive Compounds

M. Angela A. Meireles

CONTENTS

1.1 Volatile or Essential Oils 2
 1.1.1 Phase Equilibrium in Systems Containing VO Compounds 3
 1.1.2 Thermophysical Properties of Selected VO Components 4
1.2 Other Bioactive Compounds 4
1.3 References 7

Because of today's pursuit for health products, the production and purification of vegetable extracts is an area of interest to the industry and academia. In this book, extraction and purification techniques are discussed. This book deals with unit operations for which mass transfer and phase equilibria dictate the performance of the processes. For instance, in Chapter 2 the use of steam distillation is discussed as applied to the deacidification of vegetable oils and the production of volatile (essential) oils. In both cases, the knowledge of mass transfer as well as the thermodynamic behavior of the systems is required in order to optimize the process and, eventually, to bring the process to industry, an estimation of the cost of manufacturing the product by the selected technology is also needed. So, the three chapters of the book that deal with extraction techniques address the question of estimation of the cost of manufacturing.

In Chapter 2 the target substances are a mixture of esters of glycerin and fatty acids (Section 2.2), thus, a fixed or vegetable oil or a volatile or essential oil (Sections 2.3 and 2.4). The fixed oils are well known to food scientists as they play an important role in food processing. Volatile oils (VOs), on the other hand, are less known. This is because, in spite of their importance in seasoning food and in spite of being known since antiquity, the volume of their production is enormously different from vegetable oils. Therefore, their economical importance is restricted to niche areas. However, as consumers are becoming more and more aware of the importance of using bioactive compounds in either form as a food supplement or as a functional food to improve their health, vegetable extracts can in the near future gain economic importance.

The techniques for obtaining several of the bioactive compounds important in food processing are discussed in this book. VOs are a source of several bioactive compounds; this chapter provides a brief introduction to these complex mixtures denoted as volatile or essential oils. Several of the compounds found in volatile oils may be classified as bioactive compounds; nonetheless, other bioactive compounds such as flavonols, flavonoids, polyphenols, and so on are generally present in the extracts of plants obtained by extraction with an organic solvent that may or may not be environmentally friendly. In Chapters 4 and 6, obtaining antioxidants using GRAS (generally recognized as safe) or green solvents is discussed. Depending on the specific application of the bioactive compound or bioactive mixture, purification must be added to the process. Purification processes such as distillation (Chapter 3), liquid–liquid extraction (Chapter 5), and adsorption/desorption (Chapters 6 and 7) are deeply discussed and applied to production of cachaça (pronounced ca-sha-ssa), a famous spirit from sugar cane produced in Brazil, to fractionation of orange oil, and to improvement of soluble coffee aroma.

1.1 VOLATILE OR ESSENTIAL OILS

VOs are a mixture of volatile terpenoids that are produced by the plant's secondary metabolism, or the isoprenoid path [1]. Originally, VOs were defined as the volatile portion of the plant obtained by steam distillation, but volatile oils can also be produced by fractionating the oleoresin obtained by solvent extraction (at low or high pressures; see Chapters 4 and 6). VOs can have a very simple composition, as in the case of clove buds oil (eugenol, 64.3%; ß-caryophyllene, 19.6%; eugenol acetate, ~13.8%; humulene, 2.3% [2]) or can be as complex as turmeric oil (see Table 1.1). The major compounds forming a VO belong to the chemical classes of the monoterpenes ($C_{10}H_{16}$), oxygenated monoterpenes, sesquiterpenes ($C_{15}H_{24}$), and oxygenated

TABLE 1.1
Composition of the Volatile Fraction (VO) of Turmeric Extract Obtained by SFE at 20 MPa and 303 K

Compound	% (area)
Ar-curcumene	2.3
α-zingiberene	1.6
β-sesquiphellandrene	2.4
Ar-turmerol	1.2
Ar-turmerol isomer	1.3
Ar-turmerone	28.1
(Z)-γ-atlantone	24.2
(E)-γ-atlantone	20.3
6S,7R-bisabolone	1.2
Nonidentified	17.4

Source: Based on Braga, M. E. M., et al., *Journal of Agricultural and Food Chemistry*, 51:6604–6611, 2003.

TABLE 1.2
Examples of Terpenoids Found in Food

Terpenoids	Example	Food
Terpene ($C_{10}H_{16}$)	Limonene	Orange
Oxygenated terpene ($C_{10}H_{12}O_2$)	Eugenol	Clove buds
Sesquiterpene ($C_{15}H_{24}$)	α-Humulene	Black pepper
Oxygenated sesquiterpene ($C_{15}H_{26}O$)	Nerolidol	Ginger
Diterpene ($C_{20}H_{28}O_3$)	Cafestol	Coffee
Triterpene ($C_{30}H_{50}$)	Squalene	Shark liver oil
Tretaterpene ($C_{40}H_{56}$)	Lycopene	Tomato

sesquiterpenes. Other terpenoids such as the diterpenes ($C_{20}H_{32}$) and triterpenes ($C_{30}H_{48}$) are not volatile but are widely found in certain foods (see Table 1.2) and are important bioactive compounds.

1.1.1 Phase Equilibrium in Systems Containing VO Compounds

The phase equilibria of VO components and solvents is important for the process design optimization of steam distillation, supercritical fluid extraction (SFE) (with or without cosolvent), and low-pressure solvent extraction. The phase equilibria can be calculated using the activity coefficients and models for the excess Gibbs free energy [4] as discussed in Chapters 3 and 5 or by using the fugacity coefficients calculated with an equation of state (EOS) as discussed in Chapter 6; this last method is more frequently used to describe the phase equilibria at high pressures. There are two EOSs that are frequently used to describe these unconventional systems: the EOS of Peng–Robinson [5] and the EOS of Soave–Redilich–Kwong [6]. In the case of EOS the thermophysical properties (critical temperature and pressure and acentric factor) of the VO components are required; in general, experimental values of these properties are not available. To overcome this difficulty, these properties can be predicted by several different group contribution methods [4]. In choosing a method, one is faced with the difficulty that none of the available methods were developed considering the properties of terpenoids; the majority of these methods were developed considering the compounds of interest to traditional chemical industries. In spite of that, these methods have been largely used to estimate the thermophysical properties required to describe, using EOS, the phase equilibria of VO compounds with carbon dioxide [7–11].

Araújo and Meireles [12] have demonstrated that the phase equilibria is better described when the thermophysical properties are estimated by the method that is more indicated for a given class of chemical species. The systems studied by these authors were fats and fat-related substances. These systems contain compounds from homologous series, such as the fatty acids. Although VOs contain terpenoids, and several molecules with the same chemical formula can be present simultaneously in a specific VO (see Table 1.1), it can be expected that describing phase equilibria of systems containing VOs is a difficult task. Nonetheless, Moura et al. [8] were successful in describing the phase equilibria of fennel extract with carbon dioxide.

TABLE 1.3
Composition of SFE Fennel Extract Obtained at 25 MPa and 303 K

Compound	Mass fraction (%)
Fenchone	1.05
Anethole	16.5
Palmitic acid	6.63
Palmitoliec acid	1.11
Stearic acid	2.68
Oleic acid	45.26
Linoleic acid	23.04

Source: Based on Moura, L. S., et al., *Journal of Chemical and Engineering Data,* 50:1657–1661, 2005.

Fennel VO is very rich in anethole and fenchone (74 and 15%, respectively [13]), whereas the fennel extract obtained by SFE extraction, in addition to anethole and fenchone, contains fatty acids [8] (see Table 1.3). Thus, the success of Moura et al. [8] is due to the large asymmetry of the system fennel extract/CO_2, which is composed of a small molecule (CO_2) and several large molecules (terpenoids and fatty acids).

1.1.2 Thermophysical Properties of Selected VO Components

Terpenoids, in spite of being volatile, have a normal boiling point higher than that of water. In general, these molecules are thermolabile and would degrade at temperatures far below their estimated critical temperature; this behavior explains the scarcity of thermophysical data of terpenoid molecules in the literature. Thus, it would be perfectly acceptable in the case of these molecules to choose one group contribution method, for instance, the Joback and Reid method [14], and use it throughout the phase equilibrium calculations. Rodrigues [2] did a study similar to that of Araújo and Meireles [12] considering literature data of terpenoids. Because for terpenoids the database is far smaller than that of fat and fat-related substances, the success in clearly selecting a group contribution method was very limited. Additionally, for molecules with the same chemical formula, the group contribution methods tend to predict similar or even the same values for the thermophysical properties. Table 1.4 shows a compilation made by Rodrigues [2] of thermophysical properties of compounds usually found in VOs. The molecular structures of these compounds are available in Adams [29].

1.2 OTHER BIOACTIVE COMPOUNDS

Several classes of compounds display antioxidant activity and other properties that make their ingestion a good health habit. Some of these compounds are polyphenols, widely found in aromatic, condimentary, and spice plants. The actions of these substances are discussed in Chapters 4 and 6. VOs exhibit antioxidant activity, which is due to the presence of mono- and sesquiterpenes and not to the presence of large molecules. In order to obtain higher molecular mass substances from plant matrices, certain organic solvents are used, and the extract is generally denoted as an oleoresin

TABLE 1.4
Thermophysical Properties of Some VO Compounds

Compound	CAS number	MM/kg·kmol^{-1}	T_b/K	T_f/K	T_c/K	P_c/MPa	ω	D^{20}/kg·m^{-1}
Anethole ($C_{10}H_{12}O$)	104-46-1	148.20	508.45[g]	294.50[g]	723.00[g]	29.0[g]	0.4846[g]	0.9883[a]
Aromadendrene ($C_{15}H_{24}$)	489-39-4	204.36	515.72[4]	304.50[4]	706.17[8]	20.0[8]	0.434[10]	na
β-Bisabolene ($C_{15}H_{24}$)	495-61-4	204.36	529.99[4]	267.62[4]	713.36[8]	19.3[8]	0.8274[10]	0.8673[b]
Borneol ($C_{10}H_{18}O$)	507-70-0	154.24	485.15[a]	477.15[a]	675.09[1]	29.2[8]	0.6069[9]	1.1011[a]
Carvacrol ($C_{11}H_{14}O$)	499-75-2	150.21	510.85[e]	274.15[e]	723.19[1]	32.2[7]	0.5754[9]	0.9772[e]
β-Caryophyllene ($C_{15}H_{24}$)	87-44-5	204.36	529.15[b]	255.92[4]	726.54[8]	27.6[8]	0.4719[10]	0.9075[a]
Chavicol ($C_9H_{10}O$)	501-92-8	134.18	511.15[b]	288.95[b]	737.35[8]	39.3[7]	0.6163[10]	na
p-Cimeno ($C_{10}H_{14}$)	99-87-6	134.22	450.22[e]	204.25[e]	652.0[e]	28.0[e]	0.3815[9]	0.8573[e]
1,8 Cineole ($C_{10}H_{18}O$)	470-82-6	154.25	449.55[b]	274.65[b]	652.54[1]	27.8[7]	0.3674[9]	0.9267[b]
ar-Curcumene ($C_{15}H_{22}$)	644-30-4	202.34	548.12[4]	255.27[4]	739.99[8]	19.7[7]	0.6400[10]	0.8805[c]
Decanal ($C_{10}H_{20}O$)	112-31-2	156.27	488.15[g]	267.15[g]	674.2[c]	26.0[g]	0.5820[g]	na
Eugenol ($C_{10}H_{12}O_2$)	97-53-0	164.21	528.15[a]	265.65[c]	737.86[8]	32.9[7]	0.4486[10]	1.0664[b]
Eugenol acetate ($C_{12}H_{14}O_3$)	93-28-7	206.24	554.15[c]	303.65[c]	774.16[8]	31.4[7]	0.6274[10]	1.0860[c]
β-Famesene ($C_{15}H_{24}$)	18794-84-8	204.36	397.15[a]	257.00[4]	706.53[8]	19.8[8]	0.9285[10]	0.8363[c]
Fenchone ($C_{10}H_{16}O$)	4695-62-9	152.24	509.9[1]	—	742.4[1]	30.9[1]	0.4057[1]	—
Geranial ($C_{10}H_{16}O$)	141-27-5	152.24	502.15[e]	247.27[4]	699.97[1]	21.8[7]	0.4628[9]	0.8869[e]
Geraniol ($C_{10}H_{18}O$)	106-24-1	154.25	503.15[e]	258.15[e]	671.67[1]	24.0[7]	0.7799[9]	0.8894[e]
2-Hexanone ($C_6H_{12}O$)	591-78-6	100.16	400.70[g]	217.35[g]	587.6[f]	32.9[g]	0.3846[g]	0.8113[e]
α-Humulene ($C_{15}H_{24}$)	6753-98-6	204.36	523.59[4]	260.70[4]	720.87[8]	21.6[8]	0.5567[10]	0.8905[a]
Limoneno ($C_{10}H_{16}$)	5989-27-5	136.23	449.65[g]	199.00[g]	660.0[g]	27.5[g]	0.3123[g]	0.8407[e]
Linalool ($C_{10}H_{18}O$)	78-70-6	154.24	471.15[a]	258.42[4]	640.07[1]	24.4[7]	0.6674[9]	0.8622[e]
Methyl-chavicol ($C_{10}H_{12}O$)	140-67-0	148.20	488.65[e]	241.79[4]	700.43[8]	29.2[7]	0.5139[10]	0.9645[d]

continued

TABLE 1.4 *(continued)*

Compound	CAS number	MM/kg·kmol^{-1}	T_b/K	T_f/K	T_c/K	P_c/MPa	ω	D^{20}/kg·m^{-1}
Methyl-eugenol ($C_{11}H_{14}O_2$)	93-15-2	178.23	527.85(e)	269.15(e)	733.61(8)	29.9(7)	0.5447(10)	1.0396(e)
γ-Muurolene ($C_{15}H_{24}$)	30021-74-0	204.36	529.41(4)	305.48(4)	722.63(8)	20.0(8)	0.4482(10)	na
Myrcene ($C_{10}H_{16}$)	123-35-3	136.23	440.15(b)	240.40(4)	606.5(g)	23.3(g)	0.5547(9)	0.794(a)
Neral ($C_{10}H_{16}O$)	106-26-3	152.24	502.15(e)	265.48(4)	699.97(1)	22.9(7)	0.4840(9)	0.8888(e)
Nerol ($C_{10}H_{18}O$)	106-25-2	154.24	498.15(e)	258.15(e)	667.81(1)	24.0(7)	0.7498(9)	0.8756(e)
Sabinene ($C_{10}H_{16}$)	3387-41-5	136.23	437.15(b)	238.32(4)	635.56(6)	27.30(7)	0.3532(9)	0.8437(b)
β-Selinene ($C_{15}H_{24}$)	17066-67-0	204.36	543.15(b)	270.50(4)	729.59(8)	15.60(8)	0.5065(10)	0.9196(b)
γ-Terpinene ($C_{10}H_{16}$)	99-85-4	136.23	456.15(e)	227.35(4)	662.94(1)	28.28(7)	0.2725(9)	0.8490(e)
α-Terpineol ($C_{10}H_{18}O$)	98-55-5	154.25	494.00(e)	313.65(e)	676.75(1)	28.58(8)	0.8386(9)	0.9337(c)
Thujone ($C_{10}H_{16}O$)	546-80-5	152.23	478.22(4)	281.92(4)	686.5(l)	28.33(7)	0.427(9)	na
Thymol ($C_{15}H_{24}O$)	89-83-8	150.21	505.65(e)	324.65(e)	698.0(c)	33.58(7)	0.7273(9)	na
Verbenone ($C_{10}H_{14}O$)	80-57-9	150.21	500.65(e)	282.95(e)	721.59(1)	28.33(7)	0.4350(9)	0.9978(e)
Zingerone ($C_{11}H_{14}O_3$)	122-48-5	194.23	589.19(4)	313.65(e)	812.93(8)	2913(7)	0.6602(10)	na
Zingiberene ($C_{15}H_{24}$)	495-60-3	204.36	528.98(4)	261.77(4)	540.19(8)	17.05(8)	0.5434(10)	na

na: not available.

Literature values: (a)Merck Index [15]; (b)Weast et al. [16]; (c)Lide [17]; (d)Fenaroli [18]; (e)Ikan [19]; (f)DIPPR [20].

Estimated values: (1)Joback and Reid [14]; (2)Tsibanogiannis et al. [21]; (3)Willman and Teja [22]; (4)Constantinou and Gani [23]; (5)Klincewiz and Reid [24]; (6)Lin and Chao [25]; (7)Reid et al. (Lydersen method) [26]; (8)Somayjulu [27]; (9)Vetere [28]; Reid et al. (Edmister rule) [26].

if it comes from an aromatic, condimentary, or medicinal plant. The flavonoid quercetin can be obtained from the flowers of macela (*Achyrocline satureioides*) by extraction with ethanol or with CO_2 modified with ethanol.

1.3 REFERENCES

1. Fennema, O. R. 1996. *Food chemistry.* 3rd ed. New York: Marcel Dekker.
2. Rodrigues, V. M. 2001. Determinação da solubilidade em sistemas peudo-ternários: cravo-da-índia (*Eugenia caryophyllus*) + CO_2, gengibre (*Zingiber officinale*) + CO_2 e erva-doce (*Pimpinella anisum*) + CO_2. PhD diss., UNICAMP (State University of Campinas), Brazil.
3. Braga, M. E. M., P. F. Leal, J. E. Carvalho, and M. A. A. Meireles. 2003. Comparison of yield, composition, and antioxidant activity of turmeric (*Curcuma longa* L.) extracts obtained using various techniques. *Journal of Agricultural and Food Chemistry* 51:6604–6611.
4. Poling, B. E., J. M. Prausnitz, and J. P. O'Connel. 2001. *The properties of gases and liquids.* New York: McGraw-Hill.
5. Peng, D. Y., and D. B. Robinson. 1976. A new two-constant equation of state. *Industrial Engineering and Chemistry Fundamentals* 15:59–64.
6. Soave, G. 1972. Equilibrium constants from a modified Redilich-Kwong equation of state. *Chemical Engineering Science* 27:1192–1203.
7. Souza, A. T., M. L. Corazza, L. Cardozo-Filho, R. Guirardello, and M. A. A. Meireles. 2004. Phase equilibrium measurements for the system clove (*Eugenia caryophyllus*) oil + CO_2. *Journal of Chemical and Engineering Data* 49:352–356.
8. Moura, L. S., M. L. Corazza, L. Cardozo-Filho, and M. A. A. Meireles. 2005. Phase equilibrium measurements for the system fennel (*Foeniculum vulgare*) extract + CO_2. *Journal of Chemical and Engineering Data* 50:1657–1661.
9. Takeuchi, T. M., P. F. Leal, R. Favareto, L. Cardozo-Filho, M. L. Corazza, P. T. V. Rosa, and M. A. A. Meireles. 2008. Study of the phase equilibrium formed inside the flash tank used at the separation step of a supercritical fluid extraction unit. *Journal of Supercritical Fluids* 43:447–459.
10. Stuart, G. R., C. Dariva, and J. V. Oliveira. 2000. High-pressure vapor-liquid equilibrium data for CO_2–orange peel oil. *Brazilian Journal of Chemical Engineering* 17:181–189.
11. Corazza, M. L., L. Cardozo-Filho, O. A. C. Antunes, and C. Dariva. 2003. High-pressure phase equilibria of related substances in the limonene oxidation in supercritical CO_2. *Journal of Chemical and Engineering Data* 48:354–358.
12. Araújo, M. E., and M. A. A. Meireles. 2000. Improving phase equilibrium calculation with the Peng–Robinson EOS for fats and oils related compounds/supercritical CO_2 systems. *Fluid Phase Equilibria* 169:49–64.
13. Moura, L. S., R. N. Carvalho, Jr., M. B. Stefanini, L. C. Ming, and M. A. A. Meireles. 2005. Supercritical fluid extraction from fennel (*Foeniculum vulgare*): global yield, composition and kinetic data. *Journal of Supercritical Fluids* 35:212–219.
14. Joback, K. G., and R. Reid. 1987. Estimation of pure component properties from group contributions. *Chemical Engineering Communications* 57:233–243.
15. *The Merck index.* 1983. 10th ed. Rahway, NJ: Merck Co.
16. Weast, R. C., and M. J. Astle. 1987. *CRC handbook on organic compounds,* Vols. I and II. Boca Raton, FL: CRC Press.
17. Lide, D. R. 1997–1998. *Handbook of chemistry and physics.* 78th ed. Boca Raton: CRC Press.
18. Fenaroli, G. 1971. *Fenaroli's handbook of flavor ingredients.* Boca Raton: CRC Press.

19. Ikan, R. 1969. *Natural products—A laboratory guide*. Jerusalem: Israel University Press.
20. DIPPR 801. 2008. Thermophysical properties database for pure chemical compounds. http://www.aiche.org/DIPPR/ (accessed 31 March 2008).
21. Tsibanogiannis, I. N., N. S. Kalospiros, and D. P. Tassios. 1995. Prediction of normal boiling point temperature of medium/high molecular weight compounds. *Industrial and Engineering Chemical Research* 34:997–1002.
22. Willman, B., and A. S. Teja. 1985. Method for the prediction of pure component vapor pressures in range 1 kPa to the critical pressure. *Industrial and Engineering Chemical Research* 24:1033–1036.
23. Constatntinou, L., and R. Gani. 1994. Group contribution method for estimating properties of pure compounds. *AIChE Journal* 10:40–56.
24. Klincewicz, K. M., and R. C. Reid. 1984. Estimation of critical properties with group contribution methods. *AIChE Journal* 30:137–142.
25. Lin, H.-U, and K.-C. Chao. 1984. Correlation of critical properties and acentric factor of hydrocarbons and derivatives. *AIChE Journal* 30:981–983.
26. Reid, R. C., J. M. Prausnitz, and B. E. Poling. 1987. *The properties of gases and liquids*. New York: McGraw-Hill.
27. Somayajulu, G. R. 1989. Estimation procedures for critical constants. *Journal of Chemical and Engineering Data* 34:106–200.
28. Vetere, A. 1991. Predicting the vapor pressures of the pure compounds by using the Wagner equation. *Fluid Phase Equilibria* 62:1–10.
29. Adams, Robert P. 2001. *Identification of essential oil components by gas chromatography/quadrupole mass spectroscopy*. Carol Stream, IL: Allured Publishing.

2 Steam Distillation Applied to the Food Industry

Manuel G. Cerpa, Rafael B. Mato, Maria José Cocero, Roberta Ceriani, Antonio J. A. Meirelles, Juliana M. Prado, Patrícia F. Leal, Thais M. Takeuchi, and M. Angela A. Meireles

CONTENTS

2.1 Fundamentals of Steam Distillation 11
Manuel G. Cerpa, Rafael B. Mato, and Maria José Cocero

2.1.1 Definitions 11
2.1.1.1 Steam Distillation 11
2.1.2 Description of the Process 12
2.1.2.1 Advantages of SD 12
2.1.2.2 Limitations of SD 13
2.1.3 Applications 13
2.1.3.1 Deacidification and Deodorization of Edible Fats and Oils 13
2.1.3.2 Distillation of VOs or Essential Oils 14
2.1.4 Phenomenological Study of the Process 14
2.1.4.1 Oil Release 14
2.1.4.2 Vaporization 15
2.1.4.3 Mass Transfer 16
2.1.4.4 Distillate Condensation 17
2.1.5 Nomenclature 17
2.1.6 References 17

2.2 Deacidification of Vegetable Oils by Stripping 18
Roberta Ceriani and Antonio J. A. Meirelles

2.2.1 Modeling a Reactive Batch Deodorizer 19
2.2.1.1 Mathematical Equations 19
2.2.1.2 Vapor–Liquid Equilibria and Vaporization Efficiency 21
2.2.1.3 Estimation of the Oil Composition 22

2.2.2 Computational Simulation Results 23
2.2.3 Neutral Oil Loss 23
2.2.4 *Cis–Trans* Isomerization 26
2.2.5 Waxes Degradation 30
2.2.6 Nomenclature 32
2.2.7 Acknowledgments 33
2.2.8 References 34
2.3 Obtaining Volatile Oils by Steam Distillation: State of the Art 35
Juliana M. Prado, Patrícia F. Leal, and M. Angela A. Meireles
2.3.1 Steam Distillation 35
2.3.2 VOs from Aromatic, Condimentary, and Medicinal Plants 38
2.3.3 VOs from Anise Seed, Black Pepper, Chamomile, and Rosemary 43
2.3.4 Acknowledgments 45
2.3.5 References 45
2.4 Cost of Manufacturing of Volatile Oil from Condimentary Plants 47
Patrícia F. Leal, Thais M. Takeuchi, Juliana M. Prado, and M. Angela A. Meireles
2.4.1 Characteristics of the Cost Estimation Classes 48
2.4.2 Cost Estimation Classes 48
2.4.3 Cost of Manufacturing Estimation Methods 50
2.4.4 COM for VOs from Condimentary Plants 50
2.4.4.1 Scale-Up 51
2.4.4.2 Fixed Cost of Investment 51
2.4.4.3 Raw Material Cost 51
2.4.4.4 Operational Labor Cost 51
2.4.4.5 Waste Treatment Cost 52
2.4.4.6 Cost of Utilities 52
2.4.5 COM Estimation 52
2.4.5.1 Anise Seed 55
2.4.5.2 Chamomile 58
2.4.5.3 Rosemary 60
2.4.5.4 Black Pepper 63
2.4.5.5 Thyme 65
2.4.6 Comparing Estimated COMs and Market Prices 70
2.4.7 Nomenclature 72
2.4.8 Acknowledgments 73
2.4.9 References 73

In this chapter the uses of steam distillation (SD) in food processing and related industries are discussed. First, in Section 2.1 the fundamentals of the process are presented; this section gives examples of two applications of SD in food processing: (1) deacidification of fixed oils and (2) obtaining volatile oils (VOs) from aromatic, condimentary, and medicinal plants. Next, in Section 2.2 the deacidification of vegetable oils by stripping is discussed. This section is a good example of the use of simulation in predicting the behavior of a complex system. In Sections 2.3 and 2.4

the focus is on the use of SD to produce VOs from aromatic, condimentary, and medicinal plants. VOs or essential oils are mixture of terpenoids produced by the plants' secondary metabolism. The reader will notice that we prefer to use "volatile oils" instead of "essential oils"; this distinction is intended to make clear that we are dealing with substances responsible for the aroma, and in some cases also for the taste, which are characteristic of these plants. The review of the literature in Section 2.3 is entirely devoted to the applications of SD in obtaining VOs. Finally, the basis for estimation of the cost of manufacturing (COM) VOs is presented in Section 2.4. The COM VO from five aromatic plants (anise seed, chamomile, rosemary, black pepper, and thyme) was estimated using the described methodology. Because obtaining VOs is still considered an art instead of a technology, engineering data related to process design for the production of VOs by SD are scarce in literature. Therefore, in Section 2.4 a compilation of available data for the five plants previously mentioned and their use in COM estimation is extensively discussed.

2.1 FUNDAMENTALS OF STEAM DISTILLATION

Manuel G. Cerpa, Rafael B. Mato, and Maria José Cocero

Steam distillation (SD) is a modified distillation process used for the recovery of temperature-sensitive materials. It should be used in those cases where components to be separated present different volatilities but are so low that the use of ordinary distillation would lead to degradation of thermally labile compounds.

The use of boiling water reduces the temperature of the process by reducing the partial pressure of the desired components in the vapor phase. The process is also sometimes combined with vacuum operation in order to improve temperature reduction and avoid component decomposition, when materials with very low volatility are processed.

2.1.1 DEFINITIONS

2.1.1.1 Steam Distillation

SD is a modified distillation process used for the recovery of high boiling point volatile compounds, from an inert and complex matrix (solid or liquid), using steam (saturated or superheated) as a separation and energy agent. There are three variants of this process [1]: (a) direct SD, (b) water distillation, and (c) dry steam distillation.

2.1.1.1.1 Direct Steam Distillation (Steam Distillation)

The inert matrix (raw material) is supported on a perforated grid or screen inserted some distance above the bottom of the still, but it is not in direct contact with water. The boiler can be inside or outside the still. The low-pressure saturated steam flows up through the raw material matrix, collecting the evaporated components.

2.1.1.1.2 Water Distillation (Hydrodistillation)

In this case the raw material comes in direct contact with boiling water. The boiler is inside the still, and the material may be floating on the water or be completely immersed, depending on its specific gravity and the quantity of material handled per charge. In some cases, mixing is necessary because the material agglutinates and forms large compact lumps, preventing good contact with steam.

2.1.1.1.3 Dry Steam Distillation

The raw material is supported and steam flows through it, as in SD, but steam is generated outside the still. The steam is superheated at moderate pressures.

2.1.2 Description of the Process

A generalized flowsheet of the SD process is shown in Figure 2.1.1. The raw material (inert matrix) is charged to the still (distiller) in order to form a compact fixed bed. Before loading, solid materials may be milled and/or bitted. In the case of liquids, the load is usually treated in a continuous countercurrent still.

Steam is injected using an internal distributor, at the bottom of the still, with pressure enough to overcome the hydraulic resistance of the bed. The boiler can be inside or outside the still. As the steam flows up through the bed, the raw material warms up and releases the volatile solutes. These are vaporized and transported in the steam. When the steam leaves the still, it is condensed and cooled to ambient temperature. The condensed liquid mixture forms two immiscible phases that are separated in a dynamic decanter. This decanter is known as *Florentine* in essential oil distillation processes. The condensed water can be recycled to the still or to the boiler depending on the consumption of steam. With herbaceous raw materials, the residue can be used as fuel to generate steam in a special boiler. Dry steam distillation is preferred at the industrial scale over the other steam distillation variants, because standard boilers generate steam at moderate pressures. This steam is saturated, but when it is injected to the still, it suffers an isenthalpic expansion and becomes superheated.

2.1.2.1 Advantages of SD

2.1.2.1.1 Organic-Solvent-Free Products

The SD method uses water as the separation agent. It supplies natural products free of organic solvents that can be directly used in other processes, without the necessity of additional separation processes.

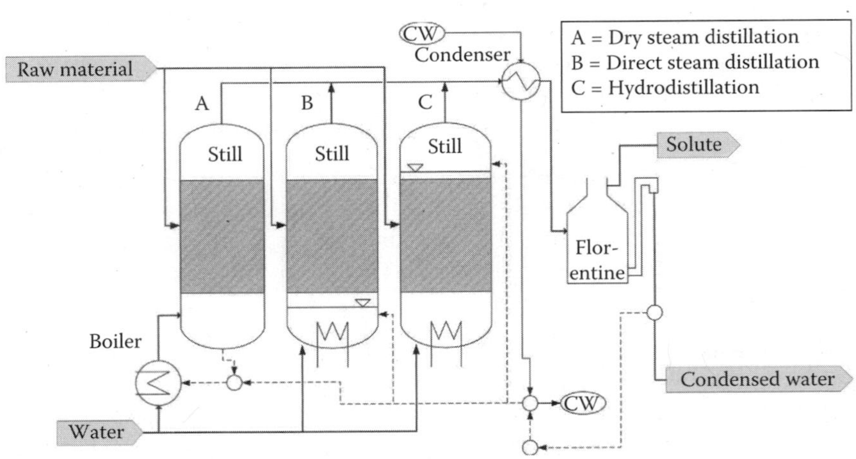

FIGURE 2.1.1 Generalized flowsheet of the different types of steam distillation.

2.1.2.1.2 High Capacity of Processing

SD can work with high loads of raw material (TM/day), with different physical characteristics. This allows a good profitability.

2.1.2.1.3 Low Costs of Investment

SD equipment is cheap, flexible, and easy to construct, and there is a big variety of materials for construction. Because SD operates at ambient conditions, it is not necessary to construct pressure vessels.

2.1.2.1.4 Know-How Available

SD is a well-known technology. The operative procedure is the same to distill herbaceous or liquid matrices. Operating conditions can be found in many books, journal articles, and Web pages or can be obtained directly with the equipment. It is not necessary to ask for licenses or permission or to buy the technology in order to distill a matrix.

2.1.2.2 Limitations of SD

2.1.2.2.1 Thermal Degradation of Products

When the solute is a natural product (volatile oils [VOs] or essential oils), thermal degradation cannot be avoided. In some cases, degradation is desirable because the solute can be enriched in main aroma compounds, but, in others cases, it generates oligomers and complex chemical compounds that decrease the shelf life of the product or change its organoleptic perception. In these cases the quality of product is affected. The hydrolysis of the solute may take place only in hydrodistillation, because in the other cases the raw material is in contact with steam. For this reason, hydrodistillation is seldom used.

2.1.2.2.2 High Consumption of Energy

As the raw material must be warmed up to boiling temperature, the consumption of energy is high. The largest contribution to energy consumption is caused by the heating of the equipment mass. Actually, the real heat duty is very large when compared to the ideal heat duty (solute vaporization), and many mechanical and operational modifications have been proposed to reduce the global energy consumption (isolation, recycle of condensed water, vacuum).

2.1.3 Applications

SD is mainly used in the food industry (1) for the removal of undesirable compounds (e.g., deacidification and deodorization of edible fats and oils) and (2) in the elaboration of VOs.

2.1.3.1 Deacidification and Deodorization of Edible Fats and Oils

Edible vegetable oils are constituted mainly by esters of glycerin (triglycerides) in which the glycerol is esterified with three fatty acids. They are usually accompanied by other products, already present in the oil or formed later during the handling of the seeds, which make them unacceptable for human consumption. These components are mainly volatile compounds, which give objectionable flavors and odors to the oil, and free fatty acids, which cause oil acidity.

Triglycerides are high-molecular-weight compounds with such low vapor pressures that they may be considered as nonvolatile. However, free fatty acids and the other odor components (aldehydes, ketones, alcohols) have higher volatilities, which make SD a suitable process for their removal.

Oil and fat deodorization of this solid raw material is carried out in batch, semicontinuous, and continuous processes, usually under reduced pressure to avoid degradation reactions. Details of this application will be discussed in Section 2.2 of this chapter.

2.1.3.2 Distillation of VOs or Essential Oils

Essential oils consist of volatile, lipophilic substances that are mainly hydrocarbons or monofunctional compounds derived from the metabolism of mono- and sesquiterpenes, phenylpropanoids, amino acids (lower mass aliphatic compounds), and fatty acids (long-chain aliphatic compounds) [2]. They are used in the food industry as flavoring. Although VOs are also obtained by other methods (solvent extraction, supercritical fluid extraction, pressing), SD is the most widespread method for their recovery in most cases [1, 3–7]. VOs are distilled from the whole plant (dill) or from separated parts: seeds (coriander, cumin, nutmeg), flowers (lavender, hyssop, spearmint), bark (cinnamon, sassafras), root (valerian), and peel (bergamot, orange).

In this case, in opposition to oil deodorization, the components must be extracted from a solid matrix before evaporation, and batch SD is used in all cases. The operation is performed close to atmospheric pressure. In Sections 2.3 and 2.4 some applications of VOs and the estimation of cost of manufacturing them, respectively, will be discussed.

2.1.4 Phenomenological Study of the Process

The goal for this section is to present a phenomenological description of the extraction process of recovered components in SD. A description of the VO distillation process is used to present the steps that occur in the model. Although this general scheme is suitable for all single-stage processes, differential remarks are presented when applied to solid or liquid raw materials.

Oil recovery from the aromatic plant takes place in four sequential stages: (1) Promoted by temperature increase, oil is released from inside the plant to its outer surface; (2) Oil vaporizes, taking vaporization heat from the steam; (3) Vapor oil molecules at the raw material surface must diffuse into the steam stream in a mass transfer process; and (4) Vapor oil molecules carried along by the steam are condensed and decanted. A simplified scheme of this sequential staged process is shown in Figure 2.1.2. A description of these four stages is detailed next.

2.1.4.1 Oil Release

When a liquid product is steam distilled, the whole load is directly accessible by the steam, and volatile compounds are ready to be vaporized as soon as they reach their boiling temperature. This is the case with oil refining and deodorizing, and under these circumstances, the oil release stage must be omitted, and vaporization should be taken as the starting point.

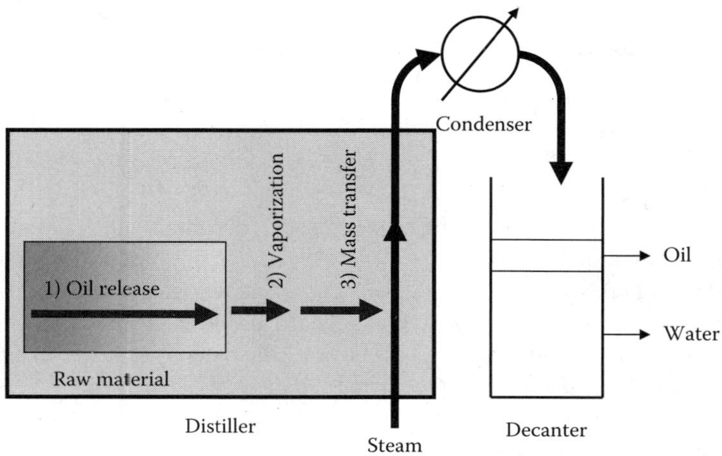

FIGURE 2.1.2 Schematic representation of extraction steps in steam distillation of essential oils.

In the case of solid materials, as it is in VO distillation, at least a portion of the recoverable components is not in contact with steam, and it must flow out of the solid before it can be vaporized. The mechanism by which this oil is released out of the plant depends on where it is located. Two main oil locations and release mechanisms are described in the literature.

2.1.4.1.1 Seeds, Fruits, or Roots

The solid shows an isotropic material behavior, with a uniform distribution of oil. Coriander seeds [8, 9] or aniseed grains [10] have been successfully described using this model, where diffusion inside the solid matrix is assumed.

2.1.4.1.2 Leaves or Flowers

Oil is deposited on the surface of the plant inside fragile glandular trichomes. In other oil extraction processes, such as supercritical CO_2 extraction [11, 12] or microwave extraction [13], the disruption of all or a significant part of the trichomes has been demonstrated. However, in SD, the integrity of the wall containing the oil inside the trichome has been verified by SEM (scanning electron microscopy) [13–15], and an exudation model has been proposed in this case where the oil slowly permeates through membranes and cuticle [8, 14, 15].

Because the oil release stage is a slow transfer mechanism, it is usually the controlling stage in the final part of the distillation, mainly in ground particles where diffusion inside the particle is the main resistance to oil recovery (see 2.1.4.1.1). This is the main reason why seeds and roots are usually crushed before distillation.

2.1.4.2 Vaporization

Vaporization occurs at the liquid–vapor interface. In this process molecules of components in the liquid phase move to the vapor phase, according to their volatilities.

The relation between compositions in both phases is regulated by the usual vapor–liquid equilibrium expression:

$$y_i P = \frac{x_i \gamma_i f_i^o}{\hat{\phi}_i}, \tag{2.1.1}$$

where P is the total or operation pressure, x_i and y_i are the molar fractions of each component in the liquid and vapor phases, respectively, γ_i is the activity coefficient of component i in the liquid phase, f_i^o the standard state fugacity of pure component i, and $\hat{\phi}_i^V$ the fugacity coefficient of component i in the vapor phase. These terms may be simplified, assuming ideal gas behavior, calculated from experimental measurements or estimated from group contribution methods.

In the case of oil refining and deodorization, the process is carried out under a vacuum (a few millibars) and high temperatures (381–543 K) in a single liquid phase. However, in VO steam distillation, the presence of condensed water wetting the plant surface, together with the flow of VO released by the plant, lead to the formation of two immiscible liquid phases, in direct contact with steam. If water and volatile (or oil) phases are considered totally immiscible, by Dalton's law, then

$$P = P_w^{vap} + P_C^{vap}, \tag{2.1.2}$$

where P is the total pressure, and P_w^{vap} and P_C^{vap} are the water and volatile substances vapor pressures, respectively. The presence of liquid water in a separated phase reduces the boiling temperature of the mixture because its contribution to the vapor pressure allows the liquid to boil at a lower temperature.

2.1.4.3 Mass Transfer

Molecules of vaporized components at the liquid–vapor interface must go into the steam stream by a mass transfer process. Mechanisms involve diffusion and convective mass transfer.

In VO distillation, steam flows through a porous bed of solid material, wetted by the liquid oil–water phases, and conventional mass transfer correlation coefficients [16, 17], K_g, may be used to calculate the molar flow of volatilized components, $\dot{m}_i$, incorporating into the global steam stream:

$$\dot{m}_i = K_g S \left(y_i - y_i^G\right), \tag{2.1.3}$$

where S is the transfer surface of contact between the porous bed and the steam, and y_i and y_i^G are the vapor phase mole fractions of component i in the liquid–vapor interface and in the global steam stream, respectively.

In oil refining and deodorization, mass transfer is usually considered as a limitation to vapor–liquid equilibrium and, instead of mass transfer coefficients, a stage efficiency parameter is used. This is the conventional practice in distillation, where Murphree efficiency is used to correct equilibrium deviations caused by mass transfer limitations and other efficiency-reducing phenomena, such as liquid droplets carried out by the steam flow. Distillation is discussed in depth in Chapter 3.

2.1.4.4 Distillate Condensation

Vapor leaving the distiller is condensed in the water cooled external condenser. In a total condenser no change in flow or composition takes place, because all vapors are condensed into a liquid phase.

2.1.5 Nomenclature

Symbol	Definition	Units in SI system	Dimensions in M, N, L, T, and θ
f_i^o	Standard state fugacity of pure component *i*	Pa	$ML^{-1}T^{-2}$
K_g	Mass transfer correlation coefficient	kmol s^{-1} m^{-2}	$NT^{-1}L^{-2}$
L	Total moles of liquid in the still	kmol	N
P	Pressure	Pa	$ML^{-1}T^{-2}$
P_i^{vap}	Vapor pressure of component *i*	Pa	$ML^{-1}T^{-2}$
P_w^{vap}	Vapor pressure of water	Pa	$ML^{-1}T^{-2}$
S	Transfer surface of contact between the porous bed and the steam	m^2	m^2
x_i	Component *i* liquid molar fraction at the vapor–liquid interphase	—	—
y_i	Component *i* vapor molar fraction at the vapor–liquid interphase	—	—
y_i^G	Component *i* vapor molar fraction at the global steam stream	—	—
Greek letter			
$\hat{\phi}_i^V$	Fugacity coefficient of component *i* in the vapor phase	—	—
γ_i	Activity coefficient of component *i* in the liquid phase	—	—

2.1.6 References

1. Günther, E. 1948. *The essential oils*. Vol. 1 of *History and origin in plants production analysis*. New York: Krieger Publishing.
2. Ullmann. 2007. Flavors and fragrances: Essential oils. In *Ullmann's encyclopedia of industrial chemistry*. Hoboken, NJ: John Wiley & Sons.
3. Di Cara, A., Jr. 1983. Essential oils. In *Encyclopedia of chemical processing and design*, Vol. 19, edited by J. J. McKetta, 352–381. New York: Marcel Dekker–Taylor & Francis–CRC.

4. Mookherjee, B. O., and R. Wilson. 2001. Oils essential. In *Kirk-Othmer encyclopedia of chemical technology*, ECT (CD) Vol. 17. New York: John Wiley & Sons.
5. Masango, P. 2005. Cleaner production of essential oils by steam distillation. *Journal of Cleaner Production* 13:833–839.
6. Muñoz, F. 2002. *Plantas medicinales y aromáticas: Estudio, cultivo y procesado.* Madrid: Ediciones Mundi-Prensa.
7. Peter, K. V. 2004. *Handbook of herbs and spices*. London: Woodhead Publishing.
8. Benyoussef, E. H., S. Hasni, R. Belabbes, and J. M. Bessiere. 2002. Modélisation du transfert de matiére lors de l`extraction de l´huile essentielle des fruits de coriandre. *Chemical Engineering Journal* 85:1–5.
9. Sovová, H., and S. A. Aleksovski. 2006. Mathematical model for hydrodistillation of essential oils. *Flavour Fragrance Journal* 21:881–889.
10. Romdhane, M., and C. Tizaoui. 2005. The kinetic modelling of a steam distillation unit for the extraction of aniseed (*Pimpinella anisum*) essential oil. *Journal of Chemical Technology and Biotechnology* 80:759–766.
11. Zizovic, I., M. Stamenić, A. Orlović, and D. Skala. 2007. Mathematical modelling of essential oil SFE on the micro-scale—Classification of plant material. 5th International Symposium on High Pressure Process Technology and Chemical Engineering, Segovia (Spain), June 24–27.
12. Mukhopadhyay, M. 2000. *Natural extracts using supercritical carbon dioxide*. New York: CRC Press.
13. Iriti, M., G. Colnaghi, F. Chemat, J. Smadja, F. Faoro, and F. A. Visinoni. 2006. Histocytochemistry and scanning electron microscopy of lavender glandular trichomes following conventional and microwave-assisted hydrodistillation of essential oils: A comparative study. *Flavour Fragrance Journal* 21:704–712.
14. Cerpa, M. G. 2007. Hidrodestilación de aceites esenciales. Doctoral diss., Department of Chemical Engineering and Environmental Technology, University of Valladolid, Spain.
15. Cerpa, M. G., R. B. Mato, and M. J. Cocero. 2008. Modeling steam distillation of essential oils: Application to lavandin super oil. *AIChE Journal* 54 (4): 909–917.
16. Knudsen, J. G., H. C. Hottel, A. F. Sarofim, et al. 1999. Heat and mass transfer. In *Perry´s chemical engineers handbook*, 7th ed., edited by R. H. Perry and D. W. Green. New York: McGraw-Hill.
17. Rexwinkel, G., A. B. M. Heesink, and W. P. M. Van Swaaij. 1997. Mass transfer in packed beds at low Peclet numbers—Wrong experiments or wrong interpretations? *Chemical Engineering Science* 52 (21–22): 3995–4003.

2.2 DEACIDIFICATION OF VEGETABLE OILS BY STRIPPING

Roberta Ceriani and Antonio J. A. Meirelles

Vegetable oils are composed mainly of triacylglycerols (TAGs), i.e., esters of glycerin and fatty acids. They also contain a wide range of minor constituents, such as sterols (phytosterols), waxes (esters of long-chain alcohols and fatty acids), tocols, pigments (carotenoids, chlorophyll), and vitamins. Due to hydrolysis, a small portion of the fatty acids attached to the glycerol is released as free fatty acids (FFAs) or oil acidity, generating also partial acylglycerols (monoacylglycerols [MAG] and diacylglycerols [DAG]). Most of these minor constituents are removed during the refining process, a series of purification steps to which the majority of vegetable oils are submitted to become edible. Some of these compounds, such as sterols and tocols, can be recovered and sold as valuable by-products.

Steam deacidification and steam deodorization are mass transfer stripping steps of the refining process that aim to remove FFAs and/or odor-causing substances by applying high temperatures and high vacuum. In these conditions of processing, the majority of the unwanted substances are largely more volatile than triacylglycerols, and their removal can be accomplished by injecting a stripping agent. Industrially, live steam is used as the stripping agent, although nitrogen was suggested as an alternative because it does not promote the hydrolysis reaction in the oil. From a thermodynamic point of view, the required amount of stripping gas is proportional to its molecular weight, which suggests the preference for a low-molecular-weight agent such as steam.

Although these processes target only the vaporization of undesirable substances, simultaneous losses of nutraceutical compounds and of acylglycerols (neutral oil loss [NOL]), due to volatilization, take place. Petrauskaitè et al. [1] studied the steam deacidification of coconut oil in a lab-scale batch deodorizer and concluded that NOL depends on the initial content of partial acylglycerols, initial oil acidity, and process conditions that influence their volatility, such as temperature, pressure, and the amount of stripping agent injected. A loss due to mechanical carryover or entrainment of the oil droplets by the rising vapor was also found in their experiments.

The high temperatures applied in the steam deacidification of vegetable oils also ease the occurrence of important chemical reactions, such as hydrolytic, oxidative, and thermal degradation reactions, which affect the final quality of refined oils. One important chemical reaction under study nowadays is the *cis–trans* isomerization of polyunsaturated fatty acids (PUFAs). The *cis*-isomer is an essential fatty acid in human metabolism. The *trans*-isomer, on the other hand, has effects similar to saturated fatty acids in human blood cholesterol. The initial content of *trans* PUFA in crude oils, which is usually lower than 0.3%, may increase to 5% during the deodorization/deacidification step. Refined edible oil should contain no more than 1.0% of *trans* PUFA to be considered as a good quality product in European countries [2].

Most of the published literature on steam deacidification and/or deodorization has been focused on quantifying experimental quality variables other than final oil acidity, such as the formation of *trans* fatty acid [3–6], waxes degradation [7], and tocopherol retention [8]. Relatively little attention has been paid to modeling and computational simulation of these processes. In this part of the chapter, the concepts underlying the appropriate modeling of steam deacidification and/or deodorization are presented. The main results of some of our published articles that deal with this subject [9, 10] are also summarized.

2.2.1 Modeling a Reactive Batch Deodorizer

2.2.1.1 Mathematical Equations

Previously in this chapter, the basic equations that describe conventional steam distillation were presented. Here, an extension of this standard model including chemical reactions is given. A scheme of a lab-scale batch deodorizer is shown in Figure 2.2.1

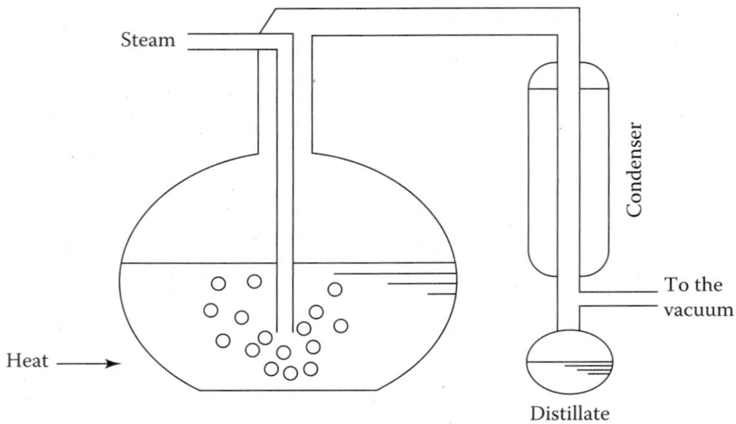

FIGURE 2.2.1 Scheme of a lab-scale batch deodorizer.

In this process, a still (batch deodorizer) is fed and then heated until the deodorization temperature is reached. Then, the injection of sparge steam begins promoting the volatilization of the undesirable substances, which are condensed and collected in a receiver. In this way, the whole deodorization time can be divided in two parts: *heating* (in absence of water) and *stripping* with sparge steam at constant temperature, which is allowed by the presence of small amounts of condensed steam that are dissolved into the oil. Despite this low level, water has a strong influence in the vapor–liquid equilibria of the whole multicomponent mixture.

The total and component molar balances for the reactive batch deodorizer are given by

$$\frac{dL}{dt} = -V + \Delta R_t, \tag{2.2.1}$$

and

$$\frac{d(L \cdot x_i)}{dt} = -V \cdot y_i + (R_i)_t, \tag{2.2.2}$$

where L is the total moles of liquid in the still, V is the molar vaporization rate in moles/time, x_i and y_i are the liquid and vapor molar fractions of component i in the liquid and vapor phases, respectively, ΔR_t is the total change of number of moles caused by reaction course (moles) at a given time, and $(R_i)_t$ is the number of moles of component i produced (or consumed) by the reaction (moles) at time t.

ΔR and $(R_i)_t$ can be calculated using the relations below:

$$(\Delta R)_t = \left(\sum_i R_i\right)_t \tag{2.2.3}$$

$$(R_i)_t = (k_i)_t . (x_i \cdot L)_t, \tag{2.2.4}$$

where $(k_i)_t$ is the constant of reaction of component i at time t.

For the distillate, the total and component molar balances are as follows:

$$\frac{dD}{dt} = V, \tag{2.2.5}$$

and

$$\frac{dD_i}{dt} = V \cdot y_i, \tag{2.2.6}$$

where D is the total moles of distillate and D_i represents the moles of component i in the distillate.

The molar vaporization rate, V, is a function of the heat supplied by the heating source to vaporize the volatiles and the vaporization enthalpy of the mixture. Ceriani and Meirelles [9] estimated an average molar vaporization value to be an input in the simulation program, based on the total amount of distillate formed during the experimental trials of Petrauskaitè et al. [1]. In this way, it was not necessary to do energy balances in their simulations.

2.2.1.2 Vapor–Liquid Equilibria and Vaporization Efficiency

The variables x_i and y_i that appear in Equation 2.2.2 are related to each other by the vapor–liquid equilibria at each instant:

$$y_i = x_i \cdot \frac{\gamma_i \cdot f_i^o}{\hat{\phi}_i \cdot P}. \tag{2.2.7}$$

For the system in discussion the total pressure is low; thus, assuming non-ideal gas behavior, the reference or standard-state fugacity f_i^o of Equation 2.2.7 is given by

$$f_i^o = P_i^{vap} \cdot \phi_i^{sat} \cdot exp\left(\frac{V_i^L \cdot \left(P - P_i^{vap}\right)}{RT}\right), \tag{2.2.8}$$

where R is the ideal gas constant, T is the absolute temperature of the system, P_i^{vap} and ϕ_i^{sat} are, respectively, the vapor pressure and the fugacity coefficient of the pure component i, and V_i^L is the liquid molar volume of component i. The exponential term corresponds to the Poynting factor.

At each time, Equation 2.2.9 is solved to determine the conditions in which the sum of the partial pressure of n compounds is equal to the system total pressure. During the *heating* period, the boiling point temperature of the fatty mixture should be determined by solving Equation 2.2.9. During the *stripping* period, the boiling temperature of the mixture is already set, and Equation 2.2.9 is solved to calculate the water concentration in the liquid phase at the chosen temperature and pressure conditions:

$$f = P - \sum_{i=1}^{n} x_i \cdot \frac{\gamma_i \cdot f_i^o}{\hat{\phi}_i}. \tag{2.2.9}$$

Ceriani and Meirelles [11] studied the vapor–liquid equilibria of fatty systems in detail. In their work, the fugacity coefficients were calculated using the virial equation truncated at the second term in combination with the appropriate mixing rules. Critical properties and acentric factors of the pure components, needed to calculate second virial coefficients, were estimated using Joback's technique for critical volumes and pressures and Fedor's group contributions for critical temperatures [12]. The V_i^L values for fatty compounds were obtained using the model developed by Halvorsen et al. [13]. The activity coefficients were determined using UNIFAC, and the vapor pressures were estimated by the group contribution equation suggested by Ceriani and Meirelles [11].

According to Ceriani and Meirelles [11], even at the low pressures that prevail in stripping units of the vegetable oil industry, it is necessary to include in the vapor–liquid calculations the fugacity coefficient ϕ_i^{sat} for water and fatty acids, because of the high values of P_i^{vap} at equilibrium temperatures in these cases. Ceriani and Meirelles [11] also found that UNIFAC $r^{3/4}$ [14] gave better predictions of vapor–liquid equilibrium data than original UNIFAC [15] and UNIFAC $r^{2/3}$ [16]. An earlier work of Fornari et al. [14] had similar conclusions for systems composed of vegetable oils and hexane.

One should note that Equation 2.2.7 assumes that the liquid and vapor phases are in equilibrium at each instant, which means that the steam becomes totally saturated with the volatiles as it passes through the oil in the still. The concept of vaporization efficiency is a measure of completeness with which the steam bubble becomes saturated with volatile substances during its passage through the oil layer. In 1941 Bailey [17] proposed a mathematical model for vaporization efficiency applied to steam (batch) deodorization that is still used today. At that time, the author discussed that a complete mathematical treatment of the phenomenon should consider two effects of the hydraulic pressure on the rising bubble: continuous variation on its surface area (the bubble expands significantly) and its internal pressure. In fact, because the pressure above the free surface of the liquid (P^o) is sufficiently low, 133 to 800 Pa for steam deodorization, the bubble formed at the orifice grows significantly as it ascends in a varying pressure field. As a consequence, the rising bubble expands with the decreasing external pressure, and the partial pressure of the solute, which is zero at the bottom, increases as the bubble moves toward the free surface. In an earlier work, Coelho Pinheiro and Guedes de Carvalho [18] modeled the stripping of pentane from sunflower seed oil using experimental results from the system at 298 K and pressures of 0.3 to 100 kPa. A detailed review about vaporization efficiency during steam distillation and deodorization can be found by referring to Ceriani and Meirelles [19].

2.2.1.3 Estimation of the Oil Composition

From computational simulation studies of steam deodorization and steam deacidification, it is possible to extract important information about the composition of the products (refined oil and distillate) throughout the stripping process, understanding the effects of the processing variables on the distribution of each compound or class of compounds. However, in order to achieve results with good quality, it is necessary to do an accurate estimation of the oil composition, in terms of its major compounds, such as TAG, and minor compounds, such as DAG, MAG, FFA, and nutraceuticals.

Oil composition is usually given in terms of fatty acids, as a result of the analysis by gas–liquid chromatography of the prepared methyl esters from the fatty acids attached to the glycerol part of TAG [20]. Statistical procedures, such as the one developed by Antoniosi Filho et al. [21], are capable of converting the fatty acid composition of the oil in its probable TAG composition with satisfactory accuracy, considering the distribution of the fatty acids in the three positions of the glycerol molecule. As inputs of this method, it is necessary to inform the percentage of tri-saturated TAG that usually appear in the oil, the mass concentration of fatty acids, and their molecular weights. The compositions in DAG and MAG can be estimated from the probable TAG composition, following the stoichiometric relations of the hydrolysis reactions in the following way: each TAG is split into 1,2- and 1,3-DAG; each DAG is then split into MAG.

Concentrations of minor compounds can be easily found in the literature [22] for a variety of oils.

2.2.2 Computational Simulation Results

For illustration, some phenomena that were originated from the simulation of the steam deacidification of coconut oil and of canola oil will be briefly summarized. Coconut oil is mainly composed of short-chain saturated fatty acids, which impart a lower boiling point (higher volatility) and a higher melting point to this vegetable oil. Its high content of FFAs (between 1 and 6%) denotes the presence of important quantities of DAG and MAG, which imply higher NOL. Canola oil, on the other hand, has important contents of mono- and polyunsaturated fatty acids, as oleic, linoleic, and linolenic acids, which imply a higher boiling point. Its initial content of FFA is low, being less than or equal to 1.2%. Considering that, the analyses of the results were focused on NOL in the study of the steam deacidification of coconut oil and on *trans* isomer formation in the case of canola oil. In both cases, the simulation results were compared with those reported in the literature [1, 3]. For further applications of our methodology, we also studied the reaction of decomposition of total aliphatic waxes during the deodorization of canola oil.

2.2.3 Neutral Oil Loss

To quantify NOL during steam deacidification of coconut oil and to study the effect of some processing variables, Petrauskaitè et al. [1] conducted some experiments in a lab-scale batch deodorizer while varying temperature, pressure, and percentage of steam in relation to the initial mass of oil. To simulate their experiments, the same fatty acid composition and initial acidity of coconut oil (3.18%, expressed as percentage of lauric acid) reported by Petrauskaitè et al. [1] were used. Petrauskaitè et al. [1] did not give the partial acylglycerol composition of their samples, giving us some discretion to vary its value and study the effect of the initial content of DAG and

MAG in the NOL. Three different compositions were considered in the simulations. Composition 1 (COC1) had 3% mass concentration of DAG and 1% of MAG, according to Loncin [23]. Composition 2 (COC2) had an intermediate content of partial acylglycerols: 0.89% mass concentration of DAG and 0.27% of MAG. Composition 3 (COC3), on the other hand, had none (0% DAG and 0% MAG).

From the fatty acid composition reported by Petrauskaitè et al. [1], the oil composition in terms of TAG, DAG, MAG, and FFAs were estimated using the procedure already discussed. As a whole, the probable coconut oil had 72 components: nine fatty acids, 36 TAG, 18 DAG, and nine MAG. More details about the oil composition are provided by Ceriani and Meirelles [9].

The first six experiments reported by Petrauskaitè et al. [1] were simulated, and the comparison of NOL and of final oil acidity is shown in Figures 2.2.2 and 2.2.3. As one can see, the experimental data were within the range of the simulation results, indicating that the coconut oil used by Petrauskaitè et al. [1] in their experiments might have had a value between COC1 and COC3, in terms of its partial acylglycerol concentration.

The simulation results show that NOL was proportional to the initial concentration of MAG and DAG, increasing as the oil composition changed from COC3 to COC1. In fact, as the concentration of MAG and DAG increased, part of these components was vaporized and collected in the distillate instead of FFAs, increasing the refined oil acidity and the NOL. A possible explanation of this fact is the similarity that exists between the volatility of short-chain MAG and long-chain fatty acids. As

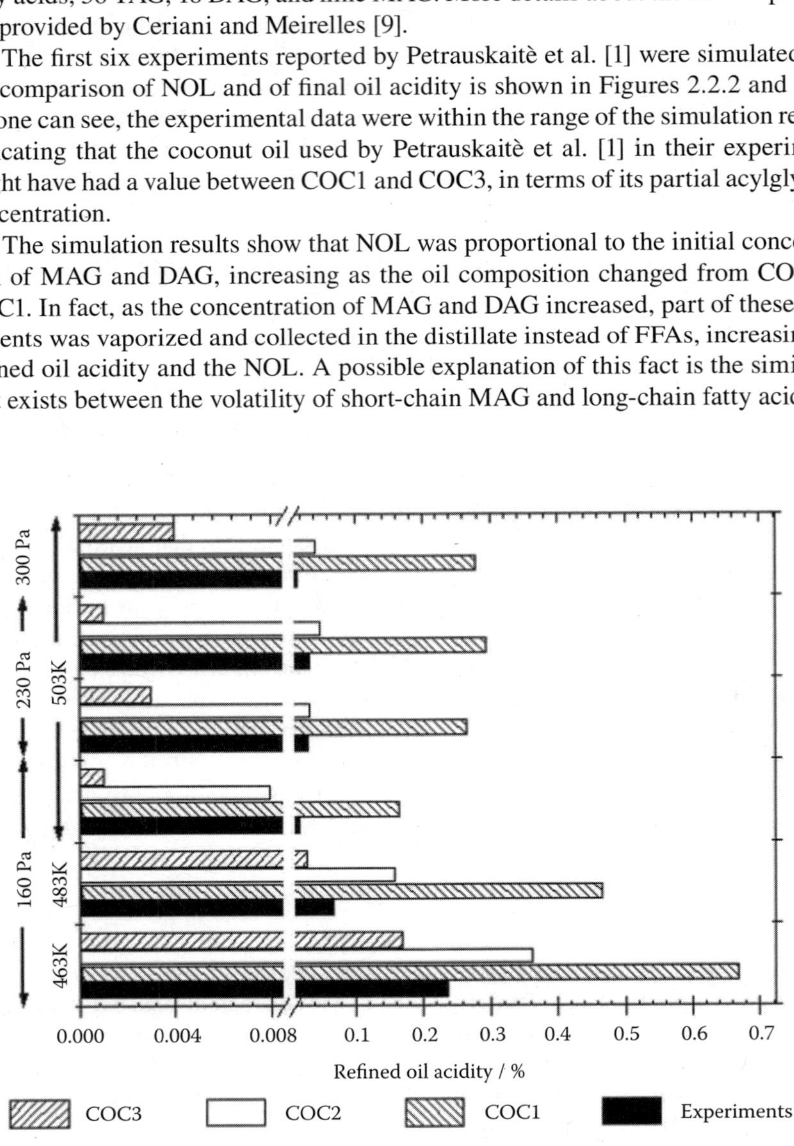

FIGURE 2.2.2 Comparison of simulation and experimental results for refined oil acidity.

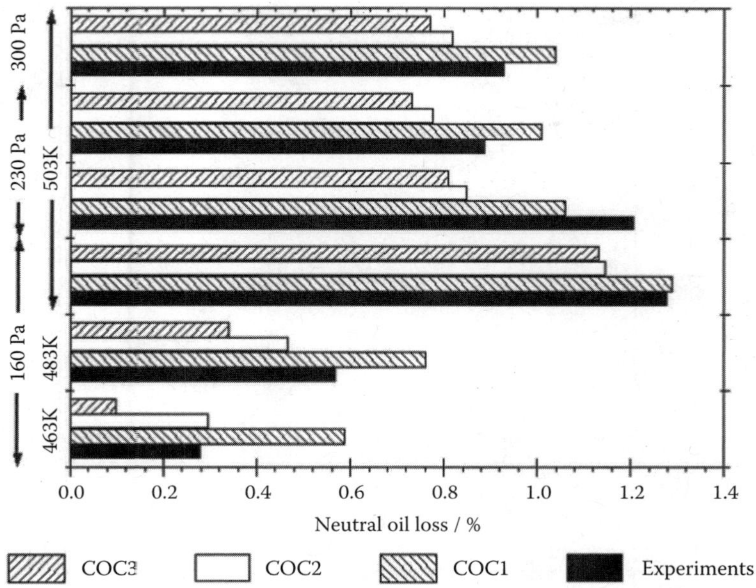

FIGURE 2.2.3 Comparison of simulation and experimental results for neutral oil loss.

expected, the refined oil acidity and NOL increased with temperature and vacuum intensity.

A further advantage of simulating batch processes is that it allows drawing the profiles of variables of interest as a function of time. To explore this tool, Figure 2.2.4 shows the profiles per time of the FFA content of the oil and of the distillate for the simulations of steam deacidification of COC2 conducted at 160 Pa and 463, 483, and 503 K.

As one can see, the oil acidity decreased with time as a consequence of the vaporization of the FFA. The profiles of the distillate acidity, on the other hand, show that an important vaporization of acylglycerols starts at 30 min of processing, reducing considerably the FFA content in the distillate. Note, in Figure 2.2.5, that this fact was more evident at 503 K, when this class of compounds was even more volatile and competitive in the vaporization process, causing also the stabilization of the oil acidity at the end of the process. As one can see in Figure 2.2.4a, in the last 10 minutes of deacidification, when the oil acidity was very close to zero, there was an important increase in the losses of TAG and DAG.

From our simulations, it is also possible to evaluate the behavior of each compound during the steam deacidification process. Figure 2.2.6 shows the profiles of the main FFAs found in coconut oil at 160 Pa and 503 K. As one can see, for the first 20 minutes, short-chain FFA was the key fraction distilled from the oil, being completely removed after 49 min of processing. At this time, the coconut oil had an oil acidity of 0.337%, formed mainly by long-chain saturated and unsaturated FFAs, such as stearic, oleic, and linoleic acids.

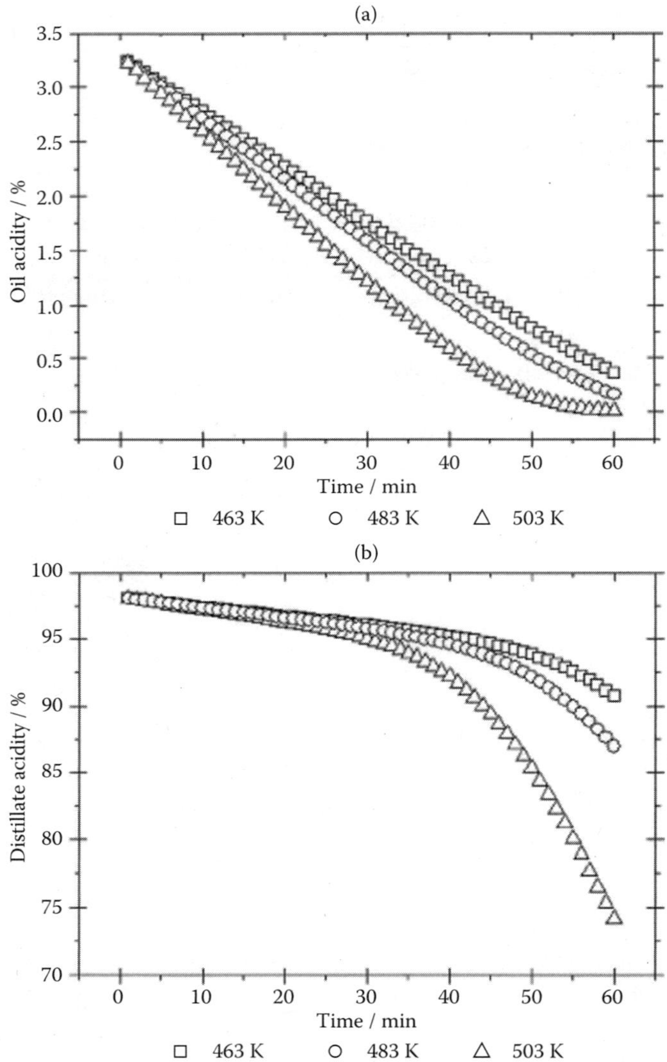

FIGURE 2.2.4 Variation of the FFA content of the oil (a) and of the distillate (b) with time at 160 Pa. 463 K (□), 483 K (○), and 503 K (Δ).

2.2.4 *Cis–Trans* Isomerization

To study the formation of *cis–trans* isomers during the steam deacidification of canola oil, it was first necessary to establish the Arrhenius type equations for the reaction of linoleic (Li) and linolenic (Ln) acids attached to the TAG. The *k* values (min^{-1}) were measured and adjusted by Hénon et al. [4], according to the equations below. For linoleic acid:

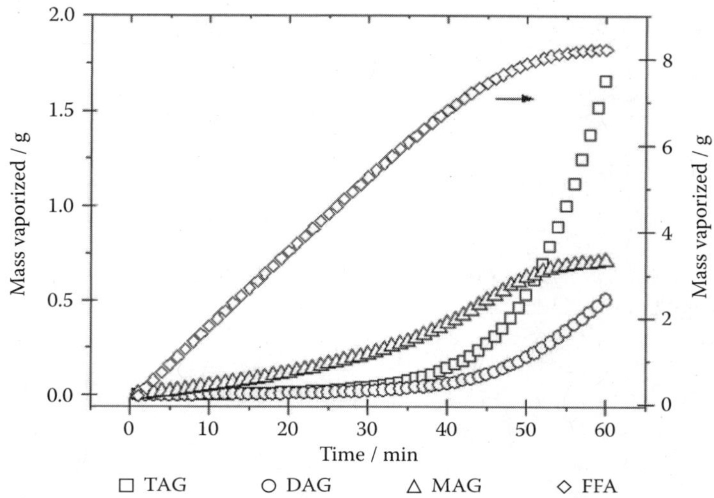

FIGURE 2.2.5 Vaporization of TAG, DAG, MAG, and FFA during the steam deacidification of COC2 at 160 Pa and 530 K. Initially, there were 8.0 g of FFA, 2.2 g of DAG, 0.7 g of MAG, and 239.1 g of TAG in the oil.

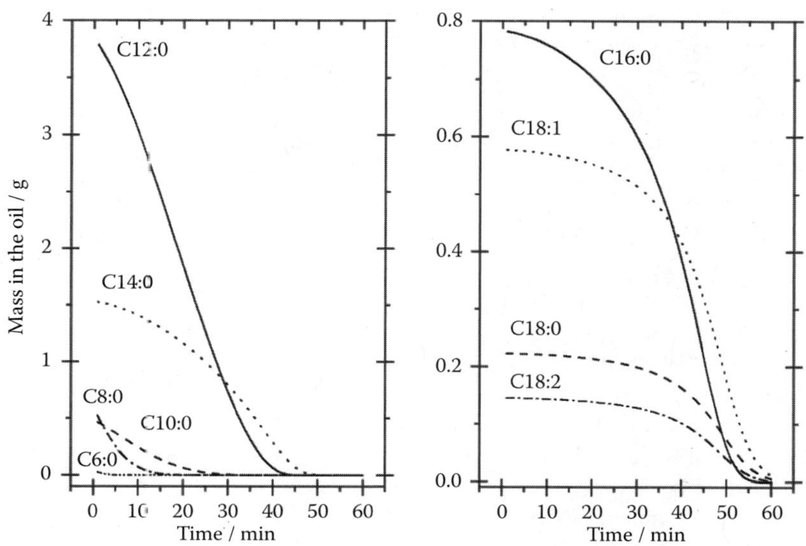

FIGURE 2.2.6 Vaporization of individual FFAs during the steam deacidification of COC2 at 160 Pa and 230°C. Initially, there were 0.03 g of C6:0, 0.52 g of C8:0, 0.47 g of C10:0, 3.79 g of C12:0, 1.53 g of C14:0, 0.78 g of C16:0, 0.22 g of C18:0, 0.58 g of C18:1, and 0.15 g of C18:2 in the oil.

$$k_{Li} = \frac{1}{3600} \cdot 10^{-7921.95/T + 12.76} \tag{2.2.10}$$

and for linolenic acid:

$$k_{Ln} = \frac{1}{3600} \cdot 10^{-6796.63/T + 11.78}, \tag{2.2.11}$$

where k_{Li} and k_{Ln} are given in seconds and temperature in Kelvin.

In Equations 2.2.3 and 2.2.4, ΔR and R_i were calculated only for TAG that contain Li acid and/or Ln acid attached. Note that the k_i values in Equation 2.2.4 should be calculated for each TAG of canola oil containing Li acid and/or Ln acid as a sum of k_{Li} and/or k_{Ln}, calculated using Equation 2.2.10 or 2.2.11 for each appearance of these fatty acids in the TAG molecule. As examples, suppose a TAG of type JWLi or JWLn (a component i of the multicomponent mixture), where J and W are types of fatty acids; then $k_i = k_{Li}$ or $k_i = k_{Ln}$, respectively. For a TAG of type JLiLn, $k_i = k_{Li} + k_{Ln}$; for one of type LiLiLi, $k_i = 3 \cdot k_L$, and so forth.

Each *cis* TAG was isomerized to its correspondent *trans,* supposing that all PUFAs attached to it isomerized at the same time. In this way, a *cis* TAG of type OLi*cis*Ln*cis* would isomerize to its correspondent *trans* TAG: OLi*trans*Ln*trans*, not OLi*cis*Ln*trans* or OLi*trans*Ln*cis*. Such simplifying assumptions allow incorporating easily the *cis–trans* reaction kinetics into the simulation algorithm.

Being a first-order reaction, the rates of formation of *trans* TAG, containing *trans* Li and/or *trans* Ln, are proportional to the concentration of the reacting substance (a *cis* TAG, in this case). In this way, it is straightforward to understand that the initial contents of Li and Ln acids in the oil influence the final amount of *trans* isomers of these fatty acids in the deacidified oil. Ceriani and Meirelles [10] analyzed, by response surface methodology and computational simulation, the effect of the composition of canola oil, in terms of Li and Ln levels, on the final *trans* content in the steam deacidified oil. More details about the canola oil compositions estimated using the statistical procedure of Antoniosi Filho et al. [21] can be found by referring to Ceriani and Meirelles [10]. In their factorial design, duration of the batch and temperature were also included as independent variables, following the same ranges given in the experimental design of Hénon et al. [3]. In the total, 25 simulations were performed by Ceriani and Meirelles [10] as a result of a factorial design composed of 2^4 trials plus a star configuration and one central point. The coded variables (designated as X_k), which ranged from –2 to +2 in the factorial design, were set within the following limits of the real variables: temperature (X_1) 463 K $\leq T \leq$ 523 K, duration of the batch (X_2) 1 h $\leq t \leq$ 5 h, initial content of *cis* Li acid (X_3) 18% $\leq$ % C18:2*cis* $\leq$ 30% and initial content of *cis* Ln acid (X_4) 6% $\leq$ % C18:3*cis* $\leq$ 14%.

Figure 2.2.7 shows the profiles of C18:2*cis* (%), C18:3*cis* (%), C18:2*trans* (%), and C18:3*trans* (%) as a function of time for the deacidification of canola oil at 220°C and for a 3-h duration. As one can see, the initial levels of C18:2*cis* (%) and C18:3*cis* (%) decreased slightly. On the other hand, the content of C18:3*trans* (%) increased even more than the content of C18:2*trans* (%), because the C18:3*cis* acid is more reactive (three unsaturations).

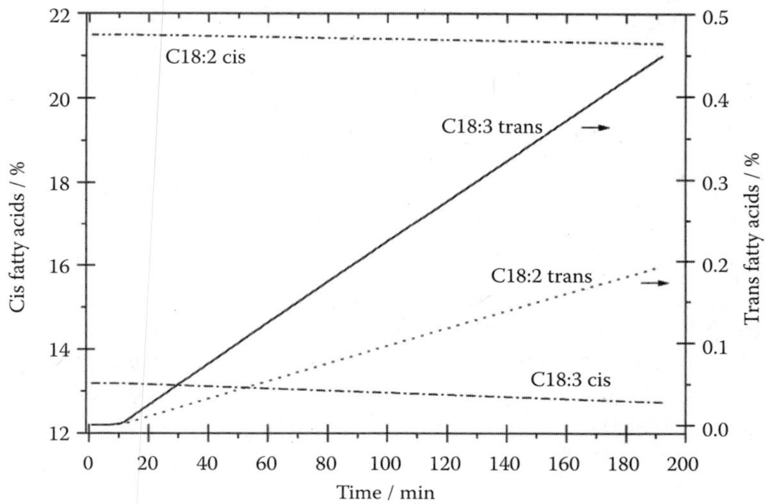

FIGURE 2.2.7 Changes in the content of C18:2*cis*, C18:2*trans*, C18:3*cis*, and C18:3*trans* (mass %) during deodorization of canola oil at 220°C and for 3-h duration.

Using the quadratic models obtained by Ceriani and Meirelles [10] from the statistical analysis of the simulation results in terms of the percentage of C18:2*trans* (%), C18:3*trans* (%), and TOTAL *trans* PUFA (%), shown in Equations 2.2.12 through 2.2.14, it was possible to compare the computational simulation tool with the experimental work of Hénon et al. [3]. Note that these models presented very high correlation coefficients, in addition to an adequate analysis of variance for the responses at 99.0% of confidence. In this way, they were capable of describing the effects of the coded variables on the three responses studied:

$$\log_{10}\left[\text{C18}:2trans\ (\%,mass)\right] = -0.7210 + 0.4272 \cdot X_1 + 0.1542 \cdot X_2 - 0.0289 \cdot X_2^2 + 0.0418 \cdot X_3 + 0.0554 \cdot X_4 \quad (2.2.12)$$

$$\log_{10}\left[\text{C18}:3trans\ (\%,mass)\right] = -0.3879 + 0.4123 \cdot X_1 + 0.1534 \cdot X_2 - 0.0256 \cdot X_2^2 + 0.0333 \cdot X_3 + 0.1021 \cdot X_4 \quad (2.2.13)$$

$$\log_{10}\left[\text{Total } trans \text{ PUFA } (\%,\ mass)\right] = -0.2212 + 0.4170 \cdot X_1 + 0.1537\, X_2 - 0.0267 \cdot X_2^2 + 0.0352\, X_3 + 0.0873 \cdot X_4 \quad (2.2.14)$$

Nine combinations of time and temperature resulted from the factorial design set by Ceriani and Meirelles [10]: 478 K and 2 h ($X_1 = -1$ and $X_2 = -1$), 508 K and 2 h ($X_1 = +1$ and $X_2 = -1$), 478 K and 4 h ($X_1 = -1$ and $X_2 = +1$), 508 K and 4 h ($X_1 = +1$ and $X_2 = +1$), 463 K and 3 h ($X_1 = -2$ and $X_2 = 0$), 523 K and 3 h ($X_1 = +2$ and $X_2 = 0$),

493 K and 1 h ($X_1 = 0$ and $X_2 = -2$), 493 K and 5 h ($X_1 = 0$ and $X_2 = +2$), and 493 K and 3 h ($X_1 = 0$ and $X_2 = 0$). Using Equations 2.2.12 and 2.2.13, it was possible to investigate the influence of the oil composition regarding the initial content of Li (18 to 30%) and Ln (6 to 14%) in the oil. The results are shown in Table 2.2.1. It is possible to note that the minimum and maximum values of C18:2*trans* and C18:3*trans* were obtained respectively for the lower and the higher temperatures (463 and 523 K), indicating the importance of this variable in the formation of *trans* fatty acids, independent of the initial content of Li and/or Ln. As one can see, in six of the nine combinations shown in Table 2.2.1, the minimum values of C18:2*trans,* calculated using Equation 2.2.12 with 18% of Li and 6% of Ln ($X_3 = -2$ and $X_4 = -2$), were very close to the experimental value. On the other hand, in seven of the nine combinations, the maximum values of C18:3*trans,* calculated using Equation 2.2.13 with 30% of Li and 14% of Ln ($X_3 = +2$ and $X_4 = +2$), were closer to the value reported by Hénon et al. [3]. These facts suggest that the composition of the canola oil used by Hénon et al. [3] might be not far from the minimum value in terms of Li and from the maximum value in terms of Ln.

The combination of computational simulation and response surface methodology allowed analysis of the influence of two factors that would be difficult to control in experimental trials of natural oils, such as their initial levels of *cis* Li and *cis* Ln acids. The relevance of these variables for an industrial plant of small size relies on the seasonality of crops and in the variation of the oils processed.

2.2.5 Waxes Degradation

The turbidity (haze, cloudiness) formation during the storage under normal warehouse conditions is a problem recently observed in bottled canola oil and can affect consumer preferences. Usually, 100–200 mg/kg of waxes, which would crystallize

TABLE 2.2.1
Comparison between the Experimental Values of *Trans* PUFA and the Minimum and Maximum Values Calculated with Equations 2.2.12 and 2.2.13

	T = 463 K	T = 478 K		T = 493 K			T = 508 K		T = 523 K
	t = 3 h	t = 2 h	t = 4 h	t = 1 h	t = 3 h	t = 5 h	t = 2 h	t = 4 h	t = 3 h
C18:2 (%)									
Mininum value	0.02	0.03	0.06	0.05	0.12	0.19	0.21	0.43	0.87
Hénon et al. [3]	0.07	0.05	0.07	0.10	0.12	0.19	0.20	0.34	0.64
Maximum value	0.04	0.07	0.15	0.12	0.30	0.46	0.52	1.06	2.13
C18:3 (%)									
Minimum value	0.03	0.06	0.11	0.09	0.22	0.35	0.38	0.76	1.47
Hénon et al. [3]	0.07	0.17	0.39	0.28	0.66	1.11	1.14	2.11	3.41
Maximum value	0.11	0.20	0.40	0.30	0.76	1.22	1.31	2.65	5.10

The comparison between the experimental values of *trans* PUFA (mass %) is from Hénon et al. [3], and the minimum and maximum values were calculated with Equations 2.2.12 and 2.2.13, considering the limits of the initial Li and Ln contents in the factorial design from Ceriani and Meirelles [10].

at room temperature, are removed by chilling the oil in a continuous heat exchanger to about 278 K, and filtering it. Wax contents lower than 50 mg/kg no longer produces a visible haze. The crystallization and/or filtration are expensive processes because of the associated neutral oil losses and energy requirements. In this context, steam deacidification and/or deodorization could be a previous step for helping in the removal of waxes from canola oil.

An aliphatic wax is a result of the esterification of a long-chain fatty acid and a long-chain fatty alcohol. Tubaileh et al. [7] established the kinetics of decomposition of waxes of 36, 38, 40, 42, 44, and 46 carbon atoms, during deodorization of olive oil. The reactions were modeled as of order "zero," with their constants following the Arrhenius' law. Tubaileh et al. [7] did not specify which fatty acid and fatty alcohol were produced during the decomposition of these waxes, but Przybylski et al. [24] found different fatty alcohol chain lengths in the analysis of sediments isolated from bottled canola oil. Among them, the main fractions were 22, 24, 26, and 28 carbon atoms.

To study the decomposition of waxes during deodorization of canola oil by computational simulation, we selected some combinations of temperature and duration from Table 2.2.1 (463 K and 1 h, 463 K and 3 h, 478 K and 1 h, 478 K and 3 h, 493 K and 1 h, 493 K and 1 h, 493 K and 3 h, and 508 K and 1 h). A canola oil with 21.0% Li and 8.0% Ln was selected for this investigation [10], including 198 mg/kg (0.0198%) of waxes (0.0033% for each type of wax) in the complete composition of the oil. It was supposed that in the beginning of steam deacidification there was no fatty alcohol in the oil.

The R_i values that appear in Equations 2.2.3 and 2.2.4 were calculated using Equation 2.2.15 for each wax and its corresponding fatty acid and fatty alcohol. The vapor–liquid equilibria were calculated according to the procedure already described, with the vapor pressure estimated inclusive for fatty alcohols and waxes.

$$(R_i)_t = (k_i \cdot V_{oil})_t, \qquad (2.2.15)$$

where V_{oil} is the calculated oil volume in m^3 for each instant t, using the method of Halvorsen et al. [13], and $(R_i)_t$ is given in kmol of $i \cdot s^{-1}$ and k_i in kmol of $i \cdot m^{-3} \cdot s^{-1}$.

The values of k_i were taken from Tubaileh et al. [7]. In Equation 2.2.15, R_i is negative for waxes and positive for fatty acids and fatty alcohols. It was supposed that wax C36 degradated in a fatty acid of type C16:0 and a fatty alcohol of type C20:0, wax C38 degradated in a fatty acid of type C16:0 and a fatty alcohol of type C22:0, wax C40 degradated in a fatty acid of type C16:0 and a fatty alcohol of type C24:0, wax C42 degradated in a fatty acid of type C18:0 and a fatty alcohol of type C24:0, wax C44 degradated in a fatty acid of type C18:0 and a fatty alcohol of type C26:0, and wax C46 degradated in a fatty acid of type C18:0 and a fatty alcohol of type C28:0.

Changes in the contents of total waxes for the selected conditions are shown in Figure 2.2.8. In general, the initial content of total waxes decreased during deodorization, and the degradation of waxes was more intense for the lower temperature studied. In fact, Tubaileh et al. [7] found that k values for the decomposition of waxes

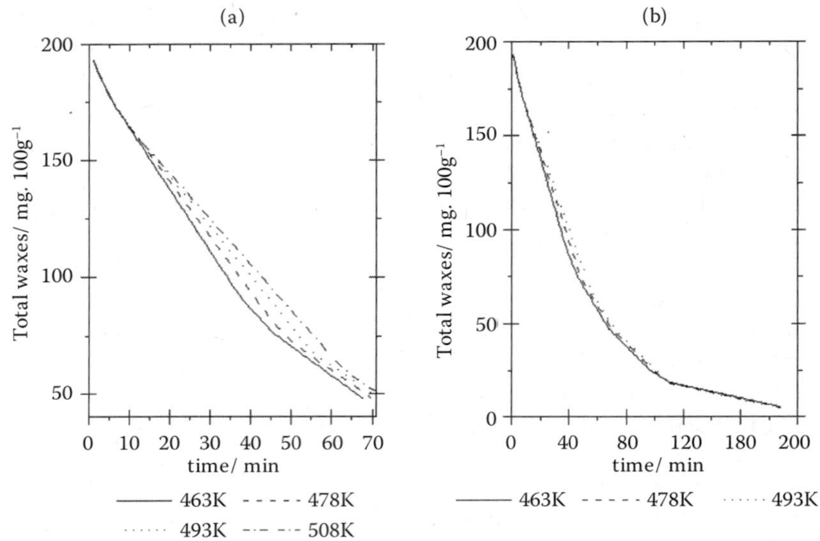

FIGURE 2.2.8 Changes in the total wax content (mg/kg) during deodorization of canola oil for selected conditions: (a) 1-h duration and (b) 3-h duration.

decreased with an increase in the temperature in the range of temperature values investigated. In all cases studied, the simulation program generated final levels of waxes lower than 50 mg/kg. For the same processing time (1 or 3 h), there were no important differences among the final content of total waxes as a consequence of temperature. As one can see in Figure 2.2.8A, 47 to 48 mg/kg of waxes were still in the deodorized oil, and in Figure 2.2.8B, only 3 to 5 mg/kg of waxes were not decomposed. The sharpest decreases were found at the beginning of the deodorization, at lower temperatures. Our simulation results showed that steam deacidification could be designed to decompose waxes, in a way that reduces the necessity of further steps for their removal.

2.2.6 Nomenclature

Acronym	Description
COC1	Composition 1 (3% mass concentration of DAG and 1% of MAG)
COC2	Composition 2 (0.89% mass concentration of DAG and 0.27% of MAG)
COC3	Composition 3 (0% DAG and 0% MAG)
DAG	Diacylglycerols
FFA	Free fatty acids
Li	Linoleic acid
Ln	Linolenic acid
MAG	Monoacylglycerols
NOL	Neutral oil loss
PUFA	Polyunsaturated fatty acids
TAG	Triacylglycerols

Symbol	Definition	Units in SI system	Dimensions in M, N, L, T, and θ
f_i^o	Standard state fugacity of pure component *i*	Pa	$ML^{-1}T^{-2}$
P_i^{vap}	Vapor pressure of component *i*	Pa	$ML^{-1}T^{-2}$
V_i^L	Liquid molar volume of component *i*	$m^3 \cdot kmol^{-1}$	$L^3 N^{-1}$
$(k_i)_t$	Constant of reaction of component *i* at time *t* in Eq. (2.2.4)	S^{-1}	T^{-1}
$(k_i)_t$	Constant of reaction of component *i* at time *t* in Eq. (2.2.15)	$kmol \cdot m^{-3}s^{-1}$	$NL^{-3}T^{-1}$
$(R_i)_t$	Moles of component *i* produced (or consumed) by the reaction (moles) at time *t*	$kmol \cdot s^{-1}$	$N \cdot T^{-1}$
ΔRt	Total change of number of moles caused by reaction course (moles) at a given time	$kmol \cdot s^{-1}$	$N \cdot T^{-1}$
D	Total moles of distillate	kmol	N
D_i	Moles of component *i* in the distillate	kmol	N
L	Total moles of liquid in the still	kmol	N
P	Pressure	Pa	$ML^{-1}T^{-2}$
R	Gas constant	$J \cdot kmol^{-1} \cdot K^{-1}$	$MN^{-1}L^2T^{-2}\theta^{-1}$
T	Absolute temperature of the system	K	θ
t	Time	s	T
V	Molar rate of vaporization	$kmol \cdot s^{-1}$	$N \cdot T^{-1}$
V_{oil}	Oil volume for each instant *t*	m^3	L^3
x_i	Molar fraction of component *i* in the liquid phase	—	—
X_k	Coded variable of factorial design	—	—
y_i	Molar fraction of component *i* in the vapor phase	—	—
Greek letter			
$\hat{\phi}_i$	Vapor-phase fugacity coefficient	—	—
ϕ_i^{sat}	Fugacity coefficient of the pure component *i*		
γ_i	Activity coefficient of component *i* in the liquid phase	—	—

2.2.7 Acknowledgments

R. Ceriani thanks Fundação de Amparo à Pesquisa do Estado de São Paulo (FAPESP) for the postdoctoral fellowship (05/02079-7). The authors thank FAPESP for financial support (05/53095-2).

2.2.8 References

1. Petrauskaitè, V., W. F. De Greyt, and M. J. Kellens. 2000. Physical refining of coconut oil: Effect of crude oil quality and deodorization conditions on neutral oil loss. *Journal of the American Oil Chemists' Society* 77:581–586.
2. Aro, A., J. Van Ameslvoort, W. Becker, et al. 1998. Trans fatty acids in dietary fats and oils from 14 European countries: The TRANSFAIR study. *Journal of Food Composition and Analysis* 11:137–149.
3. Hénon, G., P. Y. Vigneron, B. Stoclin, and J. Caigniez. 2001. Rapeseed oil deodorization study using the response surface methodology. *European Journal of Lipid Science Technology* 103:467–477.
4. Hénon, G., Z. Zeményny, K. Recseg, F. Zwobada, and K. Kövári. 1999. Deodorization of vegetable oils. Part 1: Modeling the geometrical isomerization of polyunsaturated fatty acids. *Journal of the American Oil Chemists' Society* 76:73–81.
5. Kemény, Z., K. Recseg, G. Hénon, K. Kövari, and F. Zwobada. 2001. Deodorization of vegetable oils: Prediction of trans polyunsaturated fatty acid content. *Journal of the American Oil Chemists' Society* 78:973–979.
6. León-Camacho, M., M. V. Ruiz-Méndez, M. M. Graciani-Constante, and E. Graciani-Constante. 2001. Kinetics of the cis-trans isomerization of linoleic acid in the deodorization and/or physical refining of edible oils: Prediction of trans polyunsaturated fatty acid content. *Journal of Lipid Science Technology* 103:85–92.
7. Tubaileh, R. M., M. M. Graciani Constante, M. León-Camacho, A. López López, and E. Graciani-Constante. 2002. Kinetics of the decomposition of total aliphatic waxes in olive oil during deodorization. *Journal of the American Oil Chemists' Society* 79:971–976.
8. De Greyt, W. F., M. J. Kellens, and A. D. Huyghebaert. 2001. Effect of physical refining on selected minor compounds in vegetable oils. *Fett/Lipid* 101:428–432.
9. Ceriani, R., and A. J. A. Meirelles. 2004. Simulation of batch physical refining and deodorization processes. *Journal of the American Oil Chemists' Society* 81:305–312.
10. Ceriani, R., and A. J. A. Meirelles. 2007. Formation of trans PUFA during deodorization of canola oil: A study through computational simulation. *Chemical Engineering and Processing* 46:375–385.
11. Ceriani, R., and A. J. A. Meirelles. 2004. Predicting vapor-liquid equilibria of fatty systems. *Fluid Phase Equilibria* 215:227–236.
12. Reid, Robert C., J. M. Prausnitz, and B. E. Poling. 1987. *The properties of gases and liquids*. New York: McGraw-Hill.
13. Halvorsen, J. D., W. C. Mammel, and L. D. Clements. 1993. Density estimation for fatty acids and vegetable oils based on their fatty-acid composition. *Journal of the American Oil Chemists' Society* 70:875–880.
14. Fornari, T., S. Bottini, and E. A. Brignole. 1994. Applications of UNIFAC to vegetable oils–alkanes mixtures. *Journal of the American Oil Chemists' Society* 71:391–395.
15. Fredenslund, A., J. Gmehling, and P. Rasmussen. 1977. *Vapor-liquid equilibria using UNIFAC*. Amsterdam: Elsevier.
16. Kikic, I., P. Alessi, P. Rasmussen, and A. Fredenslund. 1980. On the combinatorial part of the UNIFAC and UNIQUAC models. *Canada Journal of Chemical Engineering* 58:253–258.
17. Bailey, A. E. 1941. Steam deodorization of edible fats and oils. *Industrial Engineering and Chemistry* 33:404–408.
18. Coelho Pinheiro, M. N., and J. R. F. Guedes de Carvalho. 1994. Stripping in a bubbling pool under vacuum. *Chemical Engineering Science* 49:2689–2698.
19. Ceriani, R., and A. J. A. Meirelles. 2005. Modeling vaporization efficiency for steam refining and deodorization. *Industrial and Engineering Chemistry Research* 44:8377–8386.

20. AOCS. 1993. Preparation of methyl esters of long-chain fatty acids. In *Official methods and recommended practices of the American Oil Chemists' Society*, Ce 2-66. Champaign, IL: AOCS Press.
21. Antoniosi Filho, N. R., O. L. Mendes, and F. M. Lanças. 1995. Computer prediction of triacilglicerol composition of vegetable oils by HRGC. *Journal of Chromatography A* 40:557–562.
22. O'Brien, R. D. 2004. *Fats and oils: Formulating and processing for applications*. New York: CRC Press.
23. Loncin, M. 1962. L'hydrolyze spontanée des huiles glycéridiques et en particulier de l'huile de palme. Couillet, Hainut, Belgium: Maison-D'Edition.
24. Przybylski, R., C. G. Biliaderis, and N. A. Michael Eskin. 1993. Formation and practical characterization of canola oil sediment. *Journal of the American Oil Chemists' Society* 70:1009–1015.

2.3 OBTAINING VOLATILE OILS BY STEAM DISTILLATION: STATE OF THE ART

Juliana M. Prado, Patrícia F. Leal, and M. Angela A. Meireles

2.3.1 Steam Distillation

In spite of being widely used in food and other industries, distillation is a major energy consumer process. During the energy "crisis" of the 1970s, much effort was put into making this process more efficient. Recent developments of energy shortages have refocused attention on major industrial energy users, because there is a global trend of preserving natural resources.

The distillation process may be continuous or in batch. The idea of continuous distillation is that the amount going into the still and the amount leaving the still should always equal each other at any given point in time. The simplest example of a batch process is the old-fashioned spirit making (see Chapter 3). The distiller fills a container at the start and then heats it; then, the vaporized mixture is condensed to make the alcoholic drink. When the proper quantity of drink is made, the distiller stops the still and empties it out, being then ready for a new batch.

Fractionation systems may have different objectives: the removal of light components from heavy products (stripping, see Section 2.2), the removal of heavy components from light products (rectification), or the removal of light material from heavy product and of heavy material from light product at the same time (fractionation).

One modified distillation process is steam distillation (SD). It is widely used for recovering compounds from solid matrices, such as aromatic, condimentary, and medicinal plants. Volatile oil (VO) and the residual vegetal matrix can be separated both by hydrodistillation and SD, which are processes used in industry since antiquity [1]. Although VOs may be extracted through a hydrodistillation process, long contact time leads to degradation or hydrolysis, which can be avoided by SD [2]. Thus, batch SD is the classical process for obtaining VO from condimentary, medicinal, and aromatic plants.

On a laboratorial scale, the most used distillation method for obtaining VO is hydrodistillation. There is wide research work on the identification of the chemical composition and on the biological activity of VOs obtained by hydrodistillation. However, on an industrial scale, the most common distillation technique used is SD.

Although the phenomenon involved in both techniques is the same, the yield of each process may be different, as long as the chemical composition of the VO is subject to variation.

Different operating conditions of a single extraction method can also positively or negatively influence the quality and the yield and therefore, the cost of manufacturing the VO. The literature reports the effect of different distillation methods on the content (yield), chemical composition, and biological activity of VOs [3–10].

Comparing the SD process on laboratorial and industrial scales, some important differences should be noted. At the laboratory, for research purposes, the SD process frequently uses selected parts of the plant, while in industry the plant material is used just as it has been collected from the field. Moreover, laboratorial SD is exhaustive, leading to reproducible results for the oil chemical composition. Koedan 1982, cited by Mateus et al. [11], emphasized the contribution of the operational conditions to the variations on the oil chemical composition. Thus, the industrial operation does not have to be exhaustive, but should be carried out until the desired chemical composition of the oil is attained.

Another major point to be considered in industry is the energy consumption. It is closely related to the process or cycle time. The process time of the SD process is as important as for any other extraction process. It is strongly connected to the steam flow rate. At the end of the distillation process, the increase observed in oil yield is very low if compared to the beginning of extraction, leading to longer processing and higher energy consumption [12]. With shorter distillation periods, the chemical composition of the oil can be representative, although it will not usually be exactly the same as that of the exhaustive processing.

The SD equipment is multipurpose and, therefore, is adequate for obtaining a wide variety of active principles from aromatic and condimentary plants. However, it is less adequate or even inadequate for processing vegetal matrices that possess thermosensitive active principles or when the degradation product of a thermosensitive component is toxic. Figure 2.3.1 shows an industrial unit of a multipurpose SD process.

Distillation has always been the most commonly used method for the recovery of essential oils, because it takes advantage of their volatility. The components in VOs, however, have much higher boiling points than water; therefore, they are actually distilled with steam. The steam acts as a carrier and removes the oil vapors, which have been evaporated well below their boiling point. This is especially important because many of the VO components have high boiling points and would thermally degrade far below their normal boiling points. After condensation, the oils and water are immiscible and thus are easily separated.

In the cases where the separation is more complicated or when the amount of oil recovered is too low, there are some alternatives. One of them is increasing separation time for a few days, if it is necessary. Another possibility is dissolving salt in the emulsion, although this procedure downgrades the hydrosol. The emulsion can also be frozen and then separated. Finally, an organic solvent immiscible in water, such as dichloromethane, toluene, and hexane or petroleum ether, can be added to the emulsion. In that case, the global process can no longer be considered clean or green process.

FIGURE 2.3.1 (a) Steam distillation unit used to produce volatile oils, (b) stills, (c) condenser, and separator of oil and hydrosol. (From LINAX, Votuporanga, Brazil, www.clinax.com.br. With permission.)

During SD, two different products are obtained: VO and hydrosol (nonalcoholic condensed water). Little amounts of the aromatizing compounds are present in hydrosol, conferring to it a pleasant aroma. Many hydrosols obtained from SD of flowers and leaves have great potential for usage by the cosmetic, food, and pharmaceutical industries. They can be used in aqueous medium formulations of cosmetics, lotions, soaps, foods, and beverages and as ambient aromatizers. The usage of the hydrosols by other industries can prevent pollution, since the presence of organic compounds in wastewaters increases the chemical oxygen demand [12]. However, the hydrosols are usually discarded by companies that do not know their selling potential.

Some compounds of VOs are lost with the residual water (hydrosol). In the case where the vegetal matrix and the water are mixed in the reservoir (hydrodistillation), part of the VO may be lost with both the reservoir water and the aqueous phase condensed after the condenser. The residual oil dissolved in the wastewater does not always have a pleasant aroma and may also cause an unpleasant odor. The alternative for recovering this oil is to redistill the water (reservoir water and/or aqueous condensate). However, the redistillation process increases the cost of utilities because of the energy costs involved in that process.

Although the traditional SD for obtaining VO is not a process involving patents and the instrumentation is not critical because it is a widespread process, information related to process conditions (e.g., temperature, pressure, cycle time) is restricted. The patents found in the database of the United States Patent and Trademark Office (USPTO) are derivations of the traditional process. For example, the patent

US4319963 from March 16, 1982, suggests modifications in the equipment aiming to decrease the vapor condensation in the vegetal material and, therefore, decrease the possibility of hydrolysis that directly affects the quality of the essential oil. It also reduces the risk of degradation of the vegetal matrix by overheating due to the high steam temperature at the inlet when compared to the temperature of the vegetal material located in the extraction column. There is another patent that presents an alternative for increasing the yield of VO obtained by SD by adding surfactants to the vegetal material before the distillation process (US5891501 from April 6, 1999).

Generally, there is a temperature decrease over the column length. The temperature at the steam inlet (reboiler) is higher than at the top of the column, which causes water condensation inside the distillation column, diminishing the yield. Additionally, the presence of organic compounds in the residual water increases the chemical oxygen demand, as mentioned before. The modification proposed by Masango [12] includes a steam jacket introduced externally to the distillation column with the objective of reducing the condensation of water by heating up uniformly all the distillation column length and consequently, diminishing the volatile compound loss within the residual water (aqueous phase of the condenser, the hydrosol).

2.3.2 VOs from Aromatic, Condimentary, and Medicinal Plants

Aromatic, condimentary, and medicinal plants coming from the Middle East were valuable during the late Middle Ages. During the fifteenth and sixteenth centuries, Portugal, Spain, and Venice competed in funding maritime travels aiming to discover spice production centers.

Aromatic, condimentary, and some medicinal plants are widely employed in cooking, giving food pleasant flavors and aromas. Besides the great contribution of condiments to the improvement of palatability by enhancing the flavor of food, they present antimicrobial and antioxidant properties. Those preservative properties of condiments guarantee better conservation of food, increasing its shelf life. Black pepper added to meat formulations (e.g., bologna, sausages) is an example, because besides conferring flavor to the food, it also preserves it. Similarly, condiments have been used in bakery products and fish, among others foods. Various products have suffered formulation modifications in order to substitute synthetic food additives by powders, oleoresins, or VOs from natural sources, such as condiments.

Facing the great demand for practical, durable, and easily accessible food, processing has become inevitable. Industrialized foods conquered a visible, wide market in the nineteenth century, positively affecting the development of a wide variety of additives, among them being antioxidants and preservatives, which aim to increase the products' shelf life. These ingredients may come from natural sources or chemical processes. The use of synthetic antioxidants has been severely restricted in the food industry because of their side effects, such as allergies and possible cancer-promoting effects that have been found in studies using laboratory animals [13].

In this context, the usage of condiments in processed products has promoted the development and improvement of oleoresins and VO extraction techniques in order to potentiate their conservative and antioxidant actions.

VOs are substances of interest for the aroma industry, including beverage and food companies. This market requires products of high quality and competitive prices. The expectation for VO demand increase will come from the food industry, once there is a growing demand for processed products that include in their composition additives that can extend shelf life (antioxidant properties) or bring some benefit to health (functionality).

The VOs are located in the oil bags or in the oil cells of the plants. If the plants are kept intact, the access to the oil is more difficult and the process becomes slower because the vaporization rate is then determined by the hydrodiffusion rate. The milling process of the raw material allows the breaking of the cells, favoring the contact between the steam and the oil and increasing the vaporization rate. Seeds and fruits must be milled in order to break the maximum of cell walls, facilitating the access of the steam to the oil. Roots and stems must be cut in small pieces in order to expose a greater number of oil bags. On the other hand, flowers and leaves may be distilled without milling, if their structure is sufficiently permeable to allow the occurrence of rapid oil vaporization.

VOs represent a small fraction of plants' composition, but confer to them characteristics for which aromatic plants are used in the pharmaceutical, food, and fragrance industries [14]. The aroma of each plant is the result of the combination of the aromas provided by all the components, from the major ones to the trace ones, and these last are very important, because they give the oil a characteristic and natural odor [14]. Thus, it is very important that the natural proportion of the components is maintained during extraction of the VOs from plants, particularly if they are designated for use in the fragrance industry. On the other hand, a target compound may be desired to be in higher concentration for pharmaceutical usage. Therefore, the future application of the recovered VO dictates the best extraction process.

VOs are generally expensive (from several to several thousand US$/kg) compared to "duplicate oils" (synthetics combined with natural oils), which usually lack certain odor notes of the natural products because of the absence of trace components. This is the reason why the more "chemical" odor is popularly attributed to the combined oils [14].

In SD of tea tree, the hydrosol contains about 2% of VO emulsified in water [15], which allows its usage in other industries. The hydrosol obtained in SD of lavender and artemisia contain 0.26 and 0.24% of VO, respectively [12]. The distillated leaves can be used for organic fertilization. The possibility of usage of the waste streams in other industries, because they do not have any toxic residues, is one of the characteristics of the SD process that makes it environmentally friendly.

In his research on the theories of VO distillation, Von Rechenberg 1910, cited by Baker et al. [16], demonstrated the early appearance of oxygenated components in the distillation of oils from intact plant material. This was explained by hydrodiffusion, rather than the boiling point, and was proposed as the rate-determining step in distillation. He also concluded, by observing that it was not possible to recover 100% of oil from a plant by SD, that some volatiles were retained because of their affinity to nonvolatile substances, such as lipids. This was confirmed by Koedam et al. 1979, cited by Baker et al. [16], who extended distillation for 24 h but found that some hydrocarbon fractions of the VO were not recovered.

Other studies have shown the losses and artifact formations associated with the distillation of VO. For instance, Southweel and Stiff 1989, cited by Baker et al. [16], found that the compounds sabinene, *cis*-sabinene hydrate, and *trans*-sabinene hydrate, found in the flush leaves of tea tree, are thermally transformed to terpinen-4-ol, α-terpinene, and γ-terpinene with distillation. Therefore, to obtain the best quality of oil, it is necessary to ensure that, during distillation, the VO is maintained at a low temperature, or, at least, that it is kept at a high temperature for the shortest time possible [17].

Studies involving superheated vapor for obtaining VOs mention that temperatures superior to 303 K cause partial pyrolysis of the biomass and the decomposition of the VO. Thus, the ideal temperature for the flash distillation of the VO is between 478 and 497 K [18].

In tea tree SD (Johns et al. 1992, cited by Baker et al. [16]), in line with Von Rechenberg's hydrodiffusion theories, the oxygenated components, particularly terpinen-4-ol and 1,8-cineole, are extracted faster in spite of their higher boiling points. Those authors suggested that their recovery is controlled by the film mass transfer, whereas for the components extracted later (monoterpenes and sesquiterpenes), mass transfer is controlled by diffusion. The increased resistance of these compounds to diffusion is attributed to the hydrophobic properties of the monoterpenes plus the larger molecule size of the sesquiterpenes (Johns et al. 1992, cited by Baker et al. [16]).

As the hydrodiffusion is always a slow process, if the plants are left intact, the rate of recovery of oil will be entirely determined by the rate of diffusion [17]. Therefore, ground material tends to be less affected by the effects accompanying hydrodistillation, namely the diffusion of VOs and hot water through the plant membranes, and decomposition occasioned by heat.

Considering all the presented facts, the observation of the following principles leads to the best yields and to a high quality of VOs [17]: (1) maintenance of as low a temperature as possible, not forgetting, however, that the production rate will be determined by the temperature; (2) use of as little steam as possible in direct contact with raw material, but keeping in mind that some water should be present to promote diffusion; and (3) thorough comminution of raw material before distillation and very careful, uniform packing of the still charge, remembering that excessive comminution results in channeling of steam through the mass of raw material, reducing efficiency because of poor contact between steam and charge.

Because the SD process is very simple to carry out, most of its applications are done without the study of process conditions. Although the literature reports many studies involving SD of VOs, most of the time the operational conditions are disregarded, and sometimes SD and hydrodistillation are not even differentiated. Table 2.3.1 shows the SD recovery of some bioactive compounds that have been recently studied. The lack of information about the operational conditions is clear in most of the articles cited.

The study of Baker et al. [16] found that in SD of tea tree, although the distillation time (120 min compared to 360 min) did not have influence on the total yield, the VO composition was different for these two cycle times. Although the amount of monoterpenes was higher for 120 min of extraction, the amount of sesquiterpenes was higher for 360 min of extraction. The authors attributed this fact to the dissolution of the more hydrophobic isolates in the increased volumes of condensate with time.

Povh et al. [19] studied SD of chamomile. These authors observed that operating pressure, distillation time, and steam flow rate exerted a significant effect on yield. Among the operational conditions evaluated, they found that extraction at 98 kPa for

TABLE 2.3.1
Bioactive Compounds Obtained from Vegetal Matrices by Steam Distillation

Bioactive compound	Plant material	Operational conditions: steam flow rate ($\dot{W}$), time (t), particle size (d), pressure (P), temperature (T)	Reference
Essential oil	*Coriander sativum*	$t = 300$ min, $P \sim 90$ kPa, $T = \sim 358$ K	[14]
l-menthol, menthone, eucalyptol	*Mentha piperita* (peppermint)	$t = 100$ min	[22]
Essential oil	*Melaleuca alternifolia* (tea tree)	$\dot{W} = 3.3 \times 10^{-5}$ kg/s, $t = 120$–360 min	[16]
α-Bisabolol, chamazulene	*Chamomila recutita* (chamomile)	$\dot{W} = 5$–10×10^{-4} kg/s, $t = 45$–60 min, $P = 49$–98 kPa	[19]
Essential oil (carvacrol)	*Thymbra spicata* (thyme)	$\dot{W} = 1.8$–29.9×10^{-4} kg/s, $t = 105$–150 min, $d = 0.50$–2.05 mm	[20]
Antioxidant	*Rosmarinus officinalis* (rosemary)	$t = 120$ min	[4]
Essential oil, curcuminoids	*Curcuma longa* (turmeric)	$t = 60$–180 min, $P = 0.10$–0.15 MPa, $T = 374$–383 K	[17]
Essential oil	*Artemisia annua* (artemisia)	$\dot{W} = 4.2$–33.3×10^{-4} kg/s, $t = 15$–100 min, $T = 372$ K	[12]
Essential oil	*Lavendula angustifolia* (lavender)	$\dot{W} = 3.3$–33.3×10^{-4} kg/s, $t = 15$–150 min, $T = 372$ K	[12]
Anethole	*Pimpinella anisum* (aniseed)	$\dot{W} = 1.7 \times 10^{-3}$ kg/s, $t = 150$ min, $P = 140$–250 kPa, $T = 382$–393 K	[1]
Essential oil	*Lavendula angustifolia* (lavender)	$t = 10$–90 min, $T = 373$ K	[21]
Essential oil	Thyme	$\dot{W} = 4.4$–6.9×10^{-4} kg/s, $t = 10$–40 min, $T = 373$–523 K	[5]
Essential oil	Black pepper	$\dot{W} = 4.4 \times 10^{-4}$ kg/s, $t = 10$–40 min, d = whole or ground, $T = 373$–523 K	[5]
Eugenol	*Eugenia caryophyllata* (clove)	$\dot{W} = \sim 3.2 \times 10^{-5}$ kg/s, $t = \sim 540$ min	[6]
Essential oil	*Cordia verbenacea*	$\dot{W} = 1.6 \times 10^{-4}$ kg/s, $t = 300$ min, $T = 421$ K	[23]
Essential oil	*Pimpinella anisum* (aniseed)	$\dot{W} = 1.4 \times 10^{-4}$ kg/s, $t = 300$ min, $T = 413$ K	[23]
Essential oil	*Chamomila recutita* (chamomile)	$\dot{W} = 1.4 \times 10^{-4}$ kg/s, $t = 300$ min, $T = 430$ K	[23]
Essential oil	*Rosmarinus officinalis* (rosemary)	$\dot{W} = 1.6 \times 10^{-4}$ kg/s, $t = 300$ min, $T = 419$ K	[23]

45 min with a steam flow rate of 1×10^{-3} kg/sec was the best choice, because besides presenting a high yield, the oil obtained under those conditions presented the highest amount of α-bisabolol and chamazulene in its chemical composition.

The study of Hanci et al. [20] showed important effects of the steam flow rate and particle size on the yield and process time. The use of whole leaves (2.05 mm) and a higher steam flow rate (2.9×10^{-4} kg/sec) for 75 min of distillation was chosen as the optimum combination of conditions among the studied ones, because it provided the lowest amount of monoterpene hydrocarbons, the complete recovery of oxygenated compounds, and the highest yield (1.57%) in a shorter time. Considering the same distillation time, the yield was only 0.75% for nonoptimized conditions.

Studying SD and hydrodistillation of rosemary, Boutekedjiret et al. [4] found that after 10 min of SD, more than 80% of the VO was recovered, whereas for hydrodistillation, it took 30 min to extract 88% of the oil. In addition, the chemical composition of the VOs obtained by those methods was slightly different because of the hydrolysis of some monoterpene components that was observed in hydrodistillation. This study also presented the change in oil composition with the time of extraction. Considering all these facts, the SD was considered a better process for recovering VO from rosemary because of the higher yield, shorter process time, and improved chemical composition (according to commercial standards), when compared to the hydrodistilled oil.

In the study of Manzan et al. [17], it was concluded that among the operational conditions studied, SD of turmeric at 0.1 MPa and 374 K for 120 min provided the highest yield (0.45%) and the best chemical composition. The use of nonoptimized SD conditions resulted in only 0.15% of yield.

Masango [12] studied the effect of steam flow rate on yield. In contrast to the results obtained by Hanci et al. [20], the author concluded that lower steam flow rates led to higher yields. The author also proposed a new jacketed still for keeping the temperature constant all over the still, which would decrease the condensation inside it. This procedure also increased yield by decreasing the VO loss in the hydrosol and decreased energy and water consumption by decreasing the amount of required steam. On the other hand, Rouatbi et al. [5], for SD from thyme, observed the opposite effect: the thyme oil yield increased as steam flow rate increased, in accordance with the results obtained by Hanci et al. [20]. Those authors also found that ground black pepper SD presented a higher yield when compared to the whole fruit. This result is in disagreement with the one found by Hanci et al. [20] for thyme leaves.

In the evaluation of superheated steam temperature, Rouatbi et al. [5] observed that the increase in temperature positively affected extraction yield of both thyme and black pepper. They attributed this effect to the increase in vapor pressure and consequently, in mass transfer rate, of the VO components with temperature. These authors concluded that superheated steam at 448 K and higher steam flow rate were the best extraction conditions, considering both yield and VO composition.

In the study of aniseed SD, Romdhane and Tizaoui [1] described the influence of pressure on yield. The yield increased with pressure until a maximum (200 kPa) was reached, and the inverse effect was observed from that point on. The authors focused the explanation of this phenomenon on the increase of temperature with pressure. The temperature increase enhances the driving force for mass transfer as a result of the increase in the solutes diffusion. However, the increase of temperature also causes thermal degradation of some compounds present in the vegetal matrix, leading to a decrease in yield.

Chemat et al. [21] studied an SD process where the still was inside a microwave oven (for further details see Chapter 4). The microwave accelerated SD, resulting in similar yield, but in a shorter time (10 vs. 90 min), without alteration of the lavender VO chemical composition, when compared to simple SD. Because of the sorter extraction time, energy and water consumption were substantially reduced.

From literature data collected, it is important to note that operational conditions (steam flow rate, extraction time, particle size, pressure, and temperature) presented an impressive influence on yield and VO composition. This means that the recovery of VOs by SD could be optimized by more accurate studies. Nevertheless, literature is still scarce and divergent on that matter. Most of literature studies report hydrodistillation instead of SD data [2, 7, 24–31], even though SD is the most common process in industrial scale. This becomes an especially important point when it is considered that other extractive techniques that directly compete with SD in VOs recovery have been more deeply studied and, therefore, improved.

Even though 93% of VOs are still extracted by SD [12], especially because of the low investment costs when compared to other extractive techniques, studies have increasingly shown the disadvantages of SD compared to those other methods [6, 9, 14, 16, 17, 19, 22, 32, 33]. In most of those comparative studies, however, the SD operational conditions are not studied and optimized, as in the case of the competing methods [6, 9, 14, 16, 22, 32, 33]. On the other hand, the studies that have evaluated different SD operational conditions have found great differences on yield and/or chemical composition [1, 5, 12, 17, 19, 20], indicating that the process should be optimized in order to continue competing with the other extraction methods.

The technical evaluation of the process should always be carried out together with the economical evaluation, so that the optimization of the process can be guaranteed. This way, the cost of manufacturing (COM) estimation is an important tool to evaluate the economical viability of the process. For instance, the complete exhaustion of the VO from a determined vegetal matrix may be economically unfeasible in a first analysis, because of the energy related costs involved when long cycles are used. However, reducing the process time may make the SD process more economically attractive. For this reason, additional information concerning the COM estimation becomes relevant and should be confronted with technical information of the process (impact of process conditions such as temperature, pressure, steam flow, and cycle time on the yield and oil quality).

2.3.3 VOs from Anise Seed, Black Pepper, Chamomile, and Rosemary

In Section 2.4, methods used to estimate the cost of manufacturing of VOs from condimentary plants will be discussed. These plants were selected both because of availability of the required data and their importance in food processing. The selected plants are black pepper (*Piper nigrun*), chamomile (*Chamomilla recutita*), rosemary (*Rosmarinus officinalis*), anise seed (*Pimpinella anisum*), and thyme (*Thymus vulgaris*). Next, a brief review of the usage of their VOs is presented.

Anise seed belongs to the Umbellifera (Apiaceae) family. The fruit is industrially used for the production of VO, tincture, fluid extract, alcoholic extract, and hydrosol. The phytochemical analysis of the VO shows that anethole, which is the component responsible for its characteristic anise flavor and aroma, is its major constituent (90%–95%). Pharmacological essays have shown that the fruits' extract and the VO have antifungal and antiviral activities and can be used as insect repellents and expectorant and antispasmodic agents. Popularly, anise seeds are consumed as infusions, because of the beneficial effects against cold, cough, bronchitis, fever, colic, mouth and throat inflammation, digestive problems, and loss of appetite [34].

Chamomile belongs to the Compositae (Asteraceae) family. It is an herbal, annual, and aromatic plant. The part of the plant used for therapeutic treatments is the dry flower. It is a plant used in both scientific and popular medicines in the form of an infusion or a decocted product (cooked flowers), as a bitter tonic, digestive helper, sedative, appetite stimulator, gas eliminator, and anti-colic agent. Its phytochemical analysis shows the presence of chamazulene, chamavioline, and α-bisabolol. Among its fixed constituents there are polysaccharides with immune-stimulating properties; bicyclical ethers that under experimental condition have shown antispasmodic activity similar to that of papaverine; flavonoids with bacteriostatic and antitrichomoniasis activity; and apigenin, which presents anxiolytic and sedative properties.

The aqueous infusion of the flowers or the VO itself are still used in ointment and cream formulations and in pharmaceutical preparations of external use for healing skin lesions, for relieving gum inflammation, and as an antiviral for herpes treatment, with all these properties being attributed mainly to the α-bisabolol. Industrially, chamomile is used in the cosmetic, food, and beverage fields [34].

Rosemary is a plant native to the Mediterranean region and belongs to the Lamiaceae family. It is recognized as one of the plants possessing the highest antioxidant activity. According to Ibañez et al. [35], the compounds associated with this antioxidant activity are the phenolic diterpenes such as carnosol, rosmanol, 7-methyl-*epi*-rosmanol, isorosmanol, rosmadial, carnosic acid, and methyl carnosate and phenolic acids such as caffeic acid and rosmarinic acid. The chemical composition of the rosemary extract varies a lot, influenced, among other factors, by the local cultivation and extraction techniques (Reverchon and Sanatore 1992, cited by Carvalho [36]). The rosemary leaves and extracts are often used in food products, not only for their aroma, but also for their antioxidant properties [10, 36].

Black pepper belongs to the Piperaceae family. It is a plant native to India and is cultivated in several countries around the world; it is indicated for rheumatism, laryngitis, and chronic bronchitis treatment. [37]. The volatile compounds present in black pepper extract identified by Jirovetz et al. [38] were germacrene-D (11.01%), limonene (10.26%), β-pinene (10.02%), α-phellandrene (8.56%), β-caryophyllene (7.29%), α-pinene (6.40%), and *cis*-β-ocimene (3.19%). The VO from the seeds and leaves of black pepper, which is used as a flavoring agent in the perfume and food industries, may have more than 250 compounds [39]. The black pepper oleoresin produced by solvent extraction contains the characteristics of both pungency and aroma (Premi 2000, cited by Shaikh et al. [40]).

Thyme is rich in VO, to which several biological properties are attributed. Particularly, it possesses fungicidal, antiseptic, and antioxidant activities and is an excellent tonic. The VO from the leaves is used in perfumes, soaps, and toothpastes. Besides the applications in the cosmetic field, thyme is used as a condiment. The study of Lee et al.

[41] shows that the major components in thyme extracts, especially eugenol, thymol, and carvacrol, present higher antioxidant activity when compared to the very well-known antioxidants BHT and α-tocopherol. Thyme VO presents antibacterial activity, and Rota et al. [42] have confirmed that the VOs of the genus *Thymus*, especially *Thymus hyemalis*, *T. zygis*, and *T. vulgaris*, are potent bactericide agents that can be used in the food industry, increasing shelf life and improving food product preservation.

2.3.4 Acknowledgments

The authors thank Fundação de Amparo à Pesquisa do Estado de São Paulo (FAPESP), Conselho Nacional de Desenvolvimento Científico e Tecnológico, and Coordenação de Aperfeiçoamento de Pessoal de Nível Superior for financial support. J. M. Prado and P. F. Leal thank FAPESP for the PhD assistantships (07/03817-7, 04/09310-3).

2.3.5 References

1. Romdhane, M., and C. Tizaoui. 2005. The kinetic modelling of a steam distillation unit for the extraction of aniseed (*Pimpinella anisum*) essential oil. *Journal of Chemical Technology and Biotechnology* 80:759–766.
2. Kelkar, V. M., B. W. Geils, D. R. Becker, S. T. Overby, and D. G. Neary. 2006. How to recover more value from small pine trees: Essential oils and resins. *Biomass and Bioenergy* 30:316–320.
3. Babu, K. G. D., B. Singh, V. P. Joshi, and V. Singh. 2005. Essential oil composition of damash rose (*Rosa damascena* mill.) distilled under different pressures and temperatures. *Flavour and Fragrance Journal* 17 (2): 136–140.
4. Boutekedjiret, C., F. Bentahar, R. Belabbes, and J. M. Bessiere. 2003. Extraction of rosemary essential oil by steam distillation and hydrodistillation. *Flavour and Fragrance Journal* 18:481–484.
5. Rouatbi, M., A. Duquenoy, and P. Giampaoli. 2007. Extraction of the essential oil of thyme and black pepper by superheated steam. *Journal of Food Engineering* 78:708–714.
6. Wenqiang, G., L. Shufen, Y. Ruixiang, T. Shaokun, and Q. Can. 2007. Comparison of essential oils of clove buds extracted with supercritical carbon dioxide and other three traditional extraction methods. *Food Chemistry* 101:1558–1564.
7. Sefidkon, F., K. Abbasi, Z. Jamzad, and S. Ahmadi. 2007. The effect of distillation methods and stage of plant on essential oil content and composition of *Satureja rechingeri* Jamzad. *Food Chemistry* 100:1054–1058.
8. Sefidkon, F., M. Dabiri, and A. Rahimi-Bidgoly. 1999. The effect of distillation methods and stage of plant growth on the essential oil content and composition of *Thymus kotschanus* boiss & hohen. *Flavour and Fragrance Journal* 14:405–408.
9. Esquível, M. M., M. A. Ribeiro, and M. G. Bernardo-Gil. 1999. Supercritical extraction of savory oil: study of antioxidant activity and extract characterization. *Journal of Supercritical Fluids* 14:129–138.
10. Leal, P. F., M. E. M. Braga, D. N. Sato, J. E. Carvalho, M. O. M. Marques, and M. A. A. Meireles. 2003. Functional properties of spice extracts obtained via supercritical fluid extraction. *Journal of Agricultural and Food Chemistry* 51:2520–2525.
11. Mateus, E. M., C. Lopes, T. Nogueira, J. A. A. Lourenço, and M. J. M. Curto. 2006. Pilot steam distillation of rosemary (*Rosmarinus officinalis* L.) from Portugal. *Silva Lusitana* 14 (2): 203–217.
12. Masango, P. 2005. Cleaner production of essential oils by steam distillation. *Journal of Cleaner Production* 13:833–839.

13. Rizvi, Syed S. H. 1994. *Supercritical fluid processing of food and biomaterials*. London: Blackie Academic & Professional.
14. Anitescu, G., C. Doneanu, and V. Radulescu. 1997. Isolation of *Coriander* oil: Comparison between steam distillation and supercritical CO_2 extraction. *Flavour and Fragrance Journal* 12:173–176.
15. Castro, C., M. L. Silva, A. L. Pinheiro, and L. A. G. Jacovine. 2005. Economic analysis of the cultivation and extraction of the essential oil of *Melaleuca alternifolia* Cheel. *Journal of Brazilian Forest Science* 29:241–249.
16. Baker, G. R., R. F. Lowe, and I. A. Southwell. 2000. Comparison of oil recovered from tea tree leaf by ethanol extraction and steam distillation. *Journal of Agricultural and Food Chemistry* 48:4041–4043.
17. Manzan, A. C. C. M., F. S. Toniolo, E. Bredow, and N. Povh. 2003. Extraction of essential oil and pigments from *Curcuma longa* [L.] by steam distillation and extraction with volatile solvents. *Journal of Agricultural and Food Chemistry* 51:6802–6807.
18. Boucard, G. R., and R. W. Serth. 1998. Continuous steam distillation of essential oils. *Perfumer and Flavorist* 23 (2): 1–5.
19. Povh, N. P., C. A. Garcia, M. O. M. Marques, and M. A. A. Meireles. 2001. Extraction of essential oil and oleoresin from chamomile (*Chamomila recutita* [L.] Rauschert) by steam distillation and extraction with organic solvents: A process design approach. *Revista Brasileira de Plantas Medicinais* 4:1–8.
20. Hanci, S., S. Sahin, and L. Yilmaz. 2003. Isolation of volatile oil from thyme (*Thymbra spicata*) by steam distillation. *Nahrung/Food* 47:252–255.
21. Chemat, F., M. E. Lucchesi, J. Smadja, L. Favretto, G. Colnaghi, and F. Visinoni. 2006. Microwave accelerated steam distillation of essential oil from lavender: A rapid, clean and environmentally friendly approach. *Analytica Chimica Acta* 555:157–160.
22. Ammann, A., D. C. Hinz, R. S. Addleman, C. M. Wai, and B. W. Wenclawiak. 1999. Superheated water extraction, steam distillation and SFE of peppermint oil. *Fresenius' Journal of Analytical Chemistry* 364:650–653.
23. Leal, P. F. 2008. Estudo comparativo entre os custos de manufatura e as propriedades funcionais de óleos voláteis obtidos por extração supercrítica e arraste a vapor. PhD diss., State University of Campinas (UNICAMP).
24. Abu-Lafi, S., I. Odeh, H. Dewik, M. Qabajh, L. O. Hanuš, and V. M. Dembitsky. 2008. Thymol and carvacrol production from leaves of wild Palestinian *Majorana syriaca*. *Bioresource Technology* 99:3914–3918.
25. Al-Bayati, F. A. 2008. Synergistic antibacterial activity between *Thymus vulgaris* and *Pimpinella anisum* essential oils and methanol extracts. *Journal of Ethnopharmacology* 116:403–406.
26. Wang, W., N. Wu, Y. G. Zu, and Y. J. Fu. 2008. Antioxidative activity of *Rosmarinus officinalis* L. essential oil compared to its main components. *Food Chemistry* 108:1019–1022.
27. Babu, K. G. D., and V. K. Kaul. 2007. Variations in quantitative and qualitative characteristics of wild marigold (*Tagetes minuta* L.) oils distilled under vacuum and at NTP. *Industrial Crops and Products* 26:241–251.
28. Chyau, C., S. Tsai, J. Yang, et al. 2007. The essential oil of *Glossogyne tenuifolia*. *Food Chemistry* 100:808–812.
29. Sefidkon, F., K. Abbasi, and G. B. Khaniki. 2006. Influence of drying and extraction methods on yield and chemical composition of the essential oil of *Satureja hortensis*. *Food Chemistry* 99:19–23.
30. Kimbaris, A. C., N. G. Siatis, D. J. Daferera, P. A. Tarantilis, C. S. Pappas, and M. G. Polissiou. 2006. Comparison of distillation and ultrasound-assisted extraction methods for the isolation of sensitive aroma compounds from garlic (*Allium sativum*). *Ultrasonics Sonochemistry* 13:54–60.

31. Schanenberg, B. T., and I. A. Khan. Comparison of extraction methods for marker compounds in the essential oil of lemon grass by GC. *Journal of Agricultural and Food Chemistry* 50:1345–1349.
32. Kotnik, P., M. Škerget, and Ž. Knez. 2007. Supercritical fluid extraction of chamomile flower heads: Comparison with conventional extraction, kinetics and scale-up. *Journal of Supercritical Fluids* 43:192–198.
33. Scalia, S., L. Giuffreda, and P. Pallado. 1999. Analytical and preparative supercritical fluid extraction of Chamomile flowers and its comparison with conventional methods. *Journal of Pharmaceutical and Biomedical Analysis* 21:549–558.
34. Lorenzi, H., and J. A. Matos. 2002. *Plantas medicinais no Brasil: Nativas e exóticas cultivadas*. Nova Odessa: Instituto Plantarum de Estudos da Flora.
35. Ibañez, L., A. Kub´atov´a, F. J. Señoráns, S. Cavero, G. Reglero, and S. B. Hawthorne. 2003. Subcritical water extraction of antioxidant compounds from rosemary plants. *Journal of Agricultural and Food Chemistry* 51:375–382.
36. Carvalho, R. N., Jr. 2004. Obtenção de extrato de alecrim (*Rosmarinus officinalis*) por extração supercrítica: Determinação do rendimento global, de parâmetros cinéticos e de equilíbrio e outras variáveis do processo. PhD diss., State University of Campinas (UNICAMP).
37. Rose, J. 1999. *375 Essential oils and hydrosols*. Berkeley, CA: Frog.
38. Jirovetz, L., G. Buchbauer, M. B. Ngassoum, and M. Geissler. 2002. Aroma compounds analysis of *Piper nigrum* and *Piper guineense* essential oil from Cameroon using solid-phase microextraction-gas chromatography, solid-phase microextraction-gas chromatography-mass spectrophotometry and offactometry. *Journal of Chromatography A* 976:265–275.
39. Sumathykutty, M. A., J. M. Rao, K. P. Padmakumari, and C. S. Narayana. 1999. Essential oil constituents of some *Piper* species. *Flavour and Fragrance Journal* 14:279–282.
40. Shaikh, J., R. Bhosale, and R. Singhal. 2006. Microencapsulation of black pepper oleoresin. *Food Chemistry* 94:105–110.
41. Lee, S. J., K. Umano, T. Shibamoto, and K. G. Lee. 2005. Identification of volatile components in basil (*Ocimum basilicum* L.) and thyme leaves (*Thymus vulgaris* L.) and their antioxidant properties. *Food Chem*istry 91:131–137.
42. Rota, M. C., A. Herrera, R. M. Martinez, J. A. Sotomayor, and M. J. Jordán. 2008. Antimicrobial activity and chemical composition of *Thymus vulgaris*, *Thymus zygis* and *Thymus hyemalis* essential oils. *Food Control* 19 (7): 681–687.

2.4 COST OF MANUFACTURING OF VOLATILE OIL FROM CONDIMENTARY PLANTS

Patrícia F. Leal, Thais M. Takeuchi, Juliana M. Prado, and M. Angela A. Meireles

According to the Association for the Advancement of Cost Engineering International (AACEI) [1], the cost estimation methods that are applied to industry are arranged in five classes, namely 1, 2, 3, 4, and 5. The class 5 estimation is based on the lowest definition level of the project, whereas the class 1 estimation is closer to the complete definition of the project, which means a high level of maturity. This classification considers that the estimation of the cost of manufacturing (COM) is a dynamic process that occurs all the way through successive estimations until a final estimation provides cost information close to the real value.

Next, a brief review related to the cost estimation applied to industry will be presented. The characteristics that distinguish the five COM estimation classes will be discussed [2], along with the subdivision of the classes [3] and the estimation methodologies (Lang, Chilton). Finally, a more detailed description of the methodology used for COM estimation [4] class 5 of volatile oils (VOs) from some condimentary plants will be presented.

2.4.1 Characteristics of the Cost Estimation Classes

The following characteristics are used to distinguish the cost estimation classes from each other: level of project definition, end usage, methodology, accuracy range, and preparation effort.

The level of project definition is determined by the extent and types of input information available for the estimation. Such input information include the definition of project scope, required documents, specifications, project plans, drawings, calculations, and other information that must be developed in order to define the project. A large amount of available information is related to an advanced level of definition of the project.

The several classes, or steps, of cost estimation have different purposes. With the increase in the level of definition of the project, the purpose of the estimation progresses from a strategic evaluation to a viability study of a funding demand.

The estimation methods are divided into two broad categories: stochastic (random) and deterministic. In stochastic methods, the independent variables used in the cost estimation are not usually represented by real values, that is, the costs are often assumptions. In deterministic methods, the independent variables are represented more by definite than estimated values. As the definition level of the project increases, the cost estimation method tends to progress from the stochastic to the deterministic category, which means that as the project acquires a higher maturity level, that is, as there is more definite information available, some of the assumptions are no longer necessary. From that moment on, the cost estimation based on a more deterministic method is applied.

The accuracy range of the cost estimation measures the difference between the estimated and real costs. Accuracy is traditionally expressed as the percentage variation around the estimated point with a stated level of confidence. As the definition level of the project increases, the expected accuracy of the estimation tends to improve, which is indicated by a tighter variation range.

The effort put on the cost estimation preparation is indicated by the required cost, time, and resources. The measure of the cost of this effort is usually expressed as a percentage of the total costs of the project and varies inversely with the project size in a nonlinear fashion.

2.4.2 Cost Estimation Classes

Although the cost estimation arrangement in five classes is largely used, some companies and organizations have determined that, because of the inherent imprecision of the higher level classes, some estimation cannot be classified in a conventional or systemic way.

Class 5 estimations are usually based on very limited information, and, therefore, have large accuracy ranges. They can be prepared in a very short period of time, requiring relatively little effort. Often, little more than the type of plant, its capacity, and its location are known at the moment of this first cost estimation. For the class 5 estimation preparation, a stochastic method is virtually always used, such as cost/capacity curves and factors, which can be represented by scale of operation and Lang, Hand, Chilton, Peter-Timmerhaus, Guthrie factors, and other parametric and modeling techniques. The level of project definition required for the class 5 estimation varies between 0 and 2%.

Class 4 estimations are usually based on limited information and therefore have fairly wide accuracy ranges. They are generally used for screening, feasibility determination, project concept evaluation, and preliminary budget authorization. Class 4 estimations virtually always use stochastic methods, such as the Lang, Hand, Chilton, Peter-Timmerhaus, Guthrie, and equipment related factors, the Miller method, gross unit costs/ratios, and other parametric and modeling techniques. The level of project definition required varies between 2 and 5%.

Class 3 estimations are usually prepared to form the basis for budget authorization. As such, they typically form the initial control estimate against which all actual costs and resources will be monitored. The engineering project would at least contain the following: process flow diagrams, utility flow diagrams, preliminary piping and instrument diagrams, plot plan, developed layout drawings, complete engineered process, and utility equipment lists. Class 3 estimations usually involve more deterministic than stochastic methods. Stochastic methods may be used to estimate less-significant areas of the project. The level of project definition varies from 10 to 40%.

Class 2 estimations are generally prepared to form a detailed control baseline, in terms of cost and progress control. For contractors, this class of estimate is often used as the "bid" estimate to establish contract value. Typically, engineering is from 30 to 60% complete and would comprise at minimum the following: process flow diagrams, utility flow diagrams, piping and instrument diagrams, heat and material balances, final plot plan, final layout drawings, complete engineered process and utility equipment lists, single line diagrams for electrical installation, electrical equipment and motor schedules, vendor quotations, detailed project execution plans, and resourcing and work force plans. Class 2 estimations always involve a high degree of deterministic estimating methods, and the estimates are prepared in great detail. The level of project definition requirement varies from 30 to 60% and can sometimes be higher, depending on the project complexity.

Class 1 estimations are generally prepared for discrete parts or sections of the total project. The parts of the project evaluated with this level of detail will replace the corresponding parts of less detailed estimates. Class 1 estimations involve the highest degree of deterministic methods and require a great amount of effort. They are prepared in great detail and thus are usually performed only for the most important or critical areas of the project. All items in the estimation are usually line item costs based on actual design quantities. The level of project definition required varies from 50 to 100%.

2.4.3 Cost of Manufacturing Estimation Methods

The methodology proposed by Lang is frequently used for obtaining the order of magnitude of the cost estimation. It recognizes that the cost of a processing plant may be obtained by multiplying the cost of the basic equipment by a factor, which gives the investment needed. The Lang factors vary according to the process: solid processing plant ($F_{Lang} = 3.10$), solid–liquid processing plant ($F_{Lang} = 3.63$), and fluid processing plant ($F_{Lang} = 4.74$). These factors should be multiplied by the total cost of equipment. The equipment costs are usually based on quotations for less common items and published data for more common items. The total cost of the plant can be evaluated by the following:

$$C_{TM} = F_{Lang} \cdot \sum_{i=1}^{n} C_{Pi}, \qquad (2.4.1)$$

where C_{TM} is the total cost of the plant, C_{Pi} the cost of equipment, F_{Lang} the Lang factor, and n the total number of individual units.

The Chilton method or 0.6 rule relates the fixed cost of investment of a new plant to the cost of a previously built similar plant. For certain process configurations, the fixed cost of investment of a new plant is the same as the previously built plant multiplied by the relation between capacities elevated to an exponent. This exponent is estimated as an average between 0.6 and 0.7 for many processes if no other information is available.

The cost of manufacturing (COM) estimation proposed by Turton et al. [4] is classified as class 5 or 4, that is, the cost estimation is used for business plans according to the Association for the Advancement of Cost Engineering International [1]. This preliminary cost estimation is commonly used for strategic decisions, such as advancing or stopping a project. The COM is influenced by many factors that may be grouped into three cost categories: direct costs, fixed costs, and general expenses. The direct costs consider costs that depend directly on the production, and they include raw material, utilities, and operational cost, among others. The fixed costs do not depend directly on production, existing even when the production is stopped. They include depreciation, taxes, and insurance. The general expenses are composed of the amount needed for maintaining the business and include administration expenses, shipping expenses, and research and development.

The Turton et al. [4] methodology defines COM as the weighed sum of five main costs: fixed cost of investment (F_{CI}), cost of operational labor (C_{OL}), cost of raw material (C_{RM}), cost of waste treatment (C_{WT}) and cost of utilities (C_{UT}):

$$COM = 0.304 \times F_{CI} + 2.73 \times C_{OL} + 1.23 \times (C_{RM} + C_{WT} + C_{UT}). \qquad (2.4.2)$$

2.4.4 COM for VOs from Condimentary Plants

For the COM estimation of VO from certain condimentary plants, the methodology proposed by Turton et al. [4], previously described, was selected. Next are described the technical considerations and procedures that involve making the scale-up calculations and obtaining the costs that comprise the COM.

2.4.4.1 Scale-Up

The scale-up procedure used for SD assumed that both the yield and the extraction time of the industrial scale unit would be like those of the laboratorial scale unit if the ratio between solvent mass and feed mass (S/F) was kept constant.

Considering that the bed apparent densities for the laboratorial and industrial scale units are the same, it is possible to calculate the feed mass of raw material that must be used for each extraction cycle in the industrial column.

Using the solvent mass flow rate and the time of extraction of the laboratorial scale unit, it is then possible to calculate the steam mass used in each cycle in the industrial scale unit, and, therefore, to calculate the steam flow rate:

$$\dot{M}_{S\_ind} = \frac{M_{F\_ind} \times (M_{S\_lab} / t_{cycle})}{M_{F\_lab}}. \tag{2.4.3}$$

where $\dot{M}_{S\_ind}$ is the solvent (steam) flow rate of the industrial unit, M_{F_ind} is the feed mass of raw material in the distillation column of the industrial unit, M_{F_lab} is the feed mass of raw material in the distillation column of the laboratorial scale unit, M_{S_lab} is the solvent (steam) mass used in one cycle in the laboratorial scale unit, and t_{cycle} is the time of one distillation cycle.

2.4.4.2 Fixed Cost of Investment

The SD unit is usually composed of two distillation columns that contain inside a mobile basket for raw material accommodation. The steam is produced in a boiler, which, in Brazil, is usually fed with firewood. The steam is injected at the bottom of the column. The condenser is of the shell-and-tube type and is fed with cold or ambient temperature water. The water and oil separator is the last component of the unit. The fixed cost of investment is composed of the stills, the condenser, and the separator (Figure 2.4.1).

For the COM study, it was considered an industrial nonautomated unit, containing two 0.5-m^3 columns, a shell-and-tube condenser, and a separator, without a boiler. The cost of this unit was quoted in US$ 50,000.00 (quotation from July 2006, Votuporanga, Brazil). This value does not include the reboiler, because the steam cost was estimated using the methodology proposed by Turton et al. [4]. In this methodology, the steam cost includes all the investment cost involved in the steam production.

The annual depreciation of the plant was considered to be 10%.

2.4.4.3 Raw Material Cost

The raw material cost covers all material related to production. The cost of solid substrate covers the raw material cost and all the costs related to preprocessing it, such as drying and milling.

2.4.4.4 Operational Labor Cost

The operational labor cost was calculated using information from Ulrich [5], cited by Turton et al. [4]. For the SD process, it was considered that three operators per shift

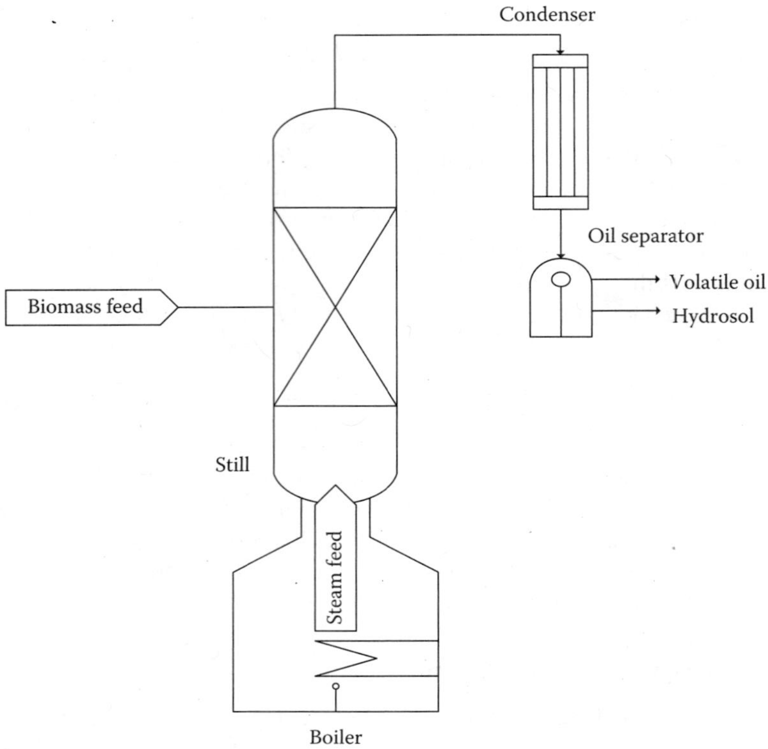

FIGURE 2.4.1 Flow diagram of a batch distillation unit used for estimation of COM.

are necessary: two of them for charging and discharging the raw material and controlling steam production, and another one for the transportation of raw material and residue. The unit considered is not automated. The operational labor was considered as US$ 3.00 h^{-1}. The estimated C_{OL} per year was US$ 47,520.00 considering 330 days of continuous operation, with three shifts per day.

2.4.4.5 Waste Treatment Cost

The residue of the SD process is the wet raw material and is therefore nonpolluting. Because usually the raw material is a plant, or part of it, it can be used as fertilizer. Thus, the C_{WT} can be neglected for this first cost estimation.

2.4.4.6 Cost of Utilities

The cost of utilities covers the steam production by the boiler destined to feed the stills and the cold water used in the condenser. The steam (US$ 16.22/ton) and cold water (US$ 14.80/10^3 ton) costs were based on the values proposed by Turton et al. [4].

2.4.5 COM Estimation

The SD process needs more studies on process operating conditions, which will guarantee superior quality for the extracts, besides a higher yield and cycle time optimization. Literature is scarce on that matter, even with SD being largely used for recovery of VOs. Other extraction techniques used for obtaining the VOs and vegetal extracts have their operating conditions widely known by the scientific community, and their processes are protected by patents, as is the case with supercritical fluid extraction. In the SD case, however, it is used as another way of protection called "know-how," which keeps information on the operating conditions as a "secret."

Because SD is a process that involves simple equipment (considered noncritical by the rules that run industrial property) and low fixed cost of investment, it is economically viable for the processing of a great variety of vegetal matrices and is accessible to a wide number of investors. However, the product that once was easily accepted by the market without any restriction has gone through a huge change with regard to product quality destined for the chemical, cosmetic, pharmaceutical, and food industries. Today, a distilled product must not only have a competitive price but must also follow strict security and standardization rules for active principles (biomarker). To satisfy all those requirements, process optimization has become a key factor for the success in market competition. Thus, in order to compete with other extractive techniques for the obtaining of VO, it is crucial that more studies on process optimization are carried out for SD of natural products.

As a result of this scenario, a simple methodology for COM estimation (class 5, according to the classification discussed previously) for some condimentary plants as a function of process time, solvent mass–to–feed mass ratio (S/F), global yield, and the major costs that comprise the COM (F_{CI}, C_{RM}, C_{OL}, C_{WT}, C_{UT}) will be presented.

Table 2.4.1 presents the operating conditions (temperature, pressure, and steam flow rate), the extraction bed characteristics (apparent bed density and mass of feed), the price rating for condimentary plants, and the steam and water costs for each case studied (data 1–4 are for anise, data 5 is for chamomile, data 6–8 are for rosemary, data 9–11 are for black pepper, and data 12–17 are for thyme). Literature data are from Romdhane and Tizaoui [6] (anise VO), Mateus et al. [7] (rosemary), and Rouatbi et al. [8] (black pepper and thyme). The experimental data (anise, black pepper, and rosemary) were obtained in the Laboratory of Supercritical Technology: Extraction, Fractionation and Identification of Vegetable Extracts (LASEFI)/FEA (College of Food Engineering)/UNICAMP (State University of Campinas) using the pilot equipment unit described by Leal [9].

The scale-up procedure used to estimate the solvent flow rate and the feed mass took the assumptions previously described (see Section 2.4.4.1).

The equipment contains a water reservoir of 15×10^{-3} m^3, a pump (model 7014-52, Cole Parmer Instrument Co., Chicago, IL) with a controller (Cole Parmer) of a heating tape that involves the tubing of the pump outlet, a steam generator (production capacity of 1.6×10^{-3} kg s^{-1}) with a heater with a recipient of capacity equal to 5 L (Labcenter, Campinas, Brazil), a temperature controller (model B144028130, Coel Controles Elétricos, São Paulo, Brazil) with two thermocouples (used for measuring the steam temperature inside the heat exchanger and the resistance temperature in order to monitor the steam superheating), a glass distillation column with 1.2×10^{-3} m^3 of capacity (diameter of 5×10^{-2} m and length of 6×10^{-1} m), a glass condenser that works with a solution of ethylene glycol (40%) in water cooled by a thermostatic bath (Marconi, model MA-184, Piracicaba, Brazil), and a glass separator of oil and hydrosol.

TABLE 2.4.1
Information for Estimation of COM of VOs: Operating Conditions and Estimated Industrial Solvent Flow Rate

	Anise				Chamomile		Rosemary		Black pepper	Thyme	
	Data 1[a]	Data 2[b]	Data 3[b]	Data 4[b]	Data 5[a]	Data 6[a]	Data 7[c]	Data 8[d]	Data 9–11[e]	Data 12–14[e]	Data 15–17[e]
P, kPa	≥100	140	200	200	≥100	≥100	190–310	140–160	N/A	N/A	N/A
T, K	140	109	393	393	430	419	401–409	395–403	373/448/523	373/338/523	373/338/523
ρ_{ap}, kg m^{-3}	425	425	425	425	135	214	130.1	99.6	318	223	223
$\dot{M}_{S\_lab}$, kg×10^{3}s^{-1}	0.14	1.7	1.7	1.7	0.14	0.16	52	62	0.444	0.444	0.694
M_{F_lab}, kg	0.108	2	2	5	0.065	0.141	29.9	23	0.01	0.007	0.007
$\dot{M}_{F\_lab}$, kg h^{-1}	997.3	637.5	637.5	637.5	538	432	409.9	484.3	424	426	666.7
M_{F_ind}, kg	212.5	212.5	212.5	212.5	67.5	107	65	49.8	159	112	112
	Quotation, US\$ ton^{-1}										
Raw material	4926[f]				5053[g]	3400[g]			3038[h]	1630[i]	
	Cost estimation (US\$ ton^{-1})[j]										
Steam	16.22										
Cold water	14.80×10^{-3}										

[a] Experimental data obtained at LASEFI/DEA/FEA/UNICAMP by Glaucia H. Carvalho; [b] data from Romdhane and Tizaouri[6]; [c] data from Mateus et al. [7] for rosemary collected from cultivation 22 days prior to distillation; [d] data from Mateus et al. [7] for rosemary collected from cultivation 1 day prior to distillation; [e] data from Rouatbi et al. [8]; [f] quotation from Hervaquímica Ind. Com., São Paulo, Brazil, 2006; [g] quotation from Herboflora Produtos Naturais Ltda, São Paulo, Brazil, 2006; [h] quotation from producer located in Northeastern Brazil, 2007; [i] quotation from CEASA (Central Supplier of Campinas), Brazil, 2007; [j] Turton et al. [4]; N/A: information not available.

2.4.5.1 Anise Seed

For the COM estimation of anise VO, two series of data were selected: (1) experimental data obtained at LASEFI/DEA/FEA/UNICAMP, designated data 1, and (2) literature data of Romdhane and Tizaoui [6], designated data 2, 3, and 4 (see Table 2.4.1).

Figure 2.4.2 shows the COM and the yield as a function of the solvent-to-feed ratio (S/F) and of distillation time (data 1). It is possible to observe that the maximum extraction time was not sufficient to achieve the exhaustion of the anise seed bed. The COM markedly decreased between 60 and 120 min of extraction, from US$ 8934.00/kg to US$ 3757.00/kg. During this period of time, the yield increased 2.5 times. The lowest COM was obtained with the longest extraction time (US$ 2822.00/kg). The low extraction yield (maximum value of 0.25%) may be due to the difficult access of the steam to the VO located inside the seed. When the S/F value is doubled from 5 to 10, a considerable reduction of the COM can be observed. Larger values of S/F could be more interesting for further exhaustion of the raw material, because increasing the amount of steam available helps to overcome the physical barrier presented by the raw material structure when the seed is not milled, which hampers the access of the solvent to the VO. Observing the distribution of the costs that comprise the COM anise VO (Figure 2.4.3; data 1), it is observed that C_{RM} is the predominant cost. The maximum value of F_{CI} was 0.6%, whereas C_{UT} and C_{OL} were not more than 7 and 8%, respectively.

Figure 2.4.4 shows the COM and the extraction yield as a function of S/F and of distillation time for anise VO (data 2). The Romdhane and Tizaoui [6] study was carried out in a plate distiller and presented a higher yield of anise seed VO when compared to the traditional distiller. The plate distiller promotes higher porosity of the bed, as well as better contact between vegetal matrix and steam. After 140 min of extraction, the yield obtained for data 2 was 10 times higher than for that of data 1. Analyzing the S/F ratio and the distillation time, it is possible to observe that for 140 min of extraction time the S/F ratio was 7 for data 2, whereas it was 22 for data 1. This information indicates that the use of higher amounts of solvent does not necessarily guarantee the increase in the extraction yield. Figure 2.4.5 presents the distribution of the costs that comprise the anise COM of anise VO (data 2). Again, the C_{RM} is predominant.

Figure 2.4.6 shows the COM and the extraction yield as a function of S/F and of distillation time for anise VO (data 3). Compared to the OEC presented in Figure 2.4.4 (data 2), it is possible to observe a slight increase in the extraction yield due to the pressure and temperature increments (Table 2.4.1). However, the estimated COM was not considerably affected. Analyzing data 2 and 3, for the same S/F, the estimated COM presented a significant variation. For data 3, considering that the solvent flow rate and the amount of raw material were kept constant, the increase in pressure and temperature directly influenced the extraction yield and, therefore, the COM. Although after 140 min of SD this phenomenon was not expressive, at the beginning of the process the extraction rate was higher for data 3 when compared to data 2, leading to lower COMs. The COM distribution

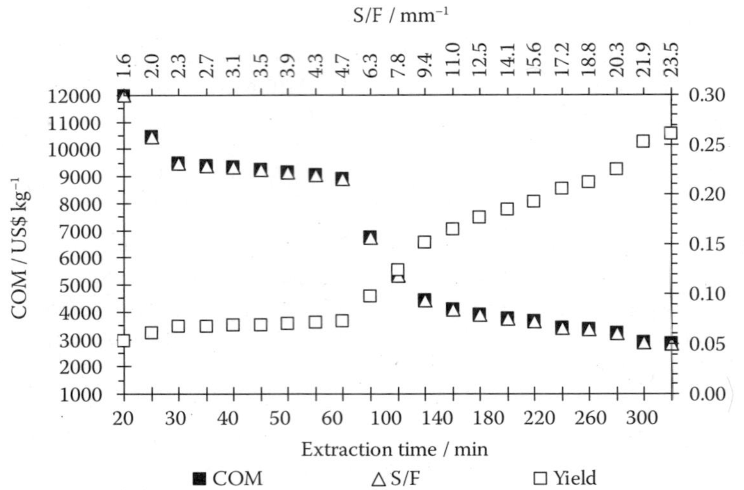

FIGURE 2.4.2 COM of anise seed VO and yield for data 1 as function of extraction time and solvent-to-feed ratio (S/F).

(Figure 2.4.7) has the same behavior as that of data 2, proving that the slight temperature variation did not exert an impact on C_{UT}.

Other data of Romdhane and Tizaoui [6] for anise VO (data 4) were also studied. In this case, the only modification when compared to data 3 was the increase of feed mass from 2 to 5 kg (Table 2.4.1), in a still of the same capacity. The yield results for data 3 and 4 were similar. Figure 2.4.8 shows that the estimated COMs were similar in both cases for 140 min of extraction. The use of higher feed mass implied in a reduction of the S/F at similar. According to Figure 2.4.6, the S/F value of 2

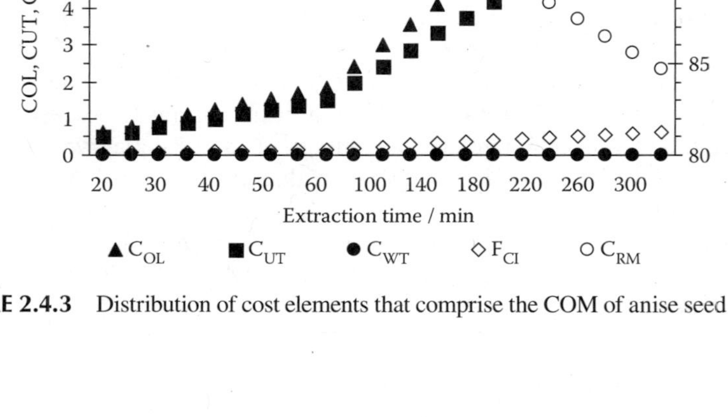

FIGURE 2.4.3 Distribution of cost elements that comprise the COM of anise seed VO (data 1).

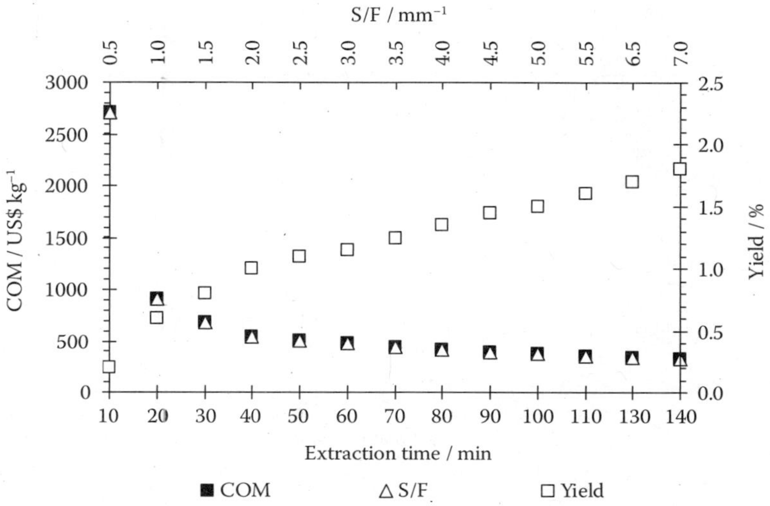

FIGURE 2.4.4 COM of anise seed VO and yield for data 2 as function of extraction time and solvent-to-feed ratio (S/F).

corresponds to a COM of US$ 450.00/kg, a distillation time of 40 min, and 1.3% of extraction yield, while Figure 2.4.8 shows that the same S/F corresponds to a COM of US$ 300.00/kg, an extraction time of 100 min, and a yield of 1.8%. Although gathering all the process information is very important in order to select the distillation time, it is also necessary to analyze the quality of the VO obtained (chemical composition of the oil and content of the bioactive compound). Once again, the C_{RM} fraction was predominant on the COM composition, as can be seen in Figure 2.4.9.

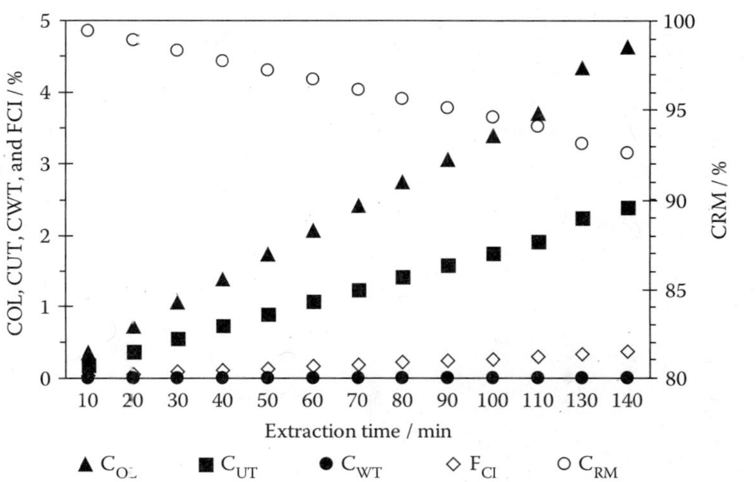

FIGURE 2.4.5 Distribution of cost elements that comprise the COM of anise seed VO (data 2).

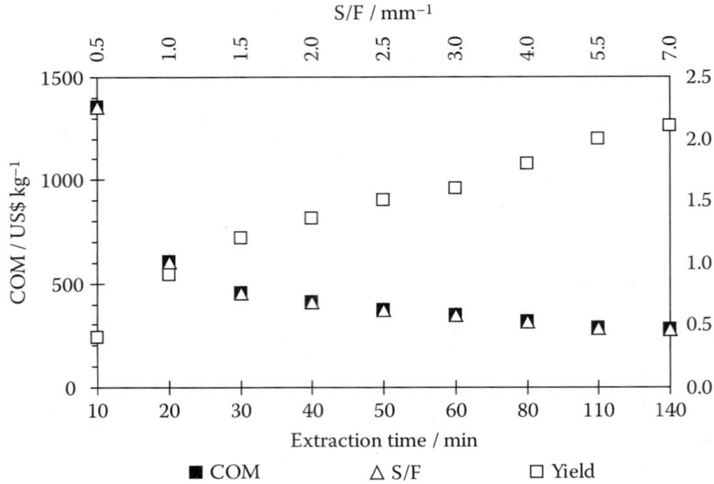

FIGURE 2.4.6 COM of anise seed VO and yield for data 3 as function of extraction time and solvent-to-feed ratio (S/F).

2.4.5.2 Chamomile

For chamomile VO, experimental data obtained at the LASEFI/DEA/FEA/UNICAMP were used (data 5 of Table 2.4.1). Figure 2.4.10 shows the COM and the yield as a function of the S/F and of the distillation time. It is observed that distillation time was not enough to exhaust the chamomile bed. The COM decreases with the distillation time from US$ 7089.00/kg to US$ 2798.00/kg for 30 and 300 min of extraction, respectively. The elevated COM of chamomile VO is related to its low extraction yield (maximum value of 0.32%). Additionally, the low apparent density

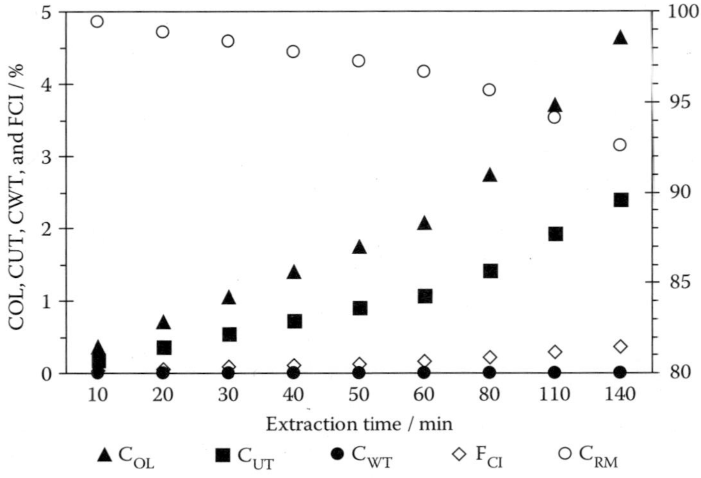

FIGURE 2.4.7 Distribution of cost elements that comprise the COM of anise seed VO (data 3).

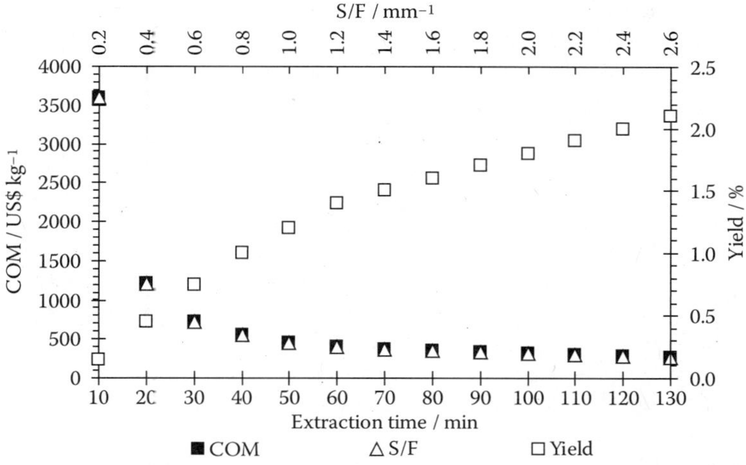

FIGURE 2.4.8 COM of anise seed VO and yield for data 4 as function of extraction time and solvent-to-feed ratio (S/F).

of the chamomile bed results in low feed mass per extraction cycle when compared to the other plants (Table 2.4.1), leading to a decrease in VO production. For 35 min of extraction cycle, the maximum productivity would be 1.2 ton/year, whereas increasing the extraction cycle to 300 min would reduce the annual productivity to 343 kg. However, it is important to observe that although the productivity is lower for longer cycle times, the estimated COM decreases with the increase in extraction time. The COM estimated for a cycle time of 300 min was less than half of the COM estimated for a cycle time of 35 min, which is why productivity and COM should be analyzed together. In this experiment the S/F ratio varied from 4 to 40. With higher

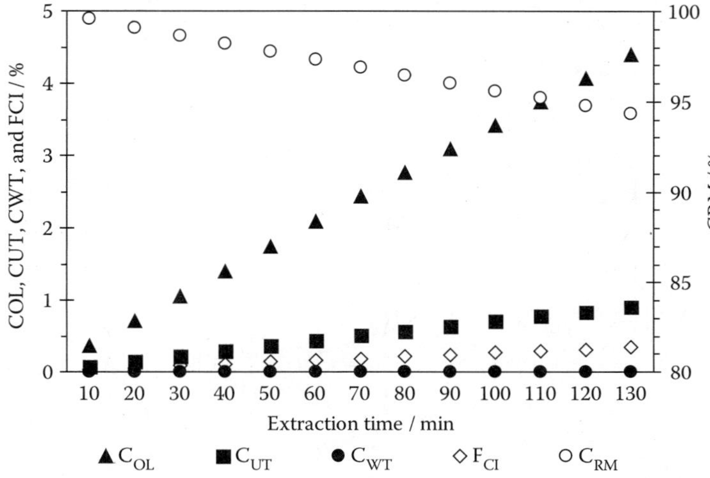

FIGURE 2.4.9 Distribution of cost elements that comprise the COM of anise seed VO (data 4).

values of S/F, higher yields were obtained; therefore, lower COMs were estimated. Figure 2.4.11 presents the cost distribution that comprises the COM. C_{RM} was predominant (69 to 99%), and C_{UT}, as expected, increased with distillation time (from 0.2 to 8.9%).

2.4.5.3 Rosemary

Experimental data (data 6) obtained at the LASEFI/DEA/FEA/UNICAMP and data obtained by Mateus et al. [7] (data 7 and 8) were selected for the COM estimation of rosemary VO. The COM estimation and the extraction yield as a function of S/F and of distillation time for rosemary VO (data 6) are presented on Figure 2.4.12. After 15 min of distillation, 91% of the VO had been extracted and the corresponding COM was US$ 375.00/kg. Figure 2.4.12 shows an atypical behavior when compared to the other raw materials discussed so far: the COM decreased up to 15 min of extraction; afterwards, it remained approximately constant up to 60 min, and after 60 min of distillation the COM increased strongly with time. The lower estimated COM was US$ 369.00/kg with an S/F of 3.7. This behavior suggests that the rosemary bed was already exhausted, and therefore, extraction cycles longer than 60 min imply a reduction of the number of cycles per year and consequent reduction of the annual production of VO. Figure 2.4.13 shows the distribution of the costs that comprise the rosemary VO COM (data 6). The C_{RM}, although predominant, decreased with extraction time, especially in the period between 60 and 300 min, whereas C_{OL} presented an increase from 7 to 20% in the same time interval.

Pereira and Meireles [10] also estimated the COM of rosemary VO. They found a COM value 4.8 times (US$ 76.50/kg) less than the lowest COM obtained for data 6 (US$ 369.00/kg). For SD COM estimations, Pereira and Meireles [10] used information

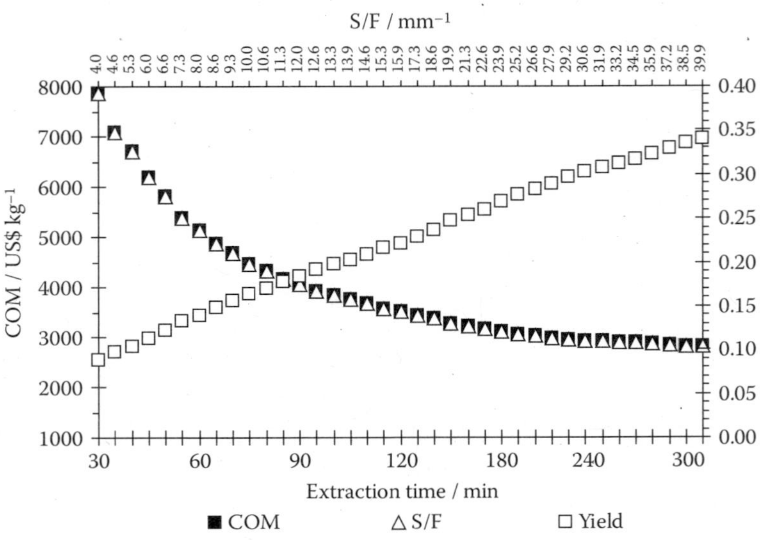

FIGURE 2.4.10 COM of chamomile VO and yield for data 5 as function of extraction time and solvent-to-feed ratio (S/F).

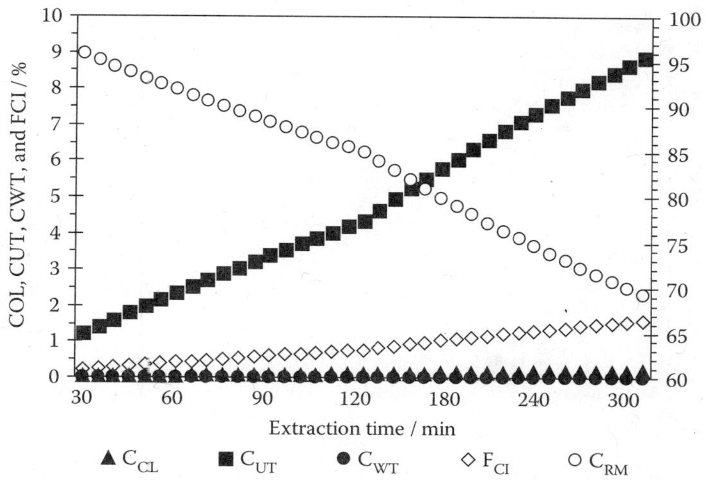

FIGURE 2.4.11 Distribution of cost elements that comprise the COM of chamomile VO (data 5).

from the study of Ondarza and Sanches [11] and made some assumptions, such as considering an S/F value of 1 and a distillation time of 2 h. The great difference between the COM estimated by Pereira and Meireles [10] and the COM estimated from data 6 is related to the difference in the raw material cost. While the Pereira and Meireles [10] study indicated that the C_{UT} was the predominant component of the COM (72.14%), the data 6 evaluation indicates that the C_{RM} cost was the predominant component (72 to 99%). IBGE (Brazilian Institute of Geography and Statistics)

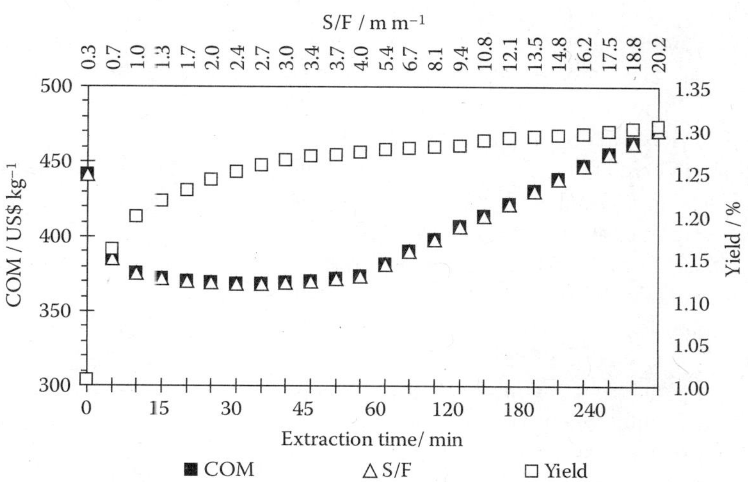

FIGURE 2.4.12 COM of rosemary VO and yield for data 6 as function of extraction time and solvent-to-feed ratio (S/F).

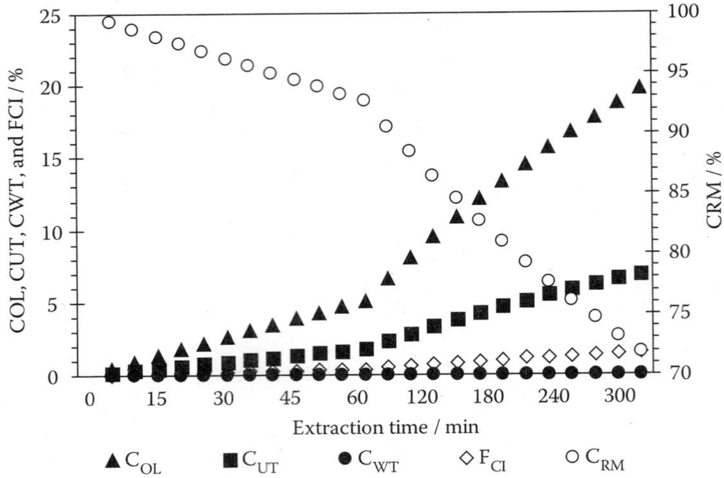

FIGURE 2.4.13 Distribution of cost elements that comprise the COM of rosemary VO (data 6).

information (2006) used by Pereira and Meireles [10] as a reference for the value of raw material cost, provides the cost of production of raw materials, not their market selling price of large quantities. Rosemary costs considered in COM estimation made by Pereira and Meireles [10] was US$ 283.19/ton, a value 12 times lower than the raw materials cost considered in the data 6 study. This way, in the Pereira and Meireles [10] estimation, although not explicitly informed, it is likely that it was considered that the industrial unit that produces the VO by SD also cultivates the raw material. For an data 6, as well as for data 7 and 8, the market selling price was considered as the raw material cost (C_{RW}). For an S/F of 1, the COM obtained by Pereira and Meireles [10] is up to five times lower than the estimated value for data 6. Using an extraction time of 2 h and raw materials cost of US$ 283.19/kg, as considered by Pereira and Meireles [10], but using the distillation conditions presented on Table 2.4.2 and the yield obtained for data 6, the COM and S/F would be US$ 83.00/kg and 8, respectively.

The COM estimation and the yield as a function of S/F and of distillation time for rosemary VO related to data 7 are presented in Figure 2.4.14. The maximum yield obtained was approximately 0.5%, and distillation periods longer than 15 min did not exert a significant impact on COM. This behavior was also observed for data 6 (Figure 2.4.12) for short cycle times. This information suggests that the rosemary VO is readily available for removal from the vegetal matrix. This way, the overestimation of the distillation time would negatively interfere in the annual productivity of VO, because of the reduction of the number of extraction cycles. Analyzing the S/F ratio for data 7, it is observed that the COM is invariant for S/F values greater than 1.5. The C_{RM} is predominant when compared to the other costs that comprise COM (Figure 2.4.15). Figure 2.4.16 shows the COM and the yield as a function of S/F and of distillation time for rosemary (data 8) VO. The study of Mateus et al. [7] reported that the lot of rosemary that was harvested 1 day prior to distillation (data 8) presented slightly higher yield (~0.65%) than the lot harvested 22 days prior to SD (~0.5%, data 7)

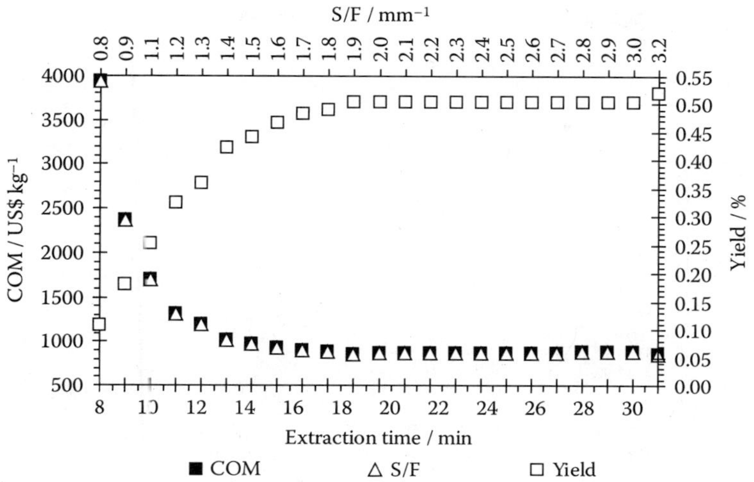

FIGURE 2.4.14 COM of rosemary VO and yield for data 7 as function of extraction time and solvent-to-feed ratio (S/F).

for the same S/F value. For S/F values greater than 3, the COM did not present large variation. Figure 2.4.17 shows the cost composition distribution for data 8. The behaviors of data 7 and data 8 were similar, with C_{RM} being the predominant cost.

2.4.5.4 Black Pepper

Experimental data obtained by Rouatbi et al. [8] were selected (data 9–11) for the COM estimation of black pepper VO. Figures 2.4.18–2.4.20 show the estimated COM and the yield as a function of S/F and of distillation time for steam temperatures of

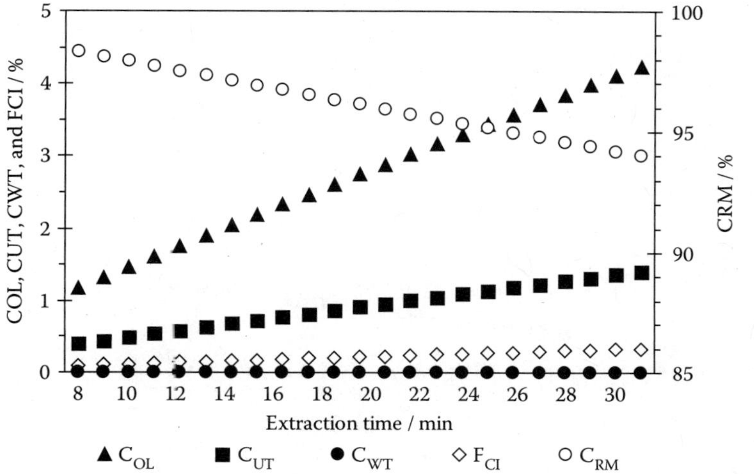

FIGURE 2.4.15 Distribution of cost elements that comprise the COM of rosemary VO (data 7).

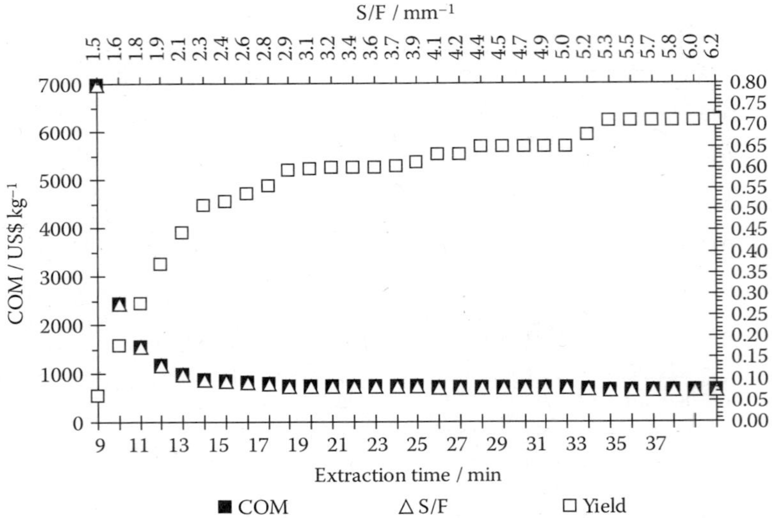

FIGURE 2.4.16 COM of rosemary VO and yield for data 8 as function of extraction time and solvent-to-feed ratio (S/F).

373 K (data 9), 448 K (data 10), and 523 K (data 11), respectively. Analyzing the OECs, it is observed that the raw material bed was not exhausted. The yield increased considerably with the increase of steam temperature. The estimated COM varied from US$ 232.00/kg to US$ 3,345.00/kg. The lowest COM was obtained with the steam temperature of 523 K. Analyzing S/F ratios and COM, the highest S/F ratio (107) corresponded to the lowest COMs. Figure 2.4.21 shows the costs distribution

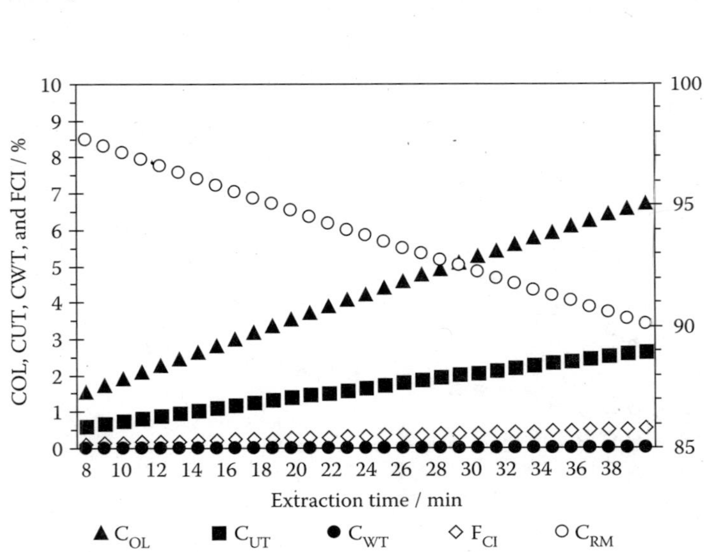

FIGURE 2.4.17 Distribution of cost elements that comprise the COM of rosemary VO (data 8).

in the COM for the three data sets of black pepper, since the temperature variation did not exert a significant effect on C_{UT}. C_{RM} was predominant when compared to the other cost components. It decreased with extraction time (from 86 to 60%), whereas the C_{UT} impact on COM increased from 12 to 34%. According to Rouatbi et al. [8], VOs extracted at 373 and 448 K presented similar quality, because the VOs obtained under both temperature conditions had similar chemical composition. However, the VO obtained at 523 K presented inferior quality when compared to the other two samples because of the degradation of some compounds and the coextraction of undesirable compounds. Rouatbi et al. [8] concluded that a steam temperature of 448 K is the more adequate temperature for obtaining black pepper VO because of the higher yield when compared to the extraction at 373 K and the superior quality regarding chemical composition when compared to the extraction at 523 K.

2.4.5.5 Thyme

Experimental data obtained by Rouatbi et al. [8] were selected (data 12–17) for the COM estimation of thyme VO. Figures 2.4.22–2.4.24 show the estimated COM and the yield as a function of S/F and distillation time steam temperatures of 373 K (data 12), 448 K (data 13), and 523 K (data 14), respectively. COM varied from US$ 79.00/kg to US$ 244.00/kg. COM decreased with temperature increase. The S/F ratio varied from 19 to 152. For data 12, the COM varied from US$ 156.00/kg to US$ 244.00/kg for S/F values of 57 and 19, respectively. When the S/F ratio was tripled (from 19 to 57), it was possible to observe a yield increase from 1 to 2.13%, reducing the manufacturing cost by 36%. The yield varied between 1 and 3.25%. For data 13, the estimated COM presented a maximum variation of 28% (from US$ 124.00/kg to US$ 172.00/kg). The yield varied from 1.5 to 4.19%. The lowest COMs were obtained for data 14 because of the higher yields obtained (from 2 to 5.25%)

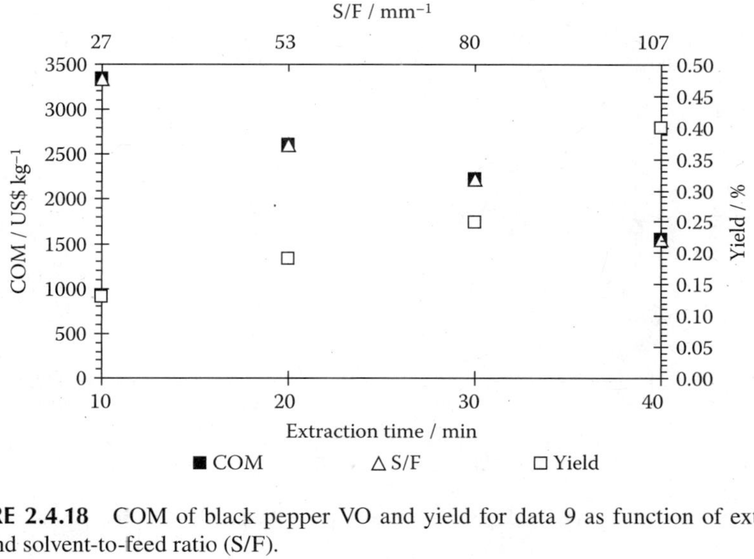

FIGURE 2.4.18 COM of black pepper VO and yield for data 9 as function of extraction time and solvent-to-feed ratio (S/F).

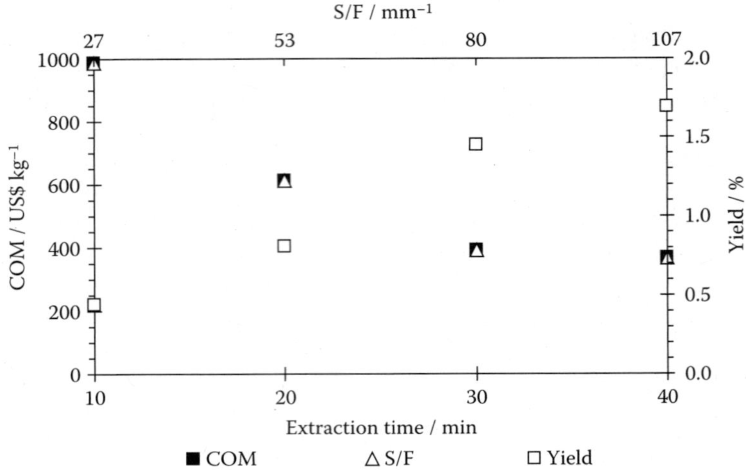

FIGURE 2.4.19 COM of black pepper VO and yield for data 10 as function of extraction time and solvent-to-feed ratio (S/F).

when compared to data 12 and 13 for the same steam flow rate. The cost distribution that comprises the COM is presented in Figure 2.4.25. A different behavior from those observed for the other condimentary plants is shown. For thyme, there was an inversion of the predominant cost. For distillation times up to 20 min the C_{RM} was predominant, whereas from 30 min on, C_{UT} represented the largest fraction of the COM. C_{RM} varied from 36.5 to 82.1%, and C_{UT} varied from 15.6 to 55.3%.

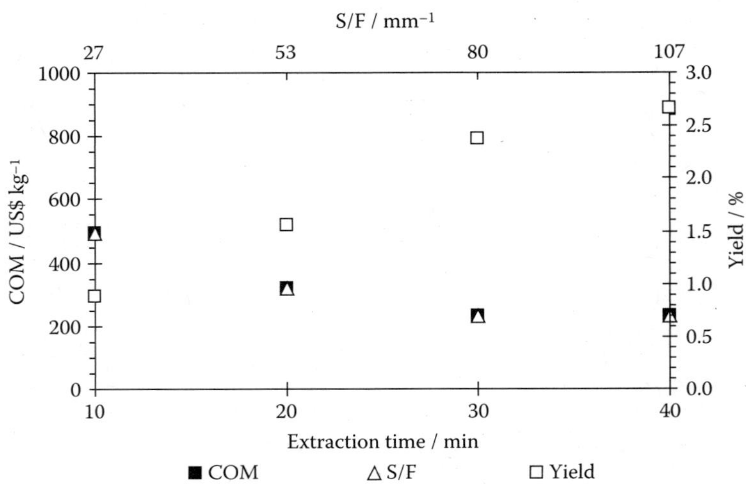

FIGURE 2.4.20 COM of black pepper VO and yield for data 11 as function of extraction time and solvent-to-feed ratio (S/F).

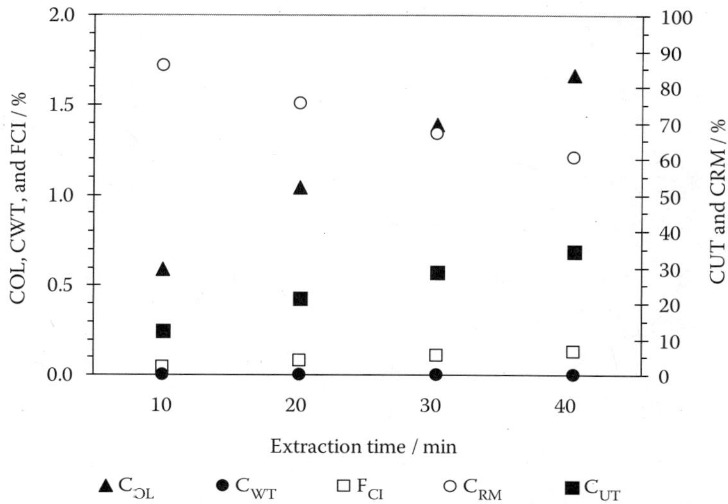

FIGURE 2.4.21 Distribution of cost elements that comprise the COM of black pepper VO (data 9, 10, and 11).

Figures 2.4.26–2.4.28 show the COM and the yield as a function of S/F and of distillation time for steam temperatures of 373 K (data 15), 448 K (data 16), and 523 K (data 17), respectively. The difference between these three data and the ones previously described (data 12–14) relies on the steam flow rate and, therefore, on the S/F ratio. The COM varied from US$ 71.00/kg to US$ 177.00/kg, the S/F ratio from 29.8 to 238.1, and the yield from 1.5 to 6.1%. The lowest COM was obtained at 523 K,

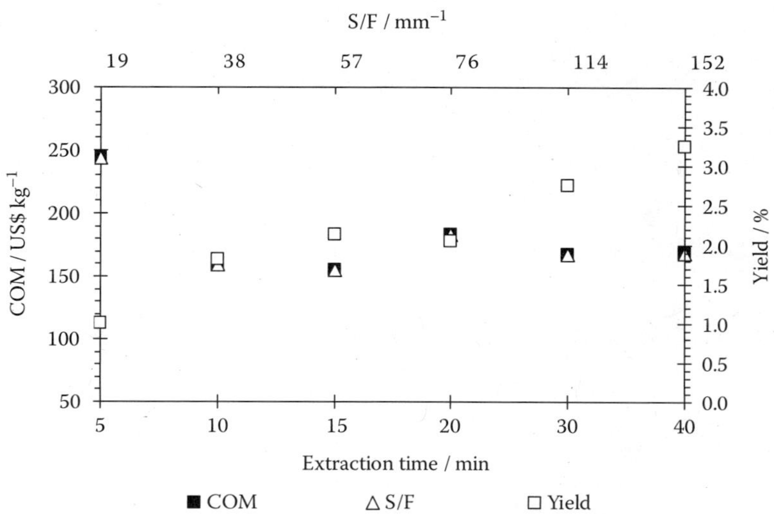

FIGURE 2.4.22 COM of thyme VO and yield for data 12 as function of extraction time and solvent-to-feed ratio (S/F).

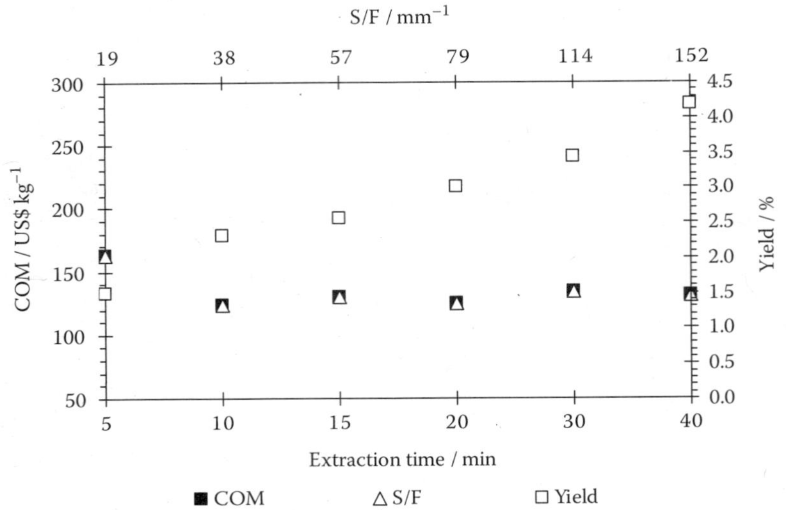

FIGURE 2.4.23 COM of thyme VO and yield for data 13 as function of extraction time and solvent-to-feed ratio (S/F).

for the distillation time of 10 min. The yield for this operating condition was 4.63%. In Figure 2.4.29 it is observed that with a distillation time of under 15 min, the C_{RM} was the predominant fraction of COM. From 20 min on, C_{UT} is responsible for the major share of the COM. C_{RM} varied between 27.8 and 75.5%, whereas CUT varied between 22.4 and 65.9%.

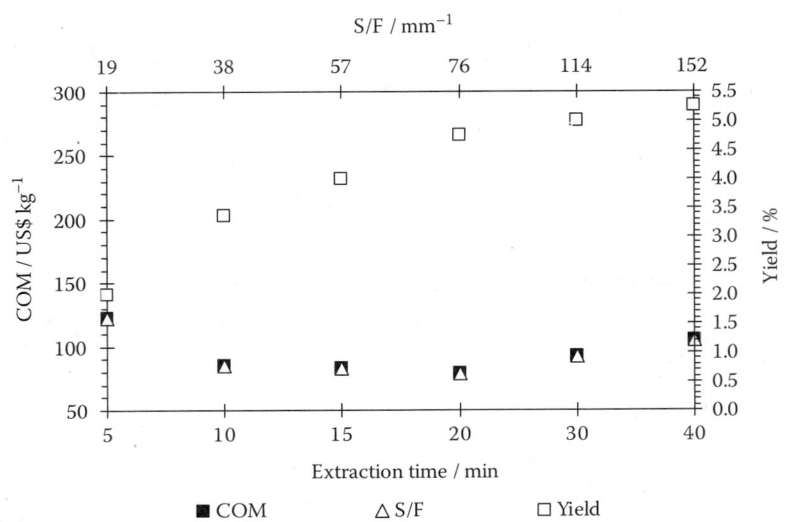

FIGURE 2.4.24 COM of thyme VO and yield from data 14 as function of extraction time and solvent-to-feed ratio (S/F).

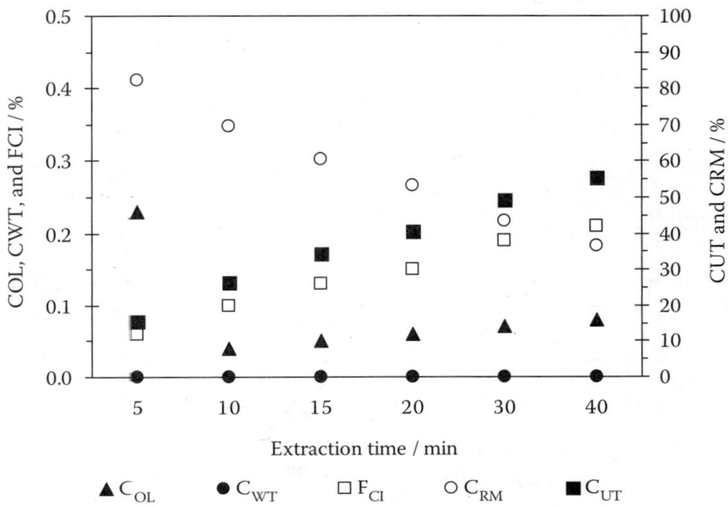

FIGURE 2.4.25 Distribution of cost elements that comprise the COM of thyme VO (data 12, 13, and 14).

In the case of thyme VO quality, the same phenomenon reported by Rouatbi et al. [8] was observed for black pepper VO. VOs extracted with steam temperatures of 273 and 448 K presented similar qualities in terms of their composition. However, the VO obtained at 523 K presented inferior quality when compared to the two other samples because of the degradation of certain compounds and coextraction of undesirable substances. Thus, thyme SD should be carried out with steam at 448 K in order to improve yield and preserve the VO quality.

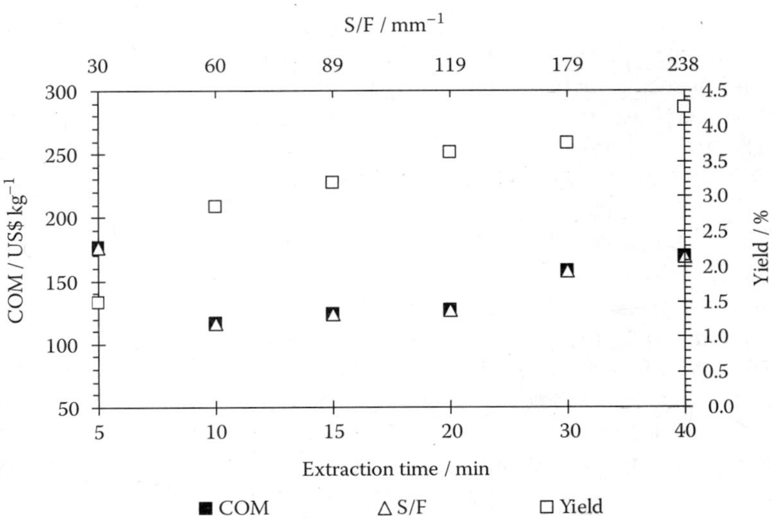

FIGURE 2.4.26 COM of thyme VO and yield for data 15 as function of extraction time and solvent-to-feed ratio (S/F).

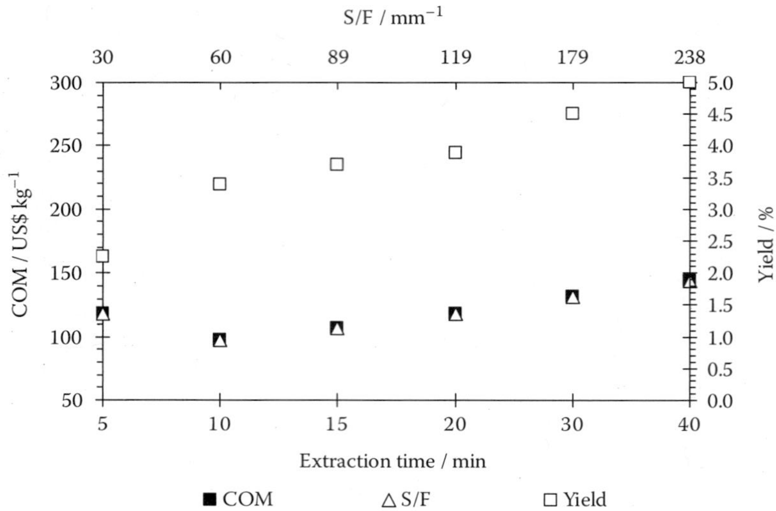

FIGURE 2.4.27 COM of thyme VO and yield for data 16 as function of extraction time and solvent-to-feed ratio (S/F).

2.4.6 Comparing Estimated COMs and Market Prices

Temperature, pressure, and solvent (steam) flow rate are operating conditions that are key factors in the COM variation, because these parameters exert influence on the extraction global yield. The selection of the processing time is another key factor for the optimization of the process, affecting its economical viability and the quality of the VO.

It is important to note that the SD results presented here indicate that the C_{RM} represents the major fraction of the COM for the majority of the raw materials evaluated. It was expected that the C_{UT} would play this role. In Chapter 6 (Section 6.2), in which the usage of supercritical fluid extraction (SFE) is discussed, it is shown that C_{RW} represents a major fraction of the COMs for producing clove bud VO and ginger oleoresin in industrial-sized equipment (two extractors of 400 L each) for the supercritical extraction process. The same behavior was reported by Leal et al. [12] for the SFE of sweet basil. Thus, this information shows that plant extract COM can be reduced by process optimization as well as by improving the agricultural techniques in order to decrease the C_{RW} cost.

Table 2.4.2 summarizes the lowest COM estimated with Turton et al. [4] methodology for each one of the condimentary plants presented in the previous sections. It also presents the estimated annual productivity considering a steam distillation unit composed of two distillation columns, each with a capacity of 0.5 m^3, operating alternately. Finally, it presents the market selling prices of some condimentary plants.

COM class 4 or 5, although based on a poor level of project definition, is a useful tool to evaluate whether the project should move forward or be abandoned.

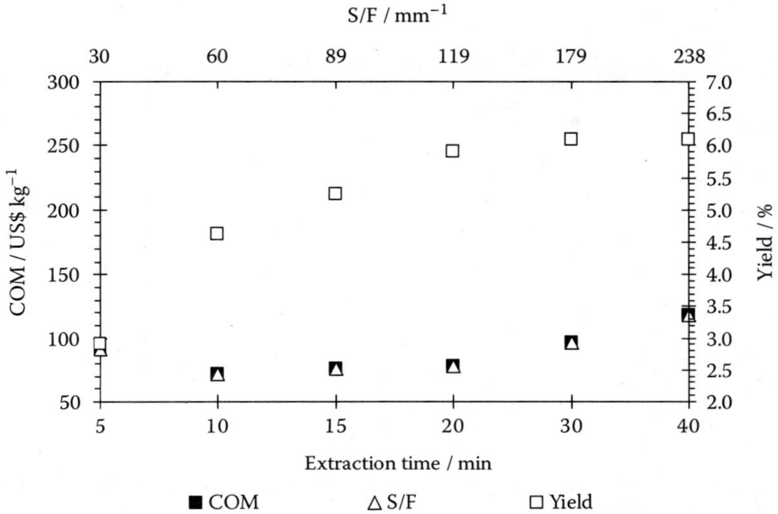

FIGURE 2.4.28 COM of thyme VO and yield for data 17 as function of extraction time and solvent-to-feed ratio (S/F).

The prices of VOs vary a lot in the market, and the main differences between the available products are their chemical composition and the quality of the raw material, which is related to its origin. For instance, the 2007 rating for VOs of rosemary, chamomile (diluted to 10%), black pepper, and thyme (Table 2.4.2) obtained from two different suppliers (a Brazilian supplier of product produced in France and a Brazilian supplier of product from different countries) indicated that the estimated

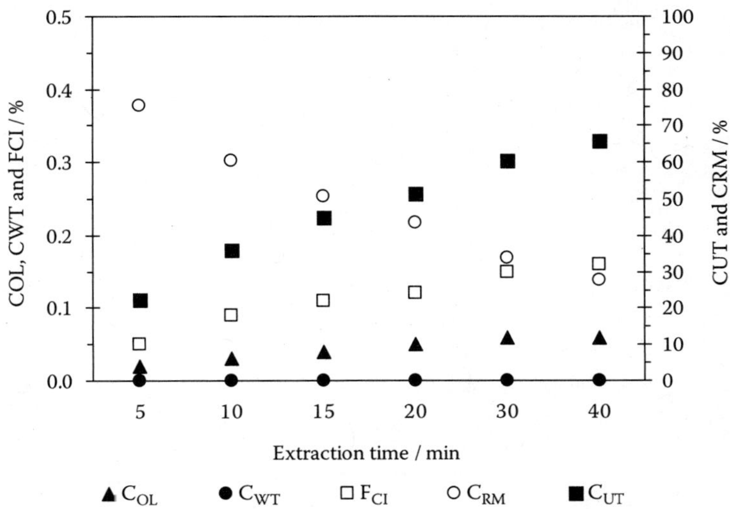

FIGURE 2.4.29 Distribution of cost elements that comprise the COM of thyme VO (data 15, 16, and 17).

TABLE 2.4.2
COM, Annual Productivity, and Market Price of VOs Obtained by SD

Raw material	COM (US$ kg^{-1})	Annual productivity (ton $year^{-1}$)	Market price (US$ kg^{-1})[a]
Anise seed	216.00	17.2	N/A
Black pepper	232.00	0.34	181.00–975.00
Chamomile	2798.00	16.3	2152.00–6625.00[b]
Rosemary	369.00	50.2	60.00–725.00
Thyme	71.00	246	155.00–428.00

N/A: information not available.
[a] Confidential source.
[b] Blue chamomile diluted to 10%.

COMs presented here are higher than the lowest selling price and lower than the highest selling price. Thus, the estimated COM presented here (which are classes 5 or 4) indicate that the SD process is attractive to investors, and the optimization of the process would certainly reduce the real COM.

2.4.7 Nomenclature

Symbol	Definition	Units	Dimensions in M, N, L, T,
M_{F_ind}	Feed mass of raw material in the distillation column of the industrial unit	kg	M
M_{F_lab}	Feed mass of raw material in the distillation column of the laboratorial scale	kg	M
M_{S_ind}	Solvent (steam) flow rate of the industrial unit	kg h^{-1}	$M{\cdot}T^{-1}$
M_{S_lab}	Solvent (steam) flow rate of the laboratorial unit	kg sec^{-1}	$M{\cdot}T^{-1}$
M_{S_lab}	Solvent (steam) mass used in one cycle in the laboratorial scale	kg	M
S/F	Ratio between solvent mass and feed mass	$kg_{solvent}\ kg_{feed}^{-1}$	$M{\cdot}M^{-1}$
t	Time	sec	T
t_{cycle}	Time of distillation	min	T
Economic variable			
C_{Pi}	Cost of equipment	US$	
C_{TM}	Total cost of an industrial plant	US$	
C_{OL}	Cost of operational labor	US$	
COM	Cost of manufacturing	US$	
C_{RM}	Cost of raw material	US$	
C_{UT}	Cost of utilities	US$	
C_{WT}	Cost of waste treatment	US$	
F_{Lang}	Lang factor	—	
FCI	Fixed cost of investment	US$	

2.4.8 Acknowledgments

The authors thank Fundação de Amparo à Pesquisa do Estado de São Paulo (FAPESP), Conselho Nacional de Desenvolvimento Científico e Tecnológico, and Coordenação de Aperfeiçoamento de Pessoal de Nível Superior for financial support. P. F. Leal, T. M. Takeuchi, and J. M. Prado thank FAPESP for the PhD assistantships (04/09310-3, 05/54544-5, 07/03817-7).

2.4.9 References

1. AACEI. 2007. Association for the Advancement of Cost Engineering International. http://www.aacei.org (accessed February, 2007).
2. Anonymous. 1997. Recommended practice (draft): Cost estimate classification system. *Cost Engineering* 39 (4): 22–25.
3. Anonymous. 1997. Recommended practice (draft): Cost estimate classification system—as applied in engineering, procurement and construction for the process industrial. *Cost Engineering* 39 (4): 15–21.
4. Turton, R., R. C. Baile, W. B. Whiting, and J. A. Shaeiwitz. 1998. *Analysis, syntesis and desing of chemical process*. Upper Saddle River, NJ: Prentice Hall.
5. Ulrich, G. D. 1984. *A guide to chemical engineering process designer and economics*. New York: John Wiley & Sons.
6. Romdhane, M., and C. Tizaoui. 2005. The kinetic modelling of a steam distillation unit for the extraction of aniseed (*Pimpinella anisum*) essential oil. *Journal of Chemical Technology and Biotechnology* 80:759–766.
7. Mateus, E. M., C. Lopes, T. Nogueira, J. A. A. Lourenço, and M. J. M. Curto. 2006. Pilot steam distillation of rosemary (*Rosmarinus officinalis* L.) from Portugal. *Silva Lusitana* 14 (2): 203–217.
8. Rouatbi, M., A. Duquenoy, and P. Giampaoli. 2007. Extraction of the essential oil of thyme and black pepper by superheated steam. *Journal of Food Engineering* 78:708–714.
9. Leal, P. F. 2008. Estudo comparativo entre os custos de manufatura e as propriedades funcionais de óleos voláteis obtidos por extração supercrítica e arraste a vapor. PhD diss., State University of Campinas (UNICAMP).
10. Pereira, C. G., and M. A. A. Meireles. 2007. Economic analysis of rosemary, fennel and anise essential oils obtained by supercritical fluid extraction. *Flavour and Fragrance Journal* 22 (5): 407–413.
11. Ondarza, M. and A. Sanchez. 1990. Steam distillation and supercritical fluid extraction of some Mexican spices. *Chromatographia* 30:16–19.
12. Leal, P. F., N. B. Maia, Q. A. C. Carmello, R. R. Catharino, M. N. Eberlin, and M. A. A. Meireles. 2008. Sweet basil (*Ocimum basilicum*) extracts obtained by supercritical fluid extraction (SFE): Global yields, chemical composition, antioxidant activity, and estimation of the cost of manufacturing. *Food and Bioprocess Technology* DOI 10.1007/s11947-007-0030-1.

3 Distillation Applied to the Processing of Spirits and Aromas

Antonio J. A. Meirelles, Eduardo A. C. Batista, Helena F. A. Scanavini, Fábio R. M. Batista, Roberta Ceriani, and Luiz F. L. Luz, Jr.

CONTENTS

3.1 Fundamentals of Distillation.......76
3.1.1 Main Concepts in the Distillation Processes.......76
3.1.2 Heat and Mass Balance Equations in Distillation Processes.......82
3.1.3 Vapor–Liquid Phase Equilibrium.......86
3.2 Recent Advances in the Simulation of Spirits and Aroma Mixtures Distillation.......97
3.3 Some Especial Applications of Distillation.......101
3.3.1 Obtaining High Quality Cachaça.......101
3.3.1.1 Batch Distillation in Alembic.......102
3.3.1.2 Continuous Distillation in Tray Columns.......109
3.3.2 Concentration and Purification of Aroma Compounds of Cashew Juice in a Batch Distillation Column.......117
3.4 Conclusion.......129
3.5 Nomenclature.......130
3.6 References.......132

In this chapter we will discuss the fundamentals of distillation and the main aspects of this process applied to the production of spirits and to the recovery and concentration of aroma compounds. The concentration and fractionation of volatile liquid mixtures are usually performed by distillation. The most important example in the food industry is the concentration of ethanol from fermented must or wine for the production of spirits, such as whisky, vodka, gin, rum, pisco, cognac, or cachaça. The recovery of aroma compounds evaporated during the concentration of fruit juices is also conducted by distillation, as is the case in the production of orange and apple concentrated juices. Essential oils and fatty acid mixtures are fractionated by distillation too, but in this case the relatively purer fractions obtained by distillation are normally used in the formulation of perfumes, fragrances, cleaning products, and cosmetics in general. In the first part of this chapter, Fundamentals of Distillation, the main concepts involved in distillation processes are discussed, the different types of equipment and the corresponding operating modes are presented, and the mathematical basis for simulating this process is indicated. In the second part, a review of the literature is presented on the topic of simulating the distillation of multicomponent mixtures found in the production of spirits and aromas. In the last part of the chapter we present our own results on the production of sugar cane spirit by alembic and continuous distillation and on the concentration and purification of cashew juice aroma by batch distillation.

3.1 FUNDAMENTALS OF DISTILLATION

3.1.1 Main Concepts in the Distillation Processes

The separation of liquid mixtures by distillation is based on the difference of the volatilities of their components, so that the light compounds (components with higher volatilities) are concentrated in the vapor phase and the heavy ones in the liquid phase. The vapor–liquid contact that characterizes distillation processes can be conducted in different ways. The simplest alternative is the differential distillation, which corresponds approximately to the operation of a batch still often used in the production of spirits on a small scale.

Figure 3.1 shows a scheme of a batch still. The heat transferred by an external source to the liquid mixture at the bottom of the equipment generates a vapor phase that flows through the liquid pool as swarms of bubbles in which the light components are concentrated. The vapor phase is condensed in the heat exchanger located at the top of the equipment and collected in the distillation pot.

A distillation process may be conducted in a batch still only when the light components have volatility much larger than the heavy ones and the required distillate concentration or purity is not very high. Both requirements are fulfilled in the case of the ethanol–water mixture found in the fermented musts used for spirit production. The ethanol concentration in spirits is usually lower than 60 °GL, a concentration expressed in Gay–Lussac, which corresponds approximately to 54.3 mass % or 31.7 mol % of ethanol in the alcoholic beverage. The volatility of ethanol is 2.9 to 12 times larger than the volatility of water for mixtures with concentration varying from much diluted ones to 60 °GL, so that the separation of ethanol from water is relatively easy in this concentration range.

When a high distillate concentration is required, a distillation column with partial reflux of the condensed vapor collected at the top of the equipment must be used. Figure 3.2 shows a scheme of a batch distillation column with reflux, containing several trays for improving the vapor–liquid contact. The vapor phase is generated by heating the liquid mixture at the bottom of the equipment, and it flows upward, bubbling through the liquid pools retained in each tray, and becomes increasingly richer in the light components as it approaches the top of the equipment. Part of the vapor phase condensed at the top of the column is refluxed to the top tray and represents the primal source of the liquid phase present on the liquid pools over the trays.

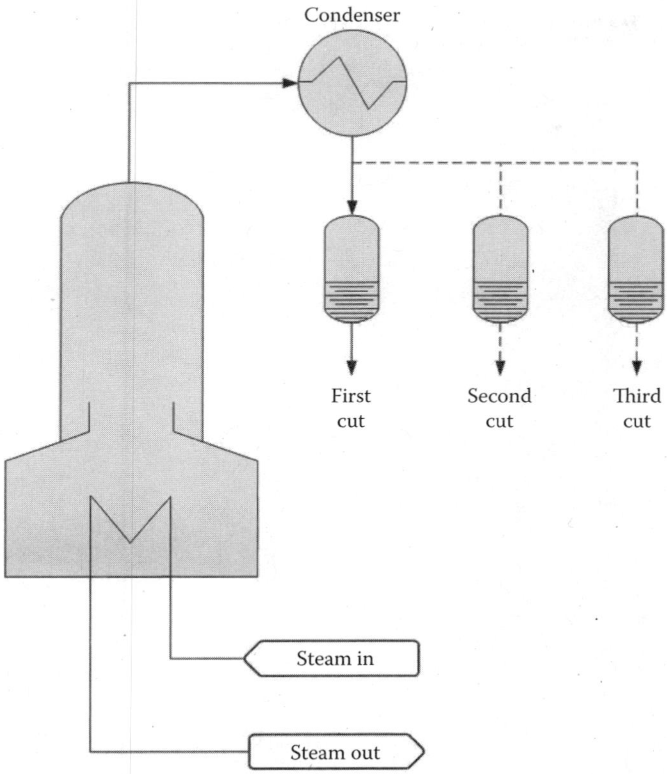

FIGURE 3.1 Scheme of a batch still.

The equipment operates as a countercurrent contactor of vapor and liquid, although on each tray the flow of both phases, vapor and liquid, is better characterized as crosscurrent. The use of reflux and of a series of distillation trays makes feasible the production of high purity distillates.

Batch distillation equipment is operated in an unsteady state, and the composition of the distillate changes continuously during the distillation run. The first portions of the distillate are the richest in the volatile compounds. As the distillation continues, the concentration of these components inside the equipment decreases and, as a consequence, the condensed vapor collected at the top becomes leaner in the volatile substances. During the process the distillate is usually separated and collected in different batches, generating a series of products of different purities that are denominated cuts.

The alembic used in the distillation of spirits in small scale is an example of a batch still. In this case the distillate is usually separated into three different cuts: The first fraction (head distillate) contains more volatile compounds, such as methanol, acetaldehyde, and ethyl acetate, in concentrations above the limits required by legislation or sensorial criteria and has an alcoholic graduation higher than 60 °GL. The second fraction (heart distillate) is the intermediate distillate portion that usually

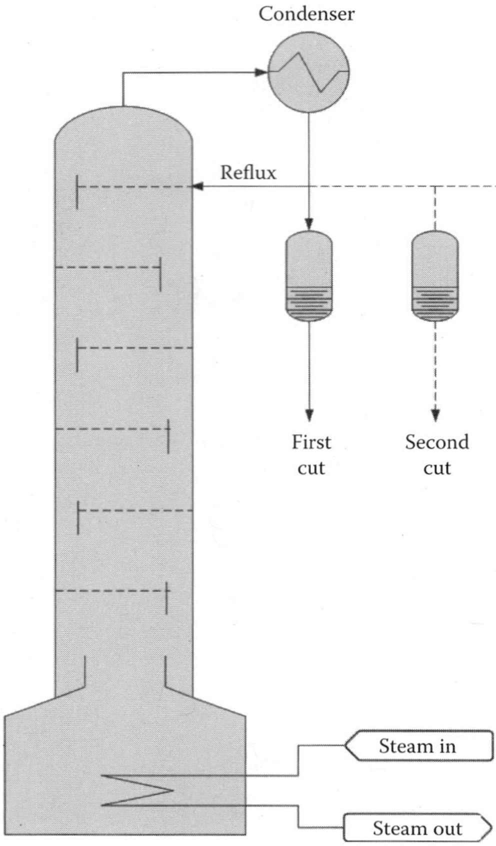

FIGURE 3.2 Scheme of a batch distillation column.

corresponds to the desired spirit. The third fraction (tail distillate), also denominated weak water, is composed mainly of water but also contains relatively lower amounts of ethanol and compounds whose boiling points are higher than 373.2 K. Batch distillation with reflux is normally used in the fractionation of essential oils. A common feature of both processes is the small scale of industrial production, with the batch of liquid processed in the still usually varying in the range of 0.5 to 1.5 m^3. The processing of high amounts of liquid mixtures by distillation requires the use of continuous equipment that is operated in steady state.

Figure 3.3 shows a typical scheme of a continuous distillation column. The liquid mixture that should be concentrated and separated is fed into the column in a tray located in the middle part of the equipment, dividing the column into two major sections: the stripping section located below the feed tray and the enriching section situated above it. At least two product streams are obtained: the distillate, which should be concentrated in the volatile components, and the bottom product, which contains mainly the heavy compounds. In some cases the column contains additional side

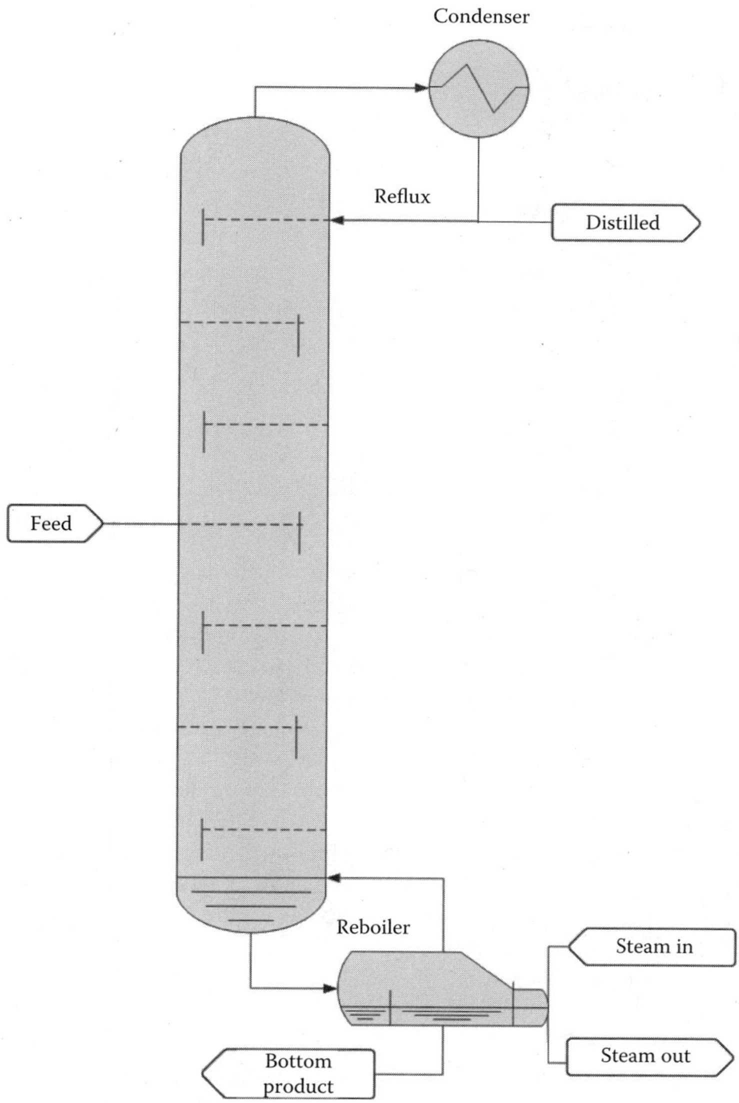

FIGURE 3.3 Continuous distillation column.

streams by means of which mixtures rich in components of intermediate volatility can be removed from the equipment. In the stripping section the volatile components should be stripped away from the liquid phase, so that the bottom product will preferentially contain only small amounts of the light substances. The enriching section must guarantee the concentration of the volatile compounds in the vapor phase and the achievement of the desired concentration at the top of the equipment. The good separation of a liquid mixture in a continuous distillation column depends mainly on the relative volatility of its components, on the number of trays of the equipment, and

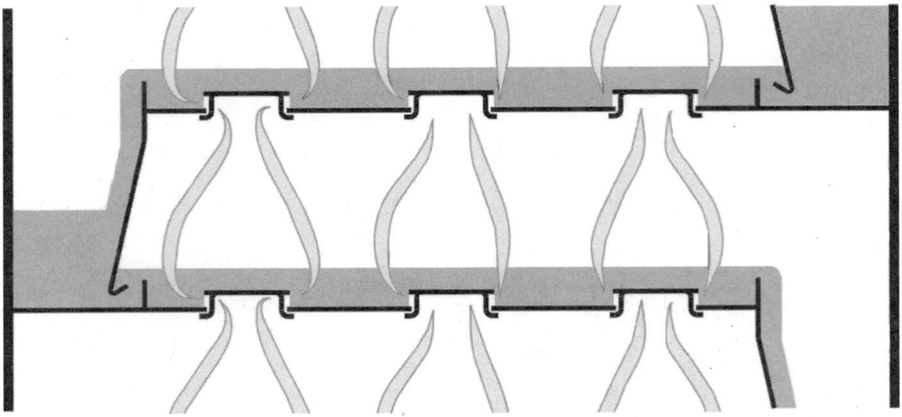

FIGURE 3.4 Valve trays in operation.

on the reflux ratio used. The reflux ratio corresponds to the ratio of the reflux stream to the distillate product stream.

In the distillation columns mentioned above, the liquid and vapor phases are contacted in a stepwise mode on each tray. The liquid passes across the tray flowing horizontally, and afterward it streams through a downcomer to the plate below. The vapor flows upward through the openings in each tray, bubbling inside the liquid pools. The froth so formed guarantees an intense contact between both phases and is usually very efficient for transferring components from one phase to another. Most parts of the mass transfer process should occur inside the froth located on each tray. Only the liquid phase should flow through the downspout, while the vapor phase, after disengaging from the froth, should stream upward without further contact with the liquid phase until it reaches the next tray above. Figure 3.4 shows a scheme of the internals of a valve tray column in operation. The mass transfer efficiency of a tray can be expressed by the Murphree efficiency:

$$\eta = \frac{y_{1,n} - y_{1,n-1}}{y_{1,n}^{*} - y_{1,n-1}}, \tag{3.1}$$

where y_1 represents the concentration, in mol fraction, of component 1 in the vapor phase, n is the index for counting the trays, from the bottom plate to the top, and y_1^* represents the concentration of component 1 in the vapor phase in equilibrium with the liquid phase, which are leaving the same tray. The denominator of Equation 3.1 indicates the maximal enriching in component 1 that the vapor phase leaving tray n–1 can attain as it passes through tray n. The tray n operates as an ideal stage when the actual concentration of the vapor leaving it, $y_{1,n}$, corresponds to the equilibrium concentration $y_{1,n}^*$, so that the Murphree efficiency equals 1.

Figure 3.5 shows the main types of internals used in tray columns. In the case of sieve trays, the openings through which the vapor must pass are perforations equally

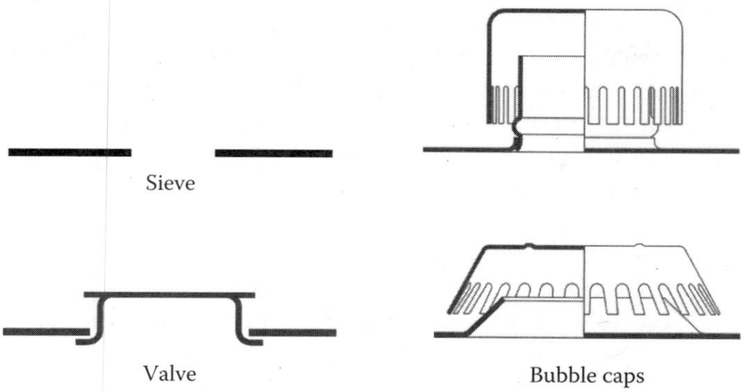

FIGURE 3.5 Main types of internals for tray columns (for the bubble caps, the inside view is on the left and the outside view is on the right).

distributed along a horizontal sheet of metal. In the parts of the metal sheet reserved for the downcomers there are no perforations. These trays have the lowest cost, but they exhibit a very limited range of appropriate operational conditions, because a lower vapor velocity allows the liquid phase to leak through the perforations, while a higher vapor velocity can easily cause an excessive entrainment of liquid and also a large increase of the liquid hold-up on the plate leading to column flooding. These effects decrease significantly the mass transfer efficiency. In the case of valve trays the openings are covered with movable caps that open wider or narrower according to the vapor phase flow, so that the effect of changes in the vapor velocity through the perforations is minimized and the above-mentioned side effects are softened. This type of tray can then operate in an extended range of operational conditions without appreciable loss of efficiency. Another type of plate is the bubble cap tray. In this tray a chimney, covered with a fixed cap, is fitted over each perforation. The chimney and the cap are connected in a way that there is free space to allow the passage of the vapor phase. The vapor flows upward through the chimney, collides with the top wall of the cap, and is directed sideward and downward by this cap. At the bottom of each cap there is a series of slots, so that the vapor is divided in a swarm of bubbles that passes through the liquid pool around the cap. Bubble caps allow a wider range of appropriate operational conditions, but they have a higher cost. Therefore, the best combination of cost and range of operational conditions is obtained by the use of valve trays.

Besides tray columns, distillation columns can also be filled with structured or random packings. In both cases the intention is to form a liquid film over a large solid surface provided by the packing, so that the liquid flows down, covering the surface of the solid structure, and the vapor flows up through the remaining empty space. Random packings are small solid pieces of regular shape, whose size should be at most one-eighth of the distillation column diameter. A very large number of those solids can be placed in a random way inside the shell of the distillation packed column. Structured packings are dense packed solid surfaces of regular shape arranged in a cylindrical way, whose diameter is slightly less than the column diameter.

Several of these structures are put inside the column shell in order to guarantee a height of solid bed and, consequently, the mass transfer area required for the specific separation that is being considered. Packed columns are especially recommended in the case of distilling under vacuum, because packings generate a lower pressure drop than the equivalent number of trays, from a mass transfer point of view. Heat-sensitive components, such as fatty acid, are usually purified in packed columns. Nevertheless, we will focus our attention on tray columns, commonly used in the distillation of spirits and aroma mixtures.

3.1.2 Heat and Mass Balance Equations in Distillation Processes

The simplest way to simulate an alembic distillation is to treat the process as a differential distillation with constant vaporization rate. The initial charge of wine is put inside the pot still and heated to the boiling point of the mixture, and then the vaporization begins. At each instant the vapor phase forms, and the liquid mixture can be assumed to be in phase equilibrium. The vapor phase, formed at the constant vaporization rate, is condensed at the top of the equipment and accumulated in the distillate receiver. This sequence of events can be described by the following set of equations:

Total and component mass balances in the still:

$$\frac{\mathrm{d}HB}{\mathrm{d}t} = -V \tag{3.2}$$

$$\frac{\mathrm{d}(HB \cdot x_i)}{\mathrm{d}t} = -V \cdot y_i \quad \text{for} \quad i = 1 \ to \ nc, \tag{3.3}$$

where HB is the total amount of liquid or liquid hold-up in the still (moles), V is the vaporization rate or vapor flow (mol/s), t is the batch time (sec) measured from the beginning of the vaporization process, x_i and y_i are liquid and vapor molar fractions of component i, respectively, and nc is the total number of components in the mixture.

Equilibrium relationships:

$$y_i = K_i \cdot x_i \quad \text{for} \quad i = 1 \text{ to } nc, \tag{3.4}$$

where K_i is the partition coefficient of component i. The calculation of the partition coefficients for a multicomponent mixture is discussed in the next section. Using the set of equilibrium relationships for the nc components in the mixture, its boiling point at the equipment pressure, and the corresponding vapor phase concentrations, can be calculated by a bubble point procedure.

Total and component mass balances in the distillate receiver:

$$\frac{\mathrm{d}HD}{\mathrm{d}t} = V \tag{3.5}$$

$$\frac{d(HD \cdot x_{Di})}{dt} = V \cdot y_i \quad \text{for} \quad i = 1 \ to \ nc, \tag{3.6}$$

where HD is the amount of distillate collected in the distillate receiver or distillate hold-up (moles) and x_{Di} is the molar fraction of component i in the distillate.

Assuming a constant vaporization rate V, the prior set of differential equations can be easily integrated, although in each and every integration step the boiling temperature and the vapor phase concentrations must be calculated by solving the system of equations by an iterative procedure (bubble point algorithm). This integration gives the complete path of boiling temperatures, mixture compositions in the still, and distillate composition. Based on the distillate composition path, decisions in terms of cutting the distillate in different products can be made.

The assumption of a constant vaporization rate is usually an acceptable approximation in the case of using a heat source with constant heat transfer rate. Considering that the molar enthalpies of vaporization of different compounds have similar values, a constant heat transfer rate means a vaporization rate, in a molar basis, that is approximately constant. Although the mixture temperature increases along the entire distillation path, the amount of energy used for keeping the mixture at the boiling point is negligible in comparison to that amount necessary for vaporizing the components. On the basis of the simplifying assumptions Scanavini et al. [1] simulated the distillation of artisan *cachaça* in an alembic.

A more rigorous approach would require the estimation of the vaporization rate via the calculation of the heat transfer rate to the mixture. To perform this calculation, information on the heat transfer area, convective coefficient, and heat source temperature is required. In the case of distilling artisanal cachaça, an additional difficulty for calculating the heat transfer rate is that the alembic is usually heated by direct fire, whose intensity is sometimes altered in order to avoid foaming and liquid entrainment that could contaminate the product. Two further aspects can also be incorporated in a more comprehensive modeling of batch distillation in a pot still. In case the alembic is not isolated, convective heat losses to the environment, occurring in the upper part of the equipment, cause internal reflux and can alter the distillation path. Chemical reactions that contribute to changing the mixture composition during distillation can also be incorporated in the approach presented above. For instance, Ceriani and Meirelles [2] investigated the formation of *trans* isomers of fatty compounds during the batch deodorization of canola oil, modeling this process as a reactive multicomponent differential distillation.

The set of equations necessary for representing the batch distillation process in a tray column is doubtless more complex and involves a series of simplifying assumptions. The following assumptions are considered in the present case: the column contains np+1 ideal stages: the first one is the reboiler and the other np stages are the column trays; the condenser is numbered as np+2 and guarantees the total condensation of the top vapor stream without subcooling of reflux and distillate; the column is perfectly isolated, components are well mixed in each tray, vapor hold-up is negligible, and molar liquid hold-up on every stage is constant.

On the basis of such assumptions the following set of equations can be formulated:

Total and component mass balance equations and enthalpy balance equations for the reboiler ($n = 1$) are as follows:

$$\frac{dHB}{dt} = L_2 - V_1 \tag{3.7}$$

$$\frac{dx_{i,1}}{dt} = \frac{1}{HB} \cdot \left[-V_1 \cdot \left(K_{i,1} \cdot x_{i,1} - x_{i,1} \right) + L_2 \cdot \left(x_{i,2} - x_{i,1} \right) \right] \quad \text{for} \quad i = 1 \text{ } to \text{ nc} \tag{3.8}$$

$$0 = Q_r - V_1 \cdot \left(H_1 - h_1 \right) + L_2 \cdot \left(h_2 - h_1 \right) - HB \cdot \frac{dh_1}{dt}. \tag{3.9}$$

Balance equations for the trays ($n = 2, np+1$):

$$0 = V_{n-1} + L_{n+1} - V_n - L_n \tag{3.10}$$

$$\frac{dx_{i,n}}{dt} = \frac{1}{HN} \cdot \left[V_{n-1} \cdot \left(K_{i,n-1} \cdot x_{i,n-1} - x_{i,n} \right) + L_{n+1} \cdot \left(x_{i,n+1} - x_{i,n} \right) - V_n \cdot \left(K_{i,n} \cdot x_{i,n} - x_{i,n} \right) \right]$$
$$\text{for} \quad i = 1 \quad \text{to} \quad nc \tag{3.11}$$

$$\frac{dH_{L_n}}{dt} = \frac{1}{HN} \cdot \left[V_{n-1} \cdot \left(H_{n-1} - h_2 \right) + L_{n+1} \cdot \left(h_{n+1} - h_n \right) - V_n \cdot \left(H_n - h_n \right) \right]. \tag{3.12}$$

Balance equations for the condenser and reflux drum ($n = np+2$):

$$0 = V_{np+1} - L_{np+2} - D \tag{3.13}$$

$$\frac{dx_{i,np+2}}{dt} = \frac{V_{np+1}}{HD} \cdot \left[K_{i,np+1} \cdot x_{i,np+1} - x_{i,np+2} \right] \quad \text{for} \quad i = 1 \text{ to } nc \tag{3.14}$$

$$0 = V_{np+1} \cdot \left(H_{np+1} - h_{np+2} \right) - HD \cdot \frac{dh_{np+2}}{dt} - Q_c. \tag{3.15}$$

where HB, HN, and HD are the reboiler, tray, and condenser plus reflux drum liquid hold-ups (mols), respectively, L is liquid flow (mol/s), V is vapor flow (mol/s), Q_r is the reboiler duty (J/mol), H and h are vapor and liquid enthalpies (J/mol), respectively, n is the stage number, D is the distillate flow (mol/s), and Q_c is the condenser duty (J/s). The reflux ratio is given by $r = L_{np+2}/D$.

In the set of equations above, the equilibrium relationships are explicitly incorporated in the component mass balances, via the K_i values. To solve these differential equations the semi-implicit method suggested by Villadsen and Michelsen [3] can be used, according to the algorithm proposed by Luz and Wolf-Maciel [4]. The integration results in the tray temperature, the liquid and vapor compositions, the liquid and vapor flows, the reboiler and condenser duties, and the distillate composition and flow as a function of batch time.

As a final set of equations for process simulation we will consider a continuous tray column operating in steady state. Three specific subsets of equations are defined, one for the trays and the other two for the reboiler and condenser, in a column with *np* trays and *np*+2 stages.

Component mass balances, enthalpy balance, and equilibrium equations for the reboiler ($n = 1$):

$$F_{1(i,1)} = b_i + v_{i,1} - l_{i,2} = 0 \quad \text{for} \quad i = 1 \text{ to } nc \tag{3.16}$$

$$F_{2(1)} = h_b + H_1 - h_2 - Q_r = 0 \tag{3.17}$$

$$F_{3(i,1)} = V_1 \cdot K_{i,1} \cdot \frac{b_i}{B} - v_{i,1} = 0 \quad \text{for} \quad i = 1 \text{ to } nc. \tag{3.18}$$

Balance and equilibrium equations for the trays ($n = 2$, $np+1$):

$$F_{1(i,n)} = \left(1 + \frac{S_n^L}{L_n}\right) \cdot l_{i,n} + \left(1 + \frac{S_n^V}{V_n}\right) \cdot v_{i,n} - v_{i,n-1} - l_{i,n+1} - f_{i,n} = 0 \quad \text{for} \quad i = 1 \text{ to } nc \tag{3.19}$$

$$F_{2(n)} = \left(1 + \frac{S_n^L}{L_n}\right) \cdot h_n + \left(1 + \frac{S_n^V}{V_n}\right) \cdot H_n - H_{n-1} - h_{n+1} - H_{f,n} = 0 \tag{3.20}$$

$$F_{3(i,n)} = \eta_{i,n} \cdot V_n \cdot K_{i,n} \cdot \frac{l_{i,n}}{L_n} - v_{i,n} + (1 - \eta_{i,n}) \cdot v_{i,n-1} \cdot \frac{V_n}{V_{n-1}} = 0 \quad \text{for} \quad i = 1 \text{ to } nc. \tag{3.21}$$

Balance and equilibrium equations for the condenser and reflux drum ($n = np+2$):

$$F_{1(i,\,np+2)} = l_{i,\,np+2} + d_i - v_{i,np+1} = 0 \quad \text{for} \quad i = 1 \text{ to } nc \tag{3.22}$$

$$F_{2(np+2)} = \text{h}_{np+2} + H_D + Q_c - H_{np+1} = 0 \tag{3.23}$$

$$F_{3(i,np+2)} = D \cdot K_{i,np+2} \cdot \frac{l_{i,np+2}}{L_{np+2}} - d_i = 0 \quad \text{for} \quad i = 1 \text{ to } nc. \tag{3.24}$$

where F_1, F_2, F_3 are the discrepancy functions, accounting for the deviation from null of each balance or equilibrium equation. B is the bottom product flow (mol/s), b_i is component i bottom product flow (mol/s), v_i and l_i are component i vapor and liquid flows (mol/s), respectively, V and L are total vapor and liquid flows (mol/s), respectively, H and h are vapor and liquid enthalpies (J/mol), respectively, S^V and S^L are vapor and liquid sidestreams (mol/s), respectively, f_i is component i feed stream (mol/s), H_f is feed stream enthalpy (J/mol), D is distillate flow (mol/s), d_i is component i distillate flow (mol/s), H_D is distillate enthalpy (J/mol), η_i is component i Murphree efficiency, and Q_r and Q_c are reboiler and condenser duties (J/s), respectively.

Note that reboiler and condenser are considered as ideal stages, but the efficiencies of the trays are taken into account. Although the distillate and feed enthalpies are indicated in capital letter, both streams can be either liquid or vapor ones. In the case of feed stream, its enthalpy, at the column pressure prevailing in the feed tray, will define the part of it fed as liquid and/or vapor. In the nomenclature above, the index n stands for the liquid or vapor stream leaving tray number n. In the cases of the bottom product and the distillate, L_l is replaced by B and V_{np+2} by D, respectively. Similar to the batch column case discussed above, the reflux ratio is given by $r = L_{np+2}/D$.

The above set of equations can be organized as a vector of discrepancy functions $\vec{F}(z)$, with $(np+2)\cdot(2nc+1)$ elements, which can be solved for the vector of variables $\vec{z}$, as indicated below:

$$\vec{F}(z) = \begin{Bmatrix} \vec{F}_1 \\ \vec{F}_2 \\ \vec{F}_3 \end{Bmatrix} = 0 \tag{3.25}$$

$$\vec{z} = \begin{Bmatrix} \vec{l} \\ \vec{v} \\ \vec{T} \end{Bmatrix}. \tag{3.26}$$

The algorithms commonly used for solving this system of equations are based on the Newton–Raphson method, and they consist in finding the solutions that minimize the errors expressed in the discrepancy functions, for instance the solutions that guarantee $\sum_{i=1}^{nc}(\bar{F}_n)^2 \le \varepsilon$, where ε corresponds to the maximum acceptable total error. Note that the total liquid and vapor streams can be directly calculated from the solution by $L = \sum_{i=1}^{nc} l_i$ and $V = \sum_{i=1}^{nc} v_i$, respectively. The same is valid for the streams' molar fractions, since $x_i = l_i/L$ and $y_i = v_i/V$.

3.1.3 Vapor–Liquid Phase Equilibrium

As indicated in the set of balance equations shown above, the design and evaluation of distillation equipment require an appropriate knowledge of enthalpies and phase equilibrium properties of the liquid and vapor phases. Several physical–chemical properties, such as heat capacity, enthalpy of vaporization, vapor pressure, and activity and fugacity coefficients, must be estimated for the mixture components. In the case of some compounds, experimental data are available at the relevant temperature and pressure ranges. Nevertheless, for some compounds such data cannot be found in the literature. Reid et al. [5] discussed in detail a series of group contribution methods that can be used for estimating these properties in the absence of appropriate experimental data.

In this section we will focus our attention on those physical–chemical properties that are the most important ones for a correct simulation and design of distillation processes, namely those properties involved in vapor–liquid phase equilibrium calculations. Consider a multicomponent system at constant absolute temperature T and pressure P, containing n different components. The thermodynamic equilibrium is described by the following condition, formulated for each component i:

$$\hat{f}_i^V = \hat{f}_i^L \tag{3.27}$$

The vapor phase fugacity of component i, $\hat{f}_i^V$, is expressed as follows:

$$\hat{f}_i^V = \hat{\phi}_i \, y_i P , \tag{3.28}$$

where y_i is the vapor phase molar fraction of component i, P is the total pressure, and $\hat{\phi}_i$ is the fugacity coefficient of component i, a variable that reflects the deviation of the ideal gas behavior in the vapor phase.

The liquid phase fugacity of component i, $\hat{f}_i^L$, is given by

$$\hat{f}_i^L = x_i \gamma_i f_i^0 , \tag{3.29}$$

where x_i is the liquid phase molar fraction of component i, f_i^0 is the standard state fugacity of component i, and γ_i is its activity coefficient, a variable that reflects the deviation from the ideal mixture behavior in the liquid phase.

Combining the prior equations, thermodynamic equilibrium can be expressed by the following new equation:

$$\hat{\phi}_i \, y_i P = x_i \gamma_i f_i^0 . \tag{3.30}$$

The standard-state fugacity, f_i^0, is the fugacity of a pure liquid, containing only molecules of component i, at the temperature and pressure of the system, and is given by

$$f_i^0 = P_i^{vap} \phi_i^S \exp \int_{P_i^{vap}}^{P} \frac{V_i^L}{RT} dP , \tag{3.31}$$

where P_i^{vap} represents the vapor pressure of component i, ϕ_i^S is the fugacity coefficient of pure component i at saturation, and the exponential term is the Poynting correcting factor. In the Poynting factor V_i^L represents the molar volume of liquid i, R is the gas constant, and T, the absolute temperature. This term expresses the influence of pressure on liquid phase fugacity. At low temperatures, a liquid is nearly incompressible, the effect of pressure on liquid phase fugacity is negligible, and the Poynting factor assumes a value very close to one. Taking this into account, the equation for phase equilibrium calculations can be expressed in the following form:

$$\hat{\phi}_i y_i P = x_i \gamma_i P_i^{vap} \phi_i^s . \tag{3.32}$$

For most distillation processes of interest in the food industry Equation 3.32 is an appropriate tool for representing vapor–liquid phase equilibrium. We will discuss the use of this equation considering, as a typical system of interest, the wine used in alcoholic distillation for cachaça production and obtained by the fermentation of sugar cane juice. This system is composed of two major components, ethanol and water, but it also contains a series of minor compounds present in very low concentrations. These minor components are called congeners, and the value of their concentration in the final distillate is usually important for the spirits' quality. Some of the main congeners present in the wine are shown in Table 3.1, as well as their concentration range.

As can be seen in Table 3.1, most of the congeners are alcohols and, except for methanol, they have volatility lower than ethanol. The other three components belong to different organic classes, such as esters, aldehydes, and acids. In fact the wine contains several other minor components, but their concentration is either lower than those reported in Table 3.1 or their influence on spirits quality is not so important.

The main objective of spirits distillation is to concentrate ethanol from the wine to the desired level and, at the same time, to keep the congeners within the levels

TABLE 3.1
Main Wine Components and Concentration Range

Component	Molar weight (kg/kmol)	Normal boiling point (K)	Concentration range (w/w)
Water	18.02	373.15	0.92–0.95 g/g[a]
Ethanol	46.07	351.55	0.05–0.08 g/g[b]
Methanol	32.04	337.85	0.0–3.2·10^{-1} mg/kg[c]
Isopropanol	60.10	355.55	N/A
Propanol	60.10	370.25	21–68 mg/kg[b]
Isobutanol	74.12	381.15	13–49 mg/kg[b]
Isoamyl alcohol	88.17	405.15	27–188 mg/kg[b]
Ethyl acetate	88.12	350.25	5.5–11.9 mg/kg[b]
Acetaldehyde	44.05	293.35	10–83 mg/kg[b]

N/A: not available.

[a] Obtained by difference.

[b] Reference [6] and [c] Reference [7].

required by legislation and/or by sensorial quality criteria. The fulfilling of this general objective depends on the volatility of the components present in the mixture. The volatility of each substance in a multicomponent mixture can be evaluated by the K_i values. They are calculated as follows:

$$K_i = \frac{y_i}{x_i} = \frac{\gamma_i \cdot \phi_i^s \cdot P_i^{vap}}{\hat{\phi}_i \cdot P}. \tag{3.33}$$

The volatility difference of two components is evaluated by the relative volatility of the light component i in relation to the heavy one, j, usually represented by the symbol α_{ij} and calculated as the ratio of the K values of both components:

$$\alpha_{ij} = \frac{\dfrac{y_i}{x_i}}{\dfrac{y_j}{x_j}} = \frac{K_i}{K_j} = \frac{\dfrac{\gamma_i \cdot \phi_i^s \cdot P_i^{vap}}{\hat{\phi}_i}}{\dfrac{\gamma_j \cdot \phi_j^s P_j^{vap}}{\hat{\phi}_j}}. \tag{3.34}$$

Values of relative volatility much larger than 1.0 indicate components that can be easily separated by distillation. When the relative volatility assumes values relatively close to 1.0, the separation by distillation requires huge numbers of ideal trays and/or extremely large reflux ratios, a situation that, from an economic point of view, is not always feasible. This kind of behavior can occur for ideal mixtures of compounds with similar vapor pressures, such as mixtures of some fatty acids. In this case, if the intention is to obtain high purity products, the separation is not feasible using only distillation processes. A relative volatility equal to 1.0 precludes the use of distillation to further concentrate a mixture, because in this case both components exhibit identical tendency to volatilize and no enriching is observed in the vapor phase obtained by distilling the liquid mixture.

In the case of spirits distillation, the relative volatility of ethanol/water is of utmost importance, but the volatility of the congeners in relation to water as well as in relation to ethanol is also a relevant factor to be considered in order to keep their concentration in the distillate within the required range of values. To have a quantitative insight into the relative volatilities of these compounds present in the spirits distillation, a further discussion of the procedures for calculating fugacity and activity coefficients is necessary.

At low pressures and relatively low densities, the interaction between molecules in the vapor phase is much weaker than the interaction between those molecules in the much denser liquid phase. It is therefore a common simplification to assume that all nonideality in vapor–liquid equilibrium calculations is concentrated in the liquid phase, attributing to the vapor phase the behavior of an ideal gas. In this case, the fugacity coefficients in the mixture, as well as for each pure component, assume the value 1.0 and the system deviation from an ideal behavior will be represented exclusively by the activity coefficients of the components in the liquid phase.

For typical mixtures, at temperatures near or slightly above the normal boiling point of the least volatile component, low pressure means pressure values restricted to few bars. However, for mixtures containing strongly associating components, such as carboxylic acids, fugacity coefficients may differ appreciably from unity even at pressures less than 1.0 bar, so that the calculation of fugacity coefficients is required for an appropriate prediction of the vapor–liquid equilibrium. Also, in the case of very light components, the fugacity coefficients, especially those calculated for pure compounds (ϕ_i^s), can be sufficiently different from unity. Very light components are those compounds whose vapor pressure is much larger than the system pressure at the equilibrium temperature. Among the substances listed in Table 3.1, acetic acid and acetaldehyde are typical compounds exhibiting the behaviors just described. This suggests that a rigorous estimation of vapor–liquid equilibrium in spirit distillation should include the calculation of the fugacity coefficients.

Usually, the fugacity coefficients are calculated using the Virial equation truncated after the second term, but for components that strongly associate, such as acetic acid, they should be estimated by means of the chemical theory. In this case the correlation of Hayden and O'Connell allows the calculation of the second Virial coefficient and the prediction of the chemical equilibrium dimerization constant. For further details see Fredenslund et al. [8].

As already mentioned, the deviation of the ideal behavior in the liquid phase can be estimated by the activity coefficients. They can be calculated using molecular models such as the NRTL (nonrandom two-liquid), Wilson, or UNIQUAC (universal quasi-chemical) equations. The NRTL model is given by the following set of equations:

$$\ln \gamma_i = \frac{\sum_j x_j \tau_{ij} G_{ji}}{\sum_k x_k G_{ki}} + \sum_j \frac{x_j G_{ij}}{\sum_k x_k G_{kj}} \left(\tau_{ij} - \frac{\sum_m x_m \tau_{mj} G_{mj}}{\sum_k x_k G_{kj}} \right) \tag{3.35}$$

$$G_{ij} = \exp(-\alpha_{ij}\, \tau_{ij}) \tag{3.36}$$

$$\tau_{ij} = \frac{A_{ij}}{RT} \tag{3.37}$$

$$A_{ij} \neq A_{ji} \tag{3.38}$$

$$\alpha_{ij} = \alpha_{ji}, \tag{3.39}$$

where A_{ij} is an interaction parameter between components i and j and α_{ij} is the nonrandom parameter.

For a binary mixture of components i and j, the NRTL model requires three parameters, A_{ij}, A_{ji}, and α_{ij}, that should be determined by fitting the model to the experimental vapor–liquid equilibrium data available for such a mixture. In the formulation presented above the model is already given for a multicomponent system, so that it can be applied for calculating the equilibrium for a mixture such as the wine given in Table 3.1. Nevertheless a whole set of interaction and nonrandom parameters for each pair of interacting components will be required. If experimental data for each binary mixture are available, these parameters can be estimated and afterwards the vapor–liquid equilibrium for the complex multicomponent system can be predicted. The possibility of using parameters estimated on the basis of experimental binary data for predicting vapor–liquid phase equilibrium of multicomponent mixtures with a usually good accuracy is one of the major advantages of activity coefficient models such as NRTL, UNIQUAC, and Wilson equations.

Unfortunately, in the case of many liquid mixtures of interest in the food industry, the corresponding experimental data are not available. For example, in the case of wine, experimental equilibrium data are available mainly for the binary mixtures containing either water or ethanol, but for binary mixtures containing a pair of congeners, the required experimental data are scarce.

In the absence of experimental data an alternative procedure is necessary. Methods, such as UNIFAC (UNIQUAC functional-group activity coefficient) and ASOG (analytical solution of groups), based on the concept of group contribution, are the best options in this case. They assume that the behavior of components in a liquid mixture can be represented by some descriptors of the components' molecule structure, such as their constituting chemical groups and the corresponding surface and volume parameters, as well as by the interaction between these chemical groups. In fact, they assume that a mixture of components can be treated as a solution of groups, so that a prediction of activity coefficients is possible even in the absence of experimental data.

The UNIFAC model is given by the following set of equations:

$$\ln \gamma_i = \ln \gamma_i^C + \ln \gamma_i^R, \tag{3.40}$$

where $\ln \gamma_i^C$ is the combinatorial contribution to the activity coefficient, related exclusively to the molecules' structure, as indicated below:

$$\ln \gamma_i^C = \ln \frac{\Phi_i}{x_i} + \frac{z}{2} \cdot q_i \cdot \ln \frac{\theta_i}{\Phi_i} + l_i - \frac{\Phi_i}{x_i} \cdot \sum_j x_j \cdot l_j, \tag{3.41}$$

where

$$\Phi_i = \frac{r_i x_i}{\sum_j r_j x_j}, \quad z = 10; \quad \theta_i = \frac{q_i x_i}{\sum_j q_j x_j}, \quad l_i = \frac{z}{2}(r_i - q_i) - (r_i - 1)$$

and

$$r_i = \sum_k \nu_k^{(i)} R_k; \qquad q_i = \sum_k \nu_k^{(i)} Q_k.$$

In this set of equations, Φ_i corresponds to a kind of volume fraction for component i, θ_i to its area fraction, and r_i and q_i represent, respectively, its van der Waals volume and surface area. The volume and surface area of component i is calculated using the number of groups of type k in its molecular structure, $v_k^{(i)}$, and the group volume and area parameters R_k and Q_k. The volume and area parameters are calculated from van der Waals group volume and surface areas given by Bondi [9], after an appropriate normalization. For further details see Fredenslund et al. [8] and Reid et al. [5].

The residual term, $\ln \gamma_i^R$, reflects the interaction between the different groups in the solution and is calculated by the following:

$$\ln\gamma_i^R = \sum_{\substack{k \\ all \\ groups}} v_k^{(i)} \cdot \left(\ln\Gamma_k - \ln\Gamma_k^{(i)}\right) \tag{3.42}$$

$$\ln\Gamma_k = Q_k * \left[1 - \ln\left(\sum_m \theta_m \Psi_{mk}\right) - \sum_m \left(\theta_m \Psi_{mk} \Big/ \sum_n \theta_n \Psi_{nm}\right)\right] \tag{3.43}$$

$$\theta_m = \frac{Q_m X_m}{\sum_j Q_j X_j} \tag{3.44}$$

$$\Psi_{mn} = \exp\left(-\frac{a_{mn}}{T}\right), \tag{3.45}$$

where Γ_k and $\Gamma_k^{(i)}$ are, respectively, the residual activity coefficient of group k in the mixture and the residual activity coefficient of the same group in a solution containing only molecules of component i, θ_m is the area fraction of group m, X_m is its mole fraction in the mixture, and a_{mn} is the interaction parameter between groups m and n. For each pair of groups there are two interaction parameters, a_{mn} and a_{nm}, with $a_{mn} \neq a_{nm}$.

The UNIFAC interaction parameters were obtained from phase equilibrium databases containing a wide range of experimental results; nevertheless, these parameters are not related to the interaction between specific molecules present in those data banks, but to the interaction between the groups that constitute those molecules, so that phase equilibrium for mixtures of other molecules composed of the same groups can also be predicted.

The original UNIFAC method was modified over time, and slightly different versions are now available, with higher accuracy for specific types of mixtures and other advantages [10–12]. Particularly in the case of mixtures occurring in the distillation of spirits, aromas, and essential oils, the UNIFAC method can be a valuable tool for process investigation and development, because the type of organic molecules present in such mixtures is very similar to those used for estimating the set of group interaction parameters available now.

Nevertheless, it should be emphasized that the UNIFAC method is a predictive procedure, useful especially in the absence of experimental data. If specific experimental data are available, the best option is always to fit one of the mentioned molecular models, because these models, with an appropriate set of parameters for each binary mixture, doubtless have higher accuracy.

Unfortunately, for complex mixtures containing several components, experimental data for each and every binary pair of compounds are usually not available. In this case a mixed strategy is probably the best option. For those pairs of components for which binary equilibrium data are available, interaction parameters of a molecular model should be adjusted. For the ones for which no experimental data were previously measured, the UNIFAC method can be used to predict the phase equilibrium data. These predicted data can then be used for fitting the remaining parameters of the selected molecular model. In this way it is possible to combine, in a coherent form, the highest possible accuracy with the available experimental data.

The most comprehensive data bank of vapor–liquid equilibrium is the DECHEMA data series [13] that contains experimental data and also the corresponding interaction parameters for the molecular models. UNIFAC parameters have been reported. Commercial software for process simulation, such as ASPEN Plus [14] and Hysis, also contain built-in data banks with interaction parameters for the molecular models as well as for the UNIFAC. When no experimental data are available, these simulation packages allow the use of the UNIFAC method to adjust interaction parameters for one of the molecular models, as explained above.

Using the ASPEN Plus [14] simulation software, we have investigated the phase equilibrium of fermented must, considering all the components given in Table 3.1. The NRTL model was selected for calculating the activity coefficients. Especially in the case of some binary mixtures of minor components, no experimental data are available, so the UNIFAC model was used for predicting the equilibrium data, according to the ASPEN Plus databank. The investigation was performed, varying the ethanol molar fraction in the whole range of interest in wine distillation, while keeping the composition of minor components at the lowest levels, so that they can be considered as infinite dilution compounds. Figure 3.6 presents the phase equilibrium in terms of the ethanol molar fraction in the liquid and vapor phases. As the minor components are present in very low concentration, this equilibrium curve is practically identical to the binary ethanol–water curve.

Most spirits have an ethanol concentration within the range 38 to 54 °GL, corresponding approximately to a maximum of 0.48 in mass fraction or 0.27 in mol fraction. In this case the relevant concentration range is restricted to the first part of the equilibrium curve given by Figure 3.6, which is exactly the part where ethanol has the highest volatility. For this reason, the distillation of spirits can be easily performed, either in a batch still without reflux or in distillation columns with low number of trays and very low reflux ratios. Hydrated ethanol, either used as biofuel or in the pharmaceutical and food industries, has a concentration close to the azeotropic point (approximately 96.5 °GL, corresponding to 95.6 in mass fraction or 89.5 in mol fraction). In this case the enriching part of the distillation

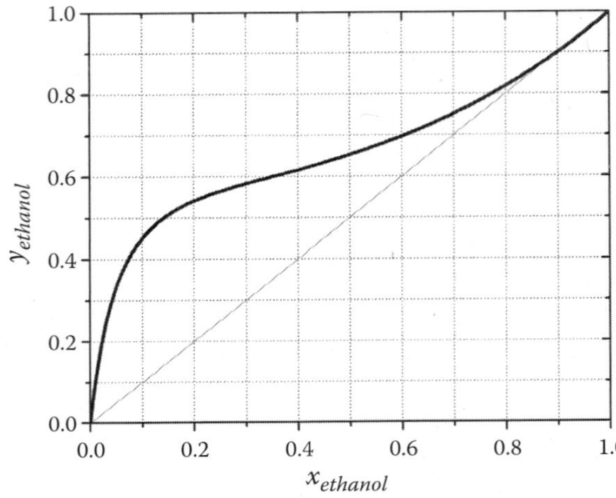

FIGURE 3.6 Ethanol equilibrium curve in wine distillation ($P = 0.1013$ MPa).

process occurs along that region of the equilibrium curve where the ethanol volatility decreases sharply and approaches the volatility of water. This is the reason why distillation columns with large number of trays and higher reflux ratios are required for producing hydrated ethanol.

Anhydrous ethanol, mainly used as an additive to gasoline, has concentrations higher than 99.6 °GL. This corresponds to a content of water lower than 0.005 in mass fraction or 0.013 in molar fraction. Anhydrous ethanol is produced from hydrated (or azeotropic) ethanol, either by especial distillation methods or by adsorption using molecular sieves. In the whole part of the equilibrium curve near the azeotropic point, ethanol volatility has a value very close to the volatility of water, requiring the addition of a third component that could change their relative volatility and allow their separation by distillation. Two main distillation methods are currently used in industrial scale for producing anhydrous ethanol: azeotropic distillation with ciclohexane, a component that enhances water volatility and allows the production of absolute ethanol as a liquid bottom product, and extractive distillation with ethylene glycol, a component that reduces the water volatility and allows the production of ethanol as distillate.

The K_i values of ethanol and the other alcoholic components of wine are shown in Figure 3.7 as a function of ethanol molar fraction in the liquid. Curves with a very similar behavior can also be obtained if one represents the relative volatility of each alcohol in relation to water ($\alpha_{alcohol\text{-}water}$) instead of the corresponding K_i values. As indicated in Figure 3.7, at very low ethanol concentrations, all the alcoholic components exhibit large volatilities. In fact, binary mixtures of water and alcohols have a positive deviation from Raoult's law ($\gamma_{alcohol} > 1.0$), indicating that repulsive interactions prevail and the alcohols' volatilities are increased in a liquid environment rich in water. This effect is significant especially in the case of the alcoholic components with larger carbon chains (more hydrophobic ones), so that

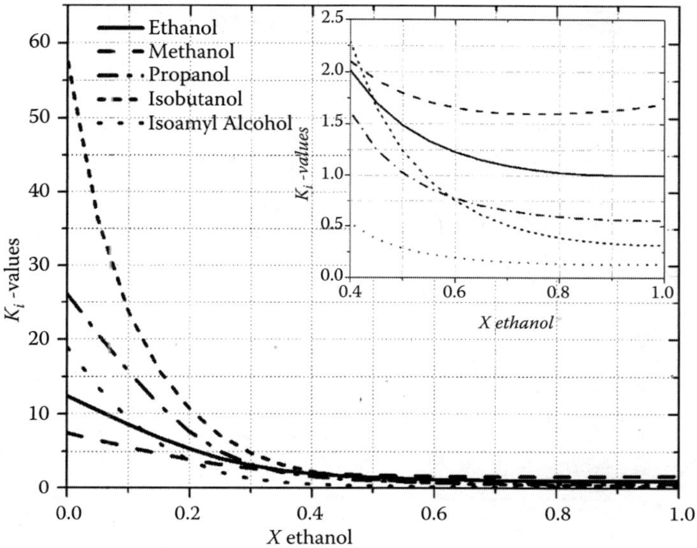

FIGURE 3.7 Volatility of alcoholic components of wine as a function of the ethanol molar fraction ($P = 0.1013$ MPa).

it predominates even upon their corresponding lower vapor pressures. As the water concentration in wine decreases, the activity coefficients of the alcoholic components also decrease and the effect of the carbon chain becomes predominant, as is indicated in Figure 3.7 for ethanol molar fractions larger than 0.4. This can be further observed in Figure 3.8, which shows the relative volatility of each minor alcoholic component in relation to ethanol. Note that, except in the region of high water concentration, ethanol has volatility greater than that of propanol, isobutanol, and isoamyl alcohol and less than that of methanol. The relative volatility of methanol–ethanol is less than 1.7, a value relatively low, which makes it difficult to decrease the level of this contaminant in distilled ethanol. Fortunately, the concentration of methanol in the wine is usually very low, except when sources of methoxylated pectin are added to the must before fermentation. In contrast, alcohols such as propanol, isobutanol, and isoamyl alcohol should be classified as wine components with intermediate volatility: they are heavier than ethanol, but they behave as light compounds in a water-rich environment.

In Figure 3.9, the K_i values of other minor components are shown. The curve profiles calculated for the aldehyde and the ester are similar to the one observed in the case of the alcoholic components, but both are lighter compounds along the whole ethanol concentration range, as the relative volatility of these components in relation to ethanol clearly indicates (see Figure 3.10). The exception is represented by acetic acid, always a heavier component in wine distillation. Its K_i values, along the entire concentration range of wine distillation, are lower than 0.1, and the relative volatilities of ethanol-acetic acid are always larger than 10.

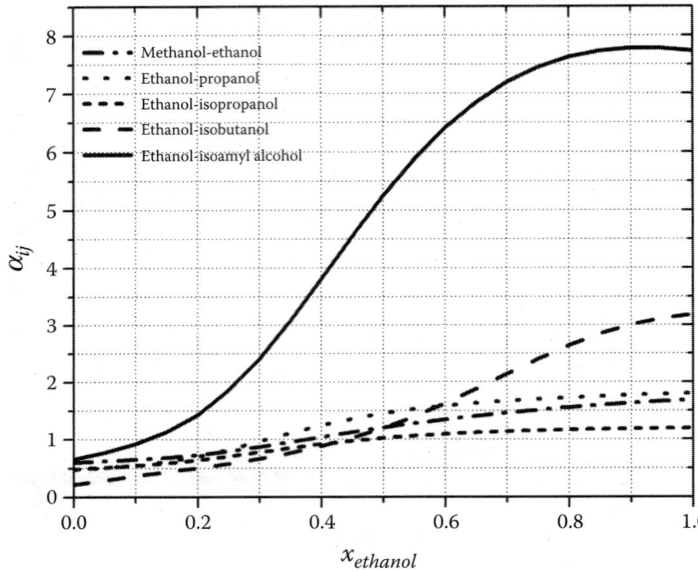

FIGURE 3.8 Relative volatility of alcoholic components of wine as a function of the ethanol molar fraction ($P = 0.1013$ MPa).

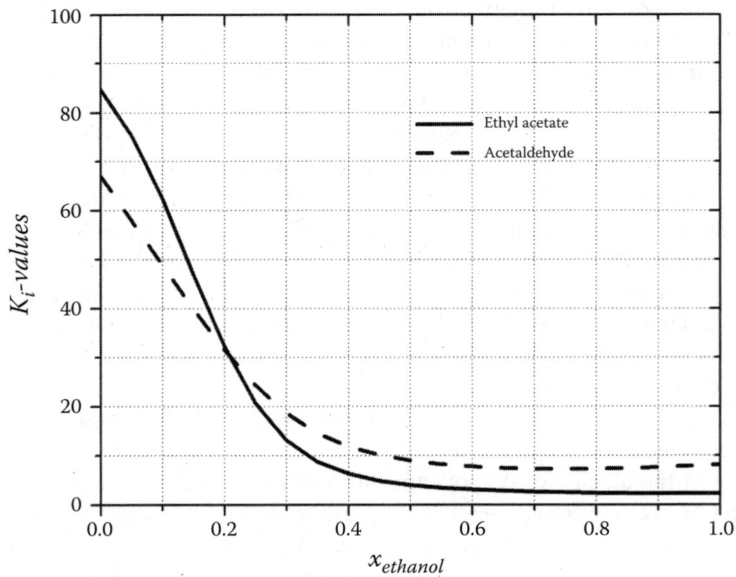

FIGURE 3.9 Volatility of volatile components of wine as a function of the ethanol molar fraction ($P = 0.1013$ MPa).

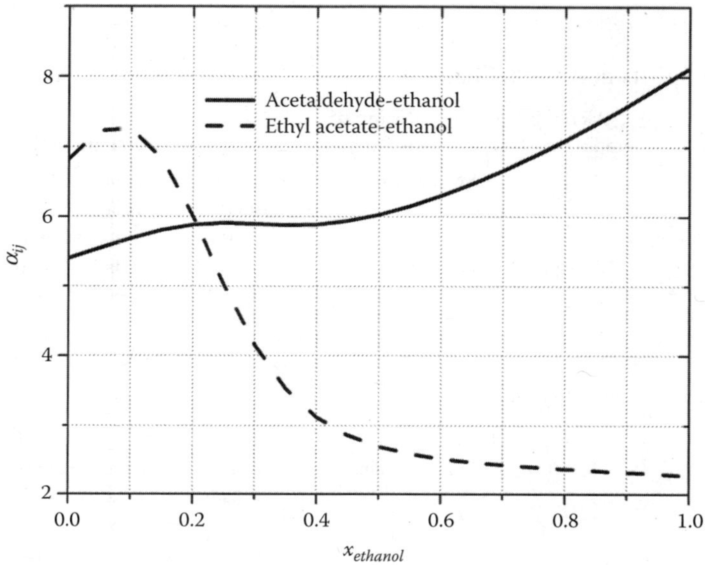

FIGURE 3.10 Relative volatility of volatile components of wine as a function of the ethanol molar fraction ($P = 0.1013$ MPa).

The characteristics of different distillation processes, the corresponding heat and mass transfer balance equations, and the vapor–liquid phase equilibrium discussed in this section represent the main fundamentals of distillation applied to the processing of liquid mixtures of interest in the food industry. Such fundamentals are often applied for the improvement and development of new processes. Recent advances in distillation processes applied to the processing of spirits and aroma mixtures will be discussed in the next section.

3.2 RECENT ADVANCES IN THE SIMULATION OF SPIRITS AND AROMA MIXTURES DISTILLATION

Table 3.2 summarizes some of the characteristics of selected spirits, including their range of ethanol graduation and particular aspects of their production. As products developed during a long period of time and in different places, there are controversies on their exact specifications, which can also vary according to each country's prevailing legal determinations. Similar spirits may also have different denominations according to the countries or regions of production. The summary presented in Table 3.2 should be considered as an overview of the general characteristics of some alcoholic beverages, without being either comprehensive or elaborated concerning the details of each spirit. As indicated in Table 3.2, ethanol graduation after distillation may be larger for absinthe, vodka, grappa, and whisky than for other spirits, but in some cases a proper dilution is performed before bottling.

Research on spirits production and technology is mainly focused on their composition and on the interplay of some aspects of their production steps and the

TABLE 3.2
Selected Spirits and Their Characteristics

Spirit denomination	Alcoholic graduation (%)	Region of production	Some characteristics
Absinthe	45–72	Switzerland, France	Obtained by distillation of alcoholic solutions of macerated herbs (wormwood, anise, fennel) [17]
Bagaceira	37.5–50.0	Portugal	Obtained by distillation of fermented grape pomace (residue from wine making after pressing); a similar spirit is denominated Orujo in Spain [18], Zivania in Cyprus [19], Tsipouro in Greece; eau-de-vie de marc in France; and rakija in Slavic countries [20–22]
Brandy	40–60	France, Spain, California, etc.	Obtained by distillation of grape wine, usually aged; denominated in France as Cognac or Armagnac, according to the corresponding French regions. Fruit brandies are obtained by distillation of fermented juices from other fruits (cider brandy, cherry brandy, etc.); some fruit brandies are not aged [15, 16, 23, 24]
Cachaça	38–48	Brazil	Obtained by distillation of fermented sugar cane juice, aged or unaged. According to the Brazilian legislation, a spirit similar to cachaça, denominated *aguardente*, contains 38 to 54% of alcoholic graduation [25–28]
Grappa	≅40–70	Italy	Obtained by distillation of fermented grape pomace (or marc), aged or unaged; wine lees can be added to grape marc in a maximal mass proportion of 1 to 4 [29, 30]
Pisco	30–50	Peru, Chile	Obtained by distillation of fermented grape mash, the product finalization may include maturation in oak casks and caramel addition [31]
Rum	≅40	Caribbean	Obtained by distillation of fermented sugar cane molasses and aged in oak barrels [32, 33]
Tequila	≅40–50	Mexico	Obtained by distillation of fermented blue agave juice and aged in oak casks [34–36]
Vodka	38–45, 50 or 56	Russia	Obtained by distillation of alimentary ethanol from grain or potato fermented must, usually distilled to higher alcoholic graduation and afterwards diluted [37]

TABLE 3.2 *(continued)*

Spirit denomination	Alcoholic graduation (%)	Region of production	Some characteristics
Whisky/ Whiskey	≥40	Scotland, Ireland, USA, etc.	Obtained by distillation of fermented grain mash and aged in wooden casks. According to the different types of whisky the following grains can be used: malted barley, barley, wheat, corn and rye. The mash is usually distilled to a higher alcoholic graduation and only diluted to the desired strength after aging [38–40]

concentration of minor components. For instance, Madrera et al. [15] investigated the influence of different aspects of the cider brandy production, such as the distillation system, oak wood type, and aging time, on the profile of volatile compounds. They tested the double distillation technique and, alternatively, a rectification column system. The distillates were matured in wood casks made of French and American oaks for 32 months. Higher levels of acetaldehyde and acetaldehyde diethyl acetal were observed in the case of the double distillation technique, whereas alcohols of higher molecular weight were better recovered in the rectification column. The distillate pH was higher for the double distillate spirit in comparison to the distillate obtained in the rectification column. They also observed that the concentration of ethanoate esters decreased during the spirits aging.

Hernández-Gomez et al. [16] investigated the distillation of fermented must from melon fruit using either a copper pot or a rectification column. They also tested the double distillation procedure. The first distillation was conducted for obtaining product with an alcoholic graduation about 17–20 °GL, and in the second step this prior distillate was separated into three fractions: a small head fraction, a heart fraction with an alcoholic content about 55 °GL, and a tail fraction, which contained the residue of alcohol recovered from the first distillate. To obtain a melon fruit spirit with an appropriate sensorial profile, the authors recommended the distillation in the copper pot. Nascimento et al. [25] investigated the influence of the alembic material on the profiles of volatile components present in sugar cane spirits. The equipment was manufactured either in copper or in stainless steel. They concluded that besides decreasing the concentration of volatile sulfur compounds whose presence can impart to the distillate an unpleasant odor, copper also participates in the formation of aldehydes. In fact, the concentration of total aldehydes in the distillate was significantly larger for the spirit produced in the copper alembic in comparison to that obtained in the stainless steel one. The investigation conducted by Cardoso et al. [41] indicates that spirits produced in stills containing either copper or aluminum as packing have lower contents of dimethylsulfite but larger ones of sulfate and methanol. As suggested by the authors, such a result is consistent with the dimethylsulfite oxidation to sulfate in the presence of either copper or aluminum, and the generation of methanol as by-product. The Brazilian legislation defines a limit of 5 mg/L of copper in the distilled beverages, and in some countries even a concentration larger than 2 mg/L is not tolerated. According to the authors, the use of copper as packing, not as a construction material of the whole alembic, allows better control of the copper contamination of the distillate, without impairing its beneficial effect on the sensorial quality of the spirit via the oxidation reaction of volatile sulfur compounds.

Boza and Horii [42] correlated the sensorial quality of sugar cane spirits and the concentration of minor components, confirming that larger propanol content and higher acidity levels impair the product quality. In a further work they observed that a larger acidity level in the distillate also corresponds to a higher copper concentration [43]. Because the distillate acidity level increases during the whole period of alembic distillation, the authors emphasized the importance of separating the heart fraction at a higher alcoholic graduation and collecting an appropriate amount of the tail fraction in order to improve the spirits quality in relation to copper and acid concentrations.

Bruno et al. [44] investigated the influence of the distillation system and procedure on the ethyl carbamate concentration of sugar cane spirits. Ethyl carbamate is a potentially carcinogenic substance, whose maximal accepted level in distilled beverages is 150 μg/L. The formation of ethyl carbamate is favored by entrainment of nitrogenous precursors and high temperatures. The influence of such factors can be diminished by a better design of the distillation equipment, by the use of an appropriate reflux rate, or by double distilling the spirits.

As indicated by this literature review, most research works on spirits processing are related to the influence of different aspects of the beverage production on the product quality. On the other hand, the use of simulation tools in order to improve the performance of the distillation process for spirits production is still a rare subject in the literature. Although simulation of ethanol distillation is a very frequent research theme, works on such a subject are usually related to the production of biofuels, focusing mainly on the energetic performance of the separation process and not taking into account the role of minor components that are important for quality and sensorial aspects of the product.

However, there are some recent works that exemplify the powerful use of simulation tools for improving spirits distillation. Osorio et al. [45] developed a model for simulating Pisco distillation as a multicomponent reactive batch distillation process with reflux. In a further work the same research group investigated, via experimental distillation runs of a model solution similar to wine, as well as via process simulation, the operating recipes for a batch column used in the production of Pisco [24]. Gaiser et al. tested the commercial software ASPEN Plus [14] through the simulation of a typical continuous distillation unit used for whisky distillation. The results obtained presented good agreement with the available experimental data.

Decloux and Coustel [46] simulated a typical production plant used for continuous distillation of neutral spirit. Neutral spirit is high purity ethanol used in the food, pharmaceutical, and chemical industries. The whole distillation plant comprises a series of seven columns for concentrating and purifying ethanol, including the decrease, to a very low value, of the presence of most contaminants such as methanol, propanol, higher alcohols, esters, aldehydes, and acidity. They used the commercial software ProSim Plus and included many congeners in order to evaluate the software capacity of correctly representing the contaminants' behavior. Their results indicated the good performance of the software and allowed them to illustrate the specific role of each column on the sequence of purification steps performed during neutral spirit production.

The use of different techniques for aroma recovery in an industrial scale, including distillation, is discussed by Karlsson and Trägårhd [47]. They showed plant schemes for integrating juice evaporation unities and aroma recovery equipment and gave some details on the vapor–liquid equilibrium involved in such processes. Yanniotis et al. [48] investigated, on a lab scale, the possibility of combining distillation and absorption techniques for aroma recovery, concluding that the combination of both techniques offers better results than the use of a simple distillation step.

The use of simulation tools for investigating the recovery of aromas by distillation is also a rare topic in the literature. In a research work similar to the ones presented above for spirits distillation, Lora et al. [49] studied the concentration of aroma compounds from wine using experimental distillation runs and simulation tools. Haypek et al. [50] simulated an industrial column for recovery of aroma compounds lost during orange juice evaporation. Because of the high concentration of terpenes in the vapor phase leaving the distillation column top tray, the distillate obtained after condensation is in fact composed of two liquid phases: an oil essence phase rich in *d*-limonene, other terpenes, and compounds with low polarity, and an aqueous essence phase, containing water, ethanol, and other polar compounds. The authors used for simulating the industrial equipment the commercial software PRO/II and concluded that the simulation results are similar to those observed in the industrial process. On the basis of the successful reproduction of the industrial column performance, the authors suggested extending their research in order to investigate, via simulation, the possibility of recovering the aroma compounds present in the oil essence phase. For this purpose, the aqueous essence phase is further concentrated, increasing its ethanol content to a value in the range of 50 to 78% (mass), so that it can be used as a solvent for recovering, by liquid–liquid extraction, the aroma compounds from the oily essence phase. The whole process was investigated by simulation, and the corresponding results exemplify appropriately the use of process simulation for evaluating, improving, and developing separation and purifying techniques of complex mixtures frequently found in the food and beverage industries.

3.3 SOME ESPECIAL APPLICATIONS OF DISTILLATION

3.3.1 Obtaining High Quality Cachaça

Cachaça, the typical Brazilian spirit, is a distilled beverage with alcoholic graduation between 38 and 48 °GL, obtained from the distillation of fermented sugar cane juice [7]. It is the world's third most consumed spirit by volume, and its consumption is increasing in the international market because of its exotic and special flavor. Currently, Brazilian production of cachaça is estimated at 1.3 billion liters per year, and government efforts will tend to increase the exported volume in the next few years.

The cachaça production process comprises fermentation of the sugar cane juice with the yeast *Saccharomyces cerevisiae*, distillation of the wine, and aging of the distillate. Fermentation exerts the main influence on the final product quality, as most of the minor components are produced in this step [51]. Several of the congener compounds are an essential part of the aroma of the distilled product. Therefore, their concentrations settle the acceptance of the product in terms of enological attributes [24].

Similar to other distilled beverages, ethanol is the main organic compound found in cachaça and is responsible for its body. Superior alcohols, such as isoamyl alcohol, isobutanol, propanol, and isopropanol, usually comprise the flavor of spirits [52], with isoamyl alcohol being responsible for half of the total amount of these alcohols. The more volatile fraction of spirits is represented by carbonilic compounds, of which the main portion (more than 90%) is constituted by acetaldehyde [52]. To obtain a good quality spirit, a very low concentration of acetaldehyde is desirable, because this compound is associated with hangover syndrome and also considered a carcinogen [53]. Two other quality parameters for spirits are low concentrations of propanol and volatile acidity. Methanol level in cachaça also concerns distillers because of severe intoxication consequences related to its ingestion [54], but this compound can be easily avoided by controlling the presence of pectin in the juice [55].

Table 3.3 gives the required limits for the minor components in cachaça according to the Brazilian legislation [27].

Artisanal cachaça is traditionally distilled in a single pot still (alembic) working as a single-step distillation unit. The wines' ethanol and minor compounds are stripped away, and part of the distillate collected during the distillation period yields the product that is then directed to the aging process. Nevertheless, the production of cachaça in larger scale is performed in distillation columns working in continuous operation. In the next sections we will discuss the cachaça distillation in alembic and in continuous columns based on results obtained by process simulation.

3.3.1.1 Batch Distillation in Alembic

Cachaça distillation in an alembic can be simulated as a differential distillation, following directions from the work of Ceriani and Meirelles [56] and Scanavini et al. [1].

TABLE 3.3
Allowable Contents of Minor Components in Cachaça According to the Brazilian Legislation

Compound	Legislation limits (mg/100 ml anhydrous ethanol)	Range of maximal values (38–54 °GL; mg/kg spirit)
Volatile acidity, in acetic acid	150	620.3–914.8
Esters, in ethyl acetate	200	827.2–1219.8
Aldehydes, in acetaldehyde	30	124.1–183.0
Superior alcohols	360	1488.9–2195.6
Methanol	20	82.7–122.0

Ceriani and Meirelles [56] simulated the steam deacidification of coconut oil in the batch process, conducted at high temperatures and low pressures. Under these conditions the more volatile fatty acids can be stripped away from crude vegetable oils, which is one of the most important steps of the edible oil refining process. Scanavini et al. [1] conducted an experimental distillation trial for cachaça production in a lab-scale pot still of 0.008 m^3 of capacity and developed a detailed algorithm for simulating the process, including the presence of several minor components. Their approach was similar to the balance equations and phase equilibrium equations described in the previous section. The model was also able to reproduce appropriately most of their experimental results.

A pot still or alembic is a type of still used for distilling spirits, such as whisky, brandy, and sugar cane spirit. It is usually made of copper and a simple scheme of the equipment is shown in Figure 3.11. Usually, heat is applied directly to the pot that contains wine. Note that the upper part of the still ("neck") is commonly not isolated, and convective heat losses might occur in this part of the equipment, causing a small reflux due to condensation of a part of the vapor phase. Usually the influence of this small reflux is negligible and the composition of the vapor phase formed at the liquid interface inside the still can be assumed to be exactly equal to the vapor phase that is condensed in the condenser.

During the traditional batch distillation of cachaça, three different fractions of distillate are usually separated by the distiller, according to the boiling temperature and/or the alcoholic graduation of the mixture [57]. The first fraction (head distillate) is composed of the more volatile compounds, such as methanol, acetaldehyde, and ethyl acetate, and has an alcoholic graduation higher than 60 °GL. The second fraction (heart distillate) is the intermediate distillate portion and corresponds to the real Brazilian sugar cane spirit. The third fraction (tail distillate), also known as *weak water,* is formed mainly by water and other compounds whose boiling points are higher than 373.2 K. The quality of the spirit depends basically on the composition

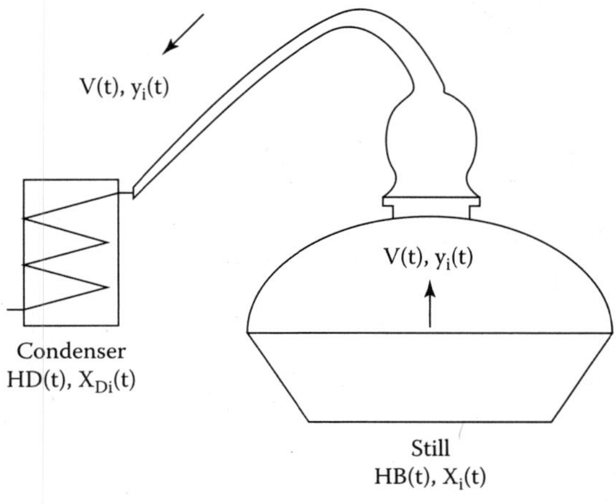

FIGURE 3.11 Scheme of alembic.

of wine, geometry of the still, and the operator's ability to do the cuts at the appropriate moments [58].

The wine is a complex mixture resulting from the fermentation of the sugar cane juice. Water and ethanol, the main components, represent more than 99% (g/g) of the total. Depending on the sugar cane and on the fermentation process, ethanol concentration usually varies from 5 to 10% in volume [57].

As said before, in a multicomponent differential distillation (batch distillation), the still is charged with wine and directly heated. Vapor flows overhead, is condensed, and then is collected in a receiver. Because the still composition is changing continuously, this process is inherently dynamic and cannot be modeled in steady state. The composition of the material collected in the receiver varies with time, so that the distillate composition of a cut is an average of all the material collected within that cut.

For simulating the multicomponent batch distillation of cachaça, typical concentrations of ethanol, water, and minor compounds were taken from the literature [6, 7, 41]. The values are shown in Table 3.4. The simulation was performed for a batch distillation of 1 m^3 of wine (52,544.7 moles) at 101.325 kPa. As indicated in Scanavini et al. [1], vaporization rates can be changed throughout the distillation process as a consequence of variations in the intensity of the heat source. This occurs especially if risks of foaming and liquid entrainment are observed during the batch period, because this could cause product contamination. Nevertheless, in order to take this into account, information either on the exact heat transfer changes or on the desired path of vaporization rate is necessary. In the absence of such information, the vaporization rate is assumed as constant and is fixed at 9.52×10^{-1} mol/sec (a value that varies around 0.09 m^3/h during the entire batch run), a reasonable value for slow distillation processes so that liquid entrainment can be better precluded.

Figure 3.12 shows the simulated profiles for the instantaneous alcoholic graduation in the still and in the condenser as well as the accumulated concentration value

TABLE 3.4
Wine Composition

Component	Composition
Water	0.9332 g/g [a]
Ethanol	0.06615 g/g [b]
Methanol	0.32 mg/kg
Isopropanol	1.02 mg/kg
Propanol	33.57 mg/kg
Isobutanol	27.75 mg/kg
Isoamyl alcohol	142.50 mg/kg
Ethyl acetate	7.685 mg/kg
Acetaldehyde	15.77 mg/kg
Acetic acid	435.10 mg/kg

[a] Obtained by difference.
[b] Corresponds to 8.2 °GL.

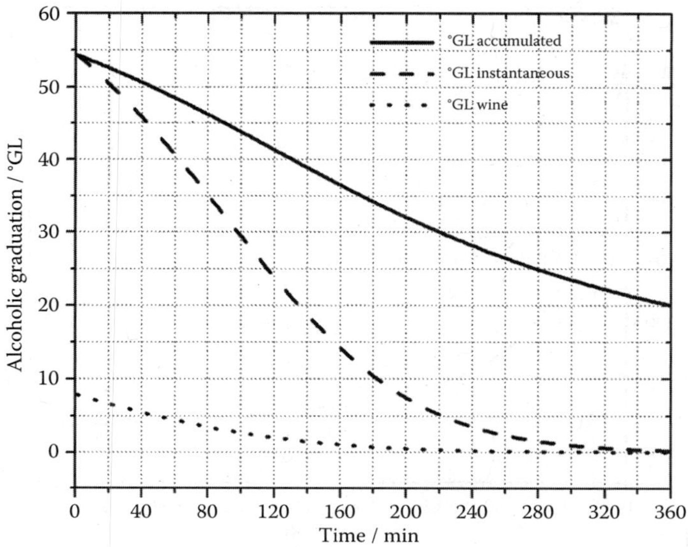

FIGURE 3.12 Alcoholic graduation profile in distillate and in still.

in the distillate receiver. As ethanol is stripped away from the wine, its concentrations in the still and in the vapor phase decrease. The accumulated concentration in distillate changes slowly because of the higher prior instantaneous concentrations observed in the vapor phase. A lower ethanol concentration in the still increases the wine boiling point, as the temperature profile in Figure 3.13 indicates.

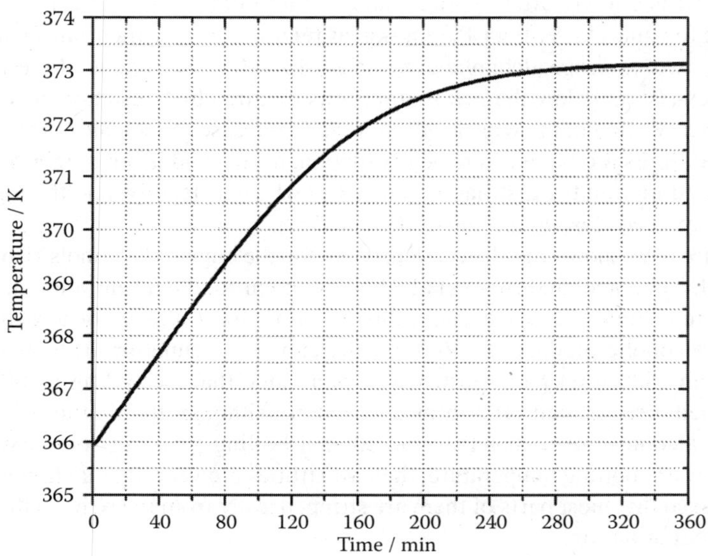

FIGURE 3.13 Temperature profile in the still.

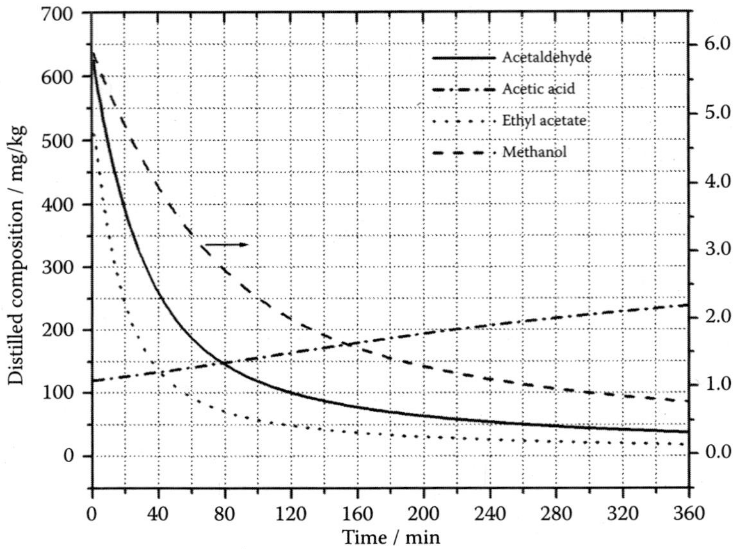

FIGURE 3.14 Profiles of minor components' composition in the distillate.

The behavior of the minor components was evaluated during the entire distillation period. Figure 3.14 shows the evolution of the distillate composition (accumulated values) with time for the light components, acetaldehyde, ethyl acetate, and methanol, and also for acetic acid. Acetaldehyde, the lightest component in the selected wine composition, is the minor component with the highest concentration in the distillate. Only after approximately 80 min does its content in the distillate decrease to values lower than 153 mg/kg, as required by the Brazilian legislation for a cachaça with average alcoholic graduation (46 °GL). The content of the other light components stays below the maximum limits required by legislation either because their concentration in the wine is very low, as is the case of methanol, or because the legislated limits are larger. In wine distillation, acetic acid is not a light component. Its concentration in the distillate increases slowly, but steadily, and the highest values are obtained close to the end of the batch run.

Figure 3.15 shows the distillate profiles for the superior alcohols (isopropanol, propanol, isobutanol and isoamyl alcohol) and their total concentration in the product. These alcohols have a strong influence on cachaça flavor. For a cachaça of an average alcoholic graduation (46 °GL), the content of superior alcohols should be lower than 1836 mg/kg of spirit, which is a value that, according to Figure 3.15, is obtained after around 60 min of distillation. Although the boiling points of the superior alcohols are higher than the ethanol boiling point, in some cases higher than the water boiling temperature, their volatilities are very high in diluted aqueous solutions, so that most parts of them are stripped away from the wine in the first part of the distillation run.

Except for acetaldehyde and superior alcohols, other minor components have distillate concentration lower than the desired maximum limits along the entire

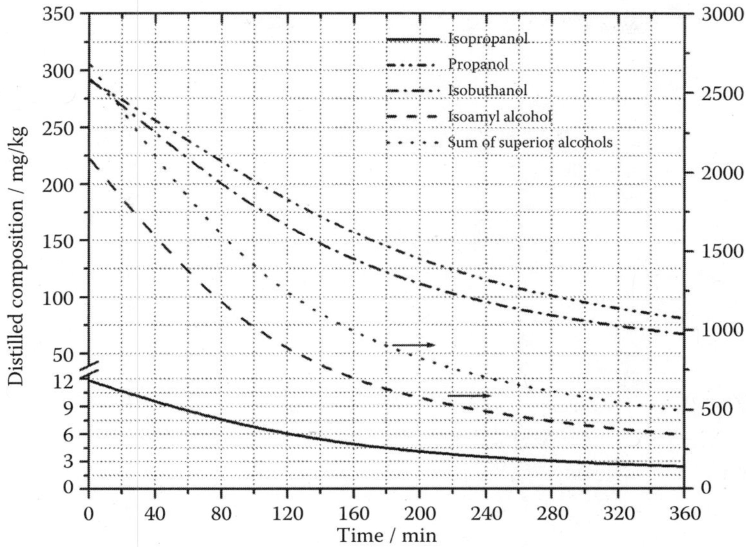

FIGURE 3.15 Profiles of superior alcohols concentration in the distillate.

distillation path. In the case of the first two classes of compounds, aldehydes and superior alcohols, the risk of outrunning the required limits is high, justifying the traditional distillation policy of cutting the alembic product in three parts: the head, heart, and tail fractions. In the first part, the head fraction, the more volatile components, mainly acetaldehyde, methanol, and superior alcohols, are concentrated, so that their residual levels in the heart cut will be, with certainty, within the required limits. The tail fraction allows the recovery of the residual ethanol still present in wine even when the alcoholic graduation in distillate is below the lowest required value. These two by-product fractions are frequently recycled in the next distillation batch, in order to improve the total ethanol recovery in alembic distillation.

Figure 3.16 shows the alcoholic graduation profiles for the three distillation cuts. The head cut corresponds to the first 5 min of distillation and represents approximately 5% of the volume of spirit produced. The heart cut or cachaça is the fraction collected until an accumulated alcoholic graduation of approximately 40 °GL is obtained. The tail fraction is the last one and is collected until the alcoholic graduation of the wine approaches a very low value, which occurs, in the present simulation case, at a batch time of about 200 min (see Figure 3.16). As Figure 3.16 indicates, the alcoholic graduation of the head cut is close to 54 °GL and that of the tail cut is close to 14 °GL. If both cuts are added to the next distillation batch of a wine with 8.2 °GL, the resulting mixture will contain a somewhat higher alcoholic content, improving the recovery of ethanol in the series of successive batches.

Figure 3.17 gives the concentration of some minor components in the distillate fraction corresponding to the three cuts shown in Figure 3.16. The second cut or heart fraction represents the cachaça spirit and can be classified as a good quality

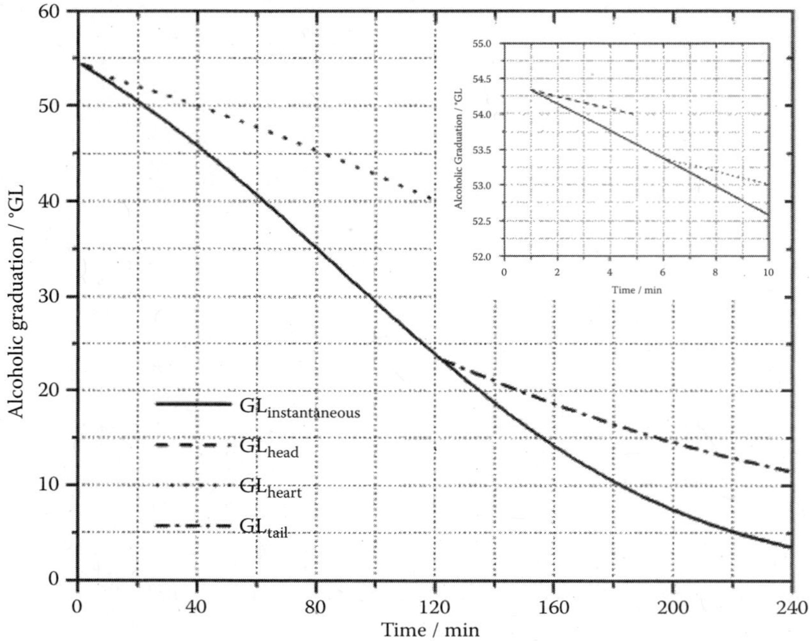

FIGURE 3.16 Alcoholic graduation of the three distillate cuts.

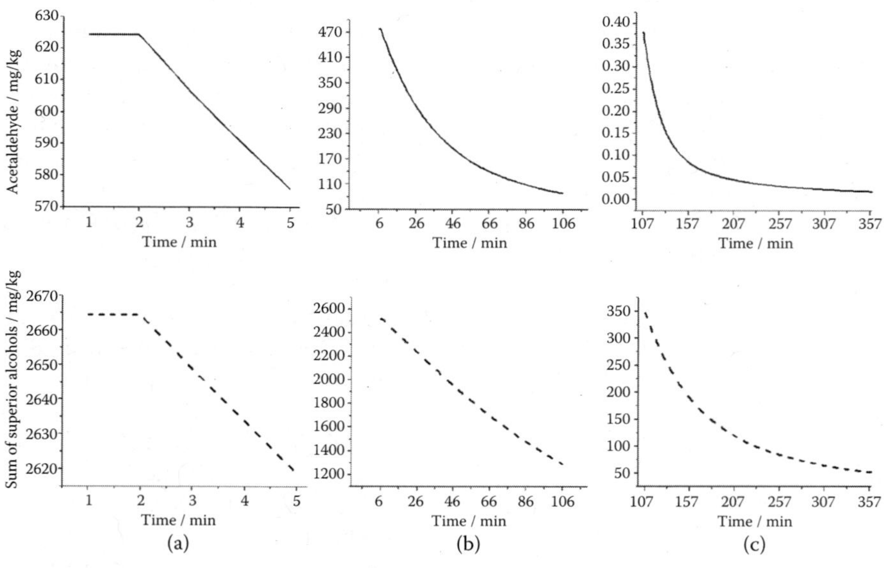

FIGURE 3.17 Minor components in three distillate cuts: (a) head fraction, (b) heart fraction, and (c) tail fraction.

product, because all legislation limits are met. For instance, its acetaldehyde composition is 100 mg/kg, and the content of superior alcohols is 1300 mg/kg, both values lower than the required maximum limits (see Table 3.3). Furthermore, its propanol and acidity levels are very low, requirements that are also very important for a product of good quality [42].

The results showed that the differential distillation model is capable of describing the distillation of cachaça in artisanal stills. A quantitative improvement could be attained if the heat loss (reflux) in the upper part of the still is considered, although such effect caused by natural convection would probably not have a large influence on the results. The proposed model could be applied to the distillation of other spirits. Other components important to the flavor of alcoholic beverages as well as chemical reactions occurring during distillation can also be considered.

3.3.1.2 Continuous Distillation in Tray Columns

A typical industrial installation for cachaça production is shown in Figure 3.18a. The column is divided in a small rectifying section, composed of two or three trays, and a stripping section, composed of 16 to 18 trays. In contrast to the production of hydrated ethanol, in cachaça distillation there is no side stream for removal of high alcohols (propanol, isopropanol, isobutanol, and isoamyl alcohol). The column is operated with a small reflux ratio, whose required value is slightly influenced by the alcoholic graduation of the wine fed into the column. A larger alcoholic concentration in the wine decreases the reflux ratio required for attaining the product specifications. The heat source is steam, which in some plants is directly injected at the bottom of the stripping section as "live" steam, so that the use of a reboiler is not

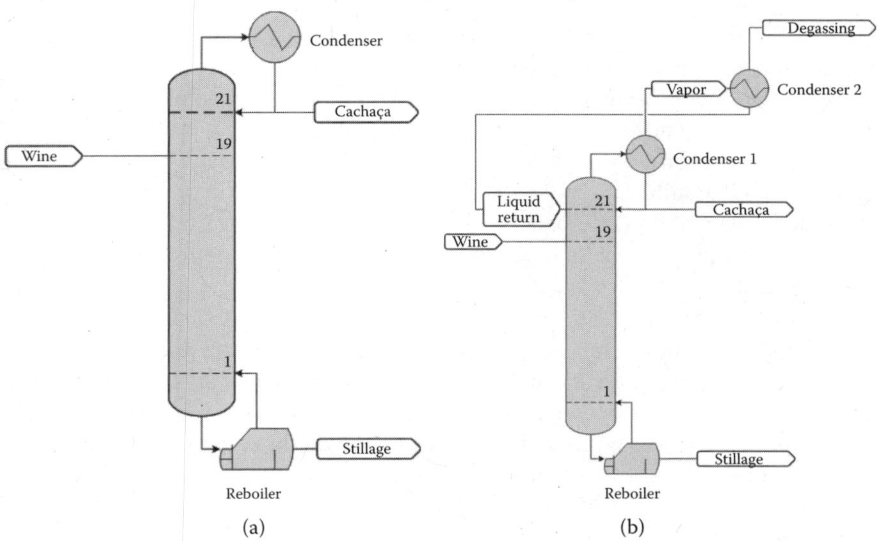

FIGURE 3.18 Typical industrial configuration for continuous cachaça production (a) without degassing and (b) with degassing.

always necessary. Nevertheless, in order to reduce the generation of waste products (stillage or vinasse), the best option is to use indirect heating with a reboiler, as is usual in conventional distillation plants.

Practically all ethanol fed into the column is recovered in the distilled stream, being admitted a maximum ethanol content of 0.02% in the bottom product, which corresponds to a loss of approximately 0.3 to 0.6% of the total ethanol amount and usually represents the main source of alcoholic loss in the process. When a stricter control of volatile components in cachaça is required, the degassing process can be a good alternative. This procedure consists in the use of a series of partial condensers in the top of the distillation column, where the vapor portion of each condenser is fed into the following condenser, and the condensed phase of each condenser is returned to the distillation column. At the last condenser of the series, the vapor portion is eliminated through the degassing stream, taking away the major part of the volatile compounds. Figure 3.18b presents the degassing scheme used for this work. As can be seen, only two condensers were used; however, the number of condensers is not limited to this number, with the possibility of using multiple condensers. It should be noted that the degassing factor can be expressed as the ratio of total flow of degassing stream to the sum of the flow of cachaça and the flow of the degassing stream.

The control of acetaldehyde concentration is a good example of the degassing function. This component can easily oxidize to acetic acid during the storage time, increasing the cachaça acidity. Knowing that the volatility of the acetaldehyde is extremely high, making possible the concentration of this component in the top of the distillation column, an increase of the degassing stream can eliminate the major part of the acetaldehyde present in cachaça, minimizing the previously mentioned problem. Because it is used only for product quality control, the value of the degassing stream is always very low in order to avoid significant ethanol losses.

The industrial process for continuous cachaça production was simulated using the commercial simulator ASPEN Plus [14]. For this simulation the wine was slightly changed, decreasing the ethanol concentration to 0.0645 g/g, an alcoholic graduation of 8.0 °GL, but keeping the concentration of all minor components to the values given in Table 3.4. The water content was increased in the exact proportion that the ethanol concentration was reduced. In a first set of simulations, without degassing (Figure 3.18a), the influence of the distillate rate and reflux ratio on the sprits' alcoholic graduation and on the ethanol loss in the stillage was investigated. The reflux ratio was varied in the range of 0.001 to 1.5 and the distillate rate from 1000 to 2000 kg/h. The feed rate was fixed at 10,000 kg/h.

According to Figure 3.19, for higher distillate flows, the alcoholic graduation is lower, but still within the range required by legislation, and the reflux ratio has no influence on the distillate concentration. For lower distillate rates, a higher reflux ratio increases the spirits' alcohol concentration, even above the required limits. The range of influence of the reflux ratio depends on the distillate rate, being the largest in the case of the lowest distillate rate. The reason for this behavior can be better understood on the basis of Figure 3.20, which shows the loss of ethanol, expressed in terms of that part of the ethanol stream fed into the column that is lost in stillage, as a function of distillate rate and reflux ratio. As can be seen in this figure, for lower distillate rates very high ethanol losses, much above the suggested limits (0.3 to 0.6% of the ethanol amount fed into the column), can be avoided only by

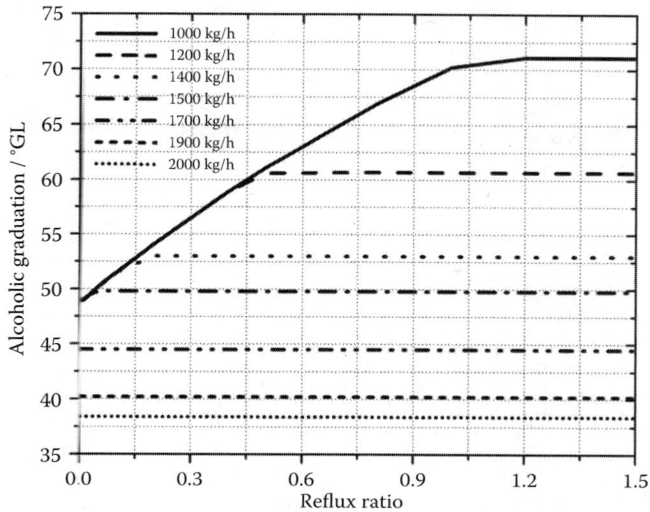

FIGURE 3.19 Cachaça alcoholic graduation as a function of reflux ratio and distillate rate.

large reflux ratios. This means that only spirits with high ethanol concentration will require higher reflux ratios in order to avoid significant ethanol losses. In fact, taking into account the alcoholic graduations required in the cachaça production, reflux ratios within the range 0.001 to 0.2 are sufficient.

Figures 3.21–3.23 show the concentration of minor compounds in the distillate (cachaça). Except for acetic acid, the reflux ratio has a very low influence on the

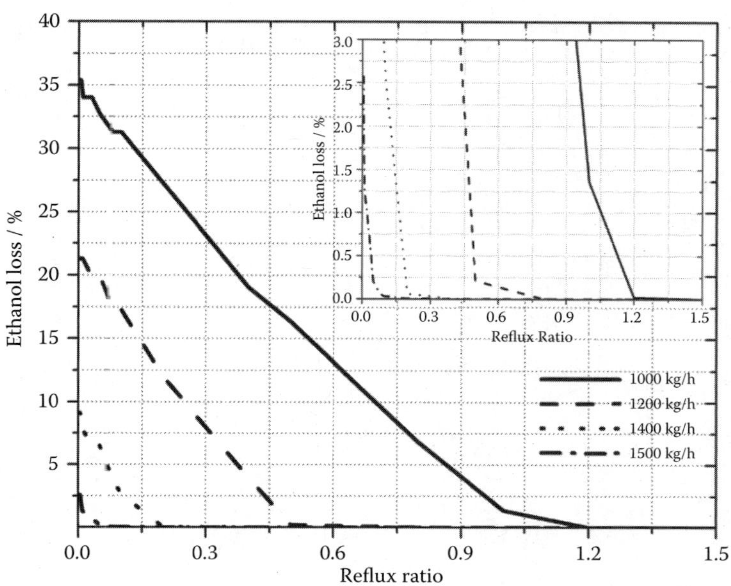

FIGURE 3.20 Ethanol loss in stillage as a function of reflux ratio and distillate rate.

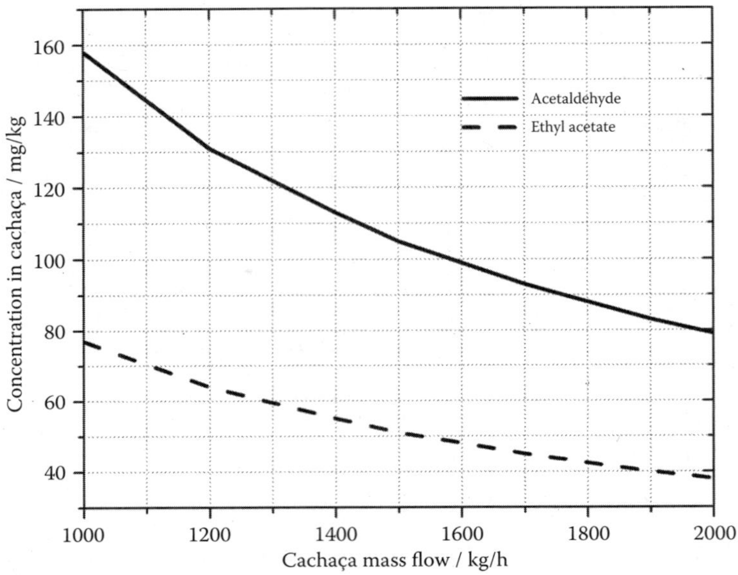

FIGURE 3.21 Acetaldehyde and ethyl acetate concentrations in cachaça as a function of distillate rate.

minor components' concentration in cachaça, and for this reason their concentration values are represented only as a function of the distillate rate. The concentrations of light components, such as acetaldehyde and ethyl acetate, decrease for large distillate rates. A similar behavior was observed for the superior alcohols.

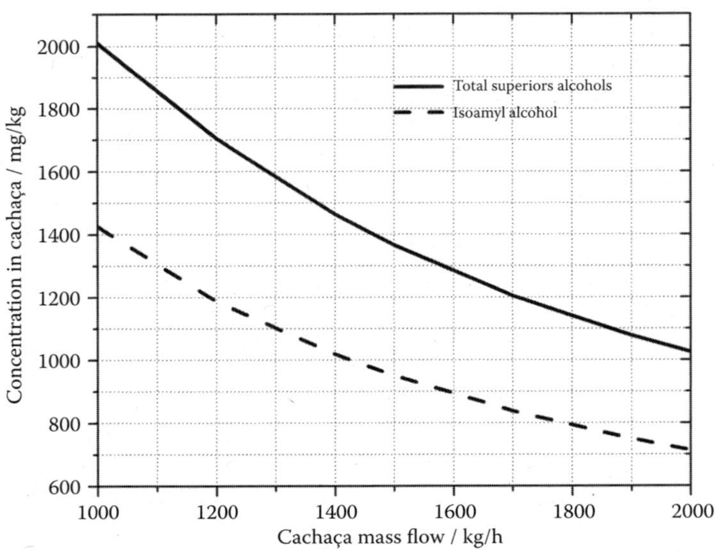

FIGURE 3.22 Isoamyl and superior alcohols concentrations in cachaça as a function of distillate rate.

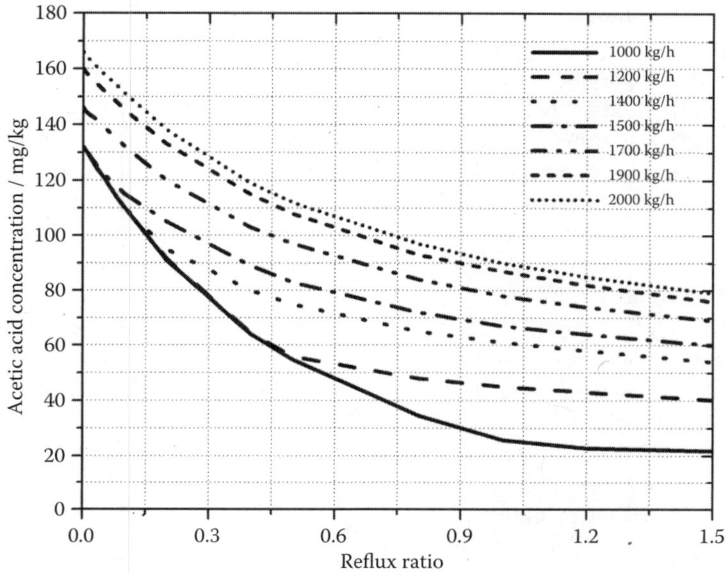

FIGURE 3.23 Acetic acid concentration in cachaça as a function of reflux ratio and distillate rate.

In the case of ethyl acetate the concentration in the distillate is always below the legislation limits (see Table 3.3), but in the cases of acetaldehyde and superior alcohols the values seem to be above the required limits for the lower distillate rates. Nevertheless, taking into account the corresponding alcoholic graduation of cachaça and the required reflux ratios in order to avoid high ethanol losses, even for low distillate rates the legislation limits are not exceeded.

Acetic acid concentration in cachaça increases with the distillate rate and decreases with the reflux ratio, a behavior usually obtained for heavier components, as is the case of this acid in spirits distillation. The limits required by legislation are easily met for this minor component in all simulated cases (see Figure 3.23).

As indicated in Table 3.3, the legislation strictly defines limits for the concentration of minor components, especially for methanol and acetaldehyde. As already explained, these limits are easily met in the case of methanol, provided that the presence of pectin is avoided during the must fermentation. For instance, in all previously simulated cases, the methanol concentration in cachaça was not higher than 1.68 mg/kg, well below the legislation limits. In the case of acetaldehyde it is surely more difficult to produce a spirit within the legislation limits. As a consequence of its very high volatility, acetaldehyde will doubtless concentrate in the distillate, so that a higher concentration of this component in the wine means necessarily a risk of exceeding the maximum allowed limit. Besides its deleterious direct effect on the product quality, acetaldehyde can also easily oxidize to acetic acid, increasing the spirits' acidity.

The effect of acetaldehyde concentration in the wine will be further investigated. A degassing (vapor phase) stream can be used for controlling the presence of light components. This was investigated for a selected case of the prior simulation set, namely for a distillate rate of 1500 kg/h and reflux ratio of 0.2. To produce different degassing flows, the temperature of condenser 2 was varied from 293.2 to 353.2 K. At the lowest temperature, little degassing was produced, and the opposite effect was observed at the highest temperature. In this way, it was possible to investigate the influence of this stream on the acetaldehyde concentration and on the ethanol loss.

Aiming to help in the control of the volatiles' content in the spirit, the degassing stream can be used when the original concentration of those compounds in the wine leads to a distillate composition in disagreement with the legislation limits. Taking into account the usual content range of acetaldehyde in the wine (see Table 3.1), we increased its content to 26 mg/kg.

In this set of simulation cases, a further component was included in the wine composition, namely carbon dioxide. This compound is important for evaluating the performance of the degassing process, represented by the degassing stream. Carbon dioxide is produced during must fermentation, and it could carry part of the generated ethanol away, increasing the product losses. In order to avoid such losses the industrial fermentation process is performed in a closed vessel and the outlet gas stream is pumped into an absorption column used for recovering the volatile component. The industrial fermentation vessel is operated at temperatures about 305.2 K and under a slightly positive manometric pressure (6.0–8.0 kPa). Assuming that the light phase inside the vessel is composed of gas saturated with ethanol and water and considering that this gas is, for practical purposes, pure carbon dioxide, the solubility concentration of CO_2 in the wine can be easily estimated. Using the NRTL parameters for ethanol–water interactions and the CO_2 Henry constants in ethanol–water solutions given by Dalmolin et al. [59], a solubility around 1100 mg CO_2/kg of wine (8.0 °GL) was estimated.

Using these values for acetaldehyde and carbon dioxide, the water content in wine (see Table 3.4) was correspondingly diminished, and the new composition was used as feed stream in this set of simulations. Figure 3.24 shows the change of acetaldehyde composition in cachaça as well as the loss of ethanol through the degassing stream as a function of the degassing percentage.

As can be seen in Figure 3.24, the degassing stream makes it possible to control the acetaldehyde concentration in cachaça, but it increases the ethanol loss in the distillation process. Taking into account the alcoholic graduation of cachaça obtained in this case (see Figure 3.25), the maximum allowed limit for acetaldehyde concentration, given in Table 3.3, corresponds approximately to 167 mg of acetaldehyde/kg spirit, a value that is obtained using a degassing stream of 0.7% (10.7 kg/h). The corresponding loss of ethanol is 0.58%, which should be added to the value of loss in the stillage. Although the corresponding impact on the product alcoholic concentration is not significant (see Figure 3.25), the estimated loss of ethanol can attain values larger than the loss obtained in the stillage. For this reason the use of a degassing stream for controlling the volatile concentration in the product is appropriate only in cases when the concentration slightly exceeds the legislation limits. Figure 3.25 indicates that the concentration of other volatile components, for instance ethyl acetate, also decreases.

If the concentration of volatiles is large, an alternative equipment configuration is required. This scheme is shown in Figure 3.26. Columns A and B correspond to the stripping and enriching sections of the prior scheme, respectively. In column A

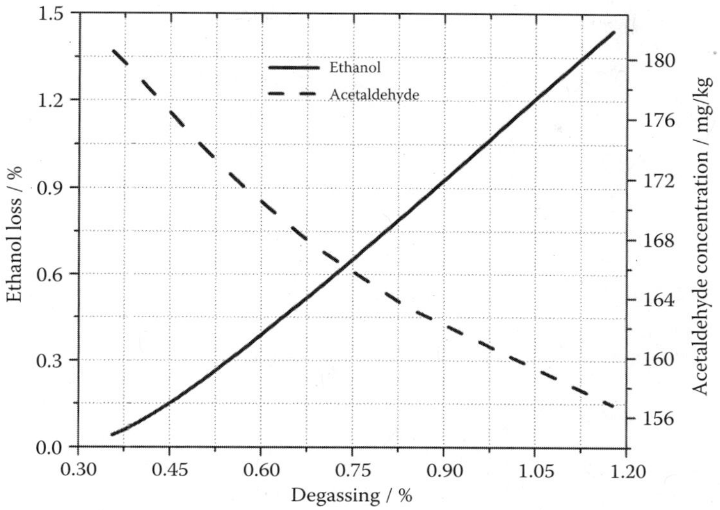

FIGURE 3.24 Acetaldehyde concentration in cachaça and ethanol loss as a function of degassing factor.

ethanol is stripped away from the liquid phase, so that the ethanol loss in the stillage is very low. In column B ethanol is concentrated up to the desired spirits graduation. Columns A_1 and D are used mainly for concentrating the light components, so that a small stream of distillate at the top of column D allows the control of volatile components' level in cachaça. This byproduct stream is named second alcohol

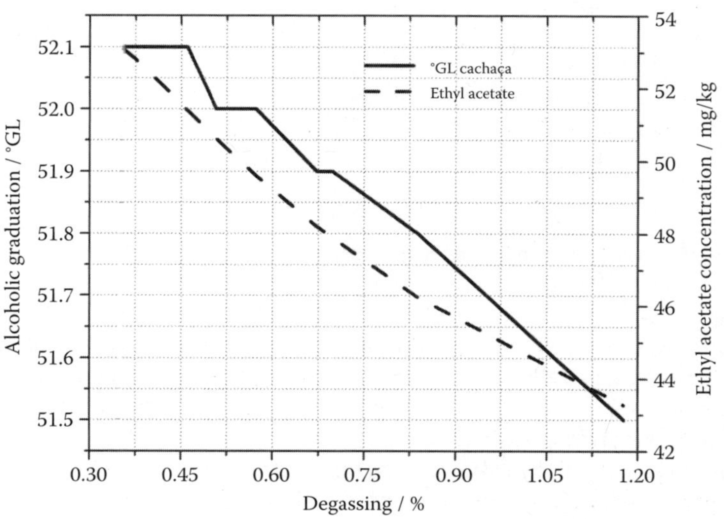

FIGURE 3.25 Cachaça alcoholic graduation and its ethyl acetate concentration as a function of degassing factor.

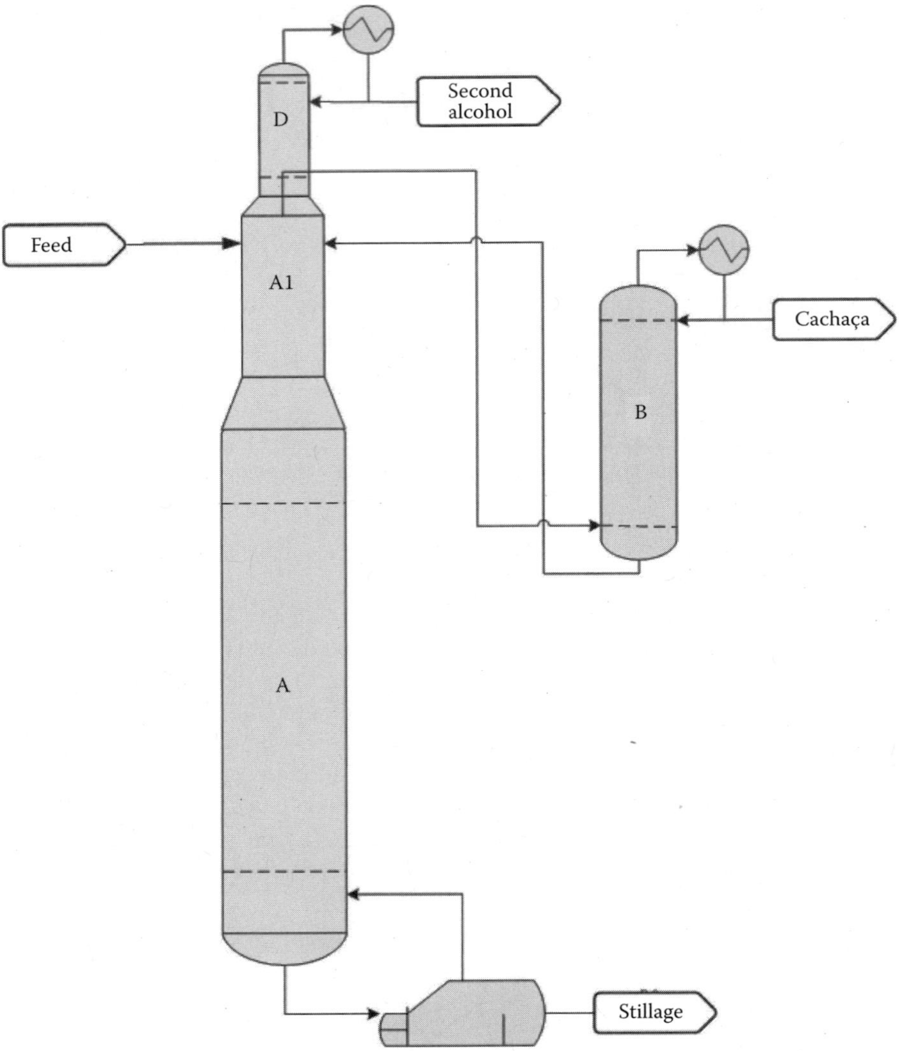

FIGURE 3.26 Alternative industrial plant for continuous cachaça production.

and corresponds to an ethanol stream rich in light components, with concentrations much larger than those allowed by legislation. This by-product stream also contains a small amount of the processed ethanol, but it has commercial value for purposes other than the spirit production.

In this configuration wine is injected at the top of column A_1, which usually contains four trays. The vapor phase of column A is directed to column D, which also contains four trays and is operated under high reflux rates. For this reason ethanol and light components are very concentrated in the distillate of this column, guaranteeing that a small stream, withdrawn from its top, will be enough to control the quality of the main product. Using such a scheme, high quality cachaça can be produced without large ethanol losses, even if the concentration of minor components in the wine is higher than usual.

3.3.2 Concentration and Purification of Aroma Compounds of Cashew Juice in a Batch Distillation Column

Fruit juice concentration reduces its natural volume and facilitates the storage, packing, and transportation of the product. However, during the conventional concentration process by evaporation, most flavor components are stripped away together with the water vapor, causing deleterious effects on the sensorial quality of the concentrated product. To minimize this consequence, specific processes are designed for recovering the juice flavor fraction lost during evaporation and reincorporating it into the concentrated juice, so that a beverage with a flavor very similar to that characteristic of the natural fruit can be obtained. This is especially the case for those juices with large international consumer markets, such as orange and apple juices.

Besides its use for recovering the natural flavor of concentrated juices, aroma compounds from juices are widely used in the food and beverage industries, either to confer a specific flavor to a product or to strengthen a characteristic flavor. A specific flavor is a consequence of the combination of several volatile substances of different chemical classes, none of them being individually responsible for that flavor.

There is a growing interest in tropical fruit juices in the international market, but the fulfilling of this increasing demand requires the adaptation of prior technologies or the development of new ones in order to preserve the fruit juices' natural flavor. Unfortunately, in the case of some tropical juices, such as cashew and acerola juices, investigations concerning flavor composition and recovery after concentration are still incomplete.

To test the use of distillation processes for recovering flavor compounds lost during tropical fruit juice evaporation, we investigated the concentration and purification of cashew juice aroma by batch distillation with reflux. Batch distillation columns are multipurpose equipment frequently used for concentrating and separating relatively small batches of mixtures on an industrial scale. In the orange juice industry, because of its very large scale, the recovery of flavor compounds from the vapor phase generated during the juice concentration is usually performed by continuous distillation. Nevertheless, the further fractionation of the recovered aroma mixtures, aqueous and orange oil essence, is often performed by batch distillation, in order to produce fractions with specific sensorial characteristics. Similarly, the batch distillation process is used for fractionating essential oils, for instance, from ginger, clove, lemon grass, eucalyptus, and citronella.

For this investigation we used an algorithm based on the dynamic model proposed by Luz and Wolf-Maciel [4], which considers mass and energy balances, and also used the vapor–liquid equilibrium relationships, as presented in Section 3.1.3. It was considered that the distillation column starts up with total reflux, that is, without any distillate withdrawal. For initializing the set of variables used in the balance and equilibrium equations, the initial composition in all plates and in the column still is assumed to be the same and equal to the initial composition of the mixture to be distilled. After a small start-up time, when the whole column is warmed up and the desired condition is achieved on the top of the equipment (condenser), the system stops to operate at total reflux, so that the product withdrawal and the separation properly begin.

The separation of compounds by batch distillation can be performed by fixing two of the following operational conditions: reflux ratio, distillation rate, boil-up rate (rate of the vapor flow leaving the reboiler), and reboiler duty. One of these specific operational conditions can be fixed during the entire batch period or a sequence of specific operational conditions, and its corresponding duration can be selected for the whole process. Alternatively the duration of a specific operational condition can be determined by a stop criterion that automatically initiates the subsequent operational condition, so that the column can operate under a sequence of different conditions. The algorithm also allows setting the moment of tank storage exchange, in other words, presetting the cuts that should be performed during the entire run. The distillate accumulated in each tank corresponds to the desired products. Each product is associated to the sequence of operational steps selected at the batch beginning and to the volatility characteristics of the mixture's components.

In a relatively recent study Garruti et al. [60] isolated the flavor compounds of the cashew fruit juice by the dynamic headspace technique. Sixty-three compounds were detected, and 49 of them were identified. Esters were the major chemical class detected, especially methyl and ethyl esters of saturated carboxylic acids from C2 to C6. According to the chromatographic and olfactometric analyses developed by Garruti et al. [60], the volatile compounds, whose identification was possible and represented the group of compounds that most intensely contribute to the formation of the characteristic cashew flavor, were the following: hexanal, 2-methyl-2-pentenal, and *cis*-3-hexenol, all with different "green" notes; ethyl isovalerate, methyl isovalerate, ethyl butanoate, and *trans*-2-ethylbutenoate, described as cashew, sweet, and fruit; and 2-methylbutanoic acid, responsible for an intense odor described as unpleasant, stinky, and reminiscent of sweat and dirty socks.

Taking the olfactometric data into account, as well as the flavor components with larger concentration in cashew juice aroma, the composition shown in Table 3.5 is assumed to correspond to the aqueous solution evaporated from cashew juice during concentration. The information on aroma composition usually reported in the literature is on a water-free basis, so that the water concentration presented in Table 3.5 must have been estimated from other sources. Haypek et al. [50] reported the composition, including the water content, of the aqueous solution generated during the industrial orange juice concentration by evaporation. The same water content was assumed as valid for the case of cashew juice evaporation.

A batch of 26,667 moles (approximately 510 kg) of a mixture with the composition given in Table 3.5 was charged into the column. Two main objectives were set for this investigation: to obtain a high recovery and concentration of the flavor volatiles, reducing to a minimum the water content in the distillate, and to purify the concentrated flavor, reducing the concentration of the undesirable volatile component (2-methylbutanoic acid) also to a minimal concentration, at least in the first cut (the first distillate product).

TABLE 3.5
Estimated Composition of the Aqueous Solution Evaporated during Cashew Juice Evaporation

Compound index	Compound	Composition (g/g)
1	2-Methyl 2-pentenal	0.0173
2	Ethyl isovalerate	0.0166
3	Hexanal	0.0127
4	Methyl isovalerate	0.0090
5	Ethyl butanoate	0.0065
6	2-Butoxyethanol	0.0039
7	2–Methylbutanoic acid	0.0040
8	Water	0.9300

In contrast to the prior case studies, there is no literature report on industrial equipment for the recovery and fractionation of aromas from cashew juice. Probably even the specific industrial know-how for this process is not yet available. For this reason we decided to investigate the process in a wide range of the main constructive and operational conditions. Although several simulation runs can usually be performed without difficulty, if the number of effects and the corresponding ranges of values to be investigated are too large, the number of required runs can increase very rapidly. An alternative is to treat the simulation runs as simulation "experiments" and to combine the approach based on simulation and the factorial design technique. Such an approach was already tested in different distillation cases with very good results [61, 62].

A complete experimental design 2^3 [63] was used, with axial points and a central point, totalizing 15 simulations runs. Three independent variables were selected: distillate rate (D, mol/h), reflux ratio (r) and number of ideal stages (np+2, number of ideal trays plus reboiler and condenser/reflux drum). The distillate rate was varied from 100 to 1100 mol/h, the reflux ratio between 4 and 40, and the number of ideal stages between 10 and 20 stages. The column operated under a pressure of 101,325 Pa.

To evaluate the simulation results, three objective functions were defined, the total recovery of the desired volatile components, R, the purification factor, F, and the productivity, P, as indicated in Equations 3.46 through 3.48 below:

$$R = \frac{\sum_{i=1}^{6} x_{i,HC} \cdot HC}{\sum_{i=1}^{6} x_{i,HI} \cdot HI} \times 100, \qquad (3.46)$$

where x_i represents the molar fraction of component i in the original mixture amount HI (moles) or in the product (distillate) amount HC (moles). Note that only the first six components are included in the summation, the seventh one being the undesirable volatile and the eighth component, water (see Table 3.5). Equation 3.46 indicates the total recovery of volatiles, except for the 2-methylbutanoic acid. A version of this equation can also be formulated for each component indicating its specific recovery. The first form was used in the process optimization, while the second one allowed evaluating the recovery of each compound in the optimized conditions.

$$F = \frac{\dfrac{\sum_{i=1}^{6} x_{i,HC}}{x_{7,HC}}}{\dfrac{\sum_{i=1}^{6} x_{i,HI}}{x_{7,HI}}}. \tag{3.47}$$

The purification factor F, calculated by Equation 3.47, is a kind of enriching factor. It indicates how many times the ratio of desired volatiles concentration to the undesired one can be increased by batch distillation.

$$P = \frac{F_m}{t_{F_m}}, \tag{3.48}$$

where F_m represents the maximal purification factor obtained in a specific simulation run and t_{F_m} is the corresponding batch time. The productivity, P, evaluated by Equation 3.48, indicates how fast a product with high purity can be obtained by batch distilling the cashew juice aroma. It should be kept in mind that batch distillation involves at least two steps: the distillation time and the period between two runs. In this last period, the prior residue, so far kept inside the equipment, is discharged and a new batch is fed into the still. Sometimes the equipment should also be cleansed between consecutive runs, to assure that flavor residues of the previous mixture do not contaminate the subsequent ones. This means that an intensive use of the batch period is an important factor in evaluating the productivity of batch distillation processes.

Figure 3.27 shows a typical result for the concentration profiles of minor components in the distillate. Product withdrawal begins after about 50 min of column start-up. Figure 3.27a shows the instantaneous concentrations and Figure 3.27b the accumulated values in the distillate receiver, calculated by integrating the instantaneous values during the entire batch period. At the very beginning, the esters exhibit the largest initial concentrations, with ethyl isovalerate reaching the maximal accumulated concentration approximately half an hour after product withdrawal. Both aldehydes reach their maximal concentration values in the collected distillate in batch times within 130–160 min from the start of distillation. During this last batch time interval, the accumulated concentrations of 2-butoxyethanol

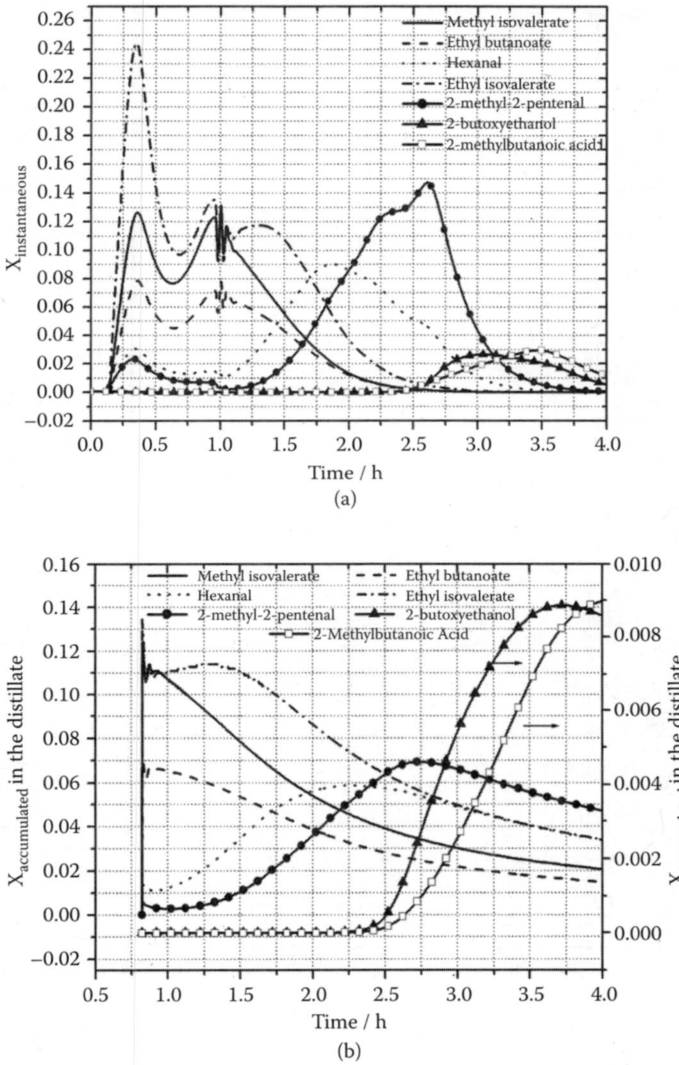

FIGURE 3.27 Concentration profiles in molar fraction, of minor components in the distillate: (a) instantaneous values and (b) accumulated concentrations in distillate receiver ($D = 600$ mol/h, $r = 22$, number of stages = 15).

and 2-methylbutanoic acid also begin to increase, but their largest concentrations were about 5–10 times lower than the simultaneous concentration obtained for the other volatiles.

Figure 3.28 shows the results for the composition of the distillate collected in the receiver, classified either according to the minor components' chemical classes or to the purification goal, in desired volatiles and 2-methylbutanoic acid. For these representations the corresponding accumulated amounts of each volatile are summed up

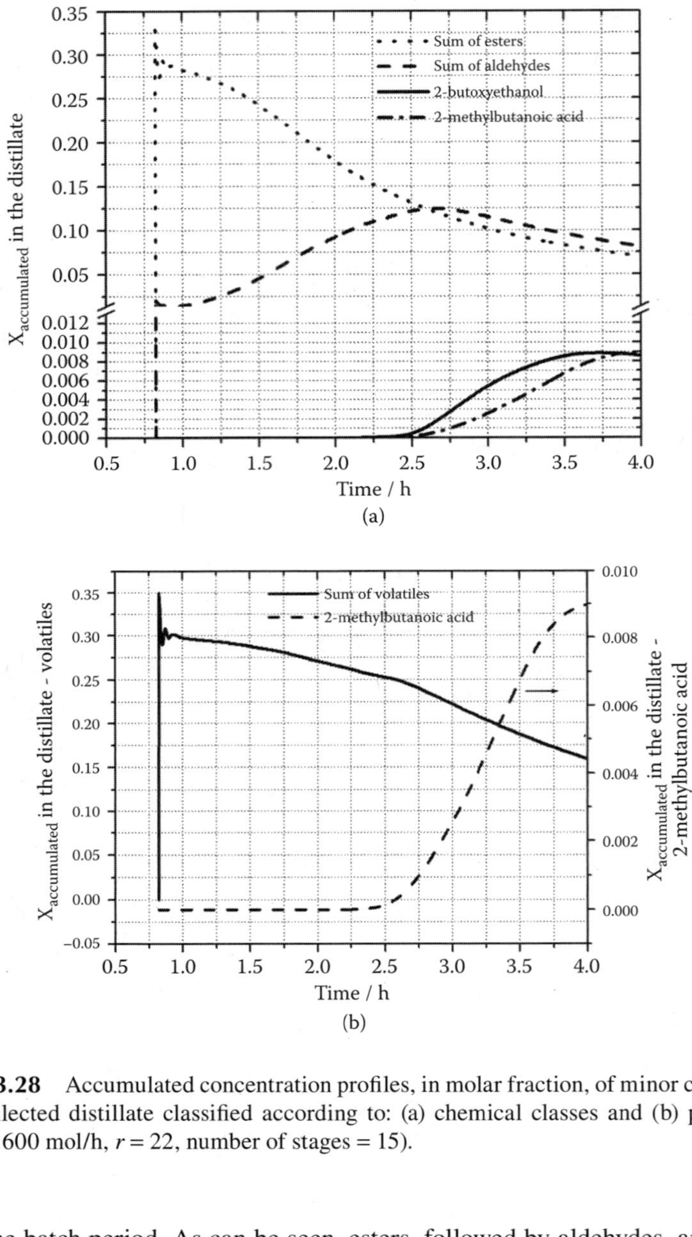

FIGURE 3.28 Accumulated concentration profiles, in molar fraction, of minor components in the collected distillate classified according to: (a) chemical classes and (b) purification goal ($D = 600$ mol/h, $r = 22$, number of stages = 15).

during the batch period. As can be seen, esters, followed by aldehydes, are the first chemical class concentrated in the distillate. The desired volatile components have high accumulated concentrations in the distillate receiver during the entire run, but their values decrease steadily, while only after about 150 min of batch distillation does the collected distillate content of the undesired volatile begin to increase. Even at its highest accumulated value, the acid concentration is approximately 12 times lower than the total concentration of desired volatiles. Furthermore, these results

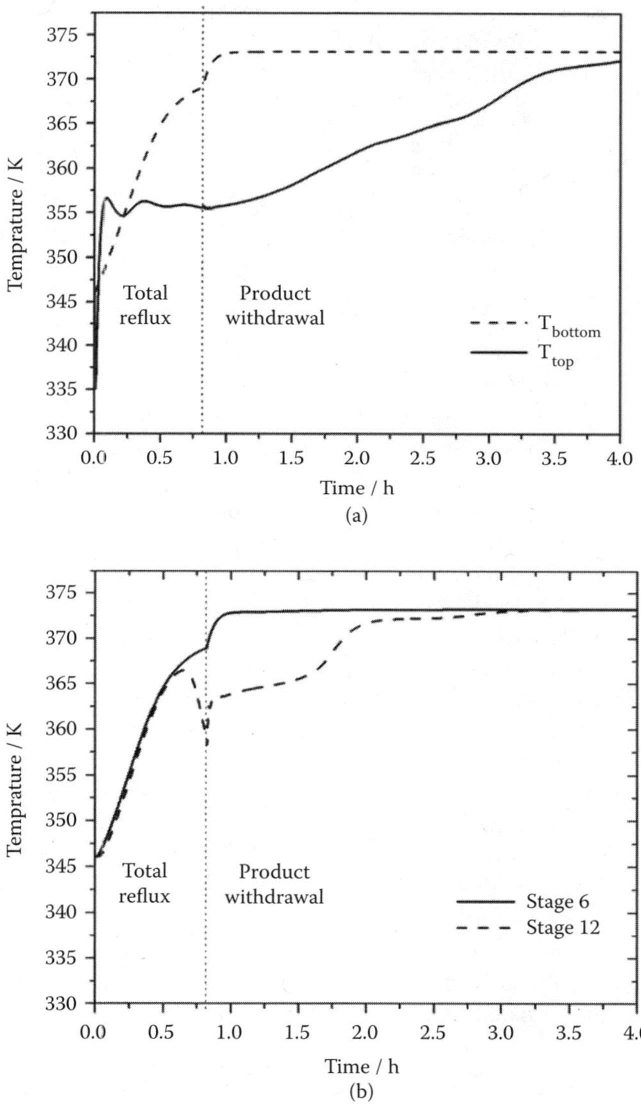

FIGURE 3.29 Temperature profile: (a) Top and bottom stages, and (b) stages 6 and 12 ($D = 600$ mol/h, $r = 22$, number of stages = 15).

indicate that a first distillate cut, performed at approximately 150 min after the distillation beginning, would generate a very pure and concentrated product, combining a good recovery of the desired volatiles, mainly esters but also part of the aldehydes, and a very low concentration of 2-methylbutanoic acid.

Figure 3.29 shows the temperature profiles of the condenser (top stage), stages 12 and 6, and reboiler (bottom stage). In the first part of the run, about 50 min, the column is operated in total reflux. It should be remembered that the algorithm used in the simulations assumes that the initial liquid concentration in all column stages is equal to the mixture's initial concentration. This means that the first part of the run corresponds to the development of a column profile inside the column operating in total reflux, with the light components concentrating in the top trays and the heavy ones in the bottom stages. The temperature profiles reflect the tendency mentioned above. As the heavy components are concentrated in the bottom stage during the operation in closed regime, the temperature in the reboiler shows a rapid increase, but its value at the very beginning corresponds approximately to the boiling point of the original mixture because of the very high liquid hold-up in the bottom stage. The initial boiling temperatures in each tray are influenced by the original concentration of the mixture, but because of the small tray liquid hold-up, they are also influenced by the vapor and liquid internal flows in the column that change the liquid tray concentration rapidly. After a very rapid increase, the top temperature oscillates around values, in most cases, lower than the reboiler temperature, and after product withdrawal it tends to increase steadily. In fact, just after the beginning of the product's withdrawal, the temperature on the top initiates a process of continuous rise after the withdrawal of the most volatile compounds. The top temperature oscillation in the first part of the run is related to the instantaneous change of the condenser/reflux drum liquid hold-up composition and to its corresponding effects on the phase equilibrium.

The start-up of an actual batch column usually involves the heating of the original mixture in the bottom stage until it reaches the boiling temperature and the formation of a vapor phase that flows upward through the trays, being cooled and condensed by the cold column shell and internals during the initial part of the start-up period. During this period the upper parts of the column are heated, and this period lasts until the vapor phase is able to get to the top of the equipment without being condensed along its way up. After this initial period, the condenser and reflux drum are filled with liquid, and the operation of the column in total reflux can be initiated. With the beginning of the reflux flow, a proper liquid hold-up is formed in each tray and the column operates in a correct way. After a further period of eventual adjustments in the boil-up rate and of distillate concentration control, product withdrawal can be initiated. Although this usual start-up procedure is not exactly what the algorithm assumes for the initialization procedure of the simulation, it should be emphasized that both procedures should give similar results at the end of the start-up period. In fact, if a closed start-up regime (total reflux) is assumed, obtaining similar results depends not on the exact way of initializing the simulation procedure, but on the algorithm capacity of representing the operation of an actual batch column with reflux after that time interval used for heating the equipment is concluded.

To get a better insight into the workings of the internal column, Figure 3.30 shows the instantaneous concentrations for selected column stages. The instantaneous concentrations of desired volatiles decrease very rapidly in reboiler and in stage 6 during the time of closed column operation (see Figure 3.30a). This decrease occurs, naturally, first at the reboiler and it is followed, with a short time delay, by the decrease observed in the sixth stage. With the beginning of product withdrawal, the desired volatile concentrations at the reboiler and stage 6 decrease even more

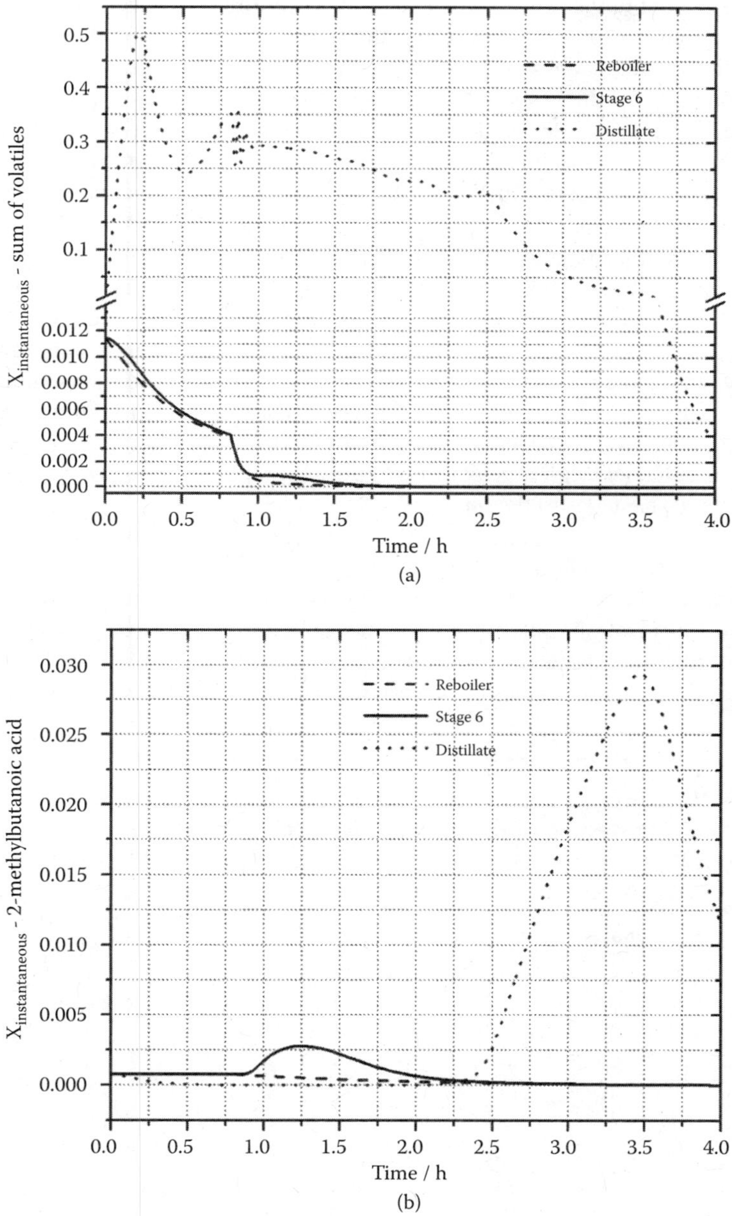

FIGURE 3.30 Concentration profiles of minor components in selected stages: (a) desired volatiles and (b) 2-methylbutanoic acid ($D = 600$ moles/h, $r = 22$, number of stages = 15).

abruptly and tend to a zero value. The instantaneous concentration profile of desired volatiles at the top of the column has a more complex pattern that is preceded by similar profiles at the column trays near the column condenser. The top compositions correspond to the instantaneous liquid concentration observed in the condenser/

reflux drum and in the reflux flow. After the beginning of product withdrawal these concentrations also correspond to the instantaneous composition of the distillate flow. If the instantaneous concentrations of the distillate flow shown in Figure 3.27a are summed up for the desired volatiles, a concentration profile equal to the top one presented in Figure 3.30a should be obtained. The oscillations of the top concentration before product withdrawal correspond to the development of the column profile during the total reflux regime: the volatiles' concentrations increase abruptly and exhibit oscillations that were damped with the product withdrawal. These damped oscillations are then related to the composition changes within the different volatiles. As the concentrations of each volatile in the distillate flow present peculiar profiles with their maximum in different batch times, summing up these component-specific profiles generates the damped oscillations observed after product withdrawal. The 2-methylbutanoic acid concentration in reboiler is very low, corresponding to its content in the original mixture, and decreases slowly and steadily during the batch time. The corresponding concentration profiles in the trays and in the distillate show a peculiar behavior, with a peak of composition propagating during the batch time from the bottom stages to the top ones.

The simulation results allowed calculating the objective functions expressed by Equations 3.46 through 3.48. The calculations were performed only for a first cut during the distillation path, which corresponds to the accumulated product until the maximum purification value F_m was obtained. The corresponding values of the objective function as well as the constructive and operational conditions tested are given in Table 3.6. The recovery of the desired volatiles varies around an average value of 46%.

TABLE 3.6
Conditions and Results of the Simulations according to the Experimental Design

Simulation run	Distillate rate (mol/h)	Reflux ratio	Number of stages	Time (h)	Recovery (%)	$F_m \times 10^{-4}$	$P \times 10^{-4}$
1	302	11	12	2.57	49.0	8.43	3.28
2	898	11	12	1.41	49.1	8.45	5.98
3	302	33	12	2.40	44.3	21.9	9.16
4	898	33	12	1.35	44.3	22.1	16.4
5	302	11	18	2.56	49.0	8.35	3.26
6	898	11	18	1.41	49.0	8.38	5.96
7	302	33	18	2.37	43.8	21.4	9.06
8	898	33	18	1.34	43.8	21.6	16.2
9	100	22	15	5.72	45.7	15.1	2.64
10	1100	22	15	1.27	45.7	15.3	12.1
11	600	4	15	1.75	53.2	3.09	1.77
12	600	40	15	1.59	42.9	25.6	16.2
13	600	22	10	1.67	46.7	15.7	9.44
14	600	22	20	1.63	45.7	15.1	9.26
15 (PC)	600	22	15	1.64	45.7	15.2	9.28

[a]Cut time corresponding to the maximum purification factor, F_m.

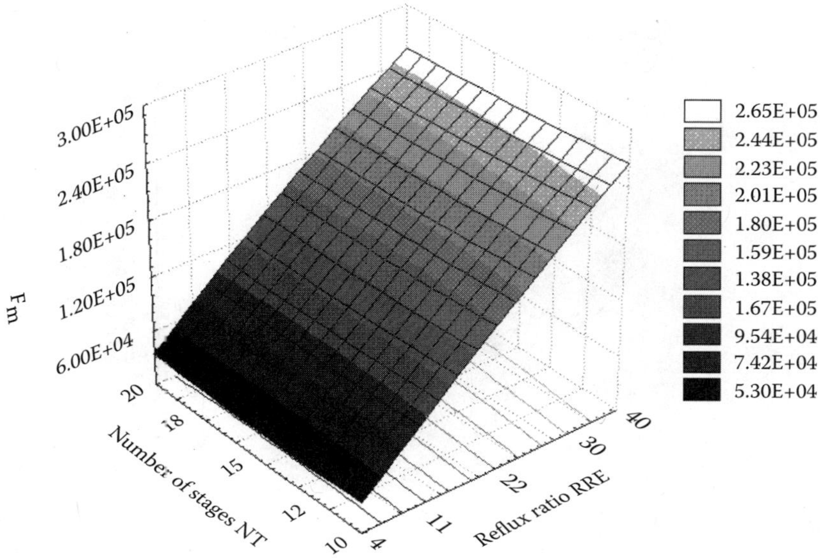

FIGURE 3.31 Response surface for the purification factor ($D = 600$ mol/h).

In fact, for a first product cut defined on the basis of the lowest contamination with the undesirable volatile component, 2-methylbutanoic acid, the operational and constructive conditions investigated do not show a large influence on the obtained recovery, which varies within the range 43 to 53%.

Very large purification factors were obtained in all simulations. A manifold enriching of the desired volatiles in the product, with a minimal concentration of 2-methylbutanoic acid, was a feasible goal for the first distillate cut. The productivity indicates that this goal could be attained at relatively short periods of batch distillation, 1.3 to 1.6 h for this first cut. A recovery of the desired volatile compounds larger than those values reported in Table 3.6 is feasible, but it implies a higher concentration of the acid in the final product.

Using the software Statistica 5.5, statistical models of the process were obtained for the purification factor and the productivity, both with high coefficients of determination, 0.9998 and 0.997, respectively. For the maximal purification factors obtained in the first cut, the statistical analysis showed that the reflux ratio and the number of stages were the significant independent variables and the corresponding response surface is represented in Figure 3.31. As expected, the reflux ratio has a large influence on the purification factor, improving the separation between the light volatile components and the heavy compound 2-methylbutanoic acid. The number of stages has only a very slight influence on the purification factor. This influence also shows an unusual behavior: a higher number of stages can have a slight but prejudicial effect on the purification factor. In the case of continuous distillation a direct relationship between a larger number of stages and a better separation of light and heavy components is valid, as a general rule. Probably the same is valid for most cases in batch distillation columns. Nevertheless, a slight but opposite effect was

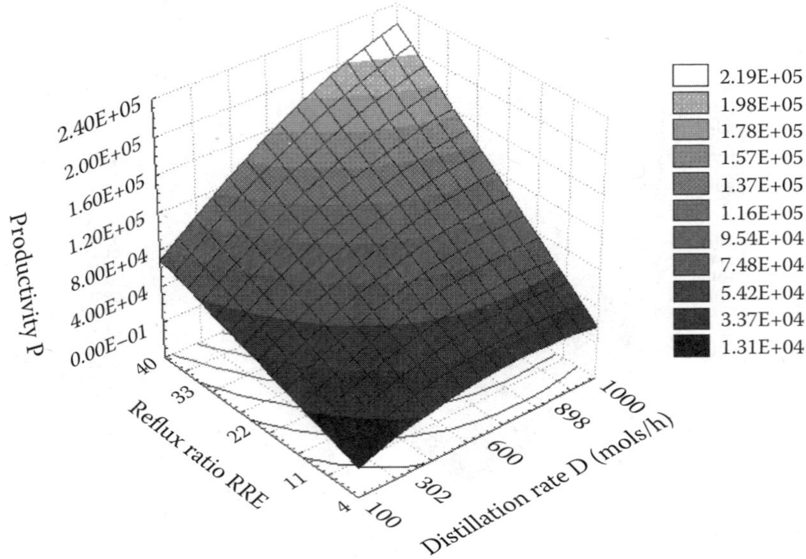

FIGURE 3.32 Response surface for productivity (number of stages = 15).

observed in the present case. This effect is possibly related to the interplay of the very low concentration of volatiles in the original aqueous solution and the dynamic behavior of a batch distillation column. It should be kept in mind that a larger number of stages corresponds to a higher total liquid hold-up inside the column trays, so that the retention time inside the equipment is larger and probably this effect can counteract the usual influence of the number of trays upon product purity.

For the productivity, the statistical analysis showed that the reflux ratio and the distillation rate were the significant independent variables, and the corresponding response surface is represented in Figure 3.32. As can be seen, the largest productivities were obtained for high reflux ratios (38–40) and high distillation rates (900–1100 mol/h). With this selection of operational conditions, a combination of higher purification factors with lower batch distillation times was accomplished. A recovery of volatile components close to 46% was also expected. Naturally other strategies for optimizing the process are also possible. In the present alternative we opted for emphasizing product purity and a short production period for the first cut. If emphasis is put on the volatiles recovery and a higher concentration of 2-methylbutanoic acid is admitted, the first cut can be postponed and other operational conditions can be tested.

On the basis of the preceding results a final simulation was conducted for the following conditions: D =1100 mol/h, r = 40, and number of stages = 10. The simulation results for the first cut, corresponding to the maximal purification factor, are shown in Table 3.7. The total recovery of volatiles was 44.2% (F_m = 2.7 ×10^5, cut time = 1.26 h, and P = 2.15 ×10^5). According to Table 3.7, the product has a high volatile concentration and low water content, much less than the 93% of the original mixture. In the case of esters, methyl isovalerate, ethyl isovalerate, and

TABLE 3.7
Concentration of the First Cut of Cashew Flavor Batch Distillation

Compound	*R* (%)	Concentration
Water	1.3	0.2769 g/g[a]
Methyl isovalerate	86.5	0.1805 g/g
Ethyl butanoate	83.1	0.1251 g/g
Hexanal	30.1	0.0886 g/g
Ethyl isovalerate	77.2	0.2970 g/g
2-methyl 2-pentenal	7.9	0.0318 g/g
2-butoxyethanol	0.0003	0.232 mg/kg
2-methylbutanoic acid	0.0002	0.152 mg/kg

[a]0.276999616.

ethyl butanoate, the recovery was larger than 70%. The distillation process also has been shown to be very efficient for reducing the content of 2-methylbutanoic acid, even below its threshold. According to the literature [64], the threshold for 2-methylbutanoic acid together with its isomer, 3-methylbutanoic acid, is 1.52 mg/kg. The threshold is defined as the lowest concentration in which an odor or flavor of a substance is capable of producing a sensation and being detected [65].

After the first cut other by-products, further cuts of lower purity can be distilled. In this case the main objective would be to concentrate the total amount of volatiles, because the purification in relation to the undesirable compound (2-methylbutanoic acid) becomes more difficult.

3.4 CONCLUSION

This chapter discussed the different types of distillation processes used in the food and beverage industries, describing the corresponding industrial equipment and their operation. The complex liquid mixtures, which very frequently occur in these industries, are multicomponent solutions containing a series of volatile compounds very important for the product sensorial quality and are often concentrated and purified by distillation procedures. Also discussed were methods for simulating different distillation techniques as well as methods for calculating and predicting the required physical–chemical properties that are now well developed, so that these mathematical tools can be a very powerful complement in the evaluation of actual separation processes and in the development of new ones. The combination of simulation studies in an extended range of constructive and operational conditions and selected experimental investigations for validation purposes allows process development and optimization with very high confidence and low cost. This surely is already contributing to improving product quality in food and beverage processing.

3.5 NOMENCLATURE

Symbol	Description	Units in SI system	Dimension in M, N, L, T, and θ
a_{mn}	Interaction parameter between the groups *m* and *n* in the residual term of UNIFAC model	K	θ
A_{ij}	Interaction parameter between components *i* and *j* for NRTL model	—	—
b_i	Bottom flow of component *i*	$Mol \cdot s^{-1}$	$N \cdot T^{-1}$
B	Total bottom product flow	$Mol \cdot s^{-1}$	$N \cdot T^{-1}$
d_i	Distillate flow of component *i*	$Mol \cdot s^{-1}$	$N \cdot T^{-1}$
D	Total distillate flow	$Mol \cdot s^{-1}$	$N \cdot T^{-1}$
$\exp \int_{P_i^{vap}}^{P} \frac{V_i^L}{RT} dP$	Poynting factor	—	—
f_i	Feed stream of component *i*	$Mol \cdot s^{-1}$	$N \cdot T^{-1}$
$\hat{f}_i^L$	Fugacity of component *i* in liquid phase	Pa	$M \cdot L^{-1} \cdot T^{-2}$
f_i^0	Standard state fugacity of component *i*	Pa	$M \cdot L^{-1} \cdot T^{-2}$
$\hat{f}_i^V$	Fugacity of component *i* in vapor phase	Pa	$M \cdot L^{-1} \cdot T^{-2}$
F	Purification factor	—	—
F_1, F_2, F_3	Discrepancy function	—	—
F_m	Maximal purification factor	—	—
h	Liquid enthalpy	$J \cdot mol^{-1}$	$(M \cdot L^2 \cdot T^{-2}) \cdot N^{-1}$
H	Vapor enthalpy	$J \cdot mol^{-1}$	$(M \cdot L^2 \cdot T^{-2}) \cdot N^{-1}$
HB	The total molar amount of liquid or liquid hold-up in the still	Moles	N
HC	Distillate amount of cashew juice	Moles	N
HD	Amount of distillate collected in the distillate receiver or distillate hold-up	Moles	N
H_D	Distillate enthalpy	$J \cdot mol^{-1}$	$(M \cdot L^2 \cdot T^{-2}) \cdot N^{-1}$
H_f	Enthalpy of feed stream	$J \cdot mol^{-1}$	$(M \cdot L^2 \cdot T^{-2}) \cdot N^{-1}$
HI	Original mixture amount of cashew juice	Moles	N
HN	Tray plus reflux drum liquid hold-up	Moles	N
K_i	Partition coefficient or volatility of component *i*	—	—
l_i	Liquid flow of component *i*	$Mol \cdot s^{-1}$	$N \cdot T^{-1}$
L	Total liquid flow	$Mol \cdot s^{-1}$	$N \cdot T^{-1}$
n	Stage number (1 to np^{+2})	—	—
nc	Total number of components in the mixture	—	—
np	Number of trays	—	—
P	Total pressure	Pa	$M \cdot L^{-1} \cdot T^{-2}$
P	Productivity	—	—

Symbol	Description	Units in SI units	Dimension in M, N, L, T, and θ
P_i^{vap}	Vapor pressure of component *i*	Pa	$M \cdot L^{-1} \cdot T^{-2}$
q_i	Surface area for component *i* at combinatorial term of UNIFAC model	—	—
Q_c	Condenser duty	$J \cdot s^{-1}$	$(M \cdot L^2 \cdot T^{-2}) \cdot T^{-1}$
Q_k	Group surface area parameter of UNIFAC model	—	—
Q_r	Reboiler duty	$J \cdot s^{-1}$	$(M \cdot L^2 \cdot T^{-2}) \cdot T^{-1}$
r	Reflux ratio	—	—
r_i	Van der Waals volume for component *i* at combinatorial term of UNIFAC model	—	—
R	Universal gas constant	$J \cdot Mol^{-1} \cdot K^{-1}$	$M \cdot L^2 \cdot T^{-2} \cdot N^{-1} \cdot \theta^{-1}$
R	Total recovery of the desired volatile compounds	—	—
R_k	Group volume of UNIFAC model	—	—
S^L	Liquid sidestream flow	$Mol \cdot s^{-1}$	$N \cdot T^{-1}$
S^V	Vapor sidestream flow	$Mol \cdot s^{-1}$	$N \cdot T^{-1}$
t	Batch time	s	T
t_{F_m}	Batch time of the maximal purification factor	s	T
T	Absolute temperature	K	θ
v_i	Vapor flow of component *i*	$Mol \cdot s^{-1}$	$N \cdot T^{-1}$
V	Vaporization rate or total vapor flow	$Mol \cdot s^{-1}$	$N \cdot T^{-1}$
V_i^L	Molar volume of liquid *i*	$m^3 \cdot mol^{-1}$	$L^3 \cdot N^{-1}$
x_{Di}	Molar fraction of component *i* in the distillate	—	—
x_i	Molar fraction of component *i* in liquid phase	—	—
x_m	Mole fraction of component *m* in the mixture in residual term of UNIFAC model	—	—
y^*	Molar fraction of vapor phase in equilibrium with liquid phase	—	—
y_i	Molar fraction of component *i* in vapor phase	—	—
γ_i	Activity coefficient of component *i*	—	—
α_{ij}	Relative volatility of the light component *i* in relation to the heavy *j*	—	—
α_{ij}	Non-random parameter for NRTL model	—	—
Γ_K	Residual activity coefficient of group *k* in the mixture	—	—
$\Gamma_k^{(i)}$	Residual activity coefficient of the group *k* in a solution containing only molecules of component *i*	—	—
γ_i^c	Combinational contribution to the activity coefficient in UNIFAC model	—	—

continued

Symbol	Description	Units in SI units	Dimension in M, N, L, T, and θ
γ_i^R	Residual contribution to the activity coefficient in UNIFAC model	—	—
ε	Maximum acceptable total error for discrepancy functions	—	—
η_i	Murphree efficiency of component i	—	—
θ_i	Area fraction for component i at combinatorial term of UNIFAC model	—	—
θ_m	Area fraction of group m at residual term of UNIFAC model	—	—
$\nu_k^{(i)}$	Number of groups of type k in molecular structure of component i	—	—
$\hat{\phi}_i$	Fugacity coefficient of component i	—	—
ϕ_i^S	Fugacity coefficient of pure component i at saturation	—	—
Φ_i	Volume fraction for component i at combinatorial term of UNIFAC model	—	—

3.6 REFERENCES

1. Scanavini, H. F. A., R. Ceriani, C. E. B. Cassini, F. M. Filho, A. J. M. Meirelles, and A. J. A. Meireles. 2005. Computer simulation of batch still for cachaça production. 4th EMPROMER, Mercosur Congress on Process Systems Engineering, Costa Verde, Rio de Janeiro, Brazil.
2. Ceriani, R., and A. J. M. Meirelles. 2007. Formation of trans PUFA during deodorization of canola oil: A study through computational simulation. *Chemical Engineering and Processing* 46:375–385.
3. Villadsen, J. and M. L. Michelsen. 1978. *Solution of differential equation models by polynomial approximation*. New Jersey: Prentice-Hall.
4. Luz, L. F. L., Jr., and M. R. Wolf-Maciel. 1995. Numerical methods for simulation of batch distillation column: Stiff system. *Brazilian Journal of Chemical Engineering* 12:111–118.
5. Reid, R. C., J. M. Prausnitz, and B. E. Poling. 1987. *The properties of gases and liquids*. 4th ed. New York: McGraw-Hill.
6. Oliveira, E. S. 2001. Características fermentativas, formação de compostos voláteis e qualidade da aguardente de cana obtida por linhagens de leveduras isoladas de destilarias artesanais. PhD Diss., Campinas State University, Brazil.
7. Boscolo, M., C. W. B. Bezerra, D. R. Cardoso, B. S. L. Neto, and D. W. Franco. 2000. Identification and dosage by Hrgc of minor alcohols and esters in Brazilian sugar-cane spirit. *Journal of the Brazilian Chemical Society* 11 (1): 86–90.
8. Fredenslund, A., J. Gmehling, and P. Rasmussen. 1977. *Vapor–liquid equilibria using UNIFAC: A group-contribution method*. New York: Elsevier Scientific Publishing.
9. Bondi, A. 1968. Physical properties of molecular crystals, liquids, and glasses. New York: Wiley.
10. Gmehling, J., J. Li, and M. Schiller. 1993. A modified UNIFAC model. 2. Present parameter matrix and results for different thermodynamic properties. *Industrial Engineering Chemical Research* 32:78–193.

11. Kikic, I., P. Alessi, P. Rasmussen, and A. Fredenslund. 1980. On the combinatorial part of the UNIFAC and UNIQUAC models. *Canadian Journal of Chemical Engineering* 58:253–258.
12. Fornari, T., S. Bottini, and E. A. Brignole. 1994. Application of UNIFAC to vegetable oils–alkane mixtures. *Journal of American Oil Chemist's Society* 71 (4): 391–395.
13. Gmehling, J., and U. Onken. 1981. *Vapor-liquid equilibrium data collection*. Frankfurt: Dechema.
14. Aspen Technology. 2003. Aspen Plus 12.1 User Guide. Cambridge, MA: MIT, Estados Unidos.
15. Madrera, R. R., D. B. Gomis, and J. J. M. Alonso. 2003. Influence of distillation system, oak wood type, and aging time on volatile compounds of cider brandy. *Journal of Agricultural and Food Chemistry* 51:5709–5714.
16. Hernández-Gómez, L. F., J. Úbeda-Iranzo, E. García-Romero, and A. Briones-Pérez. 2005. Comparative production of different melon distillates: Chemical and sensory analyses. *Food Chemistry* 90:115–125.
17. Lachenmeier, D. W., S. G. Walch, S. A. Padosch, and L. U. Kröner. 2006. Absinthe—A review. *Critical Reviews in Food Science and Nutrition* 46:365–377.
18. Silva, M. L., A. C. Macedo, and F. X. Malcata. 2000. Review: Steam distilled spirits from fermented grape pomace. *Food Science and Technology International* 6 (4): 285–300.
19. Ballabio, D., R. Kokkinofta, R. Todeschinic, and C. R. Theocharis. 2007. Characterization of the traditional Cypriot spirit zivania by means of counterpropagation artificial neural networks. *Chemometrics and Intelligent Laboratory Systems* 87 (1): 52–58.
20. Geroyiannaki, M., M. E. Komaitis, D. E. Stavrakas, M. Polysiou, P. E. Athanasopoulos, and M. Spanos. 2007. Evaluation of acetaldehyde and methanol in Greek traditional alcoholic beverages from varietal fermented grape pomaces (*Vitis vinifera* L.). *Food Control* 18 (8): 988–995.
21. Flouros, A. I., A. A. Apostolopoulou, P. G. Demertzis, and K. Akrida-Demertzi. 2003. Note: Influence of the packaging material on the major volatile compounds of tsipouro, a traditional Greek distillate. *Food Science and Technology International* 9 (5): 371–378.
22. Apostolopoulou, A. A., A. I. Flouros, P. G. Demertzis, and K. Akrida-Demertzi. 2005. Differences in concentration of principal volatile constituents in traditional Greek distillates. *Food Control* 16:157–164.
23. Léauté, R. 1990. Distillation in alembic. *American Journal of Enology and Viticulture* 41 (1): 90–103.
24. Osorio, D., J. R. Pérez-Correa, L. T. Biegler, and E. Agosin. 2005. Wine distillates: Practical operating recipe formulation for stills. *Journal of Agricultural and Food Chemistry* 53:6326–6331.
25. Nascimento, R. F, R. D. Cardoso, B. S. Lima Neto, D. W. Franco, and J. B. Faria. 1998. Influência do material do alambique na composição química das aguardentes de cana-de-açúcar. *Quimica Nova* 21 (6): 735–739.
26. Cardoso, D. R., L. G. Andrade Sobrinho, B. S. Lima-Neto, and D. W. Franco. 2004. A rapid and sensitive method for dimethylsulphide analysis in Brazilian sugar cane spirits and other distilled beverages. *Journal of the Brazilian Chemical Society* 15 (2): 277–281.
27. MAPA—Ministério da Agricultura, Pecuária e Abastecimento. Regulamento técnico para fixação dos padrões de identidade e qualidade para aguardente de cana e para cachaça. http://extranet.agricultura.gov.br/sislegis-consulta/consultarLegislacao.do?operacao=visualizar&id=12386. (accessed June 2007).
28. Dato, M. C. F., J. M. Pizauro Jr., and M. J. R. Mutton. 2005. Analysis of the secondary compounds produced by *Saccharomyces cerevisiae* and wild yeast strains during the production of "cachaça." *Brazilian Journal of Microbiology* 36 (1): 70–74.

29. Da Porto, C. 1998. Grappa and grape-spirit production. *Critical Reviews in Biotechnology* 18 (1): 13–24.
30. Da Porto, C., M. Longo, and A. Sensidoni. 1996. Effects of low pressure and a rectification column on the volatile composition of fermented grape distillate. *International Journal of Food Science and Technology* 31 (5): 403–410.
31. Peña y Lillo, M., E. Latrille, G. Casaubon, E. Agosin, E. Bordeu, and N. Martin. 2005. Comparison between odour and aroma profiles of Chilean Pisco spirit. *Food Quality and Preference* 16:59–70.
32. Pino, J. A. 2007. Characterization of rum using solid-phase microextraction with gas chromatography-mass spectrometry. *Food Chemistry* 104 (1): 421–428.
33. Da Porto, C., and S. Soldera. 2007. Behaviour of some volatile compounds during distillation of fermented marc exposed to the smoking process. *International Journal of Food Science and Technology* 43:495–500.
34. Frausto-Reyes, C., C. Medina-Gutiérrez, R. Sato-Berrú, and L. R. Sahagún. 2005. Qualitative study of ethanol content in tequilas by Raman spectroscopy and principal component analysis. *Spectrochimica Acta* Part A 61:2657–2662.
35. Mancilla-Margalli, N. A., and M. G. López. 2002. Generation of Maillard compounds from inulin during the thermal processing of *Agave tequilana* Weber var. Azul. *Journal of Agricultural and Food Chemistry* 50 (4): 806–812.
36. Peña-Alvarez, A., L. Díaz, A. Medina, C. Labastida, S. Capella, and L. E. Vera. 2004. Characterization of three agave species by gas chromatography and solid-phase microextraction–gas chromatography–mass spectrometry. *Journal of Chromatography A* 1027:131–136.
37. Legin, A., A. Rudnitskaya, B. Seleznev, and Y. Vlasov. 2005. Electronic tongue for quality assessment of ethanol, vodka and eau-de-vie. *Analytica Chimica Acta* 534:129–135.
38. Piggott, J. R., J. M. Conner, A. Paterson, and J. Clyne. 1993. Effects on Scotch whisky composition and flavour of maturation in oak casks with varying histories. *International Journal of Food Science and Technology* 28 (3): 303–318.
39. Gaiser, M., G. M. Bell, A. W. Lim, et al. 2002. Computer simulation of a continuous whisky still. *Journal of Food Engineering* 51:27–31.
40. Suomalainen, H., L. Nykanen, and K. Eriksson. 1974. Composition and consumption of alcoholic beverages—A review. *Journal of Enology and Viticulture* 25:179–187.
41. Cardoso, D. R., B. S. Lima-Neto, D. W. Franco, and R. F. Nascimento. 2003. Influência do material do destilador na composição química das aguardentes de cana—Parte II. *Química Nova* 26 (2): 165–169.
42. Boza, Y., and J. Horii. 1998. Influência da destilação sobre a composição e a qualidade sensorial da aguardente de cana-de-açúcar. *Ciência e Tecnologia de Alimentos* 18 (4): 391–396.
43. Boza, Y., and J. Horii. 2000. Influência do grau alcoólico e da acidez do destilado sobre o teor de cobre na aguardente. *Ciência e Tecnologia de Alimentos* 20 (3): 279–284.
44. Bruno, S. N. F., D. S. Vaitsman, C. N. Kunigami, and M. G. Brasil. 2007. Influence of the distillation processes from Rio de Janeiro in the ethyl carbamate formation in Brazilian sugar cane spirits. *Food Chemistry* 104 (4): 1345–1352.
45. Osório, D., R. Pérez-Correa, A. Belancic, and E. Agosin. 2004. Rigorous dynamic modeling and simulation of wine distillations. *Food Control* 15:515–521.
46. Decloux, M., and J. Coustel. 2005. Simulation of a neutral spirit production plant using beer distillation. *International Sugar Journal* 107 (1283): 628–643.
47. Karlsson, H. O. E., and G. Trägårdh. 1997. Aroma recovery during beverage processing. *Journal of Food Engineering* 34:159–178.
48. Yanniotis, S., K. Tsitziloni, G. Dendrinos, and A. Mallouchos. 2007. Aroma recovery by combining distillation with absorption. *Journal of Food Engineering* 78:882–887.

49. Lora, J., M. I. Iborra, R. Perez, and I. Carbonell. 1992. Simulation of the distillation to concentrate wine aromas. *Revista Española de Ciencia y Tecnologia de Alimentos* 32 (6): 621–633.
50. Haypek, E., L. H. M. Silva, E. A. C. Batista, D. S. Marques, M. A. A. Meireles, and A. J. A. Meirelles. 2000. Recovery of aroma compounds from orange essential oil. *Brazilian Journal of Chemical Engineering* 17:705–712.
51. Lurton, L., G. Snakkers, C. Roulland, B. Galy, and A.Versavaud. 1995. Influence of the fermentation yeast strain on the composition of wine spirits. *Journal of the Science of Food and Agriculture* 67 (4): 485–491.
52. Nykänen, L. 1986. Formation and occurrence of flavour compounds in wine and distilled alcoholic beverages. *American Journal of Enology and Viticulture* 37 (1): 84–96.
53. Nascimento, R. F., J. C. Marques, B. S. L. Neto, D. Keukeleire, and D. W. Franco. 1997. Qualitative and quantitative high-performance liquid chromatographic analysis of aldehydes in Brazilian sugar cane spirits and other distilled alcoholic beverages. *Journal of Chromatography A* 782:13–23.
54. Lamiable, D., G. Hoizey, H. Marty, and R. Vistelle. 2004. Acute methanol intoxication. *EMC-Toxicologie Pathologie* 1:7–12.
55. Zocca, F., G. Lomolino, A. Curioni, P. Spettoli, and A. Lante. 2007. Detection of pectinmethylesterase activity in presence of methanol during grape pomace storage. *Food Chemistry* 102:59–65.
56. Ceriani, R., and A. J. A. Meirelles. 2004. Simulation of batch physical refining and deodorization processes. *Journal of the American Oil Chemists' Society* 81:305–312.
57. Yokoya, F. 1995. *Fabricação de aguardente de cana*. Campinas: Série Fermentações Industriais 2, 92p.
58. Maia, A. B. R. 1994. Componentes secundários da aguardente. STAB, *Açúcar Álcool e Subprodutos* 12 (6): 29–34.
59. Dalmolin, I., E. Skovroinski, A. Biasi, M. L. Corazza, C. Dariva, and V. J. Oliveira. 2006. Solubility of carbon dioxide in binary and ternary mixtures with ethanol and water. *Fluid Phase Equilibria* 245:193–200.
60. Garruti, D. S., M. R. B. Franco, M. A. A. P. Da Silva, N. S. Janzantti, and G. L. Alves. 2003. Evaluation of volatile flavour compounds from cashew apple (*Anacardium occidentale* L) juice by Osme gas chromatography/olfactometry technique. *Journal of the Science of Food and Agriculture* 83:1455–1462.
61. Batista, E., and A. J. A. Meirelles. 1997. Simulation and thermal integration SRV in an extractive distillation column. *Journal of Chemical Engineering of Japan* 30:45–51.
62. Ceriani, R., and A. J. M. Meirelles. 2006. Simulation of physical refiners for edible oil deacidification. *Journal of Food Engineering* 76 (3): 261–271.
63. Rodrigues, M. I., and A. F. Iemma. 2005. Planejamento de experimentos e otimização de processos—Uma estratégia seqüencial de planejamentos. Campinas, SP: Ed. Casa do Pão.
64. Peinado, R. A., J. C. Mauricio, M. Medina, and J. J. Moreno. 2004. Effect of *Schizosaccharomyces pombe* on aromatic compounds in dry sherry wines containing high levels of gluconic acid. *Journal of Agricultural and Food Chemistry* 52:4529–4534.
65. Leffingwell, J. C. 2002. Chirality & odour perception. Leffingwell & Associates. http://www.leffingwell.com/index.htm (accessed October, 2005).

4 Low-Pressure Solvent Extraction (Solid–Liquid Extraction, Microwave Assisted, and Ultrasound Assisted) from Condimentary Plants

Thais M. Takeuchi, Camila G. Pereira, Mara E. M. Braga, Mário R. Maróstica, Jr., Patrícia F. Leal, and M. Angela A. Meireles

CONTENTS

4.1 Introduction 138
4.2 Fundamentals of Low-Pressure Extraction: Solid–Liquid, Microwave Assisted, and Ultrasound Assisted 139
4.2.1 Solid–Liquid Extraction 140
4.2.1.1 Mass Transfer: Balance Equations and Kinetics 142
4.2.1.2 Extractors and Operation Methods 144
4.2.1.3 Single Stage Extraction 144
4.2.1.4 Crosscurrent Extraction 147
4.2.1.5 Countercurrent Extraction 148
4.2.1.6 Thermodynamic: Phase Equilibrium 150
4.2.2 Microwave-Assisted Extraction 151
4.2.2.1 Important Factors in MAE 152
4.2.2.2 Heat Transfer: Balance Equations and Kinetics 154
4.2.3 Ultrasound-Assisted Extraction 154
4.2.3.1 Heat and Mass Transfer: Balance Equations and Kinetics ... 156
4.3 State of the Art—Mini-Review of the Literature 158
4.3.1 Solid–Liquid Extraction 158
4.3.1.1 Equipment and Process Variables 159

4.3.2 Microwave-Assisted Extraction 168
4.3.3 Ultrasound-Assisted Extraction 171
4.4 Obtaining High Quality Bioactive Compounds Using GRAS Solvents 185
4.4.1 Antioxidants 185
4.4.1.1 Solvent System 185
4.4.1.2 Temperature and Time 187
4.4.1.3 Solvent-to-Feed Ratio 188
4.4.1.4 Particle Size 189
4.4.2 Pigments 189
4.4.2.1 Solvent System 189
4.4.2.2 S/F Ratio 192
4.4.2.3 Temperature and Time 192
4.4.3 Phenolic Compounds 193
4.4.3.1 Solvent System 194
4.4.3.2 S/F Ratio 195
4.4.3.3 Temperature and Time 195
4.4.3.4 Particle Size 196
4.4.3.5 Effect of pH on Extraction Yield 196
4.5 Economical Evaluation of a Solvent Extraction Process: Sage and Macela Cases 197
4.5.1 Definition of the Solvent Extraction Process 197
4.5.2 Properties of Vegetable Materials 198
4.5.3 Equipment Sizing 198
4.5.4 Purchase Cost Estimations for Major Equipment 201
4.5.5 Capital Cost Estimation (FCI)–Lang Factor Technique (F_{Lang}) 202
4.5.6 Raw Material Costs (C_{RM}) Estimation 202
4.5.6.1 Sage Case 202
4.5.6.2 Macela Case 203
4.5.7 Costs of Utilities (C_{UT}) Estimation 203
4.5.7.1 Sage Case 204
4.5.7.2 Macela Case 205
4.5.8 Cost of Operational Labor (C_{OL}) Estimation 205
4.5.9 COM Estimation 206
4.6 Nomenclature 207
4.7 Acknowledgments 210
4.8 References 211

4.1 INTRODUCTION

Solid–liquid extraction finds numerous applications in the food industry; probably the best known example of which is the production of fixed oils (vegetable oils) from oleaginous plants. In this chapter we will discuss the process related to obtaining bioactive compounds by extraction from aromatic, condimentary, and medicinal plants. The fundamentals of solid–liquid, microwave-assisted, and ultrasound-assisted extractions will be presented. Solid–liquid extraction is discussed both ways: using analytical and graphical solutions. The review of the recent literature focuses entirely on these plants. The process parameters that must be controlled in obtaining antioxidants, pigments, and phenolic compounds are lengthily discussed, and as in Chapter 2, a methodology to estimate the cost of manufacturing (COM) is discussed using as examples the production of macela (*Achyrocline satureioides*) and sage (*Salvia officinalis*) extracts.

4.2 FUNDAMENTALS OF LOW-PRESSURE EXTRACTION: SOLID–LIQUID, MICROWAVE ASSISTED, AND ULTRASOUND ASSISTED

Condimentary plants used in daily food are known to act as an antioxidant, because of some of their pigments and polyphenolic compounds. However, this potential may be limited by industrial processes because of thermal and light degradation and low recovery of target compounds. Polyphenols, a group of chemical compounds characterized by the presence of the functional group phenol in their molecules, and widely found in every plant organ, are produced by the plant's secondary metabolism. Many antioxidants are included in this group. These compounds can be found as monomers or in polymerized forms [1] and have been classified for nutritional purposes into extractable (low and intermediate molecular weight) and nonextractable types (high molecular weight, insoluble in common organic solvents; Bravo et al. 1998, cited by Andersen et al. [2]).

Plant materials have a complex nature, and the extraction of the substances they contain is influenced by process conditions such as temperature, mechanical action (such as pressure and shaking), extraction solvent type, and solubilization of the target compounds, which effectively depend on the solvent polarity and physical conditions. In the case of antioxidants in spices such as rosemary and sage, the main polar compounds are carnosol, rosmarinic, and carnosolic acids, the latter being the most water-soluble; oregano also contains rosmarinic acid, several flavonoids, and water-extractable substances, which were proved to present high antioxidant activity [3]. For rosemary, sage, and oregano, the target antioxidant compounds are located on the leaves' surface, whereas for other species these compounds are located inside the seeds and roots. Therefore, the choice of the solvent should be combined with a pretreatment of the raw material or even with another extraction methodology, in order to reach the target compounds inside the particle and promote a high process yield.

Target compounds in the plants may vary in functionality or content, according to the degree of plant ripeness, cultivar, and edaphoclimatic conditions. Besides this natural variability, some changes may happen during the industrialization process. The chemical composition of raw material may be altered by pre- or posttransformation processes such as drying, sterilization, irradiation, extraction, evaporation, or other high temperature processes and by final storage conditions such as air or low temperature. On the other hand, coextracted substances, which have no antioxidant activity of their own, may increase the antioxidant potential of the extract [4]; among these substances (synergists) are the polyvalent organic acids, amino acids, phospholipids (lecithin), and chelating agents. As an example, some flavonoids (phenolic antioxidants), present as esters or glycosides, are partially hydrolyzed during boiling; for mushroom juice, the boiling process reduces the antiradical activity, but the boiling does not affect the activity of onions and yellow bell peppers [5]. The most common natural antioxidants, such as tocopherols, ascorbic acid, and β-carotene, were studied in model systems, but there are different unknown antioxidants from spices and essential oils. To study these antioxidants, it is necessary to monitor the retention of the target compounds throughout processing.

Therefore, the target compound and the nature of the raw material to be extracted must be known, in order to select the best process and technology, to permit a high recovery, and to guarantee the stability of the chemical compounds. Most of the extraction techniques consist of the manipulation of the physical properties of the solvent to reduce the surface tension, increase the solute's solubility, and promote a higher diffusion rate, and sometimes, a change in solvent polarity. The extraction techniques using solvents at low pressures may represent an appropriate choice for the processing of many systems. Considering the characteristics of the system, as described in the next section, the technique chosen might be the simple solid–liquid extraction, microwave-assisted extraction (MAE), or ultrasound-assisted extraction. For condimentary plants, the solvents used for extraction are mainly water and organic solvents. Besides its physical–chemical capacity in dissolving the target compound(s) and its toxicity to human beings and to the environment, the choice of the solvent should also be considered.

Various methods have been applied to extract bioactive compounds from condimentary plants. Among the extraction techniques at low pressure with solvent, there are conventional techniques, such as the solid–liquid extraction, and novel techniques, such as microwave- and ultrasound-assisted extraction. In the food industry, solid–liquid extraction has been used to recover several products, such as sugar, tea, coffee, vegetable oils, and functional compounds. This extraction technique is based on mass transfer and practical equilibrium occurrence, with or without heat application. New techniques, such as microwave- or ultrasound-assisted extractions, also have important applications. The fundamentals of these processes are different from those of conventional methods since the extraction occurs because of changes in the cell structure caused by electromagnetic or sound waves. This chapter is concerned with the fundamentals and applications of each of these low-pressure techniques.

4.2.1 Solid–Liquid Extraction

Solid–liquid extraction or solvent extraction occurs with the selective dissolution of one or more solutes from a solid matrix by a liquid solvent. This unit operation is also designated lixiviation, leaching, decoction, or elution. In fact, the terminology can be specific for a given type of extraction. For instance, lixiviation is used when the aim is to obtain alkali compounds, decoction is used when the solvent is at its boiling temperature, and elution is used when the soluble solids are at the surface of the solid matrix. Independently of the name used, this technique is one of the oldest unit operations in the chemical industry.

In the food industry, the process can be used either to obtain important substances like carotenoids or flavonoids or to remove some inconvenient compounds like contaminants or toxins. In all these cases, the extraction occurs as a result of the effect of the solvent selectivity on the soluble solute. From the industrial point of view, there are some factors that should be evaluated before the process initialization, because they influence the rate of extraction:

- *Preparation of the solid*: In food materials, the cell structure is an important factor that needs to be considered. Although the solute can be on the surface of the cell, in most of the cases it is stored in intracellular spaces, capillaries, or cell structures. This way, the success of the solvent extraction strongly depends on the solid condition. One of the pretreatment steps that must be considered is the comminuting or grinding of the raw material. Grinding before solvent extraction promotes an increase of the contact area between the solvent and the solid matrix. Besides that, in most cases this step enhances the contact between solvent and solute by breaking the cell structures. As an example, in industry, coffee grains are broken in three to five pieces. In other cases, maintaining the cell structure is required, as in the extraction of sugar from beets. In this case, the beet is cut in fine pieces, but the cell structure is preserved to avoid the extraction of undesirable compounds [6].
- *Diffusion rate*: Because of the complexity of the cell structure and the existence of porous and different compartments in the cell, the diffusivity of biological materials has a specific denomination: effective diffusivity. The effective diffusivity also depends on the composition and on the position of the solute in the solid material.
- *Temperature*: Normally, elevated temperature is attractive in terms of extraction process enhancement. Higher temperatures promote an increase of the solute's solubility in the solvent, increasing the solute diffusion rate into the solvent bulk, leading to a higher mass transfer rate. However, in the food industry, the use of elevated temperatures can generate undesirable reactions such as the degradation of thermolabile compounds. For instance, in coffee processing, elevated temperatures can cause hydrolysis.
- *Solvent choice*: The selection of the extraction solvent is based on several factors, such as its physicochemical properties, cost, and toxicity. The choice of the solvent should consider characteristics such as selectivity and capability of dissolving the solute, as well as its interfacial tension, viscosity, stability, reactivity, toxicity, and cost. Because of the toxicity of some organic solvents, there are some restrictions to their use in the food industry. In terms of human consumption, the presence of some solvents, such as acetone, ethanol, ethyl acetate, 1-propanol, 2-propanol, and propyl acetate are acceptable in small residual percentages, according to good manufacturing practice (GMP). These solvents are classified as Class 3 by the Food and Drug Administration (FDA). Others (Class 2), such as acetonitrile, chloroform, hexane, methanol, toluene, ethylmethylketone, and dichloromethane, can be used under specific conditions and present limitations concerning pharmaceutical and food products because of their inherent toxicity. The PDEs (permissible daily exposures) of the solvents in Class 2 are given to the nearest 0.1 mg/d, and concentration limits vary from 50 to 3880 ppm, depending on the organic solvent used [7]. The solvents grouped in Class 1 should not be employed in manufacturing because of their unacceptable toxicity or their deleterious environmental effects. This class includes benzene, carbon tetrachloride, 1,2-dichloroethane, 1,1-dichloroethane, and 1,1,1-trichloroethane.

- *Solid material humidity*: The water in the solid material can compete with the extraction solvent for the solute's dissolution, affecting the mass transfer. On the other hand, this humidity is necessary to permit the transport of the solute, as in coffee extraction. Nevertheless, in most of the cases the material is dried under conditions that do not cause degradation of the compounds.

4.2.1.1 Mass Transfer: Balance Equations and Kinetics

The solvent extraction is characterized by the extraction of the soluble material inside the solid matrix using a specific solvent. The extraction mechanism can be described in the following steps: First, the solvent must be transferred onto the solid surface and covered or wrapped. After that, the solvent penetrates into the solid matrix by diffusion (effective). The solute is dissolved until a concentration limited by the nature of the solid as well as the pretreatment to which it was subjected is reached. It is important to notice that the solute plus solvent mixture forms a very diluted solution; thus true equilibrium is never reached in any practical application. The solution containing the solute diffuses to the surface by effective diffusion. At the end, the solution is transferred from the surface to the bulk solution by natural or forced convection.

The rate of dissolution of a solute in the solvent of extraction is controlled by the rate of mass transfer of the solute from the solid matrix to the liquid. The transfer of the solute inside the solid particle occurs because of the concentration gradient in the solid–liquid interface, and it can be characterized by the effective diffusion. The equation that describes this phenomenon is based on the Fick's law and is given by

$$\frac{N_C}{A_T} = -D_{BC}\frac{dC_C}{dz}, \tag{4.1}$$

where N_C is the rate of dissolution of the solute C in the solution (kg/sec), A_T is the area of the solid–liquid interface (m^2), D_{BC} is the diffusivity of the solute in the solvent/inert solid (m^2/sec), C_C is the concentration of solute C in the solution (kg/m^3), and z is the distance inside the porous of the solid matrix (m).

The value of the diffusion coefficient (D_{BC}) usually is in the range 10^{-9}–10^{-10} m^2/sec; it is important and a necessary parameter in the diffusion model [8]. The mass transport in solid foods is strongly dependent on the size, shape, and porous presence. In these cases, the diffusion is expressed in terms of effective diffusivity D_{CBeff}, defined as follows:

$$D_{CBeff} = \frac{\varepsilon}{\tau} D_{BC} \tag{4.2}$$

where ε is the void fraction space or porosity of the solid, and τ is the tortuosity of the pores.

This coefficient is influenced by the nature of the solid matrix as well as by the pretreatment to which it was subjected. Values of the diffusion coefficient of various food solutes are listed in Table 4.1.

TABLE 4.1
Diffusion Coefficients and Effective Diffusion Coefficients of Food Solutes in Diverse Matrices

Food material	Solute	Solvent	Temperature (K)	D_{CB} ($\times 10^{10}$ m²/s)
Molecular diffusion coefficients				D_{CB}
Dilute solution[a]	Sucrose	Water	298	5.4
Gelatin gel[a]	Sucrose	Water	278	0.1–0.2
Dilute solution[a]	Lactose	Water	298	4.9
Effective diffusion coefficients				D_{CBeff}
Sugar cane (across grain)[a]	Sucrose	Water	348	5.1
Sugar cane (with grain)[a]	Sucrose	Water	348	3.0
Sugar beets[a]	Sucrose	Water	297	1.6–2.5
Grape pomace[b]	Polyphenols	Water	313	0.065–0.130
			323	0.010–0.211
		Ethanol	313	0.01–0.076
			323	0.011–0.048
Coffee beans[c]	Caffeine	Water	383	3.209
Milled Berries[d]	Anthocyanins	Ethanol (67%)	313	1.23
Geranium macrorhizum L.[e]	Tannins	Water	293	1.89
Nicotiana tabacum L.[e]	Crude extract	Water	293	0.395

[a] Aguilera and Stanley 1999, cited by Aguilera [9]; [b] Guerrero et al. [10]; [c] Espinoza-Perez et al. [11]; [d] Cacece and Mazza [12]; [e] Simeonov et al. [51].

On the surface of the solid particle, the transfer of the solute occurs with simultaneous molecular and turbulent transport. In this step, the rate of mass transfer can be expressed by the following equation:

$$N_C = \frac{VdC_C}{dt} = A_T k_L (C_{CS} - C_C), \tag{4.3}$$

where k_L is the mass transfer coefficient in m/sec, C_{CS} is the reference concentration of the solute C in the solution in kg/m³, and C_C is the concentration of the solute C in the solution at time t in kg/m³.

Integrating from $t = 0$ and $C_C = C_{C0}$ to $t = t$ and $C_C = C_C$, we obtain the following:

$$\int_{C_{C0}}^{C_C} \frac{dC_C}{C_{CS} - C_C} = \frac{Ak_L}{V} \int_{t=0}^{t} dt \tag{4.4}$$

$$\frac{C_{CS} - C_C}{C_{CS} - C_{C0}} = e^{-\left(k_L A / V\right)t}. \tag{4.5}$$

If pure solvent is used initially, $C_{C0} = 0$, and then

$$1 - \frac{C_C}{C_{CS}} = e^{-\left(k_L A/V\right)t} \tag{4.6}$$

$$C_C = C_{CS}\left(1 - e^{-\left(k_L A/V\right)t}\right). \tag{4.7}$$

4.2.1.2 Extractors and Operation Methods

The solvent extraction process can be carried on in batch, semi-batch (unsteady-state) or continuous (steady-state) modes. The choice of the equipment type depends on the material to be processed, the compound(s) to be extracted, and the cost. The main extractors in the food industry are shown in Table 4.2.

The methods of calculation are very similar to the one used in liquid–liquid extraction (see Chapter 5). The process can occur in single or multiple stages and it can be countercurrent or crosscurrent.

4.2.1.3 Single Stage Extraction

Consider the single stage (real) solvent extraction process shown in Figure 4.1, for which the feed, or stream F, consists of both insoluble (fiber or inert material) and soluble solids (C). Considering a single stage operation and that the extraction solvent used is pure, the stream S is constituted of pure compound B (extraction solvent). The extraction produces two outflows: the extract (the stream E), which is constituted of a relatively large amount of solvent (B) containing dissolved solute (C), and the residue (the stream R) containing the insoluble solid or inert matrix (A) and the retained solution (B + C).

From Figure 4.1 the overall mass balance and the mass balances of solute C and solvent B are, respectively, described by the following equations:

$$F + S = M = R + E \tag{4.8}$$

$$x_{iF}.F + y_{iS}.S = x_{iR}R + x_{iE}E, \tag{4.9}$$

where M is the mixture point in the single stage; $x_{iF}, x_{iS}, x_{iR}, and\, x_{iE}$ are the mass fractions of compound i in the feed, solvent, residue, and extract, respectively.

The retention index (R^*) is defined as the ratio of the mass of solution retained in the solid matrix to the mass of inert solid (A):

$$R^* = \frac{mass\ of\ adhered\ solution}{mass\ of\ inert\ solid} \tag{4.10}$$

TABLE 4.2
Characteristics and Applications of Solvent Extraction Systems

Operation	Working principle	Extraction system	Field application	Examples
Batch	Immersion extraction	Stirred vessel	Pharmacy	Alkaloids
	Static bed percolation	Single-stage percolator	Spices	Pepper
	Static bed crosscurrent percolation	Multistage percolator		
Quasi-continuous	Stationary bed, countercurrent percolation	Multistage percolator battery	Instant material, sugar	Instant coffee, sugar from beets
Continuous	Rotating cell, countercurrent percolation	Rotocel	Sugar, vegetable oil	Soybean oil
	Rotating bed, countercurrent percolation, stationary sieve tray bottom	Carrousel	Vegetable oil, spices, instant material	Soybean oil, paprika, pepper, hop
	Stationary bed, countercurrent percolation, rotating feed/discharging locations	Stationary basket	Vegetable oil, spices	Wheat germ, paprika
	Horizontal moving bed, countercurrent percolation	Sieve tray belt; sliding cell	Sugar	Sugar from beets/cane
	Horizontal moving bed, co-/countercurrent percolation	Crown loop extractor	Vegetable oil, sugar	Sugar cane/soybean oil
	Vertical moving bed, co-/countercurrent percolation	Basket elevator	Vegetable oil	Flaked oil seeds
	Moving bed, countercurrent immersion	Screw conveyer	Sugar, vegetable oil	Sugar beets, soybean oil

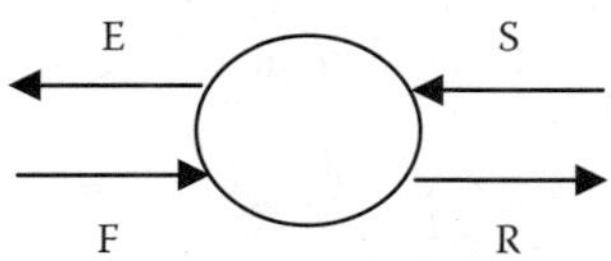

FIGURE 4.1 A single-stage extraction process.

$$R^* = \frac{x_{BR} + x_{CR}}{x_{AR}} = \frac{1 - x_{AR}}{x_{AR}}, \tag{4.11}$$

where x_{AR}, x_{BR}, *and* x_{CR} are the mass fractions of A, B, and C in the residue stream. Reorganizing:

$$x_{AR} = \frac{1}{R^* + 1} \tag{4.12}$$

$$x_{BR} + x_{CR} = \frac{R^*}{R^* + 1}. \tag{4.13}$$

The mass balance for the inert solid present in the solid matrix is as follows:

$$x_{AF}.F = x_{AR}R. \tag{4.14}$$

Then, substituting Equation 4.12 in Equation 4.14, the inert solid stream can be expressed as follows:

$$R = x_{AF}.F(R^* + 1). \tag{4.15}$$

In some cases for which the amount of retained solution is independent of the extract solution concentration, the retention index is constant. In other words, the solution retained within the solid matrix has a composition equal to that of the extract solution. In this case, there is no preferential adsorption; therefore,

$$X_{CR} = y_{CE}, \quad \text{and} \quad X_{BR} = y_{BE}, \tag{4.16}$$

where X_{CR} and X_{BR} are the mass ratio of C and B, respectively, in the retained solution expressed in inert solid free-basis (*A*).

X_{CR} can be calculated by the following:

$$X_{CR} = \frac{x_{CR}}{1 - x_{AR}}. \tag{4.17}$$

Using Equation 4.16, the practical equilibrium can be represented by the following:

$$x_{CR} = (1 - x_{AR})y_{CE}. \tag{4.18}$$

The analysis can also be made by a graphic method. The mixture point (*M*) represents the mixture stage in the equipment. The composition in this point is determined by the following:

$$x_{iF}.F + y_i.S = x_{iM}M. \tag{4.19}$$

For the solvent B and solute C, the mass fraction can be determined by Equations 4.20 and 4.21:

$$x_{BM} = \frac{x_{BF}.F + y_{BS}.S}{M} \tag{4.20}$$

$$x_{CM} = \frac{x_{CF}.F + y_{CS}.S}{M}. \tag{4.21}$$

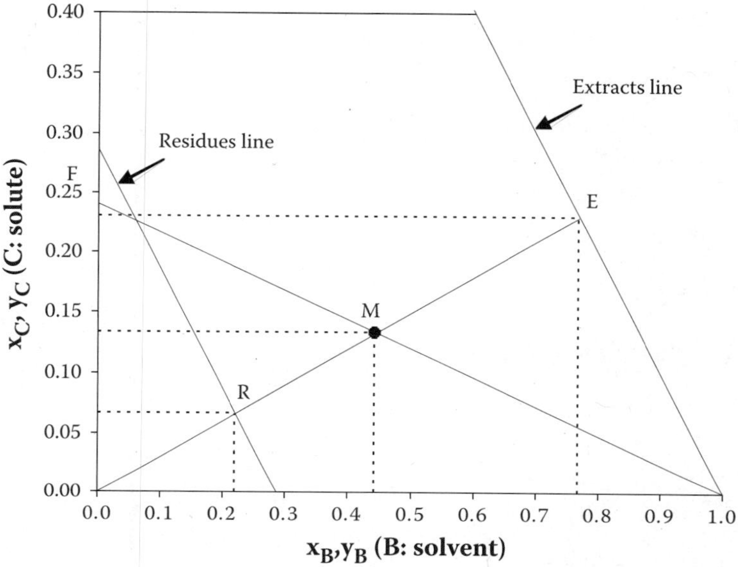

FIGURE 4.2 Graphical solution of single-stage solvent extraction.

Taking into account that the feed is solvent free and that the solvent is pure, Equations 4.20 and 4.21 can be written as follows:

$$x_{BM} = \frac{x_{BF} \cdot F}{M} \tag{4.22}$$

$$x_{CM} = \frac{S}{M}. \tag{4.23}$$

Graphically, the point (M) is represented by the intersection of the overall mass balance and practical equilibrium lines (Equations 4.8 and 4.18, respectively).

The composition of the residue can be determined by the intersection of the residue line (using Equations 4.12 and 4.13) and the practical equilibrium lines, as represented in Figure 4.2.

4.2.1.4 Crosscurrent Extraction

In this type of extraction, both the feed, at stage 1, and the residue, at the following stages, are treated in successive stages with fresh solvent. Figure 4.3 shows a crosscurrent process in two stages.

For the first stage, the solution is the same as that of the single stage extraction. For the second stage, the feed is R_1, containing the inert solid A, the unsolubilized solute C, and the retained solvent B. The overall mass balance for stage 2

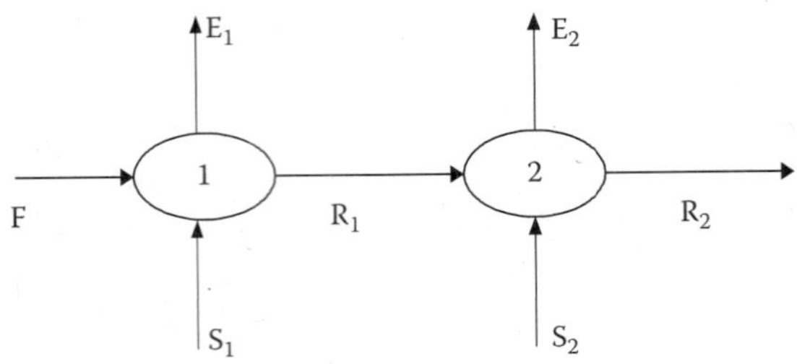

FIGURE 4.3 A crosscurrent extraction in two stages.

is given by Equation 4.24 and the mass balance for the inert solid is given by Equation 4.25:

$$R_1 + S_2 = M_2 = R_2 + E_2 \tag{4.24}$$

$$x_{AR1}.R_1 = x_{AR2}R_2. \tag{4.25}$$

If the retention index is constant, then

$$x_{AR} = \frac{1}{R^* + 1}. \tag{4.26}$$

The mixture point for the second stage is represented by Equations 4.27 and 4.28:

$$R_1 + S_2 = M_2 \tag{4.27}$$

$$x_{iF}.R_1 + y_i.S_2 = x_{iM2}M_2. \tag{4.28}$$

For solute C and solvent B, the mass fraction can be determined by the following:

$$x_{BM2} = \frac{x_{BR1}.R_1 + y_{BS2}.S_2}{M_2} \tag{4.29}$$

$$x_{CM2} = \frac{x_{CR1}.R_1 + y_{CS2}.S_2}{M_2}. \tag{4.30}$$

Similarly to the single stage extraction calculation methodology, the graphic method can be applied as shown in Figure 4.4.

4.2.1.5 Countercurrent Extraction

This operation is characterized by the enrichment of the extract solution. Both the entrance of the feed and the exit of the final extracts solution take place in the first stage (stage 1), and both the entrance of the fresh solvent and exit of the final residue take place in the last stage (stage *N* of Figure 4.5). This way, only one flow of solvent

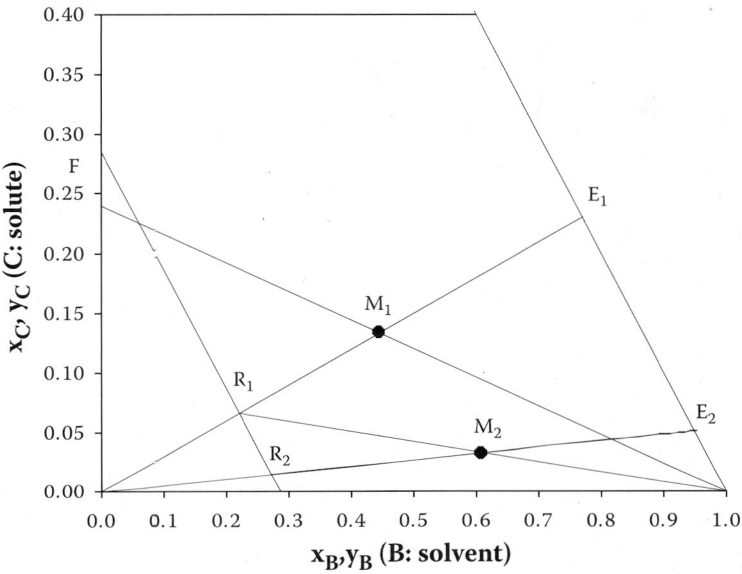

FIGURE 4.4 Graphical solution of crosscurrent extraction.

is used, and the extract solution obtained in a stage works as the extraction solvent in the next stage, as represented in Figure 4.5.

The overall mass balance for stages 1 through N is given by Equation 4.31:

$$F + E_{N+1} = R_N + E_1\,. \tag{4.31}$$

For each stage, the mass balance can be represented as follows:

Stage	Overall balance	Flow in–flow out	
1	$F + E_2 = R_1 + E_1$	$F - E_1 = R_1 - E_2 = \Delta$	(4.32)
2	$R_1 + E_3 = R_2 + E_2$	$R_1 - E_2 = R_2 - E_3 = \Delta$	(4.33)
3	$R_2 + E_4 = R_3 + E_3$	$R_2 - E_3 = R_3 - E_4 = \Delta$	(4.34)
N	$R_{N-1} + E_{N+1} = R_N + E_N$	$R_{N-1} - E_{N-1} = R_N - E_{N+1} = \Delta$	(4.35)

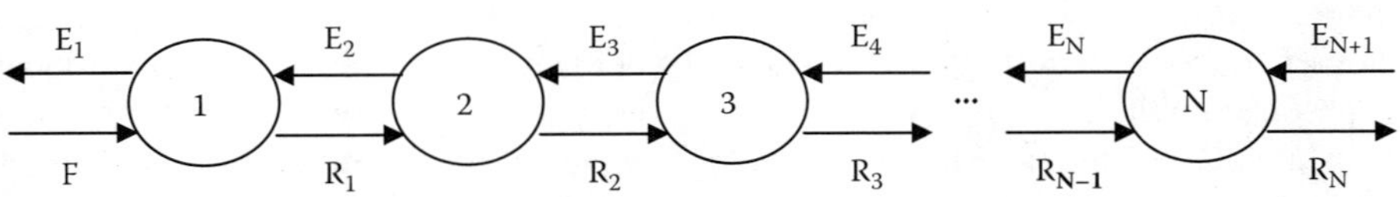

FIGURE 4.5 A countercurrent extraction process with N stages.

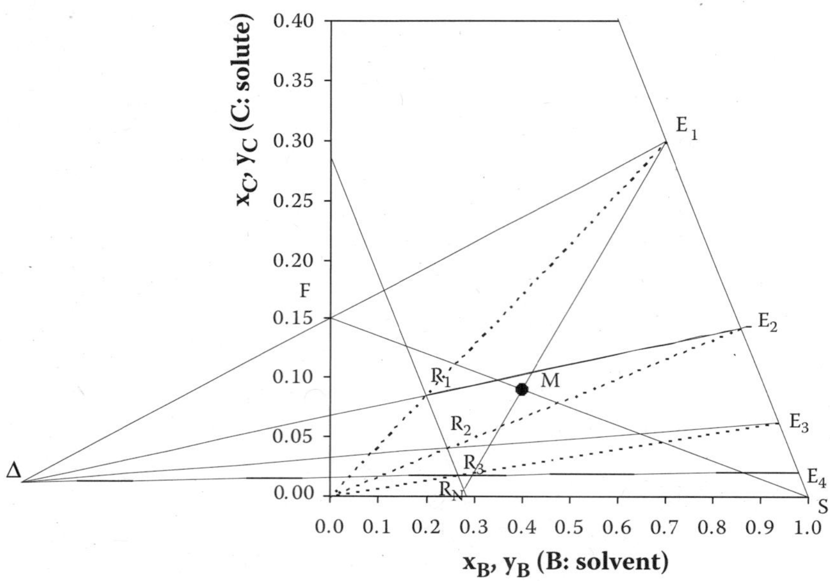

FIGURE 4.6 Graphical solution of countercurrent extraction.

The mass balance for solute C is given by Equations 4.36 and 4.37:

$$x_{CEN}.R_N - y_{CEN+1}E_{N+1} = x_{C\Delta}\Delta \quad \text{with } N \geq 1 \tag{4.36}$$

$$y_{CEN+1} = \frac{x_{CEN}.R_N - x_{C\Delta}\Delta}{E_{N+1}} \quad \text{with } N \geq 1. \tag{4.37}$$

Graphically, the solution considers the Δ-point, as can be observed in Figure 4.6.

4.2.1.6 Thermodynamic: Phase Equilibrium

The solvent extraction in the food industry is very complex because soluble material can be a complex mixture. Although the methodology of calculus is similar to the methodology in the liquid–liquid extraction, the true equilibrium in the system cannot be observed. In general, this unit operation is described empirically. In fact, the equilibrium depends not only on physicochemical conditions like temperature, pressure, and physical properties of solvent, but also on the physical conditions of the contact between the solvent and the solid matrix, such as contact time, particle size, solute mass/solid matrix mass, solute mass/solvent mass, and solvent/solid matrix interactions. Accordingly, in solvent extraction, the phase equilibrium relations are not related to true equilibrium and should be defined as practical, real, or operational equilibrium relations.

In spite of the many factors affecting the equilibrium in a solid–liquid extraction, the solute solubility is characterized by the influence of its activity coefficient, which varies with the temperature and composition of the solution, according to Equation 4.38

$$\ln \bar{x}_i = \frac{\Delta H_{fus}}{RT}\left(1 - \frac{T_m}{T}\right) - \ln \gamma_i \quad \text{for } T \le T_{m,} \tag{4.38}$$

where $\bar{x}_i$ is the molar fraction of the solute dissolved in the solvent phase at saturation, ΔH_{fus} is the molar heat of fusion (J/mol), R is the universal gas constant (J/mol·K), T_m is the melting point (K), T is the absolute temperature (K), and γ_i is the activity coefficient.

According to this expression, the solute's solubility depends on its own properties (molar heat of fusion and melting point) and on a property of the mixture (activity coefficient).

4.2.2 Microwave-Assisted Extraction

Microwaves are nonionizing electromagnetic energy with a frequency from 0.3 to 300 GHz. This energy is transmitted as waves, which can penetrate in biomaterials and interact with polar molecules inside the materials, such as water, to generate heat. MAE is a process that uses the effect of microwaves to extract biological materials. MAE has been considered an important alternative to low-pressure extraction because of its advantages: lower extraction time, lower solvent usage, selectivity, and volumetric heating and controllable heating process. Usually, domestic and industrial microwave equipment operates at 2.45 GHz, but sometimes other frequencies may be found in the United States (0.915 GHz) and Europe (0.896 GHz) [16].

Materials are classified according to their ability to absorb the microwave energy: materials like metals are conductors, and their surfaces reflect the microwaves; transparent materials, such as plastics, are insulators and are used to support the material to be heated; and materials that absorb the microwave energy, which, therefore, are easily heated, such as polar liquids, are named dielectrics (Microwave Power in Industry 1984, cited by Haque [17]).

The physical principle of this technique is based on the ability of polar chemical compounds to absorb microwave energy according to its nature, mainly the dielectric constant. This absorbed energy is proportional to the medium dielectric constant, resulting in dipole rotation in an electric field and migration of ionic species. The ionic migration generates heat as a result of the resistance of the medium to the ion flow, causing collisions between molecules because the direction of ions changes as many times as the field changes the sign. Rotation movements of the polar molecules occur while these molecules are trying to line up with the electric field, with consequent multiple collisions that generate energy and increase the medium temperature [18, 19]. The electrical component of the waves changes 4.9×10^9 times per second and the frequency of 2.45 GHz corresponds to a wavelength of 12.2 cm and energy of 0.94 J/mol [20]. Therefore, a higher dielectric constant leads to a higher absorbed energy by the molecules, promoting a faster solvent heating and extraction at higher temperatures, as from 423 to 463 K. However, other solvents with low dielectric constants are also used, and in these cases the matrix is heated and the microwave heating leads to the rupture of cell walls by expansion, promoting the delivery of the target compounds into a cooler solvent; this technique is used for the extraction of thermally labile compounds of low polarity [19, 21].

Although the microwaves penetration depth depends on the dielectric constant of target compounds, the loss factor of the matrix is also important and it is related to the transparency to microwaves and the ability to dissipate the absorbed energy. These properties depend on the moisture content, the temperature of the solid, and the frequency of the electrical field. In general, a lower loss factor and frequency promote deeper penetration. These properties (dielectric constant, loss factor, and penetration depth) were measured for some foods and materials and are listed in the literature [22].

Different from solvent extraction, MAE is improved by the presence of water. Indeed, the water contained in the solid matrix is responsible for the absorption of microwave energy. Therefore, the material undergoes internal superheating. As a result, the cell structure is disrupted, and the flow out of the chemical constituents from the solid matrix is facilitated. The phenomenology of this process is quite different from the conventional solvent extraction where the solvent diffuses in the solid matrix and dissolves the compounds.

Microwaves cause molecular motion by migration of ions and rotation of dipoles, and by solvent heating and improves its penetration. The effect of microwaves in the material is strongly dependent on the dielectric susceptibility of both the solvent and the solid matrix. The dielectric constant (ε') and dielectric loss factor (ε'') are values that express the dielectric response of materials in an applied microwave field. The dielectric constant measures the ability of the material to store microwave energy, i.e., it quantifies the capacity of the material to be polarized. In contrast, the dielectric loss factor measures the ability of a material to dissipate the stored energy into heat.

Because of this, the solvent chosen should have a high dielectric constant. Polar molecules and ionic solutions (usually acids) have a permanent dipole moment and will strongly absorb microwave energy. Solvents like ethanol, methanol, and water are sufficiently polar to be heated by microwave energy, whereas apolar solvents with low dielectric constants like hexane and toluene are not good solvents for MAE. A mixture of solvents might be considered. Although not indicated to be used in this process, hexane, when mixed with acetone, presented properties favorable to MAE.

The main solvents used in MAE are presented in Table 4.3. The higher the dielectric constant, the more energy is absorbed by the molecules and the faster the solvent heating occurs. Actually, the heat generation in the material depends not only on the dielectric constant, but also is in part dependent on the dissipation factor (ln), which is the ratio of the material dielectric loss to its dielectric costant:

$$\ln\delta = \frac{\varepsilon''}{\varepsilon'} \tag{4.39}$$

4.2.2.1 Important Factors in MAE

The great difference between MAE and convectional solvent extraction is the effect of the microwave on both the solvent and the cell structure. To optimize MAE

TABLE 4.3
Physical Constants and Dissipation Factors for Some Solvents Used in MAE

Solvent	Dielectric constant, ε' [a]	Dipole moment[b]	Dissipation factor, $\tan \delta$ (10^{-4})	Boiling point[c] (K)	Closed-vessel temperature[d] K
Hexane	1.89	0.1	—	342	—
Toluene	2.4	0.36	—	384	—
2-Propanol	19.9	1.66	6700	355	418.2
Acetone	20.7	2.69	—	329	437.2
Ethanol	24.3	1.96	2500	351	437.2
Methanol	32.6	2.87	6400	338	424.2
Acetonitrile	37.5	—	—	355	467.2
Water	78.3	2.3	1570	373	—
Hexane: Acetone (1:1)	—	—	—	325	429.2

[a] at 293 K; [b] at 298 K; [c] at 101.4 kPa: [d] at 1207 kPa.

methodology, special attention must be dedicated to factors such as temperature, pressure, solvent, volume, extraction time, and solid matrix:

- Temperature: Generally, higher temperature promotes elevated yields as a result of an increased diffusivity of the solvent into the solid material and an increase of the compound's desorption from active sites of the matrix. However, it may cause degradation in thermolabile substances.
- Pressure: It is an important factor in MAE procedures performed in closed systems. Because of the MAE dependence on temperature and its relation to the pressure of the system, the evaluation of these variables makes it possible to optimize the extraction.
- Solvent: As mentioned earlier, the choice of the solvent to be applied in MAE procedures should consider not only the related solubility of the compounds to be extracted, but also the dielectric properties that will determine the absorption of the microwave energy.
- Volume: The minimum volume of solvent necessary to immerse the solid matrix should be determined.
- Extraction time: The duration of MAE processes is very short compared to conventional extraction methodologies. For foods, the extraction times vary from 3 to 40 min, depending on the solid matrix and compounds extracted. For thermolabile compounds, a long extraction period can result in degradation.
- Solid matrix: As discussed earlier, the water content in the solid matrix is of great importance. A high dipole moment allows a strong absorption of the microwave energy.

4.2.2.2 Heat Transfer: Balance Equations and Kinetics

The general heat transfer equation can be used to estimate the heat transfer in a material that receives microwave energy. Considering a transient heat transfer in an infinite slab, for one-dimensional flux, the corresponding equation is as follows:

$$\frac{\partial^2 T}{\partial x^2} + \frac{q'''}{k} = \frac{1}{\alpha}\frac{\partial T}{\partial t}, \tag{4.40}$$

where q''' is the heat generation, k is the thermal conductivity, and α is the thermal diffusivity.

The term related to heat generation is equivalent to the power dissipation of the electromagnetic field. Microwave energy in itself is not thermal energy. The heating is a result of the electromagnetic energy generated with the dielectric properties of the material combined with the electromagnetic field applied. Assuming that the electric field is uniform throughout the volume, the conversion of the microwave energy to heat can be approximated by the expression

$$P_D = 2\pi E^2 f'\varepsilon'', \tag{4.41}$$

where P_D is the power dissipation (W/cm³), E is electrical field strength (V/cm), and f' is frequency (Hz).

The energy absorption inside the solid material causes an electric field that decreases with the distance from the material surface. The penetration depth (D_p) is the distance from the material surface where the absorbed electric field (e) is reduced to $1/e$ of the electric field at the surface. The penetration depth is inversely proportional to the frequency and the dielectric properties of the material, as shown by the expression [23]

$$D_P = \frac{c}{2\pi f' \sqrt{2\varepsilon'}\left[\sqrt{1+\tan^2\delta} - 1\right]^{1/2}}, \tag{4.42}$$

where c is the speed of light (m/sec).

If the penetration depth of the microwave is much less than the thickness of the material, only the surface is heated, and the rest of the material is heated by conduction.

4.2.3 Ultrasound-Assisted Extraction

Ultrasound has been used in different operations in chemical engineering, such as waste-water treatment, drying, sonochemistry, and extraction. In the food and pharmaceutical sectors, ultrasound has been employed to extract bioactive compounds such as flavonoids [24], essential oils and alkaloids [25], polysaccharides [26], esters and steroids [27], and others substances [28–30].

Sound waves are mechanic vibrations applied to the solid, liquid, or gas with frequencies higher than 20 kHz. Sound waves are intrinsically different from electromagnetic waves. Although the latter can pass through a vacuum, sound waves need the material presence to travel. Ultrasonic waves are elastic waves that have a frequency above the threshold of human hearing, approximately 20 kHz. They are characterized by their frequency and wavelength, and the mathematical product of these two parameters results in the wave speed through the medium. Amplitude or intensity of waves is also an important parameter and is used to classify the industrial application: low-intensity ultrasound (LIU) with less than 1 W/cm^2, and high-intensity ultrasound (HIU) with 10–1000 W/cm^2. HIU is applied at higher frequencies (up to 2.5 MHz) to modify processes or products by physical disruption of tissues, and LIU is used to monitor the quality of processes and products [31]. Waves propagate through the solid–liquid (as in food) media, moving in the longitudinal and perpendicular (as shear waves) directions of particles or close to the surface of the particle; for gases and liquids only longitudinal waves can propagate.

The effect of the sound waves in matter is the expansion and compression cycles. The expansion can create bubbles in a liquid and produce negative pressure that can reach a high local pressure of up to 50 MPa, intense heating with hot spots around 5000 K, and lifetimes of a few microseconds [32], whereas the collapse of the bubbles formed can cause cavitation. At constant ultrasound intensity, dynamic equilibrium is established between the forming and the collapsing bubbles. The collapse of cavitation bubbles near cell walls produces cell disruption. As a result, there is an enhanced solvent penetration into the cells and an intensification of the mass transfer.

These fast changes in pressure and temperature (cavitation), which cause shear disruption and thinning of cell membranes, are the phenomena that make ultrasound applicable to alter the medium state by the sonochemistry. The cavitation and consequently the mass transfer and the extraction rate, which are influenced by temperature, hydrostatic pressure, irradiation frequency, acoustic power, and ultrasonic intensity, are as important as the choice of solvent and sample preparation [33].

Another effect of this type of waves on the solid structure is that the ultrasound can facilitate swelling and hydration, causing an enlargement in the pores of the cell wall. This effect will improve the diffusion process and increase mass transfer.

Generally, the largest sonochemical effects are observed at lower temperatures, when the majority of the bubble contents is in the gas. With a decrease in the vapor pressure of the mixture, there is an increase of the implosion intensity, thus increasing the ultrasonic energy produced upon cavitation. Although the cavities are more easily formed with a solvent that has a high vapor pressure, low viscosity, and low surface tension, the cavitation intensity increases for solvents with low vapor pressure, high viscosity, and surface tension, as observed experimentally by some authors (Mason et al. 1987, cited by Thompson and Doraiswamy [33]). The ultrasonic frequency affects the cavitation process, altering the bubble critical size, with lower frequencies, producing more violent cavitation [34].

For solid–liquid systems, the most important effect of ultrasound is the mechanical effect attributed to cavitation symmetry. The hot spots are generated in the fluid by the bubble symmetrical collapse, and shock waves are produced creating a microscopic turbulence in the interfacial films that surround the solid particles. This phenomenon is named microstreaming, and results in an increased diffusion rate and enhanced mass transfer across the film [19, 32, 35, 36].

The usage of this technique is very common in wastewater treatment, and some toxicity effects can be found for systems that contain phenol composition under some conditions. Some authors studied the phenol oxidation in a NaCl medium with a high frequency (500 kHz), using a reactor at 300 K [37]. They concluded that it was necessary to optimize the ultrasound extraction with respect to frequency and time, in order to avoid the degradation of the compounds and the production of toxic substances in the medium [38].

The benefits of this method are the possibility to operate with many samples in the same equipment and short extraction times applied when compared with conventional solvent extraction. A reduction in the maceration time from 8 h to 15 min has been reported in the extraction of the alkaloid reserpine from *Rauwolfia serpentina* when this technology was applied, resulting in the same extraction yield (Bose and Sen 1961, cited by Albu et al. [39]). In another study, ultrasonic extraction promoted a yield 50% greater in 30 min than conventional extraction of berberine in 24 h (Guo et al. 1997, cited by Vinatoru et al. [40]).

As in other solvent extraction processes, the temperature and the polarity of the solvent influence the extraction procedure using ultrasound. Besides, other important factors govern the ultrasound-assisted leaching, such as frequency and sonication time.

The ultrasound frequency exerts significant influences on the extraction yield and kinetics. However, these influences are dependent on the structure of the material and on the compound to be extracted. The acceleration of the kinetics and of the extraction is obtained, probably as a result of the increase of the intraparticular diffusion of the solute that results from the disruption of the cell walls. However, in some cases, lower frequencies are required in the process to avoid degradation of bioactive compounds.

4.2.3.1 Heat and Mass Transfer: Balance Equations and Kinetics

The effects produced by ultrasound in a mass transfer process have direct relation with the intensity applied. High-intensity ultrasound enhances the mass transfer process by affecting internal and external resistance of the wall to this phenomenon. Ultrasonic intensity (UI) can be determined by calorimetric methods and can be calculated by the expression

$$UI = \frac{P_o}{A_b} = \frac{\left(\frac{dT}{dt}\right) C_p m}{A_b}, \tag{4.43}$$

where P_o is the average power, expressed in function of dT/dt that is the variation of temperature T with the time t, C_p is the heat capacity of the liquid, m is the liquid mass added into the vessel, and A_b is the area of the reaction vessel's bottom.

The few existing studies of the mechanism of extraction using ultrasound have focused on two phenomena: desorption and solid–liquid extraction. Although there are analogies between both, the interaction between solute and solvent is not the same. In the former, the action results from physical adsorption, and in the latter, there are the effects of physical and chemical adsorption, as presented in Section 4.2.1. Although both are facilitated by the effect of the sound waves in the cell structure, the mass transfer model for each extraction mechanism is different. Ji et al. [41] proposed a mass transfer model for the leaching process of geniposide from gardenia fruits using ultrasound. The model was based on the intra-particle diffusion and external mass transfer. The model applied to gardenia fruit assumed spherical particles with uniform size and density, and the instantaneous desorption of geniposide (an iridoid glycosides present in the fruit) migrating to the outer surface of the fruits into the solution adhered to the surface of the particles. The model developed is expressed by Equations 4.44 through 4.47.

1. For mass transfer in the aqueous solution,

$$\frac{dC_g}{dt} = 3\frac{k_f}{R}\frac{m}{\rho V}\left(C_{\xi=1} - C_g\right) \qquad \text{where } \xi = \frac{r}{R}, \tag{4.44}$$

where C_g is the concentration of the solute (geniposide) in the solution (mg/cm^3), t is the process time, k_f is the external mass transfer coefficient (cm/sec), R is the radius of the fruit (cm), m is the weight of the fruit, ρ is the density of the fruit (g/cm^3), V is the volume of the solution (cm^3), and $C_{\xi=1}$ is the concentration of the solute (geniposide) in the solution on the external surface of the fruit (mg/cm^3).

2. For mass transfer within the particles,

$$\frac{\partial q}{\partial t} = \frac{D_e}{R^2}\frac{1}{\xi}\frac{\partial}{\partial \xi}\left[\xi^2\left(\frac{\partial q}{\partial \xi}\right)\right], \tag{4.45}$$

where q is the remainder of the solute (geniposide) in the fruit (mg/g) and D_e is the apparent intraparticle diffusion coefficient (cm²/sec).

3. The boundary conditions

$$\frac{k_f}{R}\frac{\left(C_{\xi=1} - C_g\right)}{\rho} = \frac{D_e}{R^2}\left(\frac{\partial q}{\partial \xi}\right)_{\xi=1}. \tag{4.46}$$

4. The initial conditions are as follows: at $t = 0 \rightarrow C_g = 0$ and $q = q_0$.
5. The equilibrium equation:

$$q_{\xi=1} = \frac{KQC_{\xi=1}}{1 + KC_{\xi=1}}, \tag{4.47}$$

where K is the adsorption equilibrium constant (cm^3/mg), and Q is the adsorption capacity parameter in the Langmuir equation (mg/g).

4.3 STATE OF THE ART—MINI-REVIEW OF THE LITERATURE

4.3.1 Solid–Liquid Extraction

To obtain a high-performance extraction or a high yield of target compounds in a short process time, it is necessary to choose a selective solvent with a high solubility of the target compounds [42], and then the main factor affecting the extraction process is *solvent properties*. Related to this factor, the *viscosity of the solvent* and its *flow rate* are also important: the solvent viscosity should be sufficiently low for the liquid to go through the solid particles bed (when a packed bed is used); and higher flow rates reduce the boundary layer of concentrated solute at the particles' surface, increasing the extraction rate. Table 4.4 shows the solvent characteristics that should be considered for the extraction from natural matrices, according to Gertenbach [42]. The *solid-to-solvent ratio* and the *particle size* are other factors that influence the mass transfer. Smaller particles present higher ratios of surface area to volume, which enhance the contact between solvent and solid matrix and diminish the diffusion path of the particle to reach the surface, resulting in a faster extraction rate. On the other hand, the usage of higher liquid-to-solid ratios provides

TABLE 4.4
Solvent Characteristics for Natural Products Extraction

Characteristic	Effect in the process
Selectivity	Solvent selectivity guarantees the extract purity and solubilizes the target compounds
Compatibility with solute	The solvent should not react with the target compounds
Chemical and thermal stability	The stability of the solvent at operating extraction conditions must be assured not to alter the final extract
Low viscosity	To keep the extraction rate higher, lower viscosity is necessary to increase the diffusion coefficient
Ease of recovery	Economic aspects must be considered, and lower boiling point solvents are easily recovered and reused
Low flammability	According to the process needs and safety aspects, flammable solvents must be avoided
Low toxicity	Natural products require the absence of solvent traces and toxicity, besides the worker exposition
Regulatory issues	According to the pharmaceutical and food industries, environmental regulations should be considered so as to avoid process irregularities
Consumer acceptance	The consumer should accept the solvent usage
Low cost	Economic aspects can contribute to the final product quality

an increase in the gradient concentration of the target compounds between the particles' surfaces and their interior parts. Other factors influence the solid–liquid extraction: temperature, preparation of the solid, and humidity of the material, as presented in Section 4.1.

4.3.1.1 Equipment and Process Variables

The classification of equipment can be based on the solid–solvent contact, and generally two methods are used for the extraction from solid natural matrices: 1) slurry extraction and 2) percolation extraction.

For the slurry or dispersed-solids extraction, the solid particles are suspended in the solvent; Figure 4.7 shows an example of an extraction tank used for this technique. This method is used for finely ground raw materials, when the characteristics of the solids allow the solvent flow through the bed. The extractor consists of one or more tanks for solid–liquid mixtures and a separation step such as filtration or centrifugation to recover solvent from the extracted biomass.

For the percolation extraction, the solvent flows through a fixed bed of the solid matrix, as shown in Figure 4.8. The solvent, which may or may not fill the empty spaces between the particles, flows through the bed, taking the extract away from the particle surface. The separation between the liquid and the solids is the main advantage of this method, reducing the step of grinding the raw material into fine particles.

Some authors, such as Hu et al. [43], describe systems that use a simple extractor in batch equipment (not commercial), with a solvent mixture to obtain a bamboo leaf extract (BLE) which contains chlorogenic acid, caffeic acid, and luteolin 7-glucoside, a mixture of compounds with scavenger and antioxidant activities. Bamboo leaf powder (20–40 mesh, using a solid-to-liquid (S/L) ratio of 1:15, w/v) is kept under reflux for 1.5 h, using a hydroethanolic mixture (30%), at the mixture's boiling temperature, followed by filtration and solvent vaporization; the recovered BLE yield reaches 6%. Luteolin 7-glucoside reaches 2.8% (w/w) and chlorogenic acid 1.6%

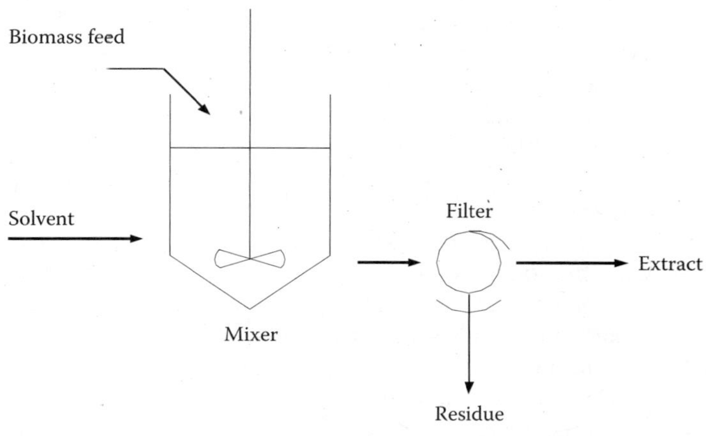

FIGURE 4.7 One-stage mixed tank for slurry extraction with filtration.

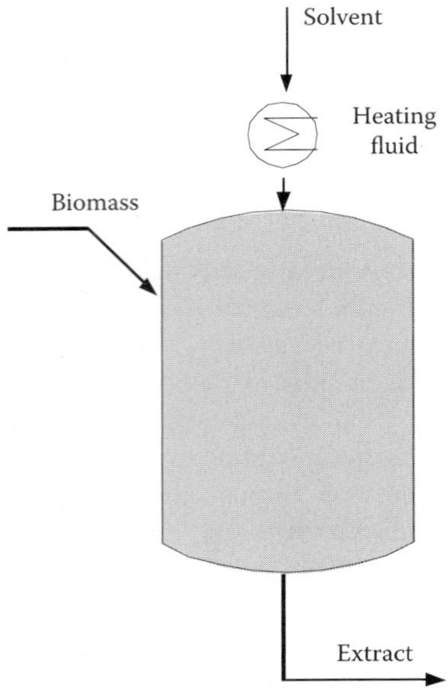

FIGURE 4.8 One-stage percolation extraction.

(w/w), quantified by HPLC and with a concentration-dependent scavenging activity measured by the DPPH (2,2-diphenil-1-picrylhydrazyl) radical.

The shiitake mushroom, widely consumed as food, has a high nutritional value and additional positive effects on health, acting as an antitumor agent and as a cholesterol-reducing agent, because it contains an alkaloid called eritadenine. The mushroom extraction is performed by methanol 80% for 3 h under reflux, using a S/L ratio of 1:20. This process was compared to methanol extraction preceded by enzymatic pretreatment (acetate buffer, pH 4.8) and followed by enzymatic hydrolysis (pH 6.0); the eritadenine was quantified by HPLC. Although the enzymatic pretreatment improved the eritadenine extraction, the difference between this process and pure methanol extraction was not statistically significant ($p > 0.05$) [44].

Methanol extraction is a very common extraction technique used for natural compounds, but generally organic solvents and water also promote the coextraction of undesired compounds. Therefore, some variations of these solvents, such as the mixture of solvents resulting in acidified or alkaline mixtures, or other solutions that may be used in raw material pretreatment or during the extraction process, have been used to improve their selectivity and the solubility of the target compounds. For example, the piperine (an alkaloid) was extracted from black pepper (*Piper nigrum*) using two hydrotrophic solutions as solvent: aromatic sulfonates and glycol sulfate substances. Hydrotrophic substances solubilize hydrophobic compounds in aqueous solutions, which present a remarkable property of disrupting the lamellar crystalline structure of surfactants in aqueous solutions, producing a continuous isotropic liquid solubility region. The authors used sodium butyl monoglycol sulfate (NaBMGS) and other hydrotropes and compared them to surfactants like sodium lauryl sulfate (SLS) and cetyltrimetylammonium bromide (CTAB) in a concentration of 0.5 mol/dm^3, at

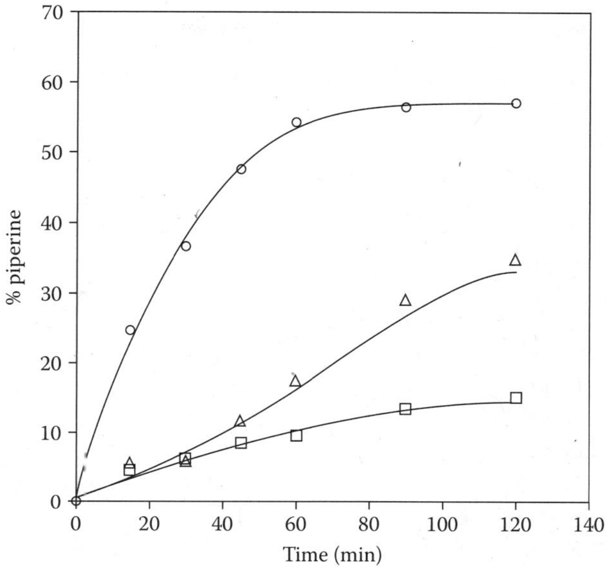

FIGURE 4.9 Extraction of piperine with surfactants SLS and CTAB (concentration = 0.5 mol/dm^3, temperature = 300 K, solid loading = 10% w/v, speed of agitation = 1100 rpm): □, SLS; Δ, CTAB; ○ NaNBBS. (Reprinted from Raman, G., and V. G. Gaikar, *Indust. Engineering Chem. Res.*, 41, 2966–2976, 2002. With permission from American Chemical Society.)

300 K in 10% (w/v) of solid (pepper fruits). The assays were performed in a fully baffled borosilicate cylindrical glass vessel (9 × 7 cm) equipped with six bladed turbine impellers, with agitation of 1100 rpm for 2 h. The hydrotropically extracted piperine (quantified by HPLC) had a higher purity than the one obtained by Soxhlet extraction [45]. Figure 4.9 shows that the piperine extraction with the NaBMGS solution is greater than that with surfactants, indicating that the hydrotropic solubilization mechanism probably involves adsorption of the hydrotrope on plant cells, penetration into the matrix, and finally, the solubilization of the target compound [45].

Low-pressure extraction through percolation was studied for rosemary (*Rosmarinus officinalis*) fresh leaves, a known spice and aromatic species from the Mediterranean region. Superheated water between 398 and 448 K was used for 30 min, at a flow rate of 2 cm^3/min and approximately 2 MPa, with a solid-to-solvent ratio of 1:15. The profile of the extract composition was compared to the profile obtained by steam distillation. For all extracted compounds, and particularly for the oxygenated compounds, their contents in the superheated water extracts were higher. Comparatively, the extraction with carbon dioxide (liquid or supercritical fluid) requires a higher solvent-to-raw-material ratio in order to extract oxygenated aroma and flavor compounds. Moreover, rosemary often needs to be previously dried for an effective extraction by CO_2 because the presence of water tends to get in the way of the desired compound solubilization. Superheated water extraction can also be considered a selective method, when compared to CO_2 extraction, because it does not extract monoterpenes, higher hydrocarbons, and lipids. In addition, it can be less expensive and does not require drying of raw material such as the rosemary system [46].

The same system was studied by Ibañez et al. [47], who performed subcritical water extraction and studied the temperature effect on the extracts composition and antioxidant activity. The maximum yield (48.6%) was obtained at the highest temperature (473 K), and although the composition profiles were different for the different temperatures tested, antioxidant activities were similar for all extracts.

For both extraction methodologies (slurry and percolation extraction) there is equipment that operates in batch or continuous modes. Because the solid matrix can be treated as a pseudo binary system containing the solute (a mixture of substances) and the inert solid (a mixture of cellulose, starch, lignins, and so on), true equilibrium between the solid matrix and the solvent is never achieved. Instead, a diluted solution is obtained and a practical equilibrium is defined as discussed in Section 4.2.1. For batch operation, the solid must be in contact with the liquid until the practical equilibrium concentrations are attained, and for continuous operation, the solvent and the solids are continuously fed to the equipment, with the recovery of extract and the removal of the residue. The process may be operated in several stages and in countercurrent, in which the fresh solvent is fed to contact the extracted biomass, and fresh biomass is fed to contact the most concentrated solvent. Figure 4.5 shows a countercurrent operation scheme, which has the main advantage of obtaining the highest rate of target compound recovery.

Commercial batch equipment for slurry extraction is generally inexpensive to install. However, a single stage produces a diluted extract; thus, multistage operation, where several tanks are assembled together (Figure 4.10), is preferred. A filtration or centrifugation step is added to remove the residue and separate the residual solvent.

The same strategies used for slurry extractions can be used for percolation extractions, using several stages and countercurrent operation. To reduce the amount of required solvent, it recirculates through the bed multiple times, until the practical equilibrium concentrations are reached. The extract is then removed, and the second charge of solvent is added into the system. These cycles of fresh solvent are

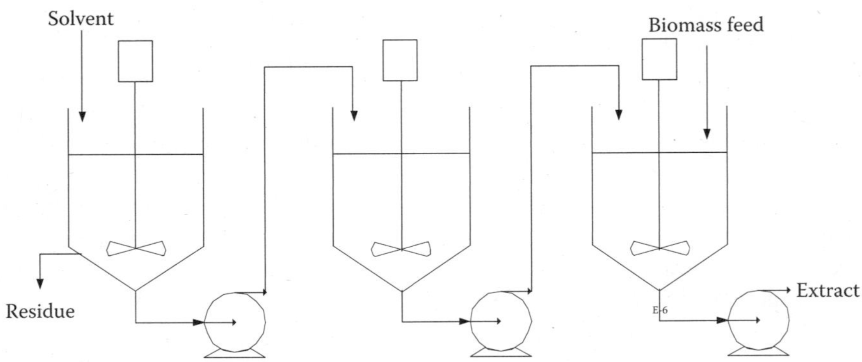

FIGURE 4.10 Countercurrent slurry extraction.

repeated until maximum recovery is attained, and after the extraction, the liquid is separated from the residue. For percolation, a uniform solvent flow that depends on bed porosity and adequate particle size to promote an acceptable extraction rate is required [42].

Some variation of this percolation process can be also obtained by operating at higher temperatures and/or pressures. An increase in temperature during the extraction changes the properties of the solvent and enhances mass transfer efficiency. Percolation extraction with increased temperature has mainly been used to obtain extracts from plants with high-molecular-weight compounds (such as oleoresins), using organic solvents. Generally, a Soxhlet apparatus, which is a laboratory scale piece of equipment that works at solvent boiling temperature, is used. Solvents used in this technique vary according to the target compounds to be extracted. Literature shows some data for Soxhlet extraction from spices, like oregano (*Oregano vulgare* L.), sage (*Salvia fruticosa*), and summer savory (*Satureja hortensis*). Exarchou et al. [48] studied the antioxidant activity and phenolic composition of extracts obtained from those plants in a Soxhlet apparatus for 6 h, using ethanol and acetone as solvents. Ethanol promoted a higher extraction yield for all tested raw materials, but acetone promoted higher total phenol contents and lower antioxidant activities by the DPPH method, which cannot be explained by the total phenol contents because they are not directly related. Therefore, other extracted compounds may have contributed to the antioxidant activity.

A heated system may be obtained by a steam jacket or by a heated solvent feed (Figure 4.11). A solid–liquid caffeine extraction from tea waste (50 g) was performed using a percolation extractor including three and five extractors each with a 500-cm^3 volume, connected in series, with steam jacket heating. The experiments were done at isothermal conditions for water and chloroform solvents, at 293 and 370 K, respectively, and a volumetric flow rate of 0.5 L/h. The highest cumulative extraction degree

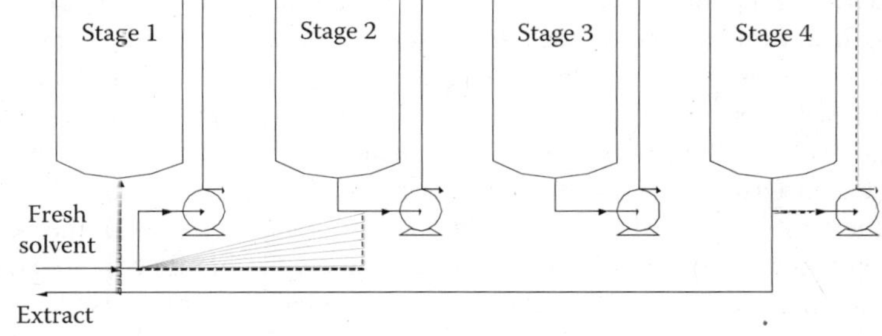

FIGURE 4.11 Four-stage percolation extraction.

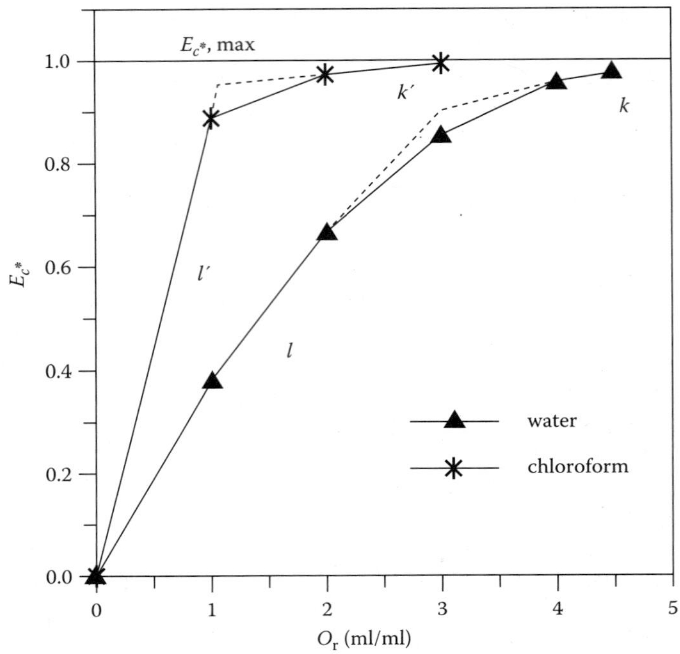

FIGURE 4.12 Variation of cumulative extraction degree with volume ratio (*Or*) for B3 battery system relative to water and chloroform solvents; Q = 0.5 L/h; $E_{C*, max} = 1$. (Reprinted from Senol, A., and A. Aydin, *J. Food Eng.*, 75, 565–573, 2006. With permission from Elsevier.)

(E_{C*}) was obtained by chloroform as compared to water. However, the significant difference observed for the first battery ($\Delta E_{C*} = 0.89–0.37$) became less pronounced with the increase in the solvent-to-solid ratio, as can be observed in Figure 4.12 [49]. The same figure shows that the water extraction performed with five extractors (B5) showed an extraction degree lower than chloroform extraction with three extractors (B3); these data reveal that the extraction degree of caffeine is notably dependent on the solvent nature and on the number of leaching stages.

A percolation extraction of virgin olive oil is a good example of natural extraction of antioxidants using only mechanical systems without chemical treatments. After the traditional discontinuous cycle of olive pressing, the percolation of crushed olives with water is followed by centrifugation in order to separate the oil from the water. These steps are common for olive processing systems, as studied by Ranalli et al. [50], with continuous percolation performed using water as solvent and a process time of 50 min and subsequent centrifugation. Three olive varieties (Leccino, Coratina, and Dritta) were tested, and the aromatic compounds that are responsible for the fruity taste and flavor were found in higher quantities in the percolation extraction. One of them was *trans*-2-hexenal, the major volatile compound found in good olive oils, which gives them a very pleasant odor and is responsible for the sensory green-fruity notes of olive oil. Although the aromatic composition was primarily affected by genetic factors, the centrifugation extraction probably removed the water-soluble volatiles from the oil [50]. Percolation produced olive oil with higher amounts of tocopherols, phenols, and aromatic compounds, which have a significant influence on the oil quality [50].

The cylindrical mixing extractor is a drying piece of equipment that has been used with success to perform plant extractions. It can use high temperatures controlled by a jacket, and this dispersed solid operation allows processing of fine particles, leading to higher concentrated extracts in relatively short cycles. Batches may be operated in countercurrent mode, and the solvent can be removed from the extractor bottom or by evaporation through the application of heating and/or a vacuum. A conical screw extractor presents the same functionality for the step of separation of the extract from the solid residue. This apparatus is equipped with an internal screw, which rotates eccentrically within the cone. The extract is drained to the bottom of the cone, where the extract is separated from the residue. Operation mode and recovery of solvent is the same as the cylindrical mixing extractor [42]. Simeonov et al. [51] studied the modeling of a screw solid–liquid conical extractor (Figure 4.13); the vertical equipment is a continuous countercurrent extractor operating with solvent recycling. *Geranium macrorhizum* L. + water extraction system was studied at 293 K, and the particles were considered as spherical. Experimental and theoretical data showed that, for the studied parameters (high volumetric solvent flow rate, long solid residence time, and diluted solutions), the kinetic curves approached the exponential curves for equilibrium under perfect mixing.

A screw extractor may be used in a batch or in a continuous mode; however, the great advantages of continuous mode over conventional batch extraction are a

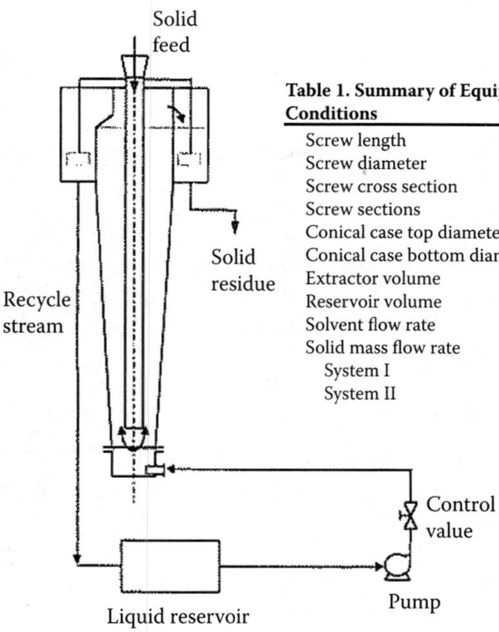

Table 1. Summary of Equipment Data and Extraction Conditions

Screw length	450×10^{-3} m
Screw diameter	44×10^{-3} m
Screw cross section	1.344×10^{-3} m^2
Screw sections	21
Conical case top diameter	90×10^{-3} m
Conical case bottom diameter	50×10^{-3} m
Extractor volume	3.15×10^{-3} m^3
Reservoir volume	$10–60 \times 10^{-3}$ m^3
Solvent flow rate	10^{-3} m^3 s^{-1}
Solid mass flow rate	
System I	1.0907×10^{-5} kg s^{-1}
System II	1.920×10^{-5} kg s^{-1}

FIGURE 4.13 Scheme of the experimental setup. (Reprinted from Simeonov et al., *Indust. Eng. Chem. Res.*, 42, 1433–1438, 2003. With permission from American Chemical Society.)

decrease of solvent consumption and of handling time. Poirot et al. [52] studied a raw material (not identified by the authors) in batch extraction to test a commercial continuous single-screw countercurrent extractor (Vatron Mau unit). This extractor was equipped with eight extraction vessels, with an average capacity of 67 L. The drainage stage was located at the last vessel and the maximum solvent flow was 10 m^3/h. The Vatron Mau unit was operated under an inert nitrogen atmosphere. Assays were performed with a raw material flow rate of 15 kg/h at ambient temperature, with a screw speed of 0.23 rpm and a solid residence time of 2 h 30 min. The countercurrent mode was not applied. Comparing kinetic assays for batch and continuous extractions, more than 90% of the extract was obtained after 1 h for batch extraction. Important information was obtained by comparing batch and continuous modes in terms of particle size, which should be large enough to avoid passing through the barrel, flying away under a strong solvent spray, or forming blocks, in order to keep a homogeneous solid flow rate and a correct solvent flow rate. However, some characteristics must be established before continuous extractions, such as the raw material swelling capacity, the solvent to be used, and the process temperature.

A scale-up of solid–liquid extraction for the screw extractor was obtained by Simeonov et al. [53] for four systems (*Geranium macrorhizum* L./water, *Amorpha fruticosa* L./petroleum ether, *Silibum marianum* L/methanol, and *Lavandula vera* L./petroleum ether). They obtained an analytical equation for the overall resistance to mass transfer, considering a linearly variable mass transfer resistance, for which the concentration profiles can be predicted from experimental data obtained from batch operation, without complementary assays from continuous extractions.

Figure 4.14 represents an immersion and a percolation type of extractor, which are examples of commercial equipment used for continuous processes. The immersion extractor is adequate for granular and powdery raw material, whereas the percolation extractor is appropriate for flakes and leaves. The Crown Iron (Model IV)

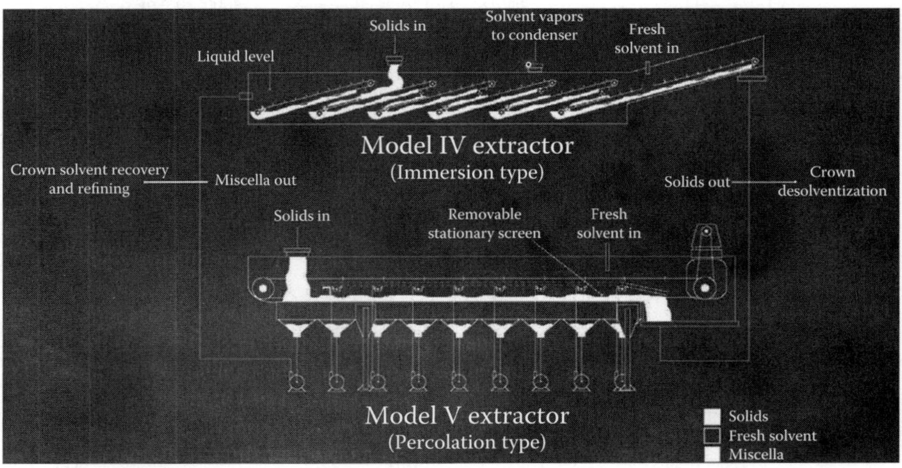

FIGURE 4.14 Crown immersion-type extractor and percolation-type (Crown iron). (Reprinted from Crown Iron, http://www.crowniron.com, 2007. With permission.)

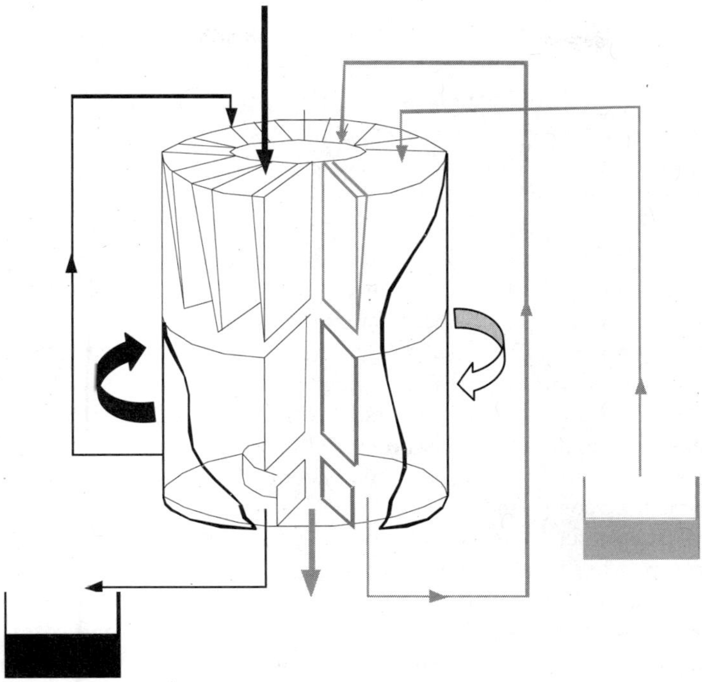

FIGURE 4.15 Continuous solid–liquid extraction pilot plant. (Reprinted from Gunt Hamburg Company, http://www.gunt.de, 2007. with permission.)

immersion extractor is not limited by screen filtration; it has a patented "en-masse"-type conveyor system that draws the material along the extractor bottom, where it is totally immersed in solvent, thus promoting a good contact between the solvent and the raw material and a low liquid velocity, in order to minimize the loss of fine particles. The percolation extractor (Model V) has also an "en-masse"-type conveyor system and a shallow bed to avoid the bed compression, with consequently less pronounced solvent channeling [54].

A continuous solid–liquid commercial extractor of Gunt Hamburg Company [55], model CE 630, is a piece of equipment that may work with up to three stages in a countercurrent flow way (Figure 4.15). It is like a carrousel extractor, with a continuously rotating extraction cell divided into compartments, with a screw feeder to feed the compartments with raw material. Control of temperature and rotation is individually performed for each stage.

Classical extractions techniques such as maceration, leaching with stirring or solvent agitation, and Soxhlet, which use solvent at its boiling temperature, have been replaced by similar industrial extraction methodologies in laboratory scale, mainly in the preparation of samples for analysis. To be effective, the selection of the extraction technique should take into consideration high extract or target compound recoveries, process time reproducibility, solvent volume, solvent removal from the extract solution and its reuse, and finally, cost.

4.3.2 Microwave-Assisted Extraction

There are two types of apparatus commercially available: closed extraction vessels under controlled pressure and temperature, and focused microwave ovens (FMASE = focused microwave-assisted solvent extraction) operating at atmospheric pressure (open vessels). These systems are schematized in Figure 4.16 as multimode and single mode. A multimode system allows random dispersion of microwave radiation within the microwave cavity, ensuring that every sample and cavity region is irradiated. A single mode or focused system permits focused microwave radiation on a restricted region in that a stronger electric field is applied on the sample.

The closed MAE system is used for extraction at high temperatures, above the solvent's boiling point. The pressure in the vessel depends on the volume and boiling point of the solvent. The great advantage of this system is that a single pressure control allows the simultaneous processing of several vessels. In the focused microwave ovens, the maximum temperature used in the apparatus is approximately the normal boiling point of the solvent. This system is mainly applied in the obtaining of organometallic compounds.

The focused microwave system can be operated using an open extraction cell under atmospheric pressure, and it can be refluxed (Figure 4.16a) with continuous irradiation and modulated power [20]. The temperature is determined by the solvent's boiling point at atmospheric pressure. To prevent the vapor losses, there is a reflux system, or, for some commercial equipment (Microwave open vessel digestion system; Milestone), a vacuum system that processes up to eight samples simultaneously in glass or quartz vessels of 250 cm^3 [56]. The diffused microwave equipment can be operated using closed extraction cells (Figure 4.16b), which allow pressure and temperature control and the application of different powers and variation of irradiation cycles in a multimode cavity [20]. For this system, the solvent can be heated above its boiling point, increasing the efficiency and accelerating the extraction speed. Additionally, the possibility of simultaneously processing several samples at the turntable can improve their homogeneity. Samples should be similar in terms of

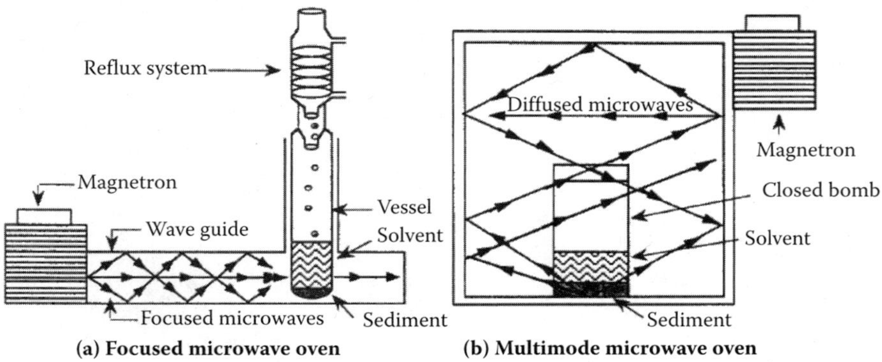

FIGURE 4.16 Schematic view of focused microwave oven (a) and multimode microwave oven (b). (Reprinted from Letellier, M., and H. Budzinski, *Analusis,* 27, 259–271, 1999. With permission from EDP Sciences and Wiley-VCH.)

both content volume and solid-to-solvent ratio because the pressure is commonly set by a single control device.

Commercial equipment supports 8–48 vessels simultaneously, with pressures of 0.4–12 MPa and vessel volumes up to 100 cm^3 (Multiwave 3000, Anton Paar [57]). Besides the extraction, this equipment can evaporate acids, preconcentrate aqueous solutions, and dry samples without carbonization or contamination.

Temperatures can be increased up to three times above the solvent's boiling point. This phenomenon is called superheating and occurs when a nonhomogeneous sample with different dielectric properties is dispersed into a homogeneous medium. This way, in order to apply this technique to obtain nonpolar target compounds, it is necessary to use solvents with dipole moments greater than zero [58].

MAE optimization of paprika (*Capsicum annum* L.) powder was obtained with different organic solvents like tetrahydrofuran, acetone, dioxane, ethanol, and methanol (90 and 15% in water). The temperature was kept under 333 K, which can be reached in 120 sec of extraction and avoids carotenoid degradation. Extraction data show that the extraction selectivity of pigments from paprika can be achieved by changing the concentration of the organic component, rather than changing the organic modifier [59].

For ginger microwave-assisted process, an improved extraction yield was observed when 1 cm^3 of a polar solvent, water acting as a modifier, was added to the system ginger–hexane. The time to obtain a maximum extraction yield was reduced from 40 to 30 seconds [60], proving that polar solvents are more appropriate to use in MAE. Considering this result, the raw material water content (humidity) may represent an improvement factor in terms of extraction yield, which might diminish, or even avoid, the drying of the raw material. Lucchesi et al. [61] studied the influence of the raw material's humidity percentage, the microwave power, and the irradiation time in the MAE of *Elletaria cardamomum* L. All variables were statistically significant (raw material humidity, extraction time, and irradiation power) with a tendency of increasing yield with the humidity and a dependency among these variables, mainly between time and power, with the power increment being associated with a reduction in the process time.

MAE of essential oil from *Laurus nobilis* L. dry leaves, which is generally obtained by hydrodistillation, was studied using a probe installed inside the Clevenger apparatus at 200 and 300 W and pulsed microwave energy at average total power of 200 W, for 1 h. MAE was selective for the phenylpropanoids compounds in both microwave power and pulsed energy, compared to the hydrodistillation. Proportionally, MAE extracts 90% more phenylpropanoids than hydrodistillation, and with the increase of the microwave power from 200 to 300 W, there was an increase of 20% in the yield [62]. The power increase in the MAE of Curcuma rhizomes leads to a pronounced increase of the main compounds of essential oil (curcumol, germacrene, and curdione; Figure 4.17) and to a reduction of the process time [63].

The same effect, a high increase of extract yield and decrease of process time as a function of power increments, was observed for other systems such as soybean, rapeseed, sunflower seeds, and olive [64, 65]. For some systems like ginger volatile oil, an increase in the microwave power from 200 to 400 W caused an enormous increase in the yield of all volatile compounds, but, at 700 W, a decrease was

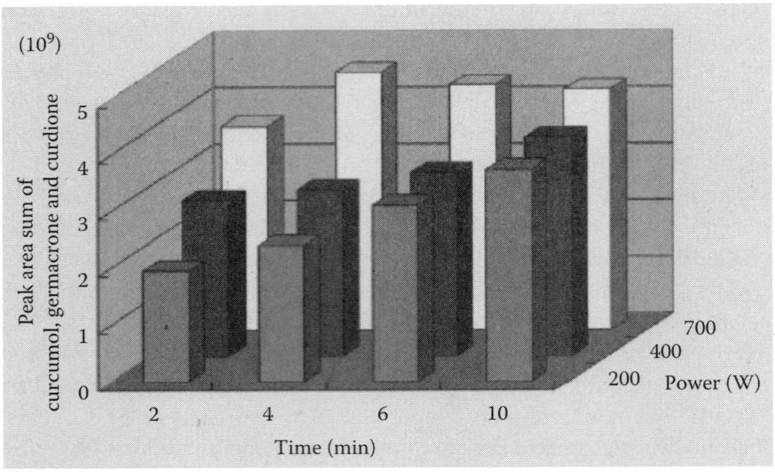

FIGURE 4.17 The effect of microwave power and irradiation time of peak area sum of curcumol, germacrone, and curdione in the TCM sample. (Reprinted from Deng, C., J. Ji, N. Li, et al., *J. Chromatogr. A,* 1117, 115–120, 2006. With permission from Elsevier.)

observed that was proportional to the increase obtained at 400 W. On the other hand, for the volatile ginger compounds, the extraction was not directly related to the microwave power [66].

Besides the interaction between power and time for many systems, temperature is directly related to the power energy absorption and should be monitored during the extraction and/or be controlled at a desired temperature to allow the recovery of larger amounts of the target compounds. The temperature of the system is related to the power energy that was used, with a sample heating as a result of the energy absorption by the polar compounds. High temperatures can be reached in short times with high irradiation power and in long times with low irradiation power or by the combination of high irradiation power and process times. Consequently, some target compounds may be favored with an increase of the solubility or disfavored with stability loss or thermal degradation.

Soy isoflavones' stability was studied in MAE at 500 W, 30 min, with extraction times that varied from 5 to 30 min, and temperatures that varied from 323 to 423 K. Higher temperatures exposed isoflavones to degradation: the temperature interval of 348–373 K mainly affected malonyl isoflavones; between 373 and 398 K the acetyl isoflavones and glucosides were affected, but the aglycones did not present degradation in this temperature interval [67]. Liazid et al. [68] studied the stability of 22 phenolic compounds during MAE, at 500 W, 20 min, and temperatures varying from 323 to 448 K. They found a relationship between the chemical structure and the stability of phenolics, where the hydroxyl-type substituents in the ring are more easily degraded than the methoxylates, for example, epicatechin, resveratrol, and myricetin.

Some advantages of MAE are shortened extraction time, reduced solvent volume, and simple extraction apparatus with easy sample heating control. An example

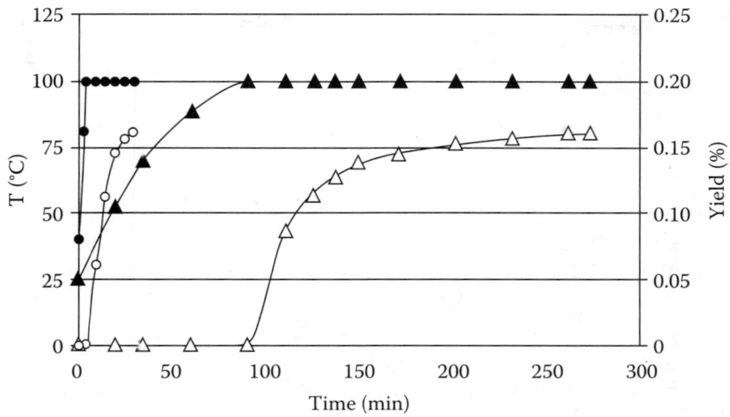

FIGURE 4.18 Temperature profiles (•, SFME; ▲, HD) and yields (○, SFME; Δ, HD) as a function of time for the SFME and HD extraction of essential oil from thyme. (Reprinted from Chemat, F., M. E. Lucchesi, J. Smadja, et al., *Analytica Chimica Acta*, 555, 157–160, 2006. With permission from Elsevier.)

is the MAE of fresh peppers to recover capsaicinoids. The assay was performed in a microwave extractor (Ethos 1600, model Milestone) at 500 W and 298 K. After 5 min of MAE, more than 95% of capsaicinoids were recovered, whereas the magnetic stirring demanded a minimum of 15 min to obtain the same content [69].

Different techniques can be applied using microwave assistance, like the solvent-free microwave extraction (SFME), which is a dry distillation combined with microwave heating to obtain, for instance, the volatile oil of basil, garden mint, and thyme. Besides the short extraction time (30 min for SFME against 4.5 h for hydrodistillation), the process saved a substantial amount of energy and was selective for some compounds. The yield of eugenol extracted from basil species increased threefold. The yield of carvone and thymol yields, extracted from garden mint and thyme, respectively, increased approximately 20%. For thyme, the extraction kinetic (Figure 4.18) indicates an important reduction of process time [70]. Microwave accelerated steam distillation of lavender essential oil resulted in the same yield of conventional steam distillation (~9%), but was three times faster [71].

In most of the studied cases, the solvent recovery was obtained by evaporation, and, consequently, if the evaporation process does not consider the degradation conditions of these compounds, this process may alter the target compound's properties. Extreme temperature conditions for a prolonged time may oxidize some antioxidants and phenolic compounds.

Table 4.5 shows a list of application of MAE to obtaining bioactive compounds.

4.3.3 Ultrasound-Assisted Extraction

Most applications of ultrasound-assisted leaching involve systems using bath or ultrasonic probe. This kind of equipment has been used for leaching organic and inorganic compounds. On the other hand, continuous apparatus has been used because

TABLE 4.5
Bioactive Compounds Obtained by Microwave- and Ultrasonic-Assisted Extraction

Process/bioactive compounds	Plant material	Operational conditions (power, frequency, solvent, time)	Reference
Microwave assisted extraction			
Terpenes	*Carum carvi* L. (caraway)	Power: 120 W, hexane, t = 60 min	[72]
Essential oil	*Elletaria cardamomum* L.	Power: 140–390 W, solvent free extraction, t =10–75 min	[73]
Essential oil	*Ocimum basilicum* L. (basil) *Mentha crispa* L. (garden mint) *Thymus vulgaris* L. (thyme)	Power: 500 W, solvent free extraction, t = 30 min	[74]
Essential oil	*Laurus nobilis* L. (laurel)	Power: 160–1600 W, probe coupled to hydrodistillation, t = 60 min	[62]
Essential oil	Rhizoma curcumae (rhizome of *Curcuma*)	Power: 200, 400, 700 W, water, t = 2–10 min	[63]
Essential oil	*Zingiber officinale* (ginger)	MD-SPME.[a] Power: 200, 400, 700 W, water, t = 1–6 min	[66]
Ginger oil and essential oils	*Zingiber officinale* (ginger)	Power: 150–300 W, hexane, ethanol, t = 3.5–4 min	[75]
Polyphenols, caffeine	*Camellia sinensis* L. (green tea)	Power: 700 W, ethanol/water (1:1), t = 4 min	[76]
Isoflavones	*Soybeans*	Power: 500 W, water, methanol, and ethanol (30–70%), t = 5–30 min	[67]
Phenolic compounds	*Vitis vinifera* (grape)	Power: n.a., methanol, t = 20 min	[68]
Bioactive compound (Artemisinin)	*Artemisia annua* L.	Power: 650 W, ethanol, trichloromethane, cyclohexane, *n*-hexane, and petroleum ether, t = 2–18 min	[77]
Edible oils	Soybean germ and seaweed	Power: n.a., open and closed vessel, hexane, t = 30–60 min	[78]
Edible oil	*Olea europaea* L. (olive)	Power: 60–120 W, hexane, 20–30 sec	[65]
Alkaloids	*Nothapodytes foetida* (slemure)	Power: 100 W, methanol (90%), t = 3 min	[79]
Pigments, carotenoids	*Capsicum annuum* L. (paprika)	Power: n.a., water: organic solvents[b] (15-90%), 30–120 sec	[80]
Saponin	*Ganoderma atrum*	Power: 800 W, closed vessel, ethanol, t = 3–30 min	[81]

Ultrasound extraction			
Antioxidants	*Rosmarinus officinalis* (rosemary)	Probe: f = 20 kHz, Bath: 40 kHz water, ethanol, and water/ethanol (1:9), t = 15–45 min	[82]
Bioflavonoid (Hesperidin)	*Citrus reticulate* (penggan)	Bath: f = 20–60–100 kHz, methanol, ethanol, and isopropanol, t = 20–160 min	[83]
Isoflavones	*Glycine max* (soybean)	Probe: f = 24 kHz, ethanol, t = 20 min	[67]
Isoflavones	*Glycine max* (soybean)	Probe and bath: f = 24 kHz, ethanol, methanol, acetonitrile (30–70%), t =10 min	[67]
Flavonoids	*Houttuynia cordata* Thunb.	Bath: f = 40 kHz, ethanol (70%), t = 50 min	[84]
Flavonoids, resin, mucilage	*Calendula officinale* (marigold)	Probe: f = 20 kHz, Bath: f = 33 kHz, ethanol/water (94%, 70% v/v), water, glycerol/water (3.5%, v/v), ethyl ether, t = 30–60 min	[85]
Edible oils	Soybean germ and seaweed	Probe: f = 19, 25, 40, 300 kHz, hexane, t = 30–60 min	[78]
Alkaloids	*Nothapodytes foetida*	Bath: f = 33 kHz, methanol (90%), t = 15, 30, 60 min	[79]
Alkaloids, oils	*Peganum harmala*	Probe: f = 20 kHz, Bath: f = 33 kHz, ethanol/water (94%, 70% v/v), water, glycerol/water (3.5%, v/v), ethyl ether, t = 30–60 min	[85]
Essential oils	*Foeniculum vulgare* (fennel) *Humulus lupulus* (hops) *Mentha piperita* (mint) *Titia cordata* (lime) *Inula helenium* (elecampane)	Probe: f = 20 kHz, Bath: f = 33 kHz, ethanol/water (94%, 70% v/v), water, glycerol/water (3.5%, v/v), ethyl ether, t = 30–60 min	[85]
Essential oils	*Laurus nobilis* L. (laurel) *Rosmarinus officinalis* L. (rosemary) *Thymus vulgaris* L. (thyme) *Oreganum majorana* (oregano) *Polianthes Tuberosa* (tuberose)	Probe: f = 20 kHz, water, t = 10 min	[86]
Essential oils	Dill, fennel, marigold, arnica, gentian, chamomile, sage, mint, coriander	Probe and bath: f = n.a., petroleum ether, ethanol (neat and aqueous) t = 15–180 min	[87]

continued

Table 4.5 (*continued*)

Process/bioactive compounds	Plant material	Operational conditions (power, frequency, solvent, time)	Reference
Virgin olive oil	*Olea europaea* L. (olive)	Probe: $f = 24$ kHz, Bath: $f = 25$ kHz,—, t = 0–30 min	[88]
Biophenols	*Olive europea* (olive)	Bath: 20 kHz, water-ethanol (50–90%), t = 6–30 min	[29]
Bioactive compounds (cineole, borneol, thujone)	*Salvia officinalis* (sage)	Probe: $f = 20$ kHz, Bath: $f = 37$–42 kHz, ethanol 65%, 1–12 h	[89]
Polysaccharides	*Salvia officinalis* (sage)	Bath: $f = 20$ kHz, ethanol, 120 min	[90]
Essential oils	*Salvia officinalis* (garden sage) *Salvia glutinosa* (glutinous sage)	Bath: $f = 40$ kHz, petroleum ether, ethanol 70%, water, 5–80 min	[91]
Biocompounds (borneol, cineole, α/β thujone)	*Salvia officinalis* (sage)	Probe: $f = 20$ kHz, ethanol 60%, variable period (broken and continuous mode for 1–4 days)	[92]
Biocompounds	*Valeriana officinalis* (valerian)	Probe: $f = 20$ kHz, ethanol 65%, variable period (broken and continuous mode for 1–4 days)	[92]
Polysaccharides	*Fagopyrum esculentum Moench* (sweet buck wheat)	Probe: $f = 20$ kHz, alkaline extractant, t = 5–10 min	[93]
Steroids, triterpenoids	*Chresta* spp.	Bath: $f = 60$ kHz, *n*-hexane, dichloromethane, and methanol, t = 30 min	[94]
Rutin	*Sophora japonica*	Probe: $f = 20$ kHz, water, methanol, t = 30min	[95]
Saponins	*Ganoderma atrum*	Bath: $f = 33$ kHz, ethanol, t = 15 min	[81]
Saponins	*Panax ginseng* (Korean and Chinese ginseng) *Panax quinquefolium* (American ginseng)	Bath: $f = 38.5$ kHz, Probe: $f = 20$ kHz, pure methanol, water-saturated *n*-butanol, water with 10% methanol, t = 60–120 min	[96]

n.a. = not available.

[a] MD-SPME: microwave distillation and simultaneous solid-phase microextraction.

[b] Acetone, dioxane, ethanol, methanol, and tetrahydrofuran (THF).

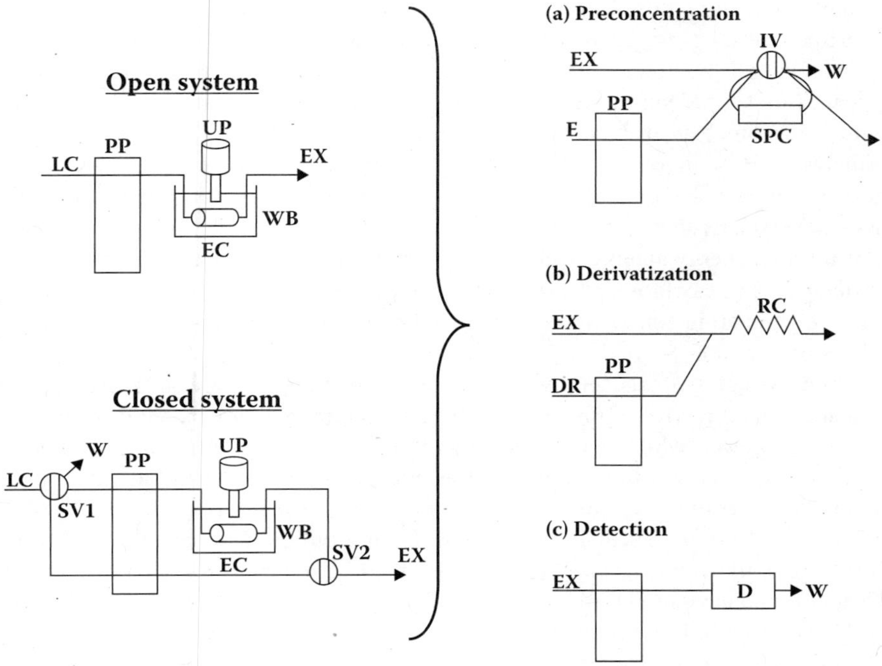

FIGURE 4.19 Experimental setup for the two modes of continuous ultrasound-assisted leaching and their coupling to other steps of an analytical process. One, two, or three steps can be used in a single method. LC: leaching carrier, PP: peristaltic pump, UP: ultrasonic probe, EC: extraction chamber, WB: water bath, W: waste, SV: selection valve, EX: extract, E: eluent, IV: injection valve, SPC: solid-phase column, DR: derivation reagent, RC: reaction coil, D: detector [36].

of the relatively reduced samples and diminished reagent consumption it allows. There are two dynamic approaches to the ultrasound-assisted leaching through continuous mode: open or closed system. The main difference in the results is that the extract obtained by a closed system is less diluted than that obtained by an open

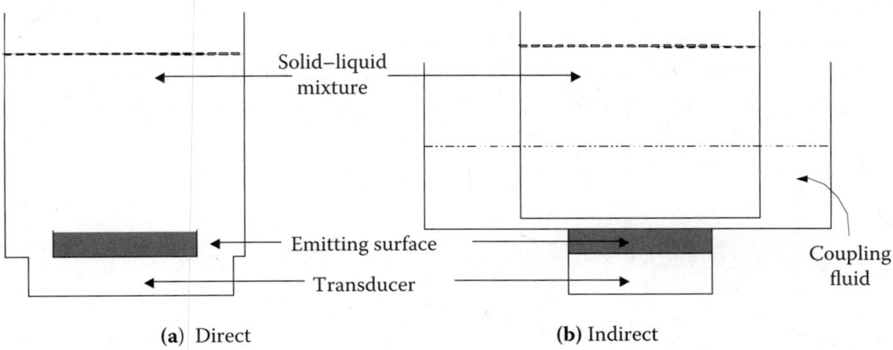

FIGURE 4.20 Methods of producing cavitation.

system. Because of this, closed systems have been used more. Figure 4.19 shows experimental setups for open and closed systems.

Among the common ultrasonic system types is the ultrasonic bath, which appeared first for cleaning purposes and is equipped with a transducer at the bottom or is submersed in a conventional tank. Because it is inexpensive and easily available, it is commonly used (Figure 4.20) in the indirect method of operation. Its disadvantage is the heating of the coupling fluid surrounding the solid–liquid mixture vessel, as shown in Figure 4.20b. The lack of uniformity in the distribution of ultrasound energy and the decline of power with time [36] are also important disadvantages. The cavitation production may be performed by direct sonication, when a device generating sound waves is placed directly inside the fluid mixture system to be processed [33].

Probe systems are generally used in the laboratory (Figure 4.21, [97]), with capacity to act directly within the solid–liquid mixture medium and delivering large amounts of power, which varies according to the variation of amplitude. The characteristic intensity distribution of an ultrasonic standing wave is in the axial direction, with higher intensity near the probe, which increasingly dissipates in the radial direction (Contamine et al. 1994, cited by Thompson and Doraiswamy [33]). The advantage of ultrasonic probes over baths is the localized energy that provides more efficient liquid cavitation [36].

Vinatoru et al. [87] obtained dry residues of the plants listed in Table 4.6 using a cleaning bath (direct sonication) at an ultrasonic power of 5 W/cm^2. The S/L ratio was 1:10 and the solvent used was ethanol 70%. The authors observed an increase of extraction yield with time for all tested plants. A probe extraction was tested to compare with ultrasonic bath for a marigold system, and the authors observed an increase in global yield for the probe system. For other systems (coriander, fennel, and dill), the ultrasonic extraction was selective for low-molecular-weight compounds.

Direct (DUSO) and indirect (IUSO) sonication of olive paste assays were performed using an ultrasound probe horn at 105 W/cm^2 and 24 kHz, and 150 W and 25 kHz, respectively, and compared to the conventional thermal treatment with respect to process yield and virgin olive oil characteristics (Table 4.7). Changes in quality parameters were not found, but, for ultrasonic assays, significant effects were found on the levels of bitterness, polyphenols, tocopherols, chlorophyll, and carotenoids for ultrasonic assays, besides the fact that off-flavor volatiles were not detected [88].

Wu et al. [96] compared ultrasonic bath and probe equipment to perform the ultrasound-assisted extraction of ginseng saponins. A cleaning bath at a frequency of 38.5 kHz and 810 W and a sonicator probe at 20 kHz and 600 W were used. For both techniques, the solvent used to extract saponins from American and Chinese ginseng was water-saturated *n*-butanol, and the S/L ratio was kept the same for all assays. Although the stabilized temperatures were different for probe and bath (at ~299 and

FIGURE 4.21 SinglePush-transducer. (Based on SinglePush-transducer of Martin Walter, Ultraschalltechnik, 2008. http://www.walter-ultraschall.de.)

TABLE 4.6
Dry Residue (g/100 g extract) Obtained by Direct Sonication in a Cleaning Bath

Sonication (time/min)	Mint	Chamomile	Marigold	Sage	Arnica	Gentian
15	0.06	1.10	0.94	0.58	0.36	—
30	0.07	1.30	0.98	0.80	0.42	1.67
60	0.25	1.43	1.14	0.92	0.67	2.66
90	0.78	1.56	1.33	0.94	1.06	2.71
120	0.82	1.79	1.75	1.13	1.20	3.24
180	—	1.80	—	—	—	—
18-h maturation	0.91	1.91	2.20	1.15	1.50	4.68
Classical 7 days + 14 days maturation	1.02	1.73	2.25	1.02	1.75	4.75

Source: Reprinted from Vinatoru, M., M. Toma, O. Radu, et al., *Ultrasonics Sonochem.*, 4, 135–139, 1997. With permission from Elsevier.

311 K, respectively) experiments, the lower frequency and power of probe (20 kHz, 600 W) affected the American ginseng extraction, leading to higher saponins content and similar total contents of Chinese ginseng (Figure 4.22).

In a study developed by Albu et al. [39] involving rosemary (*R. officinalis*) extraction, when results obtained with ultrasonic bath at 40 kHz and ultrasonic probe at 20 kHz [39] were compared, the authors concluded that the ultrasound efficiency was similar for all tested solvents. *S. officinalis* was submitted to extraction using an ultrasonic cleaning bath at 37-42 kHz and 130 W, and a probe (horn) at 20 kHz and

TABLE 4.7
Effect of Ultrasound Treatment on Sensorial Characteristics of Virgin Olive Oil

Treatment	Bitterness (K225)	Hexanal/*E*-2-hexenal (ratio)	Total volatile area (10^4 AV)	Organoleptic panel test evaluation: Positive characteristics — Fruit	Bitterness	Green	Pungent	Off-flavors — Wine
1st harvesting date								
TEST	0.28 ± 0.00[a]	2.10	99.64					
DUSO	0.24 ± 0.01[b]	1.76	99.02					
IUSO	0.25 ± 0.01[b]	1.29	95.18					
2nd harvesting date								
TEST	0.20 ± 0.00[a]	1.75	95.28	4.3	4.0	3.9	4.9	1.5
DUSO	0.21 ± 0.00[a]	1.50	93.45	4.9	3.1	4.3	5.1	0.8
IUSO	0.19 ± 0.00[b]	1.35	94.14	5.3	2.4	5.3	5.3	0.0

TEST, olive past without treatment; DUSO, direct ultrasound application by probe horn; IUSO, indirect ultrasound application by bath. Mean values ± SD ($n = 2$).

Source: Reprinted from Jiménez, A., G. Beltrán, and M. Uceda, *Ultrasonics Sonochem.*, 14(6), 725–731, 2007. With permission from Elsevier.

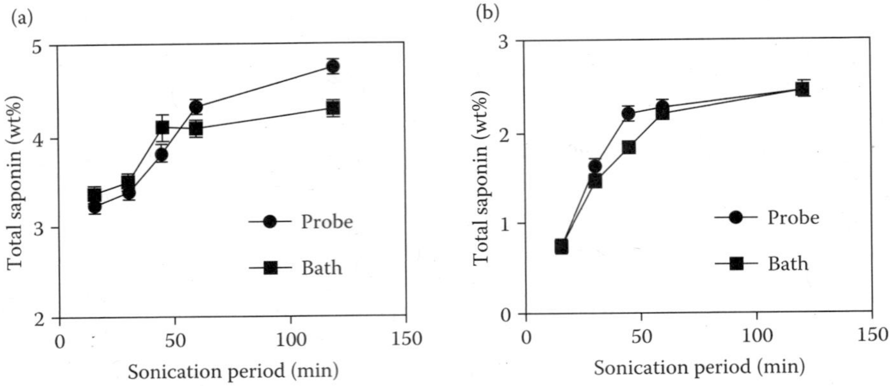

FIGURE 4.22 Saponin yields of sonication-assisted extraction for various periods of time with water-saturated *n*-butanol as the extracting solvent. (a) American ginseng root and (b) Chinese ginseng root. (Reprinted from Wu, J., L. Lin, and F. Chau, *Ultrasonics Sonochem.*, 8, 347–352, 2001. With permission from Elsevier.)

300 W, operated on a 50% cycle, with ethanol 65% as solvent. The target compounds yield obtained with the ultrasonic probe in 2 h was comparable to the result obtained using the ultrasonic bath for 5 h [89].

Transducers used for industrial applications are piezoelectric, constructed with a piezoelectric material such as quartz and based on an electric field, or magnetostrictive, based on a magnetic field and constructed with materials like nickel alloys (Hunicke 1990, cited by Thompson and Doraiswamy [33]). The piezoelectric transducers are generally used in small volume processes. They are more fragile than magnetostrictive transducers and can be damaged at temperatures higher than 423 K or by high impact. The magnetostrictive transducers are more resistant to mechanical damage and can be used in temperatures above 523 K (Hunicke 1990, cited by Thompson and Doraiswamy [33]).

Another ultrasound device is a tube reactor or sonotube, which is a stainless steel resonant tube that can be used as a flow reactor, with internal or external emission, attached to a submerged tube, working under pressure or not. Figure 4.23 shows a resonant tube constituted by a transducer of 20 kHz (C), a booster (B) with a shape that can be varied according to the wave amplitude and modular unit (M), and the resonators (R) that are fixed on both sides of the modular device. The solution flows through the tube, suffering the action of the ultrasound waves in the whole length of the reactor. Faid et al. [98] studied the effects of power ultrasound inside the resonant tube with local measurements, using three methods: a chemical dosimeter, a thermal sensor, and an electrochemical probe. Results were similar along the tube axis, but slightly different from the tube axis to the wall. A homogeneous acoustic field on a given cross section was obtained using this resonant tube, but there were large variations of effects due to standing waves in the axial direction.

Faid et al. [98] compared a cup horn to the resonant tube (Figure 4.23) and a probe (or horn) at 20 kHz and 25 W. This cup horn is constituted of a glass cylinder,

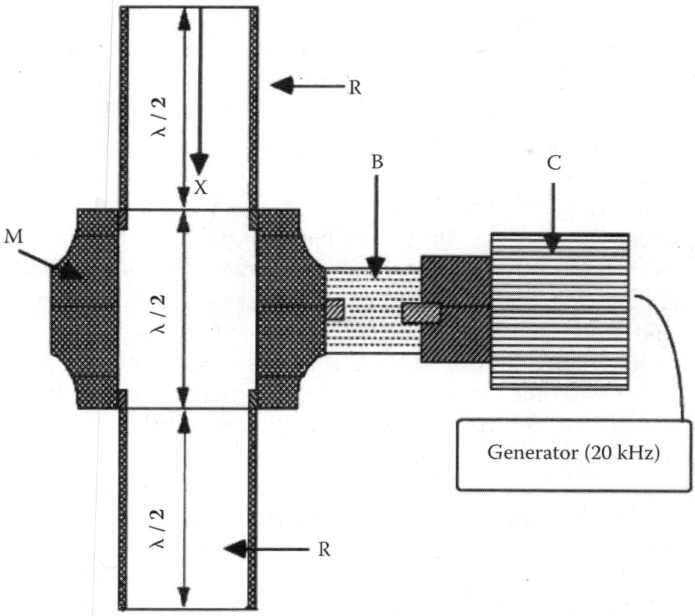

FIGURE 4.23 Scheme of sonotube. B-booster, C-transducer, M-modular unit, R-resonators. (Reprinted from Faid, F., F. Contamine, A. M. Wilhelm, et al., *Ultrasonics Sonochem.*, 5, 119–124, 1998. With permission from Elsevier.)

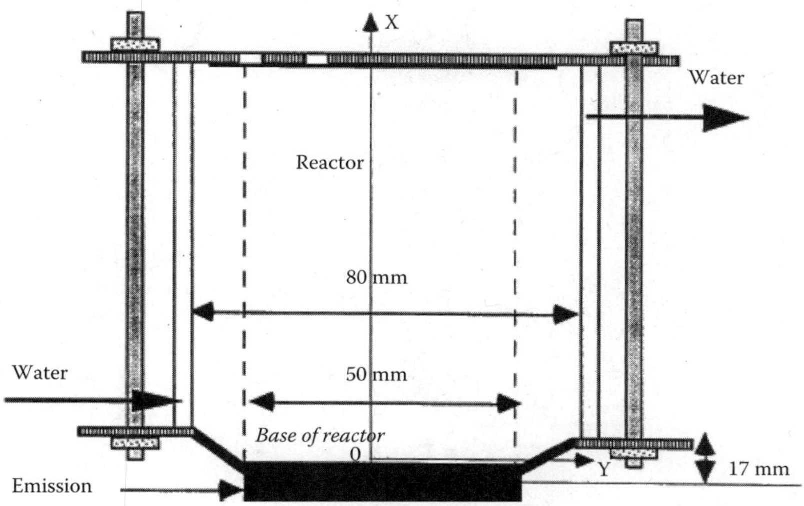

FIGURE 4.24 Scheme of the cup horn. (Reprinted from Faid, F., F. Contamine, A. M. Wilhelm, et al., *Ultrasonics Sonochem.*, 5, 119–124, 1998. With permission from Elsevier.)

with temperature control provided by a jacket, placed between two stainless steel plates, as can be seen in Figure 4.24. The comparison of performances of those devices was obtained by the intensity distribution of local cavitation effects. The extraction behavior was dependent on the equipment's potential and on the studied system, besides the complexity of the nature of vegetable matrices.

Some researchers showed that the comparison between a cleaning bath and a probe with lower frequency and similar intensity resulted in a higher extraction yield for the probe, because of the efficient cavitation it provides. In this comparison, a fixed ultrasound probe was used to perform the extraction of caraway seeds to obtain carvone and limonene. The extractor had a cooled jacket with three entries, the first for ultrasound probe, the second for cooling, and the third for sampling (Figure 4.25). The process conditions were 342 K at ultrasound power of 150 W, using a S/L ratio of 1:20 and *n*-hexane as solvent in a 60-min extraction process [99]. According to the data (Figure 4.26), the limonene extraction by ultrasound presented a pronounced increase mainly in the first 10 min. The same was observed for the carvone extraction. However, ultrasound-assisted extraction seems to be more selective at low temperatures for carvone than for limonene, because of the higher polarity of carvone and the volatility of limonene. Constant extraction rates were calculated for the obtaining of carvone and limonene in these first 10 min. Independent of temperature, the ultrasound-assisted extraction presented higher yields when compared to controls, and the extraction was 1.3 to 2 times faster.

Ultrasonic devices show heterogeneities for all equipment, which results in variation of mass transfer coefficients in axial and radial directions affected by power and power input. Some authors described the relation between the mass transfer

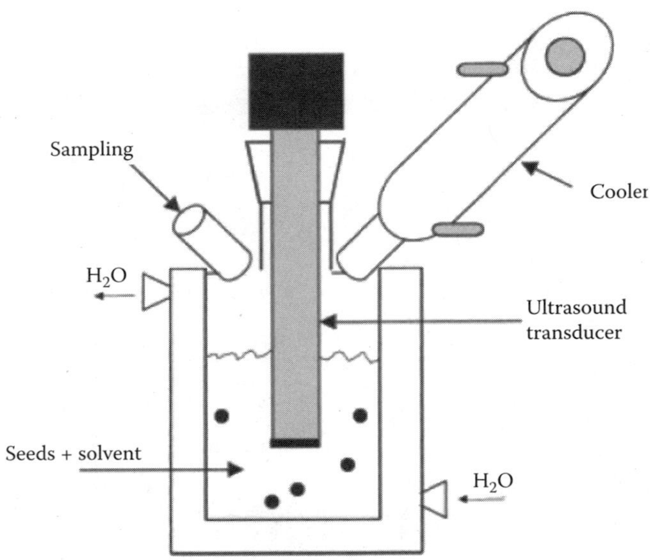

FIGURE 4.25 Ultrasound-assisted extraction experimental disposal (20 kHz). (Reprinted from Chemat, S., A. Lagha, H. AitAmar, et al., *Flav. Fragr. J.*, 19, 188–195, 2004. With permission from Wiley.)

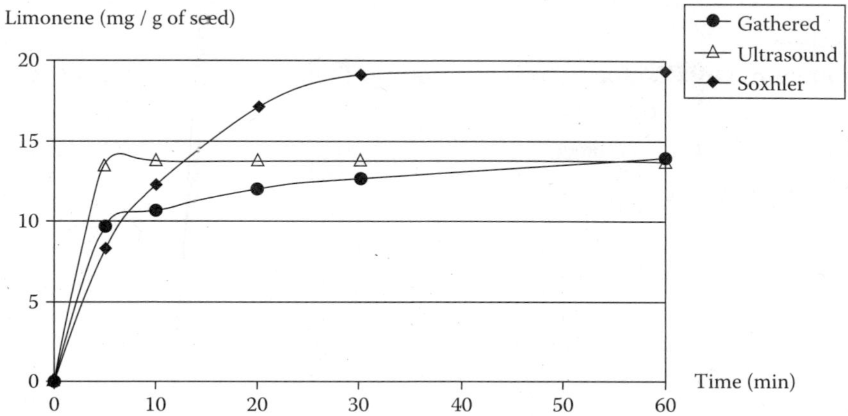

FIGURE 4.26 Comparison of hexane extraction patterns of limonene from caraway seeds with different extraction procedures. (Reprinted from Chemat, S., A. Lagha, H. AitAmar, et al., *Flav. Fragr. J.*, 19, 188–195, 2004. With permission from Wiley.)

coefficients' profile and the wave's pattern and the intensity and the cavitation effects for those three ultrasonic devices [100]. Other researchers studied and characterized the ultrasonic field propagation in ultrasonic devices by chemical and mechanical effects [101, 102].

An ultrasonic probe, similar to the one used in Slovak factories (industrial scale static extraction) was used to obtain extracts from sage (*S. officinalis* L.) and valerian (*Valeriana officinalis* L.). The probe dimensions were 79 cm of height and 5 cm of diameter. It was immersed in a stirred extraction mixture with solid-to-solvent ratio of 1:6 for sage and 1:3 for valerian, using ethanol (65 and 60%, respectively) as solvent and operating at 20 kHz and 600 W. The purpose was to sonicate by different ways, namely (1) broken mode (half-hour sonication period alternated with half-hour silent periods during 8 h, for 3 days), (2) short-time mode (2-h sonication period in the beginning of an 8-h extraction period, during 3 days), and (3) continuous mode (8 h of continuous sonication, during 3 days) [92]. A very long ultrasound contact time (continuous mode) affected the volatile substances' composition profile, with differences in cineole and α- and β-thujones contents. In terms of borneol concentrations, the difference appeared just in the second extraction day, when compared to the short-time mode. Although the yield increased for the continuous mode, the degradation risk also increased. The weak increase in yield on the third day of extraction indicates that the process need not be continued for more than 2 days (Table 4.8). Figure 4.27 indicates a direct relation between temperature and sonication time. For the continuous mode, the temperature increases quickly, and for broken mode it has a slight increase, with maximum temperature around 303 K, which indicates that, on the manufacturing scale, the extraction vessel must be cooled to avoid ethanol evaporation during the process [92]. For both systems, the shorter exposure to sonication would be expected to produce less degradation of the target compounds, when compared to the continuous mode.

TABLE 4.8
Content of Dry Residue from Sage Tinctures Prepared by Different Modes of Sonication

Time	Short time mode		Broken mode		Continuous mode	
	U (%)	C (%)	U (%)	C (%)	U (%)	C (%)
1 h	1.59	1.85	1.56	1.26	1.92	1.89
3 h	2.23	1.89	1.92	1.64	2.05	1.95
8 h	2.28	2.13	2.31	1.92	2.38	2.07
2 days	2.45	2.30	2.58	2.12	2.44	2.10
3 days	2.63	2.51	2.82	2.39	2.59	2.32
4 days	2.58	2.62	2.79	2.40		

U: extraction with ultrasound and C: control extraction.

Source: Reprinted from Valachovic, P., A. Pechova, and T. J. Mason, *Ultrasonics Sonochem.*, 8, 111–117, 2001. With permission from Elsevier.

The dynamic ultrasound-assisted extraction of oleuropein and derivatives from olive leaves was developed by Japón-Luján et al. [29]. The extraction cell was immersed in a water bath equipped with a sonifier at 20 kHz and 450 W. The optimization of olive biophenols (OBPs) obtaining was performed considering seven variables: probe position, ultrasound amplitude, percentage of ultrasound exposure duty cycle, irradiation time, solvent flow rate, solvent composition, and water bath

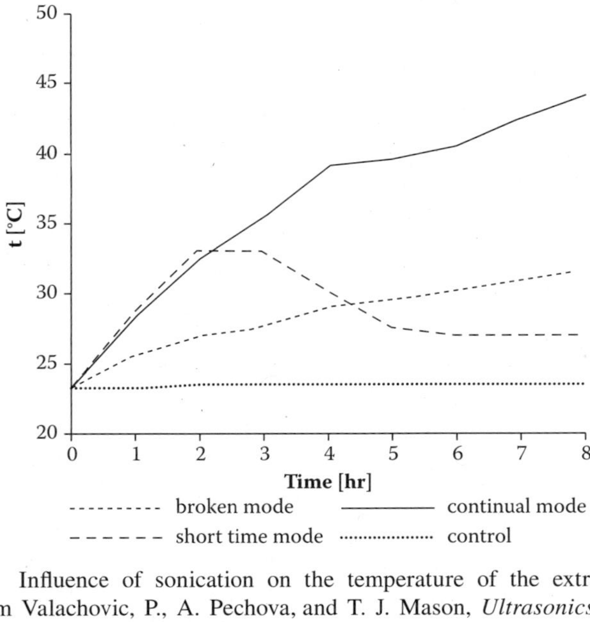

FIGURE 4.27 Influence of sonication on the temperature of the extraction mixture. (Reprinted from Valachovic, P., A. Pechova, and T. J. Mason, *Ultrasonics Sonochem.*, 8, 111–117, 2001. With permission from Elsevier.)

temperature. The best conditions to obtain higher OBPs contents were radiation amplitude of 30%, duty cycle of 70% with probe position at 4 cm, using 59% of ethanol as solvent with 5 cm^3/min at 310 K for 25 min. The researchers obtained the concentrations of 22.6, 0.48, 1.07, and 0.97 g/kg for oleuropein, verbacoside, apigenin-7-glucoside, and luteolin-7-glucoside, respectively.

Ultrasound is applied to different reactors, used for batch or continuous flow; also there are industrial systems with different methods of cavitation generation, which are described by Thompson and Doraiswamy [33]. Although these reactors have been used to promote reactions in liquid–liquid or solid–liquid systems such as oxidation, they are similar to the solid–liquid extraction units, with an additional transducer installed in the equipment. Therefore, although some of them may be adapted for solid–liquid extraction, in practice, most industrial equipment sets destined to the natural products extraction are common agitated tanks equipped with transducers, which results in relatively high equipment costs considering the improvement of extraction yield presented by researchers.

Velickovic et al. [91] studied the extraction kinetics of two sage species (*S. officinalis* L. and *Salvia glutinosa* L.) using three solvents (petroleum ether, 70% ethanol, and water) with solid-to-solvent ratio of 1:10 at 150 W and 40 kHz, 313 K, for 80 min (Figure 4.28). The extraction yield increased with solvent polarity, being higher for *S. officinalis* L. All three model equations used predicted the experimental data relatively well. The model based on the unsteady diffusion through the raw material predicted the highest diffusion coefficient values.

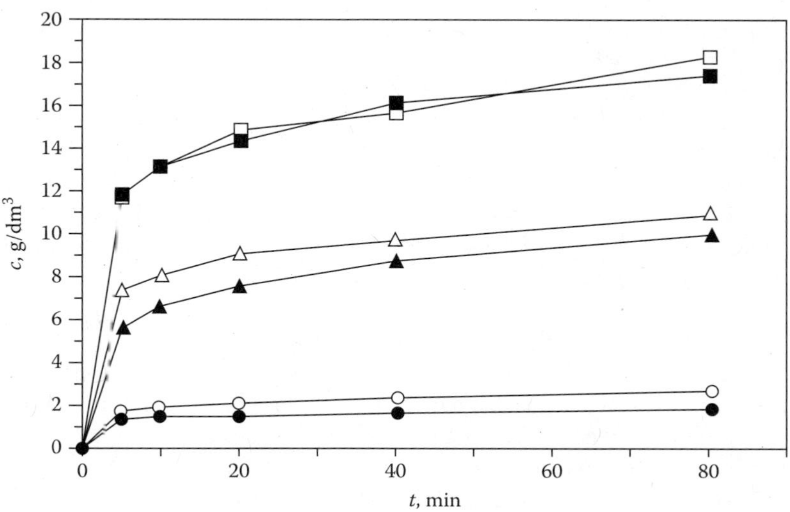

FIGURE 4.28 Variation of the concentration of ES (extractable substances) in the liquid extract with increasing sonication time during extraction (open symbols, *S. officinalis* L.; closed symbols, *S. glutinosa* L.). Extracting solvent: petroleum ether, circles; 70% ethanol, triangles; and water, squares. (Reprinted from Velickovic, D. T., D. M. Milenovic, M. S. Ristic, et al., *Ultrasonics Sonochem.*, 13, 150–156, 2006. With permission from Elsevier.)

As described before, solvent characteristics are important. The usage of ethanol as solvent was tested in respect to its instability under sonication, using gas chromatography to monitor changes in electrical conductivity of ethanol. An oxidative process was observed at a concentrations below 50% and the presence of ethanol was detected [72].

Extraction of *R. officinalis* to obtain antioxidants, like carnosoic and rosmarinic acid, was obtained using an agitated water bath and an ultrasonic bath equipped with a probe. Among the experiments performed using the agitated water bath, considering the three different solvents used (butanone, ethanol, and ethyl acetate), butanone was the most effective extraction solvent in terms of carnosoic acid yield increase. On the other hand, for the ultrasonic probe assays, the difference between results related to different extraction solvents was reduced. Similar carnosoic acid contents were obtained, at 320 K, using the ultrasonic probe with ethanol for 15 min, and the agitated water bath for 3 h [39].

Toma et al. [25] used different solvents to submit seven species to ultrasound-assisted extraction (fennel, hops, marigold, lime, mint, peganum, and elecampane). The solvents used were ethanol/water (94/70%, v/v), water, glycerol/water (3.5%, v/v), and ethyl ether. An indirect method was used with a cleaning bath at 33 kHz and 296 K. Extractions yield was determined for 30 and 60 min of process time, and the results indicated that most of the extract was obtained during the first 30 min. The solvent selectivity was specific to each species: for marigold, peganum, and mint, higher yields were obtained using water; for fennel, hops, lime, and elecampane, the solvent that improved the extraction yield were ethyl ether, ethanol/water (70%, v/v), glycerol/water (3.5%, v/v), and ethanol (94%, v/v), respectively.

To compare the temperature effect in the ultrasonic-assisted extraction of *S. officinalis* (293, 303, and 323 K), experiments were carried out in an ultrasonic cleaning bath at 37-42 kHz and 130 W, using a S/L ratio of 1:8.3 and ethanol 65% as solvent. The extraction efficiency was monitored through gas chromatography determination of the cineole, thujone, and borneol contents. At 303 K, the ultrasound effect was more pronounced, because after 12 h, the content of the active compounds was approximately 60% higher than that of the control experiment. The effect of the ultrasound was also evaluated in a system provided with mechanical stirring. After 5 h, approximately 45% more active compounds (cineole, thujone, and borneol) were obtained when ultrasound was applied when compared to the conventional stirring extraction [89].

Hromadkova et al. [90] studied a sage (*S. officinalis*) residue obtained by ethanolic ultrasonic-assisted extraction using an ultrasound probe at a frequency of 20 kHz, 600 W, and intensity of 1 W/cm^2 as an attempt at isolating polysaccharides. The usage of ultrasound-assisted extraction positively affected the polysaccharides yield, increasing the concentrations, in the extract, of glucans and arabinogalactans, as main compounds, and xylans and glucomannans, as neutral sugars.

Besides applications in food, the effect of ultrasound was studied for other extraction systems, considering the influences of solvent, solid-to-solvent ratio, particle diameter, temperature and power, frequency, and intensity of ultrasound devices. The influence of these parameters depends on the studied systems and on the ultrasonic devices chosen. Therefore, it is necessary to carefully select the equipment and the extraction solvent, as well as to study the system to be extracted and other process parameters that might exert impact on the desired result, which can be in terms of total yield or concentration/yield of target compounds [103, 104].

Table 4.5 shows a list of application of ultrasonic extraction to obtaining bioactive compounds.

4.4 OBTAINING HIGH QUALITY BIOACTIVE COMPOUNDS USING GRAS SOLVENTS

4.4.1 Antioxidants

Antioxidant compounds in food play an important role as a health-protecting factor. Antioxidants are also widely used as additives in fats and oils and in food processing to prevent or to delay spoilage of foods [105]. So there is an increased interest in the recovery of antioxidant compounds to use in the food industry. Several extraction and isolation procedures were already proposed. However, new market trends say that these compounds should be obtained using solvent with the status GRAS (Generally Recognized as Safe).

Antioxidant compounds comprise a wide variety of compounds, such as vitamins, flavonoids, terpenoids, carotenoids, and phytoestrogens. Several plant sources of antioxidant compounds have already been studied using different solvent systems, but the effects of these compounds on human health still are not well known and, therefore, will not be commented on in this chapter. Here, a brief report of some antioxidant compounds and extracts obtained from some plant matrices is presented.

Ethanol, water, and their mixtures are the preferable solvent systems currently used for natural product production. In that context, infusions (immersion in hot water) continue to be an interesting way to produce extracts with high contents of antioxidant compounds. On the other hand, the use of ethanol should also be considered, depending on the plant source and on the target compound. For instance, ethanol is widely used to recover phenolic compounds from plant matrices. Extraction of phenolics will be treated more specifically in Section 4.4.3, but some interesting results of their extraction are also pointed out in this section.

As presented in Section 4.1 several variables can influence the extraction of antioxidants from plants. Some examples are solvent system, temperature, extraction solvent-to-solid matrix (feed) ratio, time, pH, and agitation.

4.4.1.1 Solvent System

Solvent composition is always an important variable to be considered when dealing with extraction process. This variable should always be optimized in order to produce good extraction yields in an economically advantageous process.

Infusions of mate (*Ilex paraguariensis*) are well known by their antioxidant properties. For this reason, this kind of extraction system is largely used for the recovery of bioactive compounds from this species' leaves. Bastos et al. [106] extracted antioxidant compounds from mate leaves (about one-fourth of the solids present in the infusions were phenolic compounds) using water at 368 K for 5 min. The extract solution was filtered and dried, and the antioxidant composition of the resulting extract was determined by lipid oxidation inhibition. The mate extract showed antioxidant activity similar to that of the artificial antioxidant BHT.

Recently, Callemien et al. [107] studied the extraction of the well-known antioxidant resveratrol from hops (*Humulus lupulus* L.). After the removal of hydrophobic bitter compounds, the dry matter was extracted with a mixture of ethanol and water (75:25) at 333 K. The results showed good yields of resveratrol (recovery of approximately 90%). Several polyphenols, such as catechin, rutin, and quercetin, among others, were also found on hops extracts.

The extraction of antioxidant compounds from sage (*S. officinalis*) was carried out using ethanol–water mixtures [108]. Dried sage was ground in a knife mill and extracted with several mixtures of ethanol and water (from 27 to 100% of ethanol). The authors reported that the range of ~55 to 75% of ethanol was the best choice to recover the antioxidant compounds, such as rosmarinic acid- and carnosoic-type compounds. However, different ethanol–water proportions caused different behaviors in terms of the target compounds, concentrations. Rosmarinic acid was better recovered within the range of 30–70% of ethanol in the solvent mixture, but carnosoic type compounds were better extracted within the range of 70 to 100% of ethanol.

The antioxidant activities of extracts obtained from old tea leaves and black tea wastes were compared to green tea leaves [109]. Antioxidant capacity was determined by trichloroacetic acid method. The extraction was carried out in two steps: first, the dry matter was extracted with hot water at 353–378 K for 20 min to generate fraction 1; fraction 2 was produced by extracting the residue from step 1 for 30 min with hot water at temperatures varying from 373 to 403 K. The two fractions were combined and dried under vacuum. The yields were about 35, 28, and 30% for green tea leaves, old tea leaves, and black tea wastes, respectively. The antioxidant assay indicated that green tea extract was a more effective antioxidant when compared to the samples obtained from black tea wastes and old tea leaves, which presented similar results concerning antioxidant capacity.

Dormana et al. [110] studied the antioxidant activity of four herbs from the Lamiaceae family: oregano (*Origanum vulgaris* L.), rosemary (*R. officinalis* L.), sage (*S. officinalis* L.), and thyme (*Thymus vulgaris* L.). The property was assessed through four different methodologies: radical scavenging activity with DPPH, radical scavenging activity with ABTS, Fe^{3+}–EDTA/H_2O_2/ascorbate–catalyzed deoxyribose oxidative degradation assay, and ex vivo LDL oxidation inhibition. Fifty grams of the herb material was extracted twice with 500 cm^3 boiling water. The two fractions were combined, filtered, and freeze-dried. The yields (w/w) of the dry extracts were 36% for oregano, 24% for rosemary, 25% for sage, and 29% for thyme. Total phenolic contents were 149 (oregano), 185 (rosemary), 166 (sage), and 95.6 (thyme) expressed in mg GAE/g (mg of gallic acid equivalent/g of dried extract). HPLC analysis revealed that rosmarinic acid was the major constituent. Sage and rosemary extracts presented the best antioxidant activities according to all tests performed.

Ethanol proved to be an effective solvent to recover antioxidant compounds from sweet grass (*Hierochlöe odorata*) [111] when a Soxhlet apparatus is used. The crude extract, obtained after 6 h of processing, showed concentrations about 20.31% of 5, 8-dihydroxycoumarin and 2.18% of 5-hydroxy-8-*O*-β-D-glucopyranosyl-benzopyranone.

The extraction of *Cinnamomum zeylanicum* was studied by Jayaprakasha et al. [112] using hot water as solvent. Extraction was carried out with defatted matter at 393 K and ~0.1 MPa for 20 min. The antioxidant activity and radical scavenging activity were measured. Yields of extraction were almost 4% (which contained more than 44% of phenolic compounds). Antioxidant activity measured by the β-carotene-linoleate model system indicated similar results for both the water extract and BHA. The same happened for the radical scavenging activity (DPPH method) analysis.

The effect of solvent composition on the yields of phenolic compounds from wheat was demonstrated by Liyana-Pathirana and Shahidi [113]. The impact of applying different proportions of the water–ethanol mixture (30–70% of ethanol in water) was studied. The solvent composition, within the proportion of ethanol concentration evaluated, presented a quadratic relation with the phenolic compounds obtaining. The optimum proportions of the ethanol–water mixture did not widely depend on the biomass type: for soft wheat bran and soft whole wheat, the best results were obtained with a 50% ethanol aqueous solution, whereas for hard wheat bran the best ethanol concentration was 55%.

In the same way, Zhou and Yu [114] studied the antioxidant activities of wheat bran extracts obtained with aqueous ethanol (70%) and absolute ethanol. Wheat grains were cleaned, milled, extracted using Soxhlet apparatus, and concentrated. Radical scavenging capacity was determined by the DPPH method, whereas antioxidant activity (determined as trolox equivalent) was measured by 2,2' azinobis(3-ethylbenzothiazoline-6-sulfonic acid) diammonium salt) (ABTS) and Oxygen Radical Absorbance Capacity (ORAC) methods. Significant levels of antioxidant activities and phenolic compounds have been detected in wheat, indicating that it may serve as an excellent dietary source of natural antioxidants for disease prevention and health promotion. The DPPH results for the extracts obtained with 70% ethanol and absolute ethanol, respectively, were as follows: for Akron wheat bran, radical scavenging (DPPH) was about 37 and 41% (remaining radical levels) and for Trego wheat bran, it was 46 and 53%. The results for the ORAC analysis for extracts obtained with 70% ethanol and absolute ethanol, respectively, were 23 and 60 trolox equivalent for the Akron wheat bran samples and 23 and 60 trolox equivalent for the Trego wheat samples.

The release of the target compounds from plant matrices is sometimes difficult, and the solvent system is not always able to recover the compounds present on the matrix. Thus, the use of enzymes might be a good choice to break down the structures (mainly cellulose) and promote the release of some compound of interest. For instance, Kim et al. [115] studied the extraction of phenolic compounds from apple peel by combining the factors heat treatment (368 K for 20 min), acid addition (2% sulfuric acid), and pectinase addition (1 unit/10 cm^3), which resulted in a synergistic effect. After that, the peels were treated with cellulase from *Thermobifida fusca*. The results indicated that the phenolic compounds were released two times more from the treated apple peel when compared to the untreated peel.

4.4.1.2 Temperature and Time

Temperature and time are also important variables in the extraction of bioactive antioxidant compounds. Although temperature generally shows a positive effect on extraction yields, elevated temperatures might promote degradation of some target compounds.

Temperature exerted a slight influence on the extraction of antioxidant compounds from sage [108]. The authors varied temperatures from 295 to 336 K. The recovery of carnisic type compounds and rosmarinic acid was positively influenced both by the temperature increase as well as by the treatment time increase. The concentration of the target compounds in the extract obtained after a 6-h extraction time was 20% higher than that observed for the sample related to the extraction period of 1 h.

The effects of temperature and time on the extraction of phenolic compounds from wheat and their antioxidant activity were measured [113]. Temperatures varied from 313 to 353 K, and time, from 45 to 75 min. The authors found that time did not significantly affect the extraction yields, while temperature showed an important role on the phenolic compounds' recovery. In terms of temperature effect, a linear influence on the extraction of phenolics from wheat bran was observed, whereas for hard wheat bran, the relation between temperature and phenolic compounds obtained presented a quadratic nature. A marked interaction between the parameters solvent composition (aqueous ethanol) and temperature was noted. Optimum extraction temperature in terms of antioxidant capacity varied according to the kind of biomass used: for soft wheat bran extracts, the higher antioxidant results were obtained at 353 K, and for the soft whole wheat, the higher antioxidant activity was obtained at 343 K.

Temperature showed a linear correlation on the extraction of phenolic compounds from *Inga edulis* [116]. The temperature was varied from 288 to 338.4 K. The higher the temperature, the higher the phenolic compounds contents.

The extraction of aspalathin from *Aspalathus linearis* was carried out using hot water (453 K) as solvent [117]. The kinetics indicated that 30 min was enough to extract almost all the aspalathin from the dried sample. Final yields were about 12 ppm of aspalathin in wet basis.

The effect of temperature on the extraction of carnosic, ursolic, and oleanolic acids from balm leaves with ethanol as the extraction solvent was studied [118]. Temperatures from 273 to 453 K were evaluated. Surprisingly, the best choice for the obtaining of oleanolic and ursolic acids was 273 K. For carnosic acid, the authors reported that the best temperature was 293 K.

4.4.1.3 Solvent-to-Feed Ratio

The solvent-to-feed (S/F) ratio is always an important variable for the extraction of target compounds in general. This is the parameter that determines the amount of solvent used, and it is always related to economic aspects, because high S/F ratios mean higher solvent consumption. The increase of production cost due to the use of elevated amounts of solvent is not only related to the cost of the solvent itself, but also to the cost of solvent removal in case it is necessary. Despite its importance on the extraction process, few studies focused on the effect of this variable on the recovery of antioxidant compounds have been published until now. However, some investigations should be cited.

The effect of the S/F ratio variation from 6:1 to 18:1 was evaluated in terms of recovery of the antioxidant agents from sage [108]. The best extraction yields were achieved using S/F of 18:1. The authors also compared crosscurrent extraction with single-stage extraction. When sage was extracted in three stages with S/F ratio of

6:1, the yields were higher than when a S/F ratio of 18:1 was used in a single stage. On the other hand, according to another study, the S/F effect was not significant in the extraction of phenolics from *I. edulis* [116]. Ratios of 10:1, 20:1, 40:1, and 80:1 did not show significant yield differences. In this case, the conditions used to evaluate the S/F effect were 323 K for 30 min using 50% ethanol aqueous solution as the extraction solvent. The S/F ratio used for the recovery of phenolic compounds and triterpenic acids from balm (*Melissa officinalis* L.) varied from 4 to 10 L of ethanol/solid matrix kg ([118], and the best result in terms of oleanic, carnosic, and ursolic acids were observed with S/Fs of 6, 4, and 10 L/kg, respectively.

4.4.1.4 Particle Size

Sage particle with sizes varying from 1 to 3 mm were extracted using an ethanol–water (75:25) mixture [108]. As expected, the yields decreased with the increase in particle sizes. The authors suggest that the mass transfer process is limited by the joint action of two phenomena: the diffusion of the hydroalcoholic solvent into the particle and the solvent–solute diffusion out of the particle.

Depending on the target compound, different particle sizes might correspond to different patterns in terms of both extraction yield and/or composition. The effect of particle size on the extraction of antioxidant compounds from balm (*M. officinalis* L.) using ethanol as solvent was studied by Herodez et al. [118]. The recovery of carnosic acid was higher for particle sizes within the range of 0.20–0.25 mm. For ursolic and oleanolic acids, the best extraction yields were obtained for particle sizes varying from 0.315 to 0.400 mm and from 0.250 to 0.315 mm, respectively.

4.4.2 Pigments

Industries now look forward to supplying to the increasing demand of the consumers for natural product ingredients, which makes the extraction of pigments from plants an important issue. Colorants from natural sources have been used in the food industry to provide an adequate solution for consumers' needs. Additionally, because pigments are recognized for their positive role in human health, these compounds have been added to food to provide fortified versions of the products. These factors have created a demand increase for plant pigments, which have to be obtained through processes that use only GRAS solvents [119], once they are destined for human consumption.

4.4.2.1 Solvent System

The use of mixtures of ethanol and water seems to be an interesting alternative to obtain pigments from natural sources. Natural pigments comprise a wide variety of compounds, with different chemical characteristics.

Anthocyanins, which are also a phenolic compound, are water-soluble pigments. The water solubility of these pigments is attributed to the fact that their basic skeleton is often acylated with one or more polar side chains such as glucosides [119], which makes hot water, a nontoxic solvent, an interesting option for the recovery of this group of substances. Concerning that, Tsai et al. [120] reported the extraction of anthocyanins from Roselle (*Hibiscus sabdariffa* L.) petals by using boiling water as the extraction solvent. The process consisted of extracting 3 g Roselle petals (previously dried at 323 K for 36 h) with 300 cm^3 of boiling water. Then, the extract solution was immediately filtered and chilled to a temperature of 277 K. The authors also reported the role of storage time on the delphinidin-3-sambubiose content, which indicated that extracts stored for short times showed concentrations up to 80% of delphinidin-3-sambubiose, whereas extracts stored for long periods (15 weeks) presented lower concentrations of this compound (reduction of 10 to 20%, depending on the extraction conditions).

Lapornik et al. [121] studied the extraction of anthocyanins from different vegetable matrices. Water and ethanol were shown to be the best solvents for anthocyanins recovery. For all the matrices tested, 70% ethanol was better than pure water for the recovery of total anthocyanins (measured by spectrophotometric methods). For red currant, black currant, and grape, the yields obtained in 70% ethanol extraction of anthocyanins were about 2, 3, and 10 times higher, respectively, than using pure water. However, the proportion variation of the two solvents in the mixture caused different behaviors in terms of anthocyanin recovery for the different vegetable matrices. For red currant, the recovery of delphinidin-3-glucoside did not present significant variation between the results obtained with water or 70% ethanol. However, when it comes to black currant, the same compound showed to be more effectively recovered by using 70% ethanol than pure water. In a different way, water presented better recovery capacity of the compound cyanidin-3-glucoside-rutinose.

Anthocyanins are also well known by their stability at acidic conditions. Several authors have reported the extraction of anthocyanins using acidified solvents. Cacace and Mazza [12] studied the effect of ethanol concentration in water (from 50 to 84% of ethanol) for the recovery of anthocyanins from milled berries. The process consisted of extracting refrigerated black currents (which had been previously milled and sieved) with several water–ethanol mixtures. Acidified solvents (with HCl, pH ~4.0) were used. Yields varied from 10 to 15 mg/g (dry basis).

The application of acidic conditions for the obtaining of anthocyanins has also been reported by another investigation, which described their obtaining through ultrasound-assisted extraction using 1.5 M HCl–95% ethanol as solvent [122]. Another recent report [123] related the recovery of aglycons, namely, petunidin, pelargonidin, peonidin, and malvidin from a pigmented potato (*Solanum tuberosum* L.) variety. The process consisted of submitting potatoes (previously washed, cut, and blanched) to extraction with a mixture of water and hydrochloric acid (19:1, v/v). The yields obtained were in the range of 0.65–1.15 g of anthocyanins/kg of potatoes. Hu et al. [124] reported the extraction of anthocyanins from defatted wheat bran using 65% ethanol containing 0.1% HCl (pH 3.0) in a shaker (200 rpm) at room temperature (298 K). Among the compounds identified in the extract were cyanidin-3-galactoside, cyanidin-3-glucoside, pelargonidin-3-glucoside, and peonidin-3-glucoside. These examples show that acidified ethanol–water mixtures are commonly used for the recovery of anthocyanins from plant matrices.

Luque-Rodriguez et al. [125] in their study on the anthocyanins extraction from grape skin found that the use of superheated liquids could represent an attractive industrial alternative for the obtaining of this group of compounds. They reported that the extract obtained from grape skin, a by-product of the winemaking industry, using superheated mixtures of ethanol and water, presented high concentrations of anthocyanins like 3-glucosides (malvidin, peonidin, delphinidin, petunidin, and cyanidin). Some advantages of superheated liquids are that the use of temperatures above the solvent's boiling point increases diffusion rate, solubility, and mass transfer and decreases the solvent's viscosity and surface tension, and the absence of air and light reduces the possibility of degradation. The use of a superheated (393 K, 8.0 MPa) mixture of HCl acidified (0.8%, v/v) ethanol–water mixture provided the best results in terms of anthocyanins yield, which was approximately three times higher than that obtained through conventional dynamic solid–liquid extraction.

Although water is recognized as a poor solvent for carotenoids extraction, it is used for the extraction of oil from seeds and vegetables because of the fact that it is the adequate medium for the application of enzymes as a means of increasing the extraction yield. Several authors report the use of a great variety of enzymes to enhance the recovery of carotenoids from plants [126–128]. Enzymatic cell wall lyses using hydrolytic enzymes is an interesting alternative because it can degrade the cell wall constituents, thus assisting in the release of intracellular contents [126]. The major advantages of high enzyme loadings are faster rates of hydrolysis and increased sugar yields, whereas the main drawback is the high cost related to this kind of process.

As an alternative to water, ethanol showed to be a selective solvent for the recovery of carotenoids from Chili Guajillo Puya (*C. annuum* L.) flour [127]. The authors proposed a two-step process to obtain capsaicinoids and carotenoids using ethanol. Previously, an enzymatic treatment in water (pectolytic and cellulolytic enzymes were applied) would be performed. Then, the dried flour was submitted to the first extraction step with ethanol 30%, to recover a rich fraction of capsaicinoids, and to the second extraction step with industrial ethanol (96%), to recover an enriched carotenoid fraction. The authors reported that the best extraction conditions were (1) pretreatment of the flour with a solution of Viscozyme L in a concentration of 5% (120 rpm, 323 K for 7 h, with a solid-to-solvent ratio of 1:50); (2) a first extraction step using 30% v/v ethanol, obtaining a recovery of 60% of the capsaicinoids; and (3) a second extraction stage using industrial ethanol (96%), with a recovery of 83% of the carotenoids.

A recent investigation reported the optimization of the enzyme concentration for the recovery of lycopene from tomatoes [126]. Pectinase and cellulase were tested to enhance lycopene extraction. The yields of lycopene were two and almost three times higher when cellulase and pectinase were used, respectively.

Çinar [129] reported the extraction of carotenoids from orange peel, sweet potato, and carrot using different concentrations of cellulase and pectinase combinations. The process consisted of an enzymatic treatment (enzymes were used in different proportions) of the sample, filtration in celite to recover the non–water-soluble pigments, elution of the nonpolar fraction with ethanol 95%, and precipitation of the pigments with excess of water. The best extraction conditions differed for each vegetable matrix evaluated: for orange peel, the best yields were achieved with the combination of 10 and 0.5 mg/L of pectinase and cellulase, respectively, after 6 h of extraction time; for sweet potato, the best enzymatic combination was 10 mg/L of pectinase and 1 mg/L of cellulase for an 18-h treatment; and the best results for carrots were obtained with a 24-h treatment with a combination of pectinase and cellulase at concentrations of 10 and 0.5 mg/L, respectively.

Another important carotenoids characteristic that must be pointed out is their well-known instability (through oxidative degradation and isomer formation). Because of that, the extraction steps should be carried out under controlled environmental conditions. Exposure of lycopene to light should be avoided and the addition of antioxidants might be considered, depending on the specific process and application [126].

4.4.2.2 S/F Ratio

Cacace and Mazza [12] have already reported the influence of the ratio of extraction solvent-to-solid matrix on the extraction yields of anthocyanins from milled berries. They concluded that this was the most important variable compared to the others studied (temperature and solvent composition). The solvent-to-solid ratio increase was related to higher anthocyanin recoveries in an almost linear way, for all the tested solvents. These results are in accordance with the mass transfer principles. The S/F ratio varied from 6 to 74 cm^3/g, and the anthocyanins yields were 11 and 15 mg/g, in that order. Chen et al. [122] also investigated the effect of the S/F ratio on the ultrasound-assisted extraction of anthocyanins from red raspberries. An experimental optimization design with a central point was used. The S/F ratio was varied from 0.6 to 7.4 cm^3/g, and the results indicated that the quadratic term of S/F contributed significantly ($p < 0.05$) for anthocyanin recovery. The optimal S/F ratio was 4:1, resulting in approximately 31 μg of cyaniding-3-glucoside equivalent/100 g of fresh fruits.

Fan et al. [130] also reported the effect of the S/F ratio on the extraction yield of anthocyanins from purple sweet potato. Response surface methodology was used to optimize the recovery of the target compounds. The S/F ratio varied from 15 to 35 cm^3/g, and the best result was obtained with 32 cm^3/g. The extraction yield dependence on the S/F ratio could be easily observed under the experimental conditions evaluated.

4.4.2.3 Temperature and Time

Temperature seems to play an important role in the extraction of pigments. Temperature increases mass transfer and thus diminishes the extraction time. However, when dealing with thermosensitive compounds, high temperatures might lead to denaturation.

The effect of temperatures varying from 279 to 347 K on anthocyanin recovery from milled berries was evaluated [12]. The authors reported a maximum anthocyanin recovery at the temperature range of 303–308 K, and a decrease of anthocyanin yields for temperatures higher than 318 K. Therefore, between 279 and 303 K, the temperature increase was related to higher solubility and extraction yields. However, the use of even higher temperatures was either ineffective or eventually caused thermal degradation of the target compounds. The extraction time was dependent both on the temperature and on the solvent system used. The shortest extraction time (10 min) was achieved in two situations: at 328 K with dilute ethanol and at 343 K with an ethanol concentration more than or equal to 75%.

Similarly, the effects of temperature and extraction time on extraction yields of anthocyanins from purple sweet potato were described by Fan et al. [130]: evaluated temperatures were varied from 313 to 353 K, and extraction time, from 60 to 120 min. The best result in terms of anthocyanins yield (158 mg/100 g of purple sweet potato) was obtained at 353 K with 60 min of extraction time. Linear and quadratic dependence of anthocyanin yield on the temperature was observed, whereas the influence of extraction time was not as significant.

Time seemed to play an important role in the extraction of anthocyanins from different plant by-products than for the results described for purple sweet potato [121]. Using the same extraction solvent, optimal extraction time varied with the type of vegetable material submitted to processing. Higher yields were achieved after 1 h of extraction compared to 12 and 24 h for black currant and red currant when water was the solvent. However, when 70% aqueous ethanol was used as solvent, the same behavior was not observed: 24 h resulted in significantly higher yields than yields obtained after 1 and 12 h for black currant, and 12 h was the best choice when red currant was the plant material. Individual anthocyanins also showed different extraction behaviors according to extraction time for the same solvent system. Delphinidin-3-glucoside was recovered in higher yields after 1 h, with a decrease after 12 and 24 h when 70% aqueous ethanol was the solvent. However, cyaniding-3-sambubiose showed to be better recovered after 24 h compared to 1 h of extraction when water was used as solvent.

4.4.3 Phenolic Compounds

In general terms, the extraction efficiency of a target compound is usually a function of several process variables. Many authors report the influence of many variables on the extraction of phenolic compounds. The most important factors concerning the recovery of phenolic compounds from natural products are solvent type, temperature, contact time, solvent-to-solid ratio, particle size, and pH, among others. The positive or negative effect of each variable on the mass transfer phenomenon, which governs the extraction process, is specific to each type of vegetable matrix and is not always obvious. The separation of soluble phenolic compounds can be performed by promoting their diffusion from a solid matrix (plant tissue) using a liquid matrix (solvent). Several authors have reported the use of solvent extraction to recover phenolic compounds from plants. Each vegetable material possesses unique properties that might interfere in the phenolic compounds' extraction. Thus, it is important to develop optimal extraction methods for their quantification and identification [131].

Extraction is generally the first step in the isolation of phenolic compounds from plant materials. The composition and nature (simple and/or complex) of the phenolic compounds to be extracted determine the choice of the extraction conditions. Extraction is influenced by the chemical nature of the compounds (simple and complex phenolics), the extraction method employed (extraction by solvents, solid-phase extraction, and supercritical extraction), the storage time and conditions, and the presence of interfering substances [132].

The phenolic compounds in plants may vary from simple to highly polymerized substances. Some plants contain different phenolic acids, phenylpropanoids, anthocyanins, and tannins, which can interact with other plant components such as carbohydrates and proteins (these complexes might be insoluble). That is why it is difficult to develop a process capable of recovering all the phenolic compounds present in a plant matrix [132], which makes the choice of the extraction solvent a key factor.

It is the study of the solvents' nature and possible related effects that will make it possible to properly select the substance to be used in each step (extraction, fractionation, and purification) of the vegetable material processing. By understanding the properties of both the extraction solvent and the target compounds (solute), and the solvent–solute interactions, rapid fractionation and isolation of desired components might be achieved [133]. The diversity concerning the solvent's chemical characteristics and the target compounds diverse structures and compositions imply that each material–solvent system shows different behavior, which cannot be predicted, and should be investigated for each specific application [134].

Many solvents can be used to extract phenolic compounds [132]. However, in this chapter, the use of water, ethanol, and isopropanol will be discussed, as well as the influence of some other variables, such as temperature, extraction time, particles size, solvent-to-solid ratio, and pH.

In industry, the economical feasibility of the extraction process involves the search for the optimal combination of extraction conditions that will maximize the efficiency of the process and reduce costs [134].

4.4.3.1 Solvent System

Several solvent systems have been used to recover phenolic compounds from plant matrices. This discussion will be focused on the use of ethanol, isopropanol, water, and their combination; these substances are classified with the GRAS status and, for that reason, water, isopropanol, and ethanol are suitable for the recovery of nutraceuticals [132].

Ethanol is reported to be an effective solvent for the recovery of phenolic compounds and, for that reason, it is usually used for the obtaining of this group of compounds, especially when it comes to the production of nutraceuticals, which is related to its GRAS classification [132]. Some authors reported that the effectiveness of the phenolic compound recovery through solvent extraction with ethanol can be increased by the addition of different proportions of water [135–137]. Another advantage related to ethanol is that, although alcoholic solvents are not highly selective for phenols, its use is usually preferable, in view of other organic solvents, because of the possible application of the extracts in food products [138]. The acidification of the extraction solvent is a resource frequently used to improve the obtaining of anthocyanins. The positive effect of water + ethanol for the recovery of phenolic compounds was corroborated by a recent investigation developed by Markom et al. [133]. The authors compared the results obtained with ethanol, a 1:1 water–ethanol mixture, and isopropanol as the extraction solvents. The impact of other process variables such as pH, solvent-to-solid ratio and extraction time on the extraction of phenolic compounds from grape have also been evaluated. The authors concluded that the 1:1 water–ethanol was the best solvent option in terms of total phenolic compound recovery, whereas isopropanol did not provide good extraction yields.

Specifically considering the substance corilagin, the best results were obtained with a 7:3 water–ethanol mixture.

4.4.3.2 S/F Ratio

The effect of the solvent-to-solid ratio on the recovery of phenolic compounds from different plant matrices was well studied by several authors [118, 137–139]. According to mass transfer principles, the driving force during mass transfer is the concentration gradient between the solid and the bulk of the liquid, which is greater when a higher solvent-to-solid ratio is used. Therefore, according to mass transfer principles, independent of the extraction solvent used, the higher the solvent-to-solid ratio, the higher the total amount of solids obtained [138]. However, from an economical point of view, considering that the solvent consumption exerts a direct influence on the extraction process cost, this variable should be carefully analyzed and optimized.

Considering the aspects quoted above, a work developed by Bucic-Kojic et al. [140] describes a significant difference for polyphenols concentrations in grape seeds extracts due to the variation of temperature and solvent-to-solid ratio. The statistical analysis of the results indicated that the polyphenols recovery presented a significant dependence on both temperature and S/L ratio, with a clear interaction between these two variables, which means that temperature exerted different influences as the solvent-to-solid ratio used was varied. An S/F of 40 cm^3/g provided the best extraction yields at all evaluated temperatures. The highest polyphenols yield (30.243 mg GAE/g) was obtained at 353 K and an S/F of 40 cm^3/g.

4.4.3.3 Temperature and Time

Extraction time and temperature are important process parameters that should be optimized. They are closely related to the effectiveness of the process as well as playing an important role in the economical aspects of its industrial applicability.

In general aspects, there is a consensus about the roles of time and temperature in the extraction processes: increased working temperatures enhance extraction by increasing the solubility of the solutes and diffusion coefficients. However, for phenolic compounds, attention should be paid to their stability during the process; phenolic compounds, when kept above certain temperatures for certain periods of time, can suffer thermal degradation (oxidation) and activity loss [141].

These effects have been recently approached by a work developed by Spigno et al. [138]. In their study, the antioxidant activities of grape extracts were highly influenced by both time and temperature. Although the highest yield (~2.5%) was obtained at 333 K, a reduction of phenolics contents was observed after 20 h of extraction time. The authors attributed this reduction to degradation and polymerization. The same authors also studied the influence of lower temperatures (318 K) used for longer periods of time (24 h), and they observed an increase in the extraction yield (~3.0%). When working with thermosensitive compounds, the use of lower temperatures associated with longer extraction times is always preferable. Increased contact time between solvents like ethanol and solid matrices might lead to a progressive release of solute from solid matrix to solvent [142]. However, these variables have to be optimized for each specific system in order to maximize yields and satisfy economical aspects. In this context, because some phenolic compounds present thermal instability, process temperature should never exceed 323 K [142].

A recent investigation showed that the most advantageous values of phenol recovery were obtained after 3 h of extraction time. The authors commented that, because longer extraction times did not provide a significant increase of the phenols obtained, they resulted in an uneconomical final process [141].

Therefore, the choice of the extraction temperature for the obtaining of a specific group of substances should be in accordance with the target compounds' molecular structure, plant matrix characteristics, degradation tendency, and extraction time. The economical impact of these extraction variables on the process related costs should also be taken into consideration.

4.4.3.4 Particle Size

The yield of polyphenol recovery from plant materials can be strongly influenced by variations in the sample particles size. Mass transfer can be improved by the use of smaller particles to improve the penetration of solvent in the solid matrix. This effect has already been reported for the recovery of polyphenols from grape [143]. However, the particle size has to be limited because exceedingly small particles tend to agglomerate, leading to a decrease of solvent penetration in the solid matrix and, therefore, negatively affecting the mass transfer process. Particles agglomeration phenomena during extraction, leading to the appearance of preferential flow channels and offside zones, were described by Pinelo et al. [134] in their study on the extraction of grape skin.

A recent investigation reported the influence of particle size on the recovery of polyphenols from grape seeds. The smallest particles (0.16–0.125 mm) provided the best recovery of gallic acid equivalents per gram of extract (mg of GAE/g of extract). The extraction was conducted with aqueous ethanol (50%) at 353 K and S/L ratio of 40 cm^3/g [140]. The particles size increase and lower gallic acid equivalent concentrations were exponentially related to each other. Additionally, the extraction temperature influence was not the same throughout the particle size range, becoming more intense as the particle size increased.

4.4.3.5 Effect of pH on Extraction Yield

Concerning the recovery of polyphenols, pH can act according to different mechanisms and play a significant role in the extraction performance. Although the pH effect has not been as widely studied as other process variables, such as temperature and S/F ratio, the addition of acid to the extraction media as means of pH modification is frequent in the case of polyphenols recovery and provides some advantages such as increased phenol stability, including the anthocyanins [144], increased dissolution of phenolic compounds [145], and increased disintegration of cell walls, facilitation of phenolic compounds solubilization, and diffusion from the plant material [131].

In a recent investigation, the best pH conditions for the extraction of total phenols were within the range of 1.5 and 2.1 (acidified with HCl), and a decrease in the recovery of phenolic compounds was observed when the pH value of the solvent was higher than 3.0 [131].

4.5 ECONOMICAL EVALUATION OF A SOLVENT EXTRACTION PROCESS: SAGE AND MACELA CASES

When the industrial sector focuses its attention on an innovative technology, one of the first questions that emerges is: Is this process advantageous in terms of costs?

In that context, Rosa and Meireles [146], studying the economical feasibility of the supercritical technology, created a rapid COM estimation, which can be ranked to the least accurate class of estimate (Class 5) among the five classes defined in the AACE Recommended Practice No. 17R-97 [147], carefully explored by Turton et al. [148]. These authors developed a parallel between the Association for the Advancement of Cost Engineering (AACE) classification and a classification of their own, which represents the combination of other definitions found in the literature. According to their analysis, the five classes of the AACE classification can be roughly associated to the five classes of the system presented by them.

The prior purpose of the economical study that will be presented in this section was to perform a Class 5 estimate for a solvent extraction process, similar to that presented by Rosa and Meireles [146] for supercritical extraction. According to Turton et al. [148], the Class 5 estimate is an Order-of-Magnitude (also known as ratio or feasibility) Estimate, which is the one that typically relies on cost information for a complete process taken from previously built plants. This cost information is then adjusted using appropriate scaling factors, for capacity and for inflation, to provide the estimated capital cost. And normally requires only a block flow diagram. However, it was not possible to attain cost-related data for complete processes that are already installed and in current operation, which changed the focus of the examples that will be given for a Class 4 estimate.

The Class 4 estimate can be roughly associated with the Study (also known as Major Equipment or Factored) Estimate, which Turton et al. [148] define as the one that "utilizes a list of the major equipment found in the process, including all pumps, compressors and turbines, columns and vessels, fired heaters, and exchangers as the starting point. Each piece of equipment is roughly sized and the approximate cost determined. The total cost of equipment in then factored to give the estimated capital cost."

4.5.1 Definition of the Solvent Extraction Process

Once the estimate class had been chosen, it was necessary to search for a solvent extraction process in the literature. Concerning that, Rakotondramasy-Rabesiaka et al. [149] claim that although they could not find a literature reference for the batch extraction in a continuous stirred tank, it is widely used by the industry for the extraction of vegetable materials. Therefore, the examples given here were based on the extraction process studied by these authors, which basically consisted of placing a known mass vegetable material immersed in a known volume of extraction solvent inside an agitated tank.

To make it easy to visualize the main equipment involved in the process, the software SuperPro Designs 6.0 (Intelligen, Inc., Scotch Plains, NJ) was used. Concerning the software use, Takeuchi et al. [150], in their study on the performance of a supercritical extraction unit's separation tank, considered it a very important tool dedicated to the process design issue, because it is very accessible in terms of usability, which could make the communication between the scientific community and the industrial sector much simpler and faster. This software's advantageous characteristics were once again evaluated through the solvent extraction process simulation. It demands that the user select the equipment units and connect them through the process streams. The graphical representation provided by SuperPro Designs 6.0 is shown in Figure 4.29.

The reason for the addition of a second extraction vessel to the equipment set resides in the purpose of simulating a pseudo-continuous process. This means that while one of the vessels is under operation, filled with the extraction system, the other goes through the cleaning and recharging processes, with the purpose of starting its operation just as the first one finishes. Consequently, it is important to analyze whether it is possible to recondition the extraction vessel in a period of time equal to or shorter than the operation time of the other extraction vessel. On the contrary, more than two extraction vessels might be necessary.

4.5.2 Properties of Vegetable Materials

After the extraction process had been selected, it was necessary to search for solvent extraction-related information for two vegetable materials. For sage (*S. officinalis*), the necessary data were taken from a study developed by Durling et al. [108] on the use of water–ethanol mixtures to obtain phenolic compounds from this species. According to the results described by those authors, the best results were obtained with a mixture of 31% water and 69% ethanol and a solvent-to-solid ratio of 6:1 (v/w) for 3 h. Because it was not possible to find an experimental value for sage's true density in the literature, a compilation of values for other vegetable materials' true densities was carried out. The arithmetical average of all the values collected resulted in an approximate value of 1350 kg/m^3. In the case of macela flowers (*Achyrocline satureioides*), the experimental data were obtained at the Laboratory of Supercritical Technology: Extraction, Fractionation, and Identification of Vegetable Extracts (LASEFI) of the State University of Campinas (UNICAMP). The extraction solvent was ethanol, and the vegetable material's true density was 1100 kg/m^3. The extraction would be carried out for 1 h with a solvent-to-solid ratio of 25:1 (v/w).

4.5.3 Equipment Sizing

The connecting point between the COM estimation performed by Rosa and Meireles [146] for a supercritical extraction unit, and the one presented here for the solvent extraction case is an extracting vessel with a useful capacity of 0.4 m^3.

Although the software SuperPro Designs also offers the possibility of performing economical evaluation, it demands an efficient and substantial feeding of its databank with both proper estimation models and actual equipment manufacturers' information. Therefore, in terms of equipment-related data, only their dimensions (Table 4.9) were provided by the software.

Even though the agitated tanks feed streams have been carefully calculated for a 0.4 m^3 extraction vessel, the software predicts a maximum occupation volume of 90%, resulting in a tank of approximately 0.44 m^3.

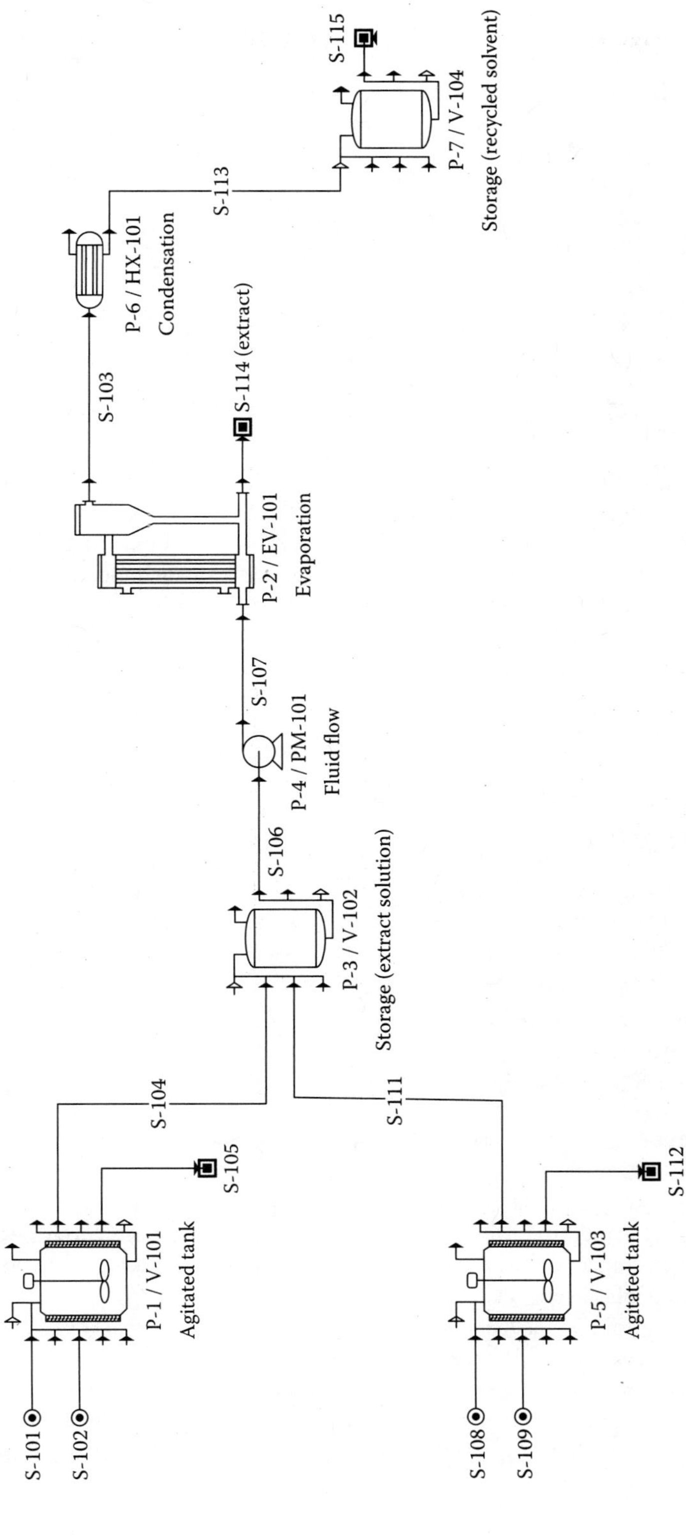

FIGURE 4.29 Graphical representation of the solvent extraction process provided by the software SuperPro Designs 6.0.

TABLE 4.9
Equipment Sizes Provided by SuperPro Designs

Equipment	Quantity	Dimension	
		S. officinalis	*A. satureioides*
Agitated tank	2	0.44 m^3	0.44 m^3
Storage tank (extract solution)	1	0.44 m^3	0.44 m^3
Storage tank (recycled solvent)	1	0.81 m^3	0.81 m^3
Centrifugal pump (4 bar of pressure increase)	1	0.02 kW and 125 L/h	0.05 kW and 305 L/h
Multiple effects evaporator	1	0.16 m^2	0.70 m^2
Condenser	1	2.18 m^2	2.5 m^2

Another important point about the agitated tanks is that the agitation power is a parameter that cannot be estimated by the software and has to be inserted by the user. According to Perry and Chilton [151], the power required to get off-bottom motion of the particles can be calculated by Equation 4.48, developed by Zweitering [152], Hirsekon and Miller [153], and Weisman and Efferding [154]. Weisman and Efferding [154] concluded that the results of the other authors agree reasonably well with their own:

$$\frac{1.74 g_C P_S}{g V_T u_S (\rho_S - \rho)} \left(\frac{1-\varepsilon_t}{\varepsilon_t} \right)^{-1/2} \frac{D_a}{D_T} = 0.16 \exp(5.3) \left(\frac{B}{D_T} \right), \tag{4.48}$$

where P_S is power to get off-bottom particle motion (ft.lb$_f$ /sec); g is acceleration due to earth's gravity (ft/sec^2); g_c is gravitational conversion factor ([32.2 lb·ft]/[lb$_f$·s^2]); ε_t is the liquid fraction based on vessel volume V_T; V_T is volume of the contents when the vessel is filled to depth equal to the diameter (ft^3); B is the distance from the impeller midplane to the vessel bottom (ft); u_s is relative velocity (ft/sec) between the particle and the fluid in a turbulent region [$1.74 \times (g\, d\, \Delta\rho/\rho)^{1/2}$]; d is particle diameter (ft); ρ_s is particle density (lb/ft^3); $\Delta\rho$ is $\rho_s - \rho$; and $0.36 < D_a/D_T < 0.43$, where D_a is the agitator or impeller diameter (ft) and D_T is the tank or vessel diameter (ft).

Comparing the two raw materials for which solvent extraction is being evaluated, sage presents the most critical characteristics in terms of the impeller design because of its higher density (1350 kg/m^3) and lower solvent-to-solid ratio (6:1, v/w). As it will be observed as this study proceeds, the purchase cost of the impeller will not significantly affect the investment cost. Therefore, the agitation power estimated for the design of an impeller for the solvent extraction of sage will be used in the study of the macela case as well.

Considering that, it was necessary to collect the sage data required by Equation 4.48 for the agitation power estimation. Thus, the following values were used: ρ = 54.054 lb/ft^3 (= 866 kg/m^3) (31% water + 69% ethanol); ρ_s = 84.265 lb/ft^3 (=1350 kg/m^3); g = 32.2 ft/sec^2 (=9.81 m/sec^2); ε_T = 0.89 for the liquid volumetric fraction; V_T = 5.2462 ft^3 (=0.1485m^3); B = 0.49 ft (=0.15 m); d = 0.0066 ft (=2 mm) [108];

$u_s = 0.6$ ft/sec (=0.18 m/sec); $D_a = 0.7532$ft (=0.2296 m); and $D_T = 1.88$ ft (=0.574 m). And the estimation of the agitation power resulted in approximately 0.55 kW.

4.5.4 Purchase Cost Estimations for Major Equipment

The next step was to estimate the purchase cost (or bare cost) of the main equipment listed in Table 4.9 and the agitators (Table 4.10).

Starting from the extraction vessels, the purchase costs for the tanks and for the propeller agitators were estimated separately. Concerning the four tanks that can be observed in Figure 4.29, their purchase costs were estimated according to the data presented in Appendix A of the book *Analysis, Synthesis, and Design of Chemical Processes* by Turton et al. [148]. The values obtained were corrected with factors for operation pressure and material of construction, presented in the same appendix. Because the equipment-related information in the book referred to the year of 2001, Equation 4.49 and the Marshall & Swift Equipment Cost Index (for the year 2005) were used to diminish the error caused by the use of dated records. The Marshall and Swift Equipment Cost Index, which is reported in the back of every issue of *Chemical Engineering*, is one of the most accepted indexes for the estimation of time effect over the equipment purchase cost:

$$C_{PC2} = C_{PC1}\left(\frac{I_2}{I_1}\right), \tag{4.49}$$

where, C_{PC} is the purchase cost and I is the cost index.

The same procedure described above was followed for the estimation of the evaporator's purchase cost. However, it was not possible to find in the same bibliographic reference adequate estimation models for the cases of the agitators, the centrifugal pump, and the condenser.

TABLE 4.10
Estimated Purchase Costs for the Main Equipment in a Solvent Extraction Facility

	Purchase Cost (US$)	
Equipment	***S. officinalis***	***A. satureioides***
Agitation tanks (extraction vessel)	42,000 (21,000 each)	42,000 (21,000 each)
Agitators	4,000 (2,000 each)	4,000 (2,000 each)
Storage tank (extract solution)	21,000	21,000
Storage tank (recycled solvent)	27,000	27,000
Centrifugal pump (4 bar of pressure increase)	5,000	5,000
Multiple effects evaporator	140,000	151,000
Condenser	28,000	77,000

In that context, the book *Plant Design and Economics for Chemical Engineers* by Peters et al. [155] was used as the alternative bibliographic reference for the cases of the centrifugal pump and the agitators. In the case of the centrifugal pump, a purchase cost of US$ 5000.00 could be estimated from Figures 12–23 of Peters et al. [155], using the correction factors for operation pressure and material of construction, whereas for the stainless steel agitator, the estimated purchase cost, attained through Figure 12–43 of Peters et al. [155], was US$ 2000.00.

The condenser's estimated purchase cost was the only value taken from the SuperPro Designs Economical Evaluation Report.

4.5.5 Capital Cost Estimation (FCI)–Lang Factor Technique (F_{LANG})

According to Turton et al. [148], the cost determined from the Lang Factor (F_{Lang}) represents the cost to build a major expansion to an existing chemical plant. The total cost is determined by multiplying the total purchase cost for all the major items of equipment by a constant. The F_{Lang} values for processing plants that operate only with fluids, only with solids, or with a combination of fluids and solids are 4.74, 3.10, and 3.63, respectively.

The case of a solvent extraction plant can also be classified as a solid–liquid extraction process. Thus, the most appropriate F_{Lang} for this case is 3.63, resulting in total capital costs of approximately US$ 970,000.00 (US$ 267,000.00 × 3.63 = US$ 969,210.00) and US$ 1,190,000.00 (US$ 327,000.00 × 3.63 = US$ 1,187,010.00) for the cases of the solvent extraction from sage and macela, respectively.

4.5.6 Raw Material Costs (C_{RM}) Estimation

Both the vegetable material and the extraction solvent are considered raw materials. To determine the quantities of each raw material component that would be required during a whole operation year, it was considered that the solvent extraction plant would operate for 330 days or 7920 h per year. This time was divided by the batch time, which is specific to each process evaluated (3 h for sage and 1 h for macela), in order to calculate the number of batches that would happen in a year, resulting in 2640 and 7920 cycles for sage and macela, respectively.

It was also important to assemble the densities and costs of every raw material component involved in the study. The ethanol (785.89 kg/m^3) and water (994.70 kg/m^3) densities at 298 K were obtained from the SuperPro Designs databank, and sage's and macela's densities had already been defined as 1350 and 1100 kg/m^3, respectively. On the other hand, according to Turton et al. [148], the costs of high purity water and ethanol are US$ 1.00/1000 kg and US$ 0.472/kg, respectively, whereas the values for sage (Hervaquímica, São Paulo, Brazil) and macela (Flor do Campo, Porto Alegre, Brazil) were provided by local producers as being approximately US$ 15.00/kg and US$ 12.00/kg, respectively.

4.5.6.1 Sage Case

In the case of sage, considering that the extraction solvent is a mixture of 31% water and 69% ethanol, that the solvent-to-solid ratio is 6:1 (v/w), and that the solvent mixture and sage densities at 298 K are 866 and 1350 kg/m^3, respectively, an extraction vessel of 0.4 m^3 of useful capacity contains 356.04 L or 308.26 kg of solvent and 43.96 L or 59.34 kg of sage. Therefore, 156,657.6 kg of sage would be submitted to extraction after 1 yr of operation, resulting in an annual cost of US$ 2,349,864.00, or approximately US$ 2,350,000.00.

As it is widely known that the mixture of ethanol and water does not present an ideal behavior, when it comes to cost estimation of sage's extraction solvent, primarily, it was considered that the mixture's final volume was produced through a weight proportion. As a second step, to avoid underestimation, the solvent cost was raised by 10%. Therefore, it was considered that 308.26 kg of solvent mixture, formed by 95.56 kg of water and 212.70 kg of ethanol, were used per extraction batch.

However, these values cannot be directly multiplied by the number of batches in a year because of the recycling of the extraction solvent. Consequently, it was decided to consider a 10% ethanol loss (more volatile) and a 5% water loss (less volatile) per batch, which means that 56,131.53 kg of ethanol and 12,609.14 kg of water would be required to replace the solvent that is lost during the extraction process.

As a result, the total amount of solvent needed for the plant operation for a year is the sum of the solvent used in the first batch plus the solvent used for replacement, resulting in 56,344.23 kg of ethanol and 12,704.7 kg of water, which correspond to costs of US$ 26,594.48 and US$ 12.70, respectively, and considering the 10% raise to account for the nonideality of the mixture, the total of US$ 26,607.18 increases up to US$ 29,267.90. Therefore, the cost related to the total extraction solvent spent in a year of operation is approximately US$ 30,000.00.

4.5.6.2 Macela Case

The same calculation procedure was followed to estimate the raw material costs related to the macela case. However, the fact that the extraction solvent is constituted only of ethanol made this process much simpler.

Once the solvent-to-solid ratio used would be 25:1 (v/w), 15.44 kg of macela and 303.33 kg could be placed in 0.4 m^3 extraction vessel. Considering that the extraction time for this species was set as 1 h, 7920 batches would be performed in a year, resulting in the requirement of 122,284.8 kg of macela flowers, with an annual cost of US$ 1,467,417.6, or approximately US$ 1,468,000.00.

When it comes to the extraction solvent, considering a 10% solvent loss per batch, the amount of ethanol required for solvent replacement would be 240,207.03 kg. This value added to the amount of solvent that would be used in the first batch results in a total of 240,510.36 kg of ethanol requirement with an annual cost of US$ 113,520.89, or approximately US$ 114,000.00.

4.5.7 Costs of Utilities (C_{UT}) Estimation

The evaporator is the equipment responsible for the solvent elimination from the extract solution, and, consequently, for the extract concentration. On the other hand, the condenser is responsible for the condensation of the solvent vapor originated in the evaporator, closing the solvent recycling process.

TABLE 4.11
Thermodynamic Properties for Water and Ethanol in SuperPro Designs Databank

Property	Water	Ethanol
Molar weight	18.02 g/gmol	46.07 g/gmol
Density (298 K)	994.70 kg/m^3	785.89 kg/m^3
Normal boiling point (T_b)	373.15 K	351.40 K
Heat capacity of the liquid (C_p)	75.25 J/(gmol·K)	113.00 J/(gmol·K)
Heat of vaporization (ΔH_v) at T_b	42,306.67 J/gmol	38,930.56 J/gmol
Heat capacity of the gas ($C_{p,gas}$ [=] J/(gmol·K))	$32.24 + 0.1924 \cdot 10^{-2} \cdot T + 0.1055 \cdot 10^{-4} \cdot T^2 - 0.6596 \cdot 10^{-8} \cdot T^3$	—

The thermodynamic properties used to estimate both the amount of vapor that feeds the evaporator and the amount of cooling water that feeds the condenser were obtained from the SuperPro Designs databank and are listed in Table 4.11.

It was considered that the vapor that would feed the evaporator would enter the equipment at 423 K and leave it at 323 K. Therefore, the estimation of the energy-to-mass factor for this vapor, according to data in Table 4.11 resulted in approximately 47.8 kJ/g, which was calculated including the heat the vapor provided when it is cooled from 423 to 373 K and when it goes through the phase change at 373 K, and the heat the liquid provides when it is cooled from 373 to 323 K.

In the case of the cooling water required by the condenser, it was considered to enter the equipment at 303 K and leave it at 313 K. Thus, the estimation of the energy to mass factor for this cooling water, according to data in Table 4.11, resulted in approximately 41.8 kJ/kg, which was calculated considering the heat absorbed by water when its temperature rises from 303 to 313 K.

It is important to point out that the methods of estimation for both the vapor and the cooling water consumption cannot be considered accurate, because of the fact that some thermodynamic principles have been neglected. One of the thermodynamic concepts that has been neglected is that both the solvent's boiling point and its heat of vaporization increase as its concentration in the solution decreases. Nevertheless, in terms of preliminary cost estimation, the effect on the approximations made should not be significant.

According to Turton et al. [148], the costs of these utilities are US$ 16.22/1000 kg and US$ 14.80/1000 m^3 for the vapor and the cooling water, respectively. Additionally, the same authors present a value of US$ 0.06/kWh for the cost of electrical power, which will have to be considered as being for the operation of both the pumps and the agitators.

4.5.7.1 Sage Case

In the case of sage, a mixture of 95.56 kg of water and 212.7 kg of ethanol is used as extraction solvent in each extraction batch. However, it will be considered that the amount of solvent used is 10% higher, in a way similar to that described in the case of the cost of raw materials estimation, in order to diminish the risk of underestimation because of the nonideality of the water–ethanol mixture.

The heat necessary to evaporate both ethanol and water from the extract solution was calculated considering both the increase of their temperatures from 298 K to their normal boiling points and their heat of vaporization related to the phase change. Consequently, it was considered that 105.12 kg of water and 233.97 kg of ethanol would demand, together, approximately 507,976 kJ or 191.6 kg of vapor per batch, which implies a total of 505,653 kg per year. As a result, the approximate vapor cost would be US$ 8200.00.

Additionally, the estimated amount of cooling water required to condensate the same quantities of water and ethanol originated by the evaporator was 10,646 kg per batch. It results in a consumption of approximately 28,104,825 kg of cooling water per year, with a related cost in the order of US$ 416,000.00.

The estimated total amount of electrical power required by the pump and the agitators was 102.96 and 3960 kWh/year, respectively, which implies a cost of approximately US$ 250.00.

Therefore, the estimated total utilities cost for the case of the solvent extraction from sage was US$ 424,450.00.

4.5.7.2 Macela Case

The amount of solvent required in each batch of solvent extraction of macela is approximately 303.33 kg of ethanol. Considering that it would have to be heated from 298 K to its normal boiling point and that additional heating would be necessary to promote the phase change, the estimated amount of vapor used per extraction batch would be on the order of 111.6 kg. Thus, approximately 883,765 kg of vapor, with a related cost of around US$ 14,350.00, would be necessary.

In terms of cooling water demand, the estimated amount of water required to condense the solvent vapor was on the order of 6139 kg per batch or 48,620,455 kg per year, with a related cost of approximately US$ 720,000.00.

When it comes to the electrical power required by both the pump and the agitators, the estimated values for annual consumption and cost were 4260.96 kWh and US$ 260.00, respectively.

Therefore, in this case, the estimated total utilities cost was on the order of US$ 734,610.00.

4.5.8 Cost of Operational Labor (C_{OL}) Estimation

According to Turton et al. [148], the technique used to estimate operating labor requirements is based on data obtained from five chemical companies and was correlated by Alkayat and Gerrard [156]. According to this method, the operating labor requirement for chemical processing plants is given by Equation 4.50:

$$N_{OL} = \left(6.29 + 31.7P^2 + 0.23N_{np}\right)^{0.5}, \tag{4.50}$$

where N_{OL} is the numbers of operators per shift, P is the number of processing steps involving the handling of particulate solids, and N_{np} is the number of nonparticulate handling processing steps.

Rosa and Meireles [146] mention a cost of US$ 3.00/h of operating labor.

Considering Equation 4.50, in the case of a solvent extraction process, as the cases of sage and macela being described here, there are two processing steps involving the handling of particulate solids: the charging and discharging of the extraction vessels. On the other hand, there are also two nonparticulate handling processing steps: evaporation and condensation.

Therefore, according to Equation 4.50, 12 operators would be necessary for the adequate performance of a solvent extraction plant. Consequently, three shifts of 8 h/day for 330 days result in 95,040 h of operating labor, with a related cost of US$ 285,120.00 per year.

4.5.9 COM Estimation

At this point of the analysis, a summary of all the cost elements that contribute to the COM estimation can be observed in Table 4.12.

The annual COM in Table 4.12 was calculated according to Equation 4.51 used by Rosa and Meireles [146] and proposed by Turton et al. [148]:

$$COM = 0.304F_{CI} + 2.73C_{OL} + 1.23 \times (C_{UT} + C_{WT} + C_{RM}). \quad (4.51)$$

The cost of waste treatment (C_{WT}) was neglected because of a conclusion very similar to that of Rosa and Meireles [146] on their study on the COM analysis for a supercritical extraction unit. They stated that the exit streams of kind of process are the exhausted solid and the CO_2 (extraction solvent) that may leak from the system. Thus, the only accumulated waste is the exhausted solid, which, being constituted of vegetable material, can be incorporated to the soil. As a result, there is no harmful waste to be treated and the CRW can be neglected.

Considering that the extraction yields are 14.5% for sage and 3.8% for macela flowers, the corresponding extracts' annual production would be 22,715 and 4647 kg. Therefore, the estimated COM was US$ 199.11/kg of extract and US$ 858.53/kg of extract for sage and macela, respectively.

TABLE 4.12
Summary of the Cost Elements in American Dollars (US$)

	S. officinalis	*A. satureioides*
Fixed capital investment (F_{CI})	970,000.00	1,190,000.00
Cost of operating labor (C_{OL})	285,120.00	285,120.00
Cost of utilities (C_{UT})	424,450.00	734,610.00
Cost of waste treatment (C_{WT})	—	—
Cost of raw material (C_{RM})	2,380,000.00	1,582,000.00
Cost of manufacturing (COM)	4,522,731.1	3,989,567.9

4.6 NOMENCLATURE

Symbol	Definition	Units	Dimensions in M, N, L, T, and θ
A	Insoluble solid or inert matrix		
A_T	Area of the solid–liquid interface	m^2	L^2
A_b	Area of the reaction vessel bottom	m^2	L^2
B	Extraction solvent		
C	Speed of light	$m \cdot s^{-1}$	LT^{-1}
C	Solute		
C_g	Concentration of solute in the solution	$mg \cdot cm^{-3}$	ML^{-3}
C_C	Concentration of the solute C in the solution	$kg \cdot m^{-3}$	ML^{-3}
C_{CS}	Reference concentration of the solute C in the solution	$kg \cdot m^{-3}$	ML^{-3}
C_{C0}	Concentration of the solute C in the solution at time t = 0	$kg \cdot m^{-3}$	ML^{-3}
CP	Heat capacity of the liquid	$J \cdot kg^{-1} \cdot K^{-1}$	$L^2 T^{-2} \theta^{-1}$
$C_{P,gas}$	Heat capacity of the gas	$J \cdot kg^{-1} \cdot K^{-1}$	$L^2 T^{-2} \theta^{-1}$
C	Concentration of solute in the solution at the external surface	$mg \cdot cm^{-3}$	ML^{-3}
D_e	Apparent intraparticle diffusion coefficient	$cm^2 \cdot s^{-1}$	L^2T^{-1}
D_{CBeff}	Effective diffusivity of the solute in the solvent/inert solid	$m^2 \cdot s^{-1}$	L^2T^{-1}
D_P	Penetration depth	m	L
E	Electrical field strength (Equation 4.41)	V·cm	$M^1L^3T^{-3}I^{-1}$
E	Extract solution stream	kg or $kg \cdot s \cdot s^{-1}$	M or $M \cdot T^{-1}$
E_{C*}	Cumulative extraction degree	—	—
E_{N-1}	Extract solution stream of the (*N*–1)th extraction stage	kg or $kg \cdot s^{-1}$	M or $M \cdot T^{-1}$
E_{N+1}	Extract solution stream of the (*N*+1)th extraction stage	kg or $kg \cdot s^{-1}$	M or $M \cdot T^{-1}$
E_1	Extract solution stream of the 1st extraction stage	kg or $kg \cdot s^{-1}$	M or $M \cdot T^{-1}$
E_2	Extract solution stream of the 2nd extraction stage	kg or $kg \cdot s^{-1}$	M or $M \cdot T^{-1}$
E_3	Extract solution stream of the 3rd extraction stage	kg or $kg \cdot s^{-1}$	M or $M \cdot T^{-1}$
E_4	Extract solution stream of the 4th extraction stage	kg or $kg \cdot s^{-1}$	M or $M \cdot T^{-1}$
f'	Frequency	Hz	T^{-1}
F	Feed stream consisted of both insoluble and soluble solids (*C*)	kg or $kg \cdot s^{-1}$	M or $M \cdot T^{-1}$
F_2	Feed stream of the 2nd extraction stage	kg or $kg \cdot s^{-1}$	M or $M \cdot T^{-1}$
k	Thermal conductivity	$W \cdot m^{-1}K^{-1}$	$M \cdot T^{-3} \cdot \theta^{-1}$
k_f	External mass transfer coefficient	$cm \cdot s^{-1}$	LT^{-1}
k_L	Mass transfer coefficient	$m \cdot s^{-1}$	LT^{-1}
k	Adsorption equilibrium constant	$cm^3 \cdot mg^{-1}$	L^3M^{-1}
$ln\ \delta$	Dissipation loss factor	—	—
M	Mass	kg	M
m	Mixture point in the single stage	kg or $kg \cdot s^{-1}$	M or $M \cdot T^{-1}$
M_2	Mixture point of the 2nd extraction stage	kg or $kg \cdot s^{-1}$	M or $M \cdot T^{-1}$
N_{np}	Number of nonparticulate handling processing steps	—	—
N_C	Rate of dissolution of the solute C in the solution	$kg \cdot s^{-1}$	MT^{-1}

continued

Symbol	Definition	Units	Dimensions in M, N, L, T, and θ
ppm	Parts per million	10^{-6}·kg·kg^{-1}	M·M^{-1}
P	Number of processing steps involving the handling of particulate solids		
P_D	Power dissipation	W·cm^{-3}	M·T^{3}·L^{-1}
P_o	Average power	W·cm^{-3}	M·T^{3}·L^{-1}
Q	Concentration of solute in the solid matrix	mg·g^{-1}	M·M^{-1}
q_0	Initial concentration of solute in the solid matrix	mg·g^{-1}	M·M^{-1}
q'''	Heat generation	W·cm^{-3}	M·T^{-3}L^{-1}
Q	Adsorption capacity parameter in Langmuir equation	mg·g^{-1}	M·M^{-1}
R	Universal gas constant (Equation 4.38)	8.314 J·gmol^{-1}·K^{-1}	ML^{2}T^{-2}N^{-1}θ^{-1}
R	Radium (Equations 4.44–4.46)	cm	L
R	Residue stream	kg or kg·s^{-1}	M or M·T^{-1}
R_N	Residue stream of the Nth extraction stage	kg or kg·s^{-1}	M or M·T^{-1}
R_{N-1}	Residue stream of the (N–1)th extraction stage	kg or kg·s^{-1}	M or M·T^{-1}
R_1	Residue stream of the 1st extraction stage (= F_2)	kg or kg·s^{-1}	M or M·T^{-1}
R_2	Residue stream of the 2nd extraction stage	kg or kg·s^{-1}	M or M·T^{-1}
R_3	Residue stream of the 3rd extraction stage	kg or kg·s^{-1}	M or M·T^{-1}
R^*	Retention index	kg or kg·s^{-1}	M or M·T^{-1}
S	Extraction solvent stream	kg or kg·s^{-1}	M or M·T^{-1}
S_2	Extraction solvent stream of the 2nd extraction stage	kg or kg·s^{-1}	M or M·T^{-1}
S/F	Solvent-to-feed ratio	kg·kg^{-1}	M·M^{-1}
t	Time	s	T
T	Absolute temperature	K	θ
T_b	Normal boiling point	K	θ
T_m	Melting point	K	θ
V	Volume	m^{3}	L^{3}
x_{AF}	Mass fraction of inert solids (A) in the feed stream (F)	kg·kg^{-1}	M·M^{-1}
x_{AR}	Mass fraction of inert solids (A) in the residue stream (R)	kg·kg^{-1}	M·M^{-1}
x_{BF}	Mass fraction of solvent B in the feed stream (F)	kg·kg^{-1}	M·M^{-1}
x_{BM}	Mass fraction of solvent B in the mixture point (M)	kg·kg^{-1}	M·M^{-1}
x_{BM2}	Mass fraction of solvent B in the 2nd extraction stage mixture point (M_2)	kg·kg^{-1}	M·M^{-1}
x_{BR}	Mass fraction of solvent B in the residue stream (R)	kg·kg^{-1}	M·M^{-1}
x_{BR1}	Mass fraction of solvent B in the 1st extraction stage residue stream (R_1)	kg·kg^{-1}	M·M^{-1}
x_b	Mass fraction of solvent in the extract solution stream	kg·kg^{-1}	M·M^{-1}
x_c	Mass fraction of solute in the extract solution stream	kg·kg^{-1}	M·M^{-1}
x_{CEN}	Mass fraction of solute C in the extract solution (E_N) of the Nth stage	kg·kg^{-1}	M·M^{-1}
x_{CF}	Mass fraction of solute C in the feed stream (F)	kg·kg^{-1}	M·M^{-1}
x_{CM}	Mass fraction of solute C in the mixture point (M)	kg·kg^{-1}	M·M^{-1}

Symbol	Definition	Units	Dimensions in M, N, L, T, and θ
x_{CM2}	Mass fraction of solute C in the 2nd extraction stage mixture point (M_2)	kg·kg^{-1}	M·M^{-1}
x_{CR}	Mass fraction of solute C in the residue stream (R)	kg·kg^{-1}	M·M^{-1}
x_{CR1}	Mass fraction of solute C in the 1st extraction stage residue stream (R_1)	kg·kg^{-1}	M·M^{-1}
$\overline{x}_i$	Molar fraction of the solute dissolved in the solvent phase at saturation	kmol·kmol^{-1}	N·N^{-1}
x_{iE}	Mass fraction of the compound i in the extract solution stream (E)	kg·kg^{-1}	M·M^{-1}
x_{iF}	Mass fraction of the compound i in the feed stream (F)	kg·kg^{-1}	M·M^{-1}
x_{iM}	Mass fraction of the compound i in the mixture point (M)	kg·kg^{-1}	M·M^{-1}
x_{iM2}	Mass fraction of the compound i in the 2nd extraction stage mixture point (M_2)	kg·kg^{-1}	M·M^{-1}
x_{iR}	Mass fraction of the compound i in the residue stream (R)	kg·kg^{-1}	M·M^{-1}
x_{iS}	Mass fraction of the compound i in the extraction solvent stream (S)	kg·kg^{-1}	M·M^{-1}
X_{BR}	Mass fraction of solvent B in the retained solution (in stream R) expressed in inert solids (A) free-basis	kg·kg^{-1}	M·M^{-1}
X_{CR}	Mass fraction of solute C in the retained solution (in stream R) expressed in inert solids (A) free-basis	kg·kg^{-1}	M·M^{-1}
y_b	Mass fraction of solvent in the residue stream	kg·kg^{-1}	M·M^{-1}
y_c	Mass fraction of solute in the residue stream	kg·kg^{-1}	M·M^{-1}
y_{BE}	Mass fraction of the solvent B in the extract solution stream (E)	kg·kg^{-1}	M·M^{-1}
y_{BS}	Mass fraction of solvent B in the extraction solvent stream (S)	kg·kg^{-1}	M·M^{-1}
y_{BS2}	Mass fraction of solvent B in the extraction solvent stream (S_2) of the 2nd extraction stage	kg·kg^{-1}	M·M^{-1}
y_{CE}	Mass fraction of solute C in the extract solution stream (E)	kg·kg^{-1}	M·M^{-1}
y_{CEN+1}	Mass fraction of solute C in the extract solution stream (E_{N+1}) of the (N+1)th extraction stage	kg·kg^{-1}	M·M^{-1}
y_{CS}	Mass fraction of solute C in the extraction solvent stream (S)	kg·kg^{-1}	M·M^{-1}
y_{CS2}	Mass fraction of solute C in extraction solvent stream (S_2) of the 2nd extraction stage	kg·kg^{-1}	M·M^{-1}
y_i	Mass fraction of the compound i in the extraction solvent stream (S) in the mixture point (M)	kg·kg^{-1}	M·M^{-1}
Z	Distance inside the porous of the solid matrix	m	L
Greek letter			
α	Thermal diffusivity	J·m^{-1}s^{-1}	L^2·T^{-1}
Δ	Flow in–flow out (in each extraction stage)	kg	M
τ	Tortuosity	—	—

continued

Symbol	Definition	Units	Dimensions in M, N, L, T, and θ
ε	Porosity of the solid	—	—
ξ	Dimensionless radial coordinate	—	—
ε'	Dielectric constant	—	—
ε''	Dielectric loss factor	—	—
ΔH_{fus}	Molar heat of fusion	$J \cdot gmol^{-1}$	$ML^2T^{-2}N^{-1}$
ΔH_V	Molar heat of vaporization at T_b	$J \cdot gmol^{-1}$	$ML^2T^{-2}N^{-1}$
γi	Activity coefficient of the compound i	—	—
ρ	Density or solvent density	$g \cdot cm^{-3}$	ML^{-3}
Variables in equations with numerical constant			
B	Distance from impeller midplane to vessel bottom	ft	L
D	Particle diameter	ft	L
D_a	Agitator or impeller diameter	ft	L
D_T	Tank or vessel diameter	m	L
g	Acceleration due to earth's gravitation	$ft \cdot s^{-2}$	LT^{-2}
g_C	Gravitational conversion factor	32.2 $lb \cdot ft \cdot lb_f^{-1} \cdot s^{-2}$	ML
P_S	Power to get off-bottom particle motion	$ft \cdot lb_f \cdot s^{-1}$	L^2MT^{-3}
u_S	Relative velocity between particle and fluid in turbulent region	$ft \cdot s^{-1}$	LT^{-1}
V_T	Volume contents when vessel is filled to depth equal to diameter	ft^3	L^3
εt	Liquid fraction based on vessel volume V_T	—	—
ρS	Particle density	$lb \cdot ft^{-3}$	ML^{-3}
Economical variables			
C_{OL}	Cost of operational labor	US$	
C_{PC}	Purchase cost	US$	
COM	Cost of manufacturing	US$	
C_{RM}	Cost of raw material	US$	
C_{UT}	Cost of utilities	US$	
C_{WT}	Cost of waste treatment	US$	
F_{CI}	Fixed capital investment	US$	
F_{Lang}	Lang factor	—	
I	Cost index	—	
N_{OL}	Number of operators per shift	—	

4.7 ACKNOWLEDGMENTS

M. E. M. Braga acknowledges Fundação para a Ciência e a Tecnologia Ministério da Ciência, Tecnologia e Ensino Superior (FCT-MCES) for the postdoctoral fellowship (SFRH/BPD/21076/2004). M. A. A. Meireles thanks Fundação de Amparo à Pesquisa do Estado de São Paulo (FAPESP), Conselho Nacional de Desenvolvimento Cientifico e Technológico, and Coordenação de Aperfeiçoamento de Pessoal de Nível Superior (CAPES) for financial support. P. F. Leal, T. M. Takeuchi, and

J. M. Prado thank FAPESP for the PhD assistantships (04/09310-3, 05/54544-5, 06/01777-5).

4.8 REFERENCES

1. Schofield, P., D. M. Mbugua, and A. N. Pell. 2001. Analysis of condensed tannins: A review. *Animal Feed Science and Technology* 91:21–40.
2. Andersen, M. L., R. K. Lauridsen, and L. H. Skibsted. 2003. Optimizing the use of phenolic compounds in foods. In *Phytochemical functional foods*. Boca Raton, FL: CRC Press, e-book.
3. Moller, J. K. S., H. L. Madsen, T. Aaltonen, et al. 1999. Dittany (*Origanum dictamnus*) as a source of water-extractable antioxidants. *Food Chemistry* 64:215–219.
4. Pokorny, J., and S. Schmidt. 2003. The impact of food processing in phytochemicals: The case of antioxidants. In *Phytochemical functional foods*. Boca Raton, FL: CRC Press, e-book.
5. Racchi, M., M. Daglia, C. Lanni, et al. 2002. Antiradical activity of water soluble components in common diet vegetables. *Journal of Agricultural and Food Chemistry* 50:1272–1277.
6. Treybal, R. E. 1981. *Mass transfer operations*. 3rd ed., 784. Singapore: McGraw Hill.
7. U.S. Food and Drug Administration. Guidance for Industry, 2003. http://www.fda.gov/cder/guidance (accessed July 11, 200).
8. Hou, K., Q. Zheng, Y. Li, et al. 2000. Modeling and optimization of herb leaching processes. *Computers and Chemical Engineering* 24:1343–1348.
9. Aguilera, J. M. 2003. Solid-liquid extraction. In *Extraction optimization in food engineering*, ed. C. Tzia and G. Liadakis, 35–55. New York: Marcel Dekker.
10. Guerrero, M. G., J. S. Torres, and M. J. Nunez. 2008. Extraction of polyphenols from white distilled grape pomace: Optimization and modeling. *Bioresource Technology* 99:1311–1318.
11. Espinoza-Pérez, J. D., A. Vargas, V. J. Robles-Olvera, et al. 2007. Mathematical modeling of caffeine kinetic during solid-liquid extraction of coffee beans. *Journal of Food Engineering* 81:72–78.
12. Cacace, J. E., and G. Mazza. 2003. Mass transfer process during extraction of phenolic compounds from milled berries. *Journal of Food Engineering* 59:379–389.
13. Simeonov, E., I. Seikova, I. Pentchev, and A. Mintchev. 2003. Modeling of a screw solid-liquid extractor through concentration evolution experiments. *Industrial & Engineering Chemistry Research* 42:1433–1438.
14. Eggers, R., and F. T. Taeger. 2003. Extraction systems. In *Extraction optimization in food engineering*, ed. C. Tzia and G. Liadakis, 95–136. New York: Marcel Dekker.
15. Frank, T. C., J. R Downey, and S. K. Gupta. 1999. Quickly screen solvent for organic solids. *Chemical Engineering Progress* 95:41–61.
16. Wang L., and C. L. Weller. 2006. Recent advances in extraction of nutraceuticals from plants. *Trends in Food Science & Technology* 17:300–312.
17. Haque, K. E. 1999. Microwave energy for mineral treatment processes—A brief review. *International Journal of Mineral Processing* 57:1–24.
18. Luque de Castro, M. D., M. M. Jiménez-Carmona, and V. Fernández-Pérez. 1999. Towards more rational techniques for the isolation of valuable essential oils from plants. *Trends in Analytical Chemistry* 18:708–716.
19. Romanik, G., E. Gilgenast, A. Przyjazny, et al. 2007. Techniques of preparing plant material for chromatographic separation and analysis. *Journal of Biochemical and Biophysical Methods* 70:253–261.
20. Letellier, M., and H. Budzinski. 1999. Microwave assisted extraction of organic compounds. *Analusis* 27:259–271.

21. Eskilsson, C. S., and E. Bjorklund. 2000. Analytical-scale microwave-assisted extraction. *Journal of Chromatography A* 902:227–250.
22. Camel, V. 2000. Microwave assisted solvent extraction of environmental samples. *Trends in Analytical Chemistry* 19 (4): 229–248.
23. Al-Harahshed, M., and S. W. Kingman. 2004. Microwave-assisted leaching—A review. *Hydrometallurgy* 73:189–203.
24. Paniwnyk, L., E. Beaufoy, J. P. Lorimer, et al. 2001. The extraction of rutin from flower buds of *Sophora japonica*. *Ultrasonics Sonochemistry* 8:299–301.
25. Toma, M., M. Vinatoru, L. Paniwnyk, et al. 2001. Investigation of effects of ultrasound on vegetal tissues during solvent extraction. *Ultrasonics Sonochemistry* 8:137–142.
26. Hromádková, Z., A. Ebringerová, and P. Valachovic. 2002. Ultrasound-assisted extraction of water-soluble polysaccharides from the roots of.valerian (*Valeriana officinalis* L.). *Ultrasonics Sonochemistry* 9:37–44.
27. Schinor, E. C., M. J. Salvador, I. C. C. Turatti, et al. 2004. Comparison of classical and ultrasound-assisted extractions of steroids and triterpenoids from three Chresta spp. *Ultrasonics Sonochemistry* 11:415–421.
28. Herrera, M. C., and M. D. Luque de Castro. 2005. Ultrasound-assisted extraction of phenolic compounds from strawberries prior to liquid chromatographic separation and photodiode array ultraviolet detection. *Journal of Chromatography A* 1100:1–7.
29. Japón-Luján, R., J. M. Luque-Rodrıguez, and M. D. Luque de Castro. 2006. Dynamic ultrasound-assisted extraction of oleuropein and related biophenols from olive leaves. *Journal of Chromatography A* 1108:76–82.
30. Ruiz-Jimenez, J., F. Priego-Capote, and M. D. Luque de Castro. 2004. Identification and quantification of *trans* fatty acids in bakery products by gas chromatography–mass spectrometry after dynamic ultrasound-assisted extraction. *Journal of Chromatography A* 1045:203–210.
31. Mulet, A., A. J. Carcel, J. Benedito, et al. 2002. Applications of low-intensity ultrasonics in the dairy industry. In *Engineering food for the 21st century*. Boca Raton, FL: CRC Press, e-book.
32. Suslick, K. S. 1990. Sonochemistry. *Science* 247:1439–1445.
33. Thompson, L. H., and L. K. Doraiswamy. 1999. Sonochemistry: Science and engineering. *Industrial & Engineering Chemistry Research* 38:1215–1249.
34. Crum, L. A. 1995. Comments on the evolving field of sonochemistry by a cavitation physicist. *Ultrasonics Sonochemistry* 2:S147–S152.
35. Elder, S. A. 1959. Cavitation microstreaming. *Journal of the Acoustical Society of America* 31:54–64.
36. Luque-Garcia, J. L., and M. D. Luque de Castro. 2003. Ultrasound: A powerful tool for leaching. *Trends in Analytical Chemistry* 22:41–47.
37. Trabelsi, F., H. Ait-Lyazidi, B. Ratsimba, et al. 1996. Oxidation of phenol in wastewater by sonoeletrochemistry. *Chemical Engineering Science* 51:1857–1865.
38. Adewuyi, Y. G. 2001. Sonochemistry: Environmental science and engineering applications. *Industrial & Engineering Chemistry Research* 40:4681–4715.
39. Albu, A., E. Joyce, L. Paniwnyk, et al. 2004. Potential for the use of ultrasound in the extraction of antioxidants from *Rosmarinus officinalis* for the food and pharmaceutical industry. *Ultrasonics Sonochemistry* 11:261–265.
40. Vinatoru, M. 2001. An overview of the ultrasonically assisted extraction of bioactive principles from herbs. *Ultrasonics Sonochemistry* 8:303–313.
41. Ji, J., X. Lu, M. Cai, et al. 2006. Improvement of leaching process of geniposide with ultrasound. *Ultrasonics Sonochemistry* 13:455–462.
42. Gertenbach, D. D. 2002. Solid-liquid extraction technologies for manufacturing nutraceuticals. In *Functional foods,* vol. 2. Boca Raton, FL: CRC Press, e-book.

43. Hu, C., Y. Zhang, and D. D. Kitts. 2000. Evaluation of antioxidant and prooxidant activities of bamboo *Phyllostachys nigra* Var. *Henonis* leaf extract in vitro. *Journal of Agricultural and Food Chemistry* 48:3170–3176.
44. Enman, J., U. Rova, and K. A. Berglund. 2007. Quantification of the bioactive compound eritadenine in selected strains of shiitake mushroom (*Lentinus edodes*). *Journal of Agricultural and Food Chemistry* 55:1177–1180.
45. Raman, G. and V. G. Gaikar. 2002. Extraction of piperine from *Piper nigrum* (black pepper) by hydrotropic solubilization. *Industrial & Engineering Chemistry Research* 41:2966–2976.
46. Basile, A., M. M. Jiménez-Carmona, and A. A. Clifford. 1998. Extraction of rosemary by superheated water. *Journal of Agricultural and Food Chemistry* 46:5205–5209.
47. Ibañez, E., A. Kubátová, F. Javier-Señoráns, et al. 2003. Subcritical water extraction of antioxidant compounds from rosemary plants. *Journal of Agricultural and Food Chemistry* 51:375–382.
48. Exarchou, V., N. Nenadis, M. Tsimidou, et al. 2002. Antioxidant activities and phenolic composition of extracts from Greek oregano, Greek sage, and summer savory. *Journal of Agricultural and Food Chemistry* 50:5294–5299.
49. Senol, A., and A. Aydin. 2006. Solid-liquid extraction of caffeine from tea waste using battery type extractor: Process optimization. *Journal of Food Engineering* 75:565–573.
50. Ranalli, A., M. L. Ferrante, G. De Mattia, et al. 1999. Analytical evaluation of virgin olive oil of first and second extraction. *Journal of Agricultural and Food Chemistry* 47:417–424.
51. Simeonov, E., I. Seikova, I. Pentchev, et al. 2003. Modeling of a screw solid-liquid extractor through concentration evolution experiments. *Industrial & Engineering Chemistry Research* 42:1433–1438.
52. Poirot, R., L. Prat, C. Gourdon, et al. 2006. Fast batch to continuous solid-liquid extraction from plants in continuous industrial extractor. *Chemical Engineering & Technology* 1:46–51.
53. Simeonov, E., I. Seikova, I. Pentchev, et al. 2004. Scale-up of the solid-liquid extraction using characteristic function technique. *Industrial & Engineering Chemistry Research* 43:4903–4907.
54. Crown Iron. 2007. http://www.crowniron.com/ (accessed July 11, 2008).
55. Gunt Hamburg Company. 2007. http://www.gunt.de/ (accessed August 10, 2007).
56. Milestone. 2007. http://www.milestonesrl.com/ (accessed July 11, 2008).
57. Paar, Anton. 2007. http://www.anton-paar.com/ (accessed August 10, 2007).
58. Pan, X., G. Niu, and H. Liu. 2001. Microwave-assisted extraction of tanshiones from *Salvia militariza bunge* with analysis by high-performance liquid chromatography. *Journal of Chromatography A* 992:371–375.
59. Csiktusnádi Kiss, G. A., E. Forgács, T. Cserháti, et al. 2000. Optimisation of the microwave-assisted extraction of pigments from paprika (*Capsicum annum* L.) powders. *Journal of Chromatography A* 889:41–49.
60. Alfaro, M. J., J. M. R. Bélanger, F. C. Padilla, and J. R. Jocelyn Paré. 2003. Influence of solvent, matrix dielectric properties, and applied power on the liquid-phase microwave-assisted processes (MAP™)[1] extraction of ginger (*Zingiber officinale*). *Food Research International* 36:499–504.
61. Lucchesi, M. E., J. Smadja, S. Bradshaw, et al. 2007. Solvent free microwave extraction of *Elletaria cardamomum* L.: A multivariate study of a new technique for the extraction of essential oil. *Journal of Food Engineering* 79:1079–1086.
62. Flamini, G , M. Tebano, P. L. Cioni, et al. 2007. Comparison between the conventional method of extraction of essential oil of *Laurus nobilis* L. and a novel method which uses microwaves applied in situ, without resorting to an oven. *Journal of Chromatography A* 1143:36–40.

63. Deng, C., J. Ji, N. Li, et al. 2006. Fast determination of curcumol, curdione and germacrone in three species of Curcuma rhizomes by microwave-assisted extraction followed by headspace solid-phase microextraction and gas chromatography-mass spectrometry. *Journal of Chromatography A* 1117:115–120.
64. García-Ayuso, L. E., and M. D. Luque de Castro. 1999. A multivariate study of the performance of a microwave-assisted Soxhlet extractor for olive seeds. *Analytica Chimica Acta* 382:309–316.
65. García-Ayuso, L. E., J. Velasco, M. C. Dobarganes, et al. 2000. Determination of the oil content of seeds by focused microwave-assisted Soxhlet extraction. *Chromatographia* 52:103–108.
66. Yu, Y., T. Huang, B. Yang, et al. 2007. Development of gas chromatography-mass spectrometry with microwave distillation and simultaneous solid-phase microextraction for rapid determination of volatile constituents in ginger. *Journal of Pharmaceutical and Biomedical Analysis* 43:24–31.
67. Rostagno M. A., M. Palma, and C. G. Barroso. 2003. Ultrasound-assisted extraction of soy isoflavones. *Journal of Chromatography A* 1012:119–128.
68. Liazid, A., M. Palma, J. Brigui, et al. 2007. Investigation on phenolic compounds stability during microwave-assisted extraction. *Journal of Chromatography A* 1140:29–34.
69. Barbero, G. F., M. Palma, and C. G. Barroso. 2006. Determination of capsaicinoids in peppers by microwave-assisted extraction-high-performance liquid chromatography with fluorescence detection. *Analytica Chimica Acta* 578:227–233.
70. Lucchesi, M. E., F. Chemat, and J. Smadja. 2004. Solvent-free microwave extraction of essential oil from aromatic herbs: Comparison with conventional hydro-distillation. *Journal of Chromatography A* 1043:323–327.
71. Chemat, F., M. E. Lucchesi, J. Smadja, et al. 2006. Microwave accelerated steam distillation of essential oil from lavender: A rapid, clean and environmentally friendly approach. *Analytica Chimica Acta* 555:157–160.
72. Chemat, S., H. AitAmar, A. Lagha, and D. C. Esveld. 2005. Microwave-assisted extraction kinetics of terpenes from caraway seeds. *Chemical Engineering and Processing* 44:1320–1326.
73. Lucchesi, M. E., J. Smadja, S. Bradshaw, W. Louw, and F. Chemat. 2007. Solvent free microwave extraction of *Elletaria cardamomum* L.: A multivariate study of a new technique for the extraction of essential oil. *Journal of Food Engineering* 79:1079–1086.
74. Lucchesi, M. E., F. Chemat, and J. Smadja. 2004. Solvent-free microwave extraction of essential oil from aromatic herbs: Comparison with conventional hydro-distillation. *Journal of Chromatography A* 1043:323–327.
75. Alfaro, M. J., J. M. R. Belangera, F. C. Padilla, and J. R .J. Pare. 2003. Influence of solvent, matrix dielectric properties, and applied power on the liquid-phase microwave-assisted processes (MAP) extraction of ginger (*Zingiber officinale*). *Food Research International* 36:499–504.
76. Pan, X., G. Niu, and H. Liu 2003. Microwave-assisted extraction of tea polyphenols and tea caffeine from green tea leaves. *Chemical Engineering and Processing* 42:129–133.
77. Hao, J., W. Han, S. Huang, B. Xue, and X. Deng. 2002. Microwave-assisted extraction of artemisinin from *Artemisia annua* L. *Separation and Purification Technology* 28:191–96.
78. Cravotto, G., L. Boffa, S. Mantegna, P. Perego, M. Avogadro, and P. Cintas. 2008. Improved extraction of vegetable oils under high-intensity ultrasound and/or microwaves. *Ultrasonics Sonochemistry* 15 (5): 898-902.
79. Fulzele, D. P., and R. K. Satdive. 2005. Comparison of techniques for the extraction of the anti-cancer drug camptothecin from *Nothapodytes foetida*. *Journal of Chromatography A* 1063:9–13.

80. Csiktusnádi Kiss, G. A. E. Forgacs, T. Cserhati, T. Mota, H. Morais, and A. Ramos. 2000. Optimisation of the microwave-assisted extraction of pigments from paprika (*Capsicum annuum* L.) powders. *Journal of Chromatography A* 889:41–49.
81. Chen, Y., M. Y. Xi, and X. F. Gong. 2007. Microwave-assisted extraction used for the isolation of total triterpenoid saponins from *Ganoderma atrum*. *Journal of Food Engineering* 81:162–170.
82. Albu, S., E. Joyce, L. Paniwnyk, J. P. Lorimer, and T. J. Mason. 2004. Potential for the use of ultrasound in the extraction of antioxidants from Rosmarinus officinalis for the food and pharmaceutical industry. *Ultrasonics Sonochemistry* 11:261–265.
83. Ma, Y., X. Ye, Y. Hao, G. Xu, and D. Liu. 2008. Ultrasound-assisted extraction of hesperidin from Penggan (Citrus reticulata) peel. *Ultrasonics Sonochemistry* 15:227–232.
84. Zhang, Y., S. Li, and X. Wu. 2008. Pressurized liquid extraction of flavonoids from *Houttuynia cordata* Thunb. *Separation and Purification Technology* 58:305–310.
85. Toma M., M. Vinatoru, L. Paniwnyk, and T. J. Mason. 2001. Investigation of the effects of ultrasound on vegetal tissues during solvent extraction *Ultrasonics Sonochemistry* 8:137–142.
86. Gutierrez, J. M. R., J. R. Jimenez, and M. D. Luque de Castro. 2008. Ultrasound-assisted dynamic extraction of valuable compounds from aromatic plants and flowers as compared with steam distillation and superheated liquid extraction. *Talanta* 75 (5): 1369–1375.
87. Vinatoru, M., M. Toma, O. Radu, et al. 1997. The use of ultrasound for the extraction of bioactive principles from plant materials. *Ultrasonics Sonochemistry* 4:135–139.
88. Jiménez, A., G. Beltrán, and M. Uceda. 2007. High-power ultrasound in olive paste pretreatment. Effect on process yield and virgin olive oil characteristics. *Ultrasonics Sonochemistry* 14 (6): 725–731.
89. Salisova, M., S. Toma, and T. J. Mason. 1997. Comparison of conventional and ultrasonically assisted extractions of pharmaceutically active compounds from *Salvia officinalis*. *Ultrasonics Sonochemistry* 4:131–134.
90. Hromadkova, Z., A. Ebringerová, and P. Valachovic. 1999. Comparison of classical and ultrasound-assisted extraction of polysaccharides from *Salvia officinalis* L. *Ultrasonics Sonochemistry* 5:163–168.
91. Velickovic, D. T., D. M. Milenovic, M. S. Ristic, et al. 2006. Kinetics of ultrasound extraction of extractive substances from garden (*Salvia officinalis* L.) and glutinous (*Salvia glutionosa* L.) sage. *Ultrasonics Sonochemistry* 13:150–156.
92. Valachovic, P., A. Pechova, and T. J. Mason. 2001. Towards the industrial production of medicinal tincture by ultrasound assisted extraction. *Ultrasonics Sonochemistry* 8:111–117.
93. Hromádková, Z., and A. Ebringerov. 2003. Ultrasonic extraction of plant materials-investigation of hemicellulose release from buckwheat hulls. *Ultrasonics Sonochemistry* 10:127–133
94. Schinor, E. C., M. J. Salvador, I. C. C. Turatti, O. L. A. D. Zucchi, and D. A. Dias. 2004. Comparison of classical and ultrasound-assisted extractions of steroids and triterpenoids from three Chresta spp. *Ultrasonics Sonochemistry* 11:415–421.
95. Paniwnyk, L., E. Beaufoy, J. P. Lorimer, and T. J. Mason. 2001. The extraction of rutin from flower buds of *Sophora japonica*. *Ultrasonics Sonochemistry* 8:299–301.
96. Wu, J., L. Lin, and F. Chau. 2001. Ultrasound-assisted extraction of ginseng saponins from ginseng roots and cultures ginseng cells. *Ultrasonics Sonochemistry* 8:347–352.
97. SinglePush-transducer. Die Produckte. 2008. Based on SinglePush-transducer of Martin Walter Ultraschalltechnik. http://www.walter-ultraschall.de (accessed July 11, 2008).
98. Faıd, F., F. Contamine, A. M. Wilhelm, et al. 1998. Comparison of ultrasound effects in different reactors at 20 kHz. *Ultrasonics Sonochemistry* 5:119–124.

99. Chemat, S., A. Lagha, H. AitAmar, et al. 2004. Comparison of conventional and ultrasound-assisted extraction of carvone and limonene from caraway seeds. *Flavour and Fragrance Journal* 19:188–195.
100. Faid, F., M. Romdhane, C. Gourdon, et al. 1998. A comparative study of local sensors of power ultrasound effects: Electrochemical, thermoelectrical and chemical probes. *Ultrasonics Sonochemistry* 5:63–68.
101. Saez, V., A. Frias-Ferrer, J. Iniesta, et al. 2005. Characterization of a 20 kHz sonoreactor. Part I: Analysis of mechanical effects by classical and numerical methods. *Ultrasonics Sonochemistry* 12:59–65.
102. Saez, V., A. Frias-Ferrer, J. Iniesta, et al. 2005. Characterization of a 20 kHz sonoreactor. Part II: Analysis of chemical effects by classical and electrochemical methods. *Ultrasonics Sonochemistry* 12:67–72.
103. Romdhane, M., and C. Gourdon. 2002. Investigation in solid-liquid extraction: Influence of ultrasound. *Chemical Engineering Journal* 87:11–19.
104. Ma, Y., X. Ye, Y. Hao, et al. 2008. Ultrasound-assisted extraction of hesperidin from Penggan (*Citrus reticulata*) peel. *Ultrasonics Sonochemistry* 15 (3): 227–232.
105. Suhaj, M. 2006. Spice antioxidants isolation and their antiradical activity: A review. *Journal of Food Composition and Analysis* 19:531–537.
106. Bastos, D. H. M., E. Y. Ishimoto, M. O. M. Marques, et al. 2006. Essential oil and antioxidant activity of green mate and mate tea (*Ilex paraguariensis*) infusions. *Journal of Food Composition and Analysis* 19:538–554.
107. Callemien, D., V. Jerkovic, R. Rozenberg, et al. 2005. Hop as an interesting source of resveratrol for brewers: Optimization of the extraction and quantitative study by liquid chromatography/atmospheric pressure chemical ionization tandem mass spectrometry. *Journal of Agricultural and Food Chemistry* 53:424–429.
108. Durling, N. E., O. J. Catchpole, J. B. Grey, et al. 2007. Extraction of phenolics and essential oil from dried sage (*Salvia officinalis*) using ethanol–water mixtures. *Food Chemistry* 101:1417–1424
109. Farhoosh, R., G. A. Golmovahhed, and M. H. H. Khodaparast. 2007. Antioxidant activity of various extracts of old tea leaves and black tea wastes (*Camellia sinensis* L.). *Food Chemistry* 100:231–236.
110. Dormana, H. J. D., A. Peltoketoa, R. Hiltunena, et al. 2003. Characterisation of the antioxidant properties of de-odourised aqueous extracts from selected Lamiaceae herbs. *Food Chemistry* 83:255–262.
111. Grigonisa, D., P. R. Venskutonisa, B. Sivik, et al. 2005. Comparison of different extraction techniques for isolation of antioxidants from sweet grass (*Hierochlöe odorata*). *Journal of Supercritical Fluids* 33:223–233.
112. Jayaprakasha G. K., P. S. Negi, B. S. Jena, and L. Jagan Mohan Rao. 2007. Antioxidant and antimutagenic activities of *Cinnamomum zeylanicum* fruit extracts. *Journal of Food Composition and Analysis* 20:330–333.
113. Liyana-Pathirana, C. and F. Shahidi. 2005. Optimization of extraction of phenolic compounds from wheat using response surface methodology. *Food Chemistry* 93 (1): 47–56.
114. Zhou, K. and L. Yu. 2004. Effects of extraction solvent on wheat bran antioxidant activity estimation. *Lebensmittel-Wissenschatf und Technologie* 37:717–721.
115. Kim, Y. J., O.-K. Kim, O. K. Chung, et al. 2005. Phenolic extraction from apple peel by cellulases from *Thermobifida fusca*. *Journal of Agriculture and Food Chemistry* 53:9560–9565.
116. Silva, E. M., H. Rogez, and Y. Larondelle. 2007. Optimization of extraction of phenolics from *Inga edulis* leaves using response surface methodology. *Separation and Purification Technology* 55:381–387.
117. Jaganyi, D., and P. J. Wheeler. 2003. Rooibos tea: Equilibrium and extraction kinetics of aspalathin. *Food Chemistry* 83:121–126.

118. Herodez, S.S., M. Hadolin, M. Skerget, et al. 2003. Solvent extraction study of antioxidants from Balm (*Melissa officinalis* L.) leaves. *Food Chemistry* 80:275–282.
119. Schoefs, B. 2004. Review: Determination of pigments in vegetables. *Journal of Chromatography A* 1054:217–226.
120. Tsai, P.-J., J. McIntosh, P. Pearce, et al. 2002. Anthocyanin and antioxidant capacity in Roselle (*Hibiscus sabdariffa* L.) extract. *Food Research International* 35:351–356.
121. Lapornik, B., M. Prosek, and A. G. Wondra. 2005. Comparison of extracts prepared from plant by-products using different solvents and extraction time. *Journal of Food Engineering* 71:214–222.
122. Chen, F., Y. Sun, G. Zhao, et al. 2007. Optimization of ultrasound-assisted extraction of anthocyanins in red raspberries and identification of anthocyanins in extract using high-performance liquid chromatography–mass spectrometry. *Ultrasonics Sonochemistry* 14 (6): 767–778.
123. Eichhorn, S. and P. Winterhalter. 2005. Anthocyanins from pigmented potato (*Solanum tuberosum* L.) varieties. *Food Research International* 38:943–948.
124. Hu, C., Y.-Z. Cai, W. Li, et al. 2007. Anthocyanin characterization and bioactivity assessment of a dark blue grained wheat (*Triticum aestivum* L. cv. Hedong Wumai) extract. *Food Chemistry* 104 (3): 955–961.
125. Luque-Rodriguez, J. M., M. D. Luque de Castro, and P. Perez-Juan. 2007. Dynamic superheated liquid extraction of anthocyanins and other phenolics from red grape skins of winemaking residues. *Bioresource Technology* 98:2705–2713.
126. Choudhari, S. M., and L. Ananthanarayan. 2007. Enzyme aided extraction of lycopene from tomato tissues. *Food Chemistry* 102:77–81.
127. Santamaria, R. I., M. D. Reyes-Duarte, E. Barzana, et al. 2000. Selective enzyme-mediated extraction of capsaicinoids and carotenoids from chili guajillo puya (*Capsicum annuum* L.) using ethanol as solvent. *Journal of Agricultural and Food Chemistry* 48:3063–3067.
128. Sadler, G., J. Davies, and D. Dezman. 1990. Rapid extraction of lycopene and beta-carotene from reconstituted tomato paste and pink grape fruit homogenates. *Journal of Food Science* 55:1460–1461.
129. Çinar, I. 2005. Effects of cellulase and pectinase concentrations on the colour yield of enzyme extracted plant carotenoids. *Process Biochemistry* 40:945–949.
130. Fan, G., Y. Han, Z. Gu, et al. 2007. Optimizing conditions for anthocyanins extraction from purple sweet potato using response surface methodology (RSM). *LWT—Food Science and Technology* 41 (1): 155–160.
131. Chirinos, R., H. Rogez, D. Camposa, et al. 2007. Optimization of extraction conditions of antioxidant phenolic compounds from mashua (*Tropaeolum tuberosum* Ruız & Pavon) tubers. *Separation and Purification Technology* 55:217–225.
132. Naczk, M., and F. Shahidi. 2006. Phenolics in cereals, fruits and vegetables: Occurrence, extraction and analysis. *Journal of Pharmaceutical and Biomedical Analysis* 41:1523–1542.
133. Markom, M., M. Hasan, W. Ramli, et al. 2007. Extraction of hydrolysable tannins from *Phyllanthus niruri* Linn.: Effects of solvents and extraction methods. *Separation and Purification Technology* 52:487–496.
134. Pinelo, M., M. Rubilar, M. Jerez, et al. 2005. Effect of solvent, temperature, and solvent-to-solid ratio on the total phenolic content and antiradical activity of extracts from different components of grape pomace. *Journal of Agriculture and Food Chemistry* 53:2111–2117.
135. Bourzeix, M., and E. Revilla. 1991. Suitability of water/ ethanol mixtures for the extraction of catechins and proanthocyanidins from *Vitis vinifera* seeds contained in a winery by-product. *Seed Science and Technology* 19:542–552

136. Yilmaz, Y., and R. T. Toledo. 2006. Oxygen radical absorbance capacities of grape/wine industry byproducts and effect of solvent type on extraction of grape seed polyphenols. *Journal of Food Composition and Analysis* 19:41–44.
137. Pinelo, M., P. Del Fabbro, L. Marzocco, et al. 2005. Optimization of continuous phenol extraction from *Vitis vinifera* byproducts. *Food Chemistry* 92:109–117.
138. Spigno, G., L. Tramelli, and D. M. De Faveri. 2007. Effects of extraction time, temperature and solvent on concentration and antioxidant activity of grape marc phenolics. *Journal of Food Engineering* 81:200–208.
139. Cacace, J. E., and G. Mazza. 2003. Optimization of extraction of anthocyanins from black currants with aqueous ethanol. *Journal of Food Science* 68:240–248.
140. Bucic-Kojic, A., M. Planinic, S. Tomas, et al. 2007. Study of solid-liquid extraction kinetics of total polyphenols from grape seeds. *Journal of Food Engineering* 81:236–242.
141. Lafka, T.-I., V. Sinanoglou, and E. S. Lazos. 2007. On the extraction and antioxidant activity of phenolic compounds from winery wastes, *Food Chemistry* 104 (3): 1206–1214.
142. Pinelo, M., M., A. Arnous, and A. S. Meyer. 2006. Upgrading of grape skins: Significance of plant cell-wall structural components and extraction techniques for phenol release. *Trends in Food Science and Technology* 17:579–590.
143. Bonilla, F., M. Mayen, J. Merida, et al. 1999. Extraction of phenolic compounds from red grape marc for use as food lipid antioxidants, *Food Chemistry* 66:209–215.
144. Escribano-Bailon, M. T., and C. Santos-Buelga. 2003. Polyphenol extraction from foods. In *Methods in polyphenol analysis*, ed. C. Santos-Buelga and G. Williamson, 1–12. London: The Royal Society of Chemistry.
145. Naczk, M., and F. Shahidi. 2004. Extraction and analysis of phenolics in food. *Journal of Chromatography A* 1054:95–111.
146. Rosa, P. T. V., and M. A. A. Meireles. 2005. Rapid estimation of the manufacturing cost of extracts obtained by supercritical fluid extraction. *Journal of Food Engineering* 67:235–240.
147. Cost Estimate Classification System. In *AACE International Recommended Practice* No. 17R–97.
148. Turton, R., R. C. Bailie, W. B. Whiting, et al. 2003. *Analysis, synthesis, and design of chemical processes.* 2nd ed. Upper Saddle River, NJ: Prentice Hall.
149. Rakotondramasy-Rabesiaka, L., J.-L. Havet, C. Porte, et al. 2007. Solid-liquid extraction of protopine from *Fumaria-officinalis* L.—Analysis determination, kinetic reaction and model building. *Separation and Purification Technology* 54:253–261.
150. Takeuchi, T. M., P. F. Leal, R. Favareto, et al. 2008. Study of the phase equilibrium formed inside the flash tank used at the separation step of a supercritical fluid extraction unit. *Journal of Supercritical Fluids* 43:447–459.
151. Perry, R. H., and C. H. Chilton. 1973. *Chemical engineers' handbook.* 5th ed. 19–11. Tokyo: McGraw-Hill Kogakuṣha.
152. Zwietering, T. N. 1958. Suspending of solid particles in liquid by agitators. *Chemical Engineering Science* 8 (3–4): 244–253.
153. Hirsekorn, F. S., and S. A. Miller. 1953. Agitation of viscous solid-liquid suspensions. *Chemical Engineering Progress* 49:459–467.
154. Weisman, J., and L. E. Efferding. 1960. Suspension of slurries by mechanical mixers. *AIChE Journal* 6 (3): 419–426.
155. Peters, M. S., K. D. Timmerhaus, and R. E. West. 2003. *Plant design and economics for chemical engineers.* 5th ed., 485–591. New York: McGraw-Hill.
156. Alkayat, W. A., and A. M. Gerrard. 1984. Estimating manning levels for process plants. *AACE Transactions* I.2.1–I.2.4.

5 Liquid–Liquid Extraction Applied to the Processing of Vegetable Oil

Eduardo A. C. Batista, Antonio J. A. Meirelles, Christianne E. C. Rodrigues, and Cintia B. Gonçalves

CONTENTS

5.1 Fundamentals of Liquid–Liquid Extraction 220
5.1.1 Equipment 221
5.1.1.1 Equipment for Liquid–Liquid Extraction 221
5.1.1.2 Equipment for Stagewise Contact 222
5.1.1.3 Equipment for Continuous Contact 222
5.1.1.4 Centrifugal Extractors 223
5.1.2 Liquid–Liquid Equilibrium Diagram for Fatty System and Short-Chain Alcohol Systems 224
5.1.3 Mass Transfer: Mass Balance Equations 225
5.1.3.1 Lever-Arm Rule 225
5.1.3.2 Single-Stage Equilibrium Extraction 227
5.1.3.3 Multistage Crosscurrent Extraction 228
5.1.3.4 Continuous Multistage Countercurrent Extractor 232
5.1.4 Thermodynamic: Phase Equilibrium 234
5.1.5 Group Contribution Models 236
5.1.5.1 UNIFAC Model 237
5.1.5.2 ASOG Model 237
5.1.5.3 Minor Component 238
5.1.6 Simulation of a Liquid–Liquid Extraction Column 239
5.2 State of the Art—Mini-Review of the Literature 241
5.3 Applications 247
5.3.1 Deacidification of Vegetable Oils 247
5.3.1.1 Effect of Temperature 247
5.3.1.2 Length Chain of Alcohols 247
5.3.1.3 Addition of Water in the Solvent 249

5.3.2 Deacidification of Vegetable Oils Retaining Bioactive Compounds 249
5.4 Nomenclature 258
5.5 References 259

In this chapter, we will discuss the fundamentals of the liquid–liquid extraction process applied to deacidification of vegetable oils with some special attention to the retention of bioactive compounds. Deacidification is the removal of free fatty acids from vegetable oils, and it is the most difficult step in oil refining, mainly because of its impact on productivity. Deacidification of oils is usually performed by chemical, physical, or miscella methods. Liquid–liquid extraction is a quite promising process for deacidification of vegetable oils that minimizes the loss of neutral oil and retains bioactive compounds. In the first part of this chapter, fundamentals of liquid–liquid extraction, the main concepts of the equipment for stagewise and continuous contact types, the liquid–liquid equilibrium diagram for fatty components and short-chain alcohol systems, distribution coefficients and selectivity of the solvent, mass transfer and some graphical methods for solving the equilibrium and mass balances, the most important thermodynamic models for description or prediction of liquid–liquid equilibrium, and the mathematical basis for simulating a stagewise column are presented and discussed. In the second part, a review of the literature in applying liquid–liquid extraction in the food and food-related processes are presented. In the last part of this chapter, we present our own results in the deacidification of vegetable oils and the retention of bioactive compounds.

5.1 FUNDAMENTALS OF LIQUID–LIQUID EXTRACTION

Crude vegetable oils are a mixture of triacylglycerols, partial acylglycerols, free fatty acids, phosphatides, pigments, sterols, and tocopherols. Refining procedures have been developed over decades to make the vegetable oil suitable for edible use. Some of the minor components are valuable and should be retained in the refined oil or recovered from the stream generated in the refining processes.

Fatty acids are almost straight chain aliphatic carboxylic acids. The most natural fatty acids are C4 to C22, with varying chain length and unsaturation. Systematic names for fatty acids are complicated for casual use. Two numbers separated by a colon represent the number of carbons and number of double bounds. The position of double bounds could be indicated from the carboxyl end of the chain, shown as Δx, where x is the number of carbons from the carboxyl end. The double-bound geometry *cis* and *trans* is represented by abbreviations c and t, respectively. Some fatty acids have common names that facilitate their identification. Nomenclatures and formulas for some fatty acids are presented in Table 5.1.

Triacylglycerols are triesters of glycerol (1,2,3-trihydroxypropane) with fatty acids. Most triacylglycerols do not have a random distribution of fatty acids on the glycerol backbone. In vegetable oils, unsaturated fatty acids predominate at position 2 of the glycerol backbone. Simplified structures and abbreviations are used to identify the fatty acids esterified to glycerol; e.g., 1-stearoyl-2-oleoyl-3-stearoyl-*sn*-glycerol is abbreviated to SOS.

The removal of free fatty acids, deacidification, is the most difficult step in oil refining, mainly because of its impact on the productivity. Deacidification of oils is

TABLE 5.1
Nomenclature and Formulas for Some Fatty Acids

Fatty acid	Common name	Symbol	Formula
8:0	Caprylic		$CH_3(CH_2)_6COOH$
10:0	Capric		$CH_3(CH_2)_8COOH$
12:0	Lauric	La	$CH_3(CH_2)_{10}COOH$
14:0	Myristic	M	$CH_3(CH_2)_{12}COOH$
16:0	Palmitic	P	$CH_3(CH_2)_{14}COOH$
18:0	Stearic	S	$CH_3(CH_2)_{16}COOH$
18:1, 9c	Oleic	O	$CH_3(CH_2)_7CH{=}CH(CH_2)_7COOH$
18:2, 9c12c	Linoleic	L	$CH_3(CH_2)_4(CH{=}CHCH_2)_2(CH_2)_6COOH$
18:3, 9c12c15c	Linolenic	Ln	$CH_3CH_2(CH{=}CHCH_2)_3(CH_2)_6COOH$
22:1, 13c	Erucic	E	$CH_3(CH_2)_7CH{=}CH(CH_2)_{11}COOH$

performed by chemical, physical, and miscella methods. Most edible oils are produced by chemical refining [1] because it is a highly versatile process applicable for all crude oil. However, for oils with high acidity, chemical refining causes high losses of neutral oil as a result of saponification and emulsification. For highly acidic oils, the physical method is also a feasible process for deacidification that results in a lower loss of neutral oil than the chemical method, but more consumption of energy is required, and the refined oil is subject to undesirable alteration in color and to a reduction of stability with regard to resisting oxidation. The miscella method is the deacidification of crude oil prior to solvent stripping. In this process, the neutralization reaction of free fatty acids with sodium hydroxide occurs in the miscella, which is a mixture of 40%–60% oil in hexane. Bhosle and Subramanian [2] present some new approaches that may be used as alternatives to current industrial deacidification, such as biological deacidification, reesterification, supercritical fluid extraction, membrane technology, and liquid–liquid extraction.

Liquid–liquid extraction is an alternative process carried out at room temperature and atmospheric pressure. According to Thomopoulos [3], this process is based on the difference in the solubility of free fatty acids and triacylglycerols in the solvent, as well as on the difference of boiling points of triacylglycerols, free fatty acids, and solvent during the subsequent separation.

Currently, cleaner processes have been developed because of environmental issues, and there is a demand for new products retaining minor compounds with bioactive properties. Liquid–liquid extraction is a quite promising process that minimizes the loss of neutral oil and retains bioactive compounds. The streams leaving the extract column, raffinate and extract, will be separated by other unity operations and a nonpolluting stream is generated.

5.1.1 Equipment

5.1.1.1 Equipment for Liquid–Liquid Extraction

The rate of mass transfer between two liquid phases is described by $N = KA\Delta c$, where N is the mass transfer rate, K is the overall mass transfer coefficient, A is the interfacial area, and Δc is the composition difference driving force. The rate may be increased by dispersing one of the liquids into smaller droplets, which are immersed into the other, with resulting large interfacial area. This favors eddy diffusion rather than molecular diffusion, which is slow.

Equipment for liquid–liquid extraction provides the direct contact of two immiscible liquids that are not in equilibrium, which involves dispersing one liquid in the form of small droplets (the dispersed phase) into the other liquid (continuous phase) in attempting to bring the liquids to equilibrium, and these resulting liquids are mechanically separated.

5.1.1.2 Equipment for Stagewise Contact

The typical and oldest extraction equipment is known as mixer-settler, in which each stage presents two well-defined and delimited regions: the first, the mixer, involves dispersing one of the liquids to the other and the second, the settler, involves the mechanical separation. Such an operation may be carried out in batch or continuous flow. If batch, the same vessel will be used for both mixing and settling; if continuous, the mixer and settler usually are in different vessels. The mixing vessel uses some form of rotating impeller placed on its center, which provides an effective dispersion of phases. The simplest settler is a decanter, and a baffle may be used to protect the vessel from the disturbance caused by the flow entering the dispersion. This basic unity of mixer-settler may be connected to form a cascade for cross-flow or, more often, countercurrent flow.

The perforated-plate (sieve-plate) column is similar to a tray distillation column. The plates contain downspouts in their free extremity, which allow the downward flow of the heavy liquid (continuous phase). Below each plate and outside the downspout, the droplets of the light phase (dispersed one) coalesce and accumulate in a liquid layer. This layer of liquid flows through the holes of the plate and is dispersed in a large number of droplets within the continuous phase located above the plate.

5.1.1.3 Equipment for Continuous Contact

In this equipment, two immiscible liquids flow countercurrently in continuous contact as a result of the difference in density of the liquid streams without settling. The force of gravity acts to provide the flows, and the equipment is usually a vertical column, with the light liquid entering at the bottom and the heavy one at the top. The complete separation of phases occurs only in one extremity of the equipment, in the top, if the dispersed phase is the light liquid, or in the bottom, if the heavy liquid is dispersed.

The simplest equipment for differential contact is the spray column, which consists basically of an empty shell with provision for introducing and removing the liquids.

If the light liquid is dispersed, the heavy liquid enters at the top through the distributor and fills the column, flows downward as a continuous phase, and leaves at the bottom. The light liquid enters at the bottom of the column by a distributor, which disperses it into small droplets. These droplets flow upward through the continuous phase, coalesce, and form an interface at the top of the column, and the light liquid leaves the equipment. Although this column is easily constructed, its use is not recommended because of its low efficiency in mass transfer as a result of absence of accessories that improve the dispersion or high axial mixture.

In packed columns, the shell of the column may be filled with a random or a structural packing arrangement. In the first case, the packing is constituted of elements one-eighth of the diameter of the column, which is made for a gas–liquid system such as Raschig, Lessing, and Paul rings, and Berl and Intalox saddles, disposed in random arrangement with intermediate support grids. The packing is made of ceramic, metal, or polymeric materials. Structured packing is formed from vertical corrugated thin sheets of ceramic, metal, or plastic with the angle of the corrugations reversed in adjacent sheets to form a very open honeycomb structure with inclined channels and a high surface area. To simplify installation, the packing is found in segments of diameter near to that of the diameter of the column. Liquid distribution is crucial for a proper distribution of the liquids in the column. The material of packing must be chosen to ensure that the continuous phase will wet it preferentially and the droplets will not coalesce.

Extractors could also be mechanically agitated in a fashion somewhat similar to that of the mixer-settler. There is a great variety of mechanically agitated columns for continuous contact.

The first example is the Rotating Disk Contactor column or simply RDC column, which has a number of horizontal stator rings fixed in the shell that divides the extractor into a number of chambers. A series of circular flat disks is fixed on a rotating central shaft and is centered in each chamber. In the literature, we could find modifications of the original RDC column, such as the ones that use perforated disks (PRDC) or columns without stators.

The Khüni column has a rotating shaft with impellers that are fixed in the center of a compartment delimited by two adjacent perforated plates. These plates help to control the volumetric fraction of the dispersed phase held inside the column. In the York–Scheibel column, the agitation is similar to the Khüni column, but each compartment with impellers is separated from the others by packing sections.

Pulsed columns are a variation of agitated columns, where perforated plates move up and down or the liquids are pulsed in a stationary column by an outside mechanism. This type of agitation is compatible with other extractors, like packed or perforated-plate columns.

5.1.1.4 Centrifugal Extractors

The most important centrifugal extractor is the Podbielniak extractor, which has a horizontal shaft that rotates a cylindrical drum rapidly (30–85 rps). There are perforated concentric plates inside the drum. The two liquids are fed into the equipment by the shaft, and the centrifugal force moves the light liquid to the center and the heavy to the wall of the drum countercurrently. Both phases leave the equipment through the shaft in the opposite sides of their feed. These extractors are important when short residence times are necessary and for liquids with a small density difference. Continuous centrifuges can also be used connected to a settler to accelerate the separation of the phases.

More information about equipment for liquid–liquid extraction can be found in Treybal [4] and Godfrey and Slater [5].

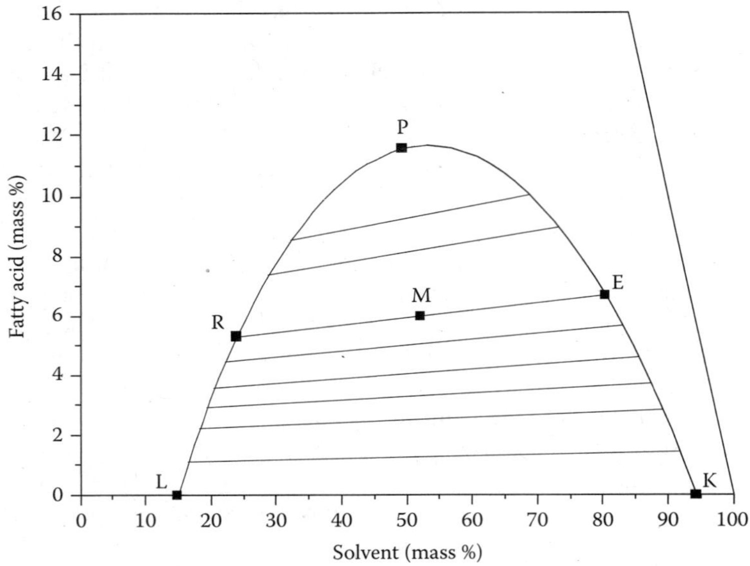

FIGURE 5.1 Liquid–liquid equilibrium diagram (K to L, base line; R to E, tie line; M, overall composition; P, plait point).

5.1.2 Liquid–Liquid Equilibrium Diagram for Fatty System and Short-Chain Alcohol Systems

In the system of vegetable oil (1) + free fatty acids (2) + short-chain alcohol (3), only the pair (1) + (3) is partially soluble. The diagrams in triangular coordinates are used at constant temperature and pressure. In a rectangular coordinate, abscissa and ordinate present the composition of the short-chain alcohol (component 3) and the free fatty acid (component 2), respectively.

Figure 5.1 presents an example of a liquid–liquid equilibrium diagram of this fatty system, of which the components 1 (vegetable oil) and 3 (short-chain alcohols) are partially miscible.

The component 2, the free fatty acid, dissolves completely in vegetable oil (1) and short-chain alcohol (3), but 1 and 3 dissolve only to a limited extend, and they are represented in the diagram by the saturated liquid binary solutions at L (rich in oil, 1) and at K (rich in short-chain alcohols, 3). Any binary mixture between L and K will separate into two immiscible liquids with composition at L and K. The point L represents the solubility of the short-chain alcohol in the vegetable oil, and the point K, the solubility of the vegetable in the short-chain alcohols.

The LRPEK curve is the binodal curve and represents the change in solubility of the phase rich in the vegetable oil (oil phase) and the phase-rich short-chain alcohol (alcoholic phase). Outside this curve, any ternary mixture will be a solution of one phase. Underneath this curve, any ternary mixture, such as mixture M, will form two immiscible mixtures of equilibrium composition indicated at R (oil phase) and E (alcoholic phase). The line RE is a tie line and must necessarily pass through point M, which represents the overall composition.

The point P, known as the plait point, is the last tie line where the binodal curve converges and the composition of the oil and alcoholic phases are equal.

The distribution coefficient (*ki*) of component *i* is defined as the ratio of its composition in phase II (alcoholic phase) to its composition in phase I (oil phase):

$$k_i = \frac{w_i^{II}}{w_i^{I}}. \tag{5.1}$$

In the example presented in Figure 5.1, the composition of free fatty acid (2) in phase II is larger than in phase I and hence the distribution coefficient will be larger than 1.

The capacity of short-chain alcohols (3) for separating the free fatty acid (2) from vegetable oil (1) is measured by the ratio of the distribution coefficient of the free fatty acid (2) to the distribution coefficient of the vegetable oil (1). This factor of separation is known as selectivity and represents the effectiveness of a short-chain alcohol in extracting the free fatty acid from the vegetable oil. Then the selectivity must exceed unity, and the greater values are the better, that is, the separation is easier:

$$\beta_{ij} = \frac{k_i}{k_j}. \tag{5.2}$$

5.1.3 Mass Transfer: Mass Balance Equations

In this section, we present the mass balances for an extractor of the stagewise type. Each stage is a theoretical stage, such that the extract and raffinate streams that are leaving are in equilibrium. In the next topic, we discuss the lever-arm rule for graphical addition in rectangular coordinates that will be useful for understanding the solutions.

5.1.3.1 Lever-Arm Rule

If a mixture with R kg is added to another E kg, both containing A, B, and C components, a new ternary mixture is generated with M kg. This mixing process is represented in Figure 5.2 and the lever-arm rule in Figure 5.3.

We can write the global mass and mass balance for components B and C as follows:

Global mass balance:

$$\mathrm{R} + \mathrm{E} = \mathrm{M}, \tag{5.3}$$

Mass balance for component B:

$$\mathrm{R}x_{\mathrm{B,R}} + \mathrm{E}y_{\mathrm{B,E}} = \mathrm{M}x_{\mathrm{B,M}}, \tag{5.4}$$

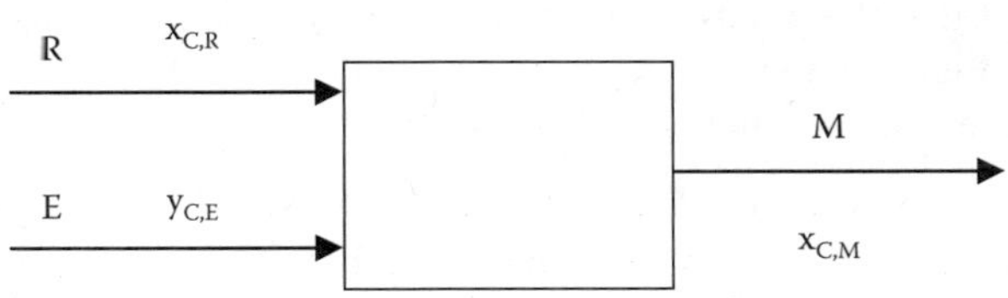

FIGURE 5.2 Mixing process.

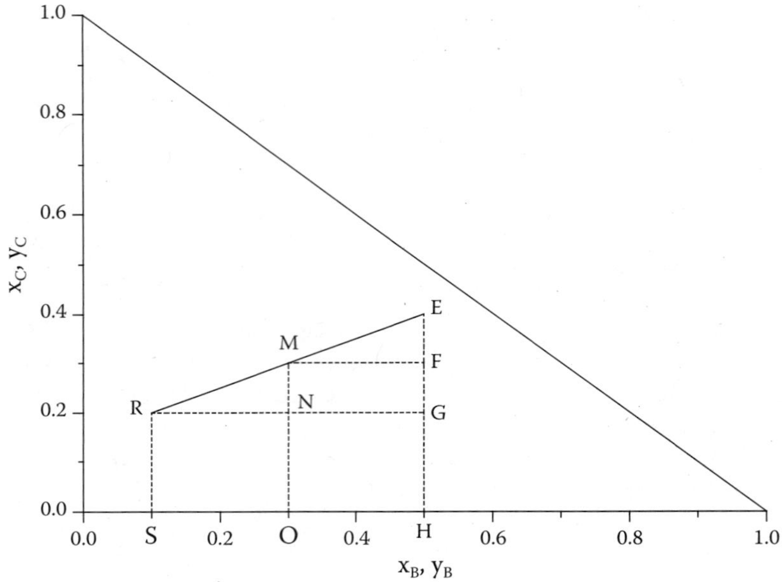

FIGURE 5.3 Lever-arm rule in rectangular coordinates.

Mass balance for component C:

$$Rx_{C,R} + Ey_{C,E} = Mx_{C,M}, \tag{5.5}$$

substituting Equation 5.3 into 5.4 and rearranging,

$$\frac{R}{E} = \frac{y_{B,E} - x_{B,M}}{x_{B,M} - x_{B,R}}, \tag{5.6}$$

substituting Equation 5.3 into 5.5 and rearranging,

$$\frac{R}{E} = \frac{y_{C,E} - x_{C,M}}{x_{C,M} - x_{C,R}}, \tag{5.7}$$

combining Equations 5.6 and 5.7 and rearranging,

$$\frac{x_{C,M} - x_{C,R}}{x_{B,M} - x_{B,R}} = \frac{y_{C,E} - x_{C,M}}{y_{B,E} - x_{B,M}}. \tag{5.8}$$

This shows that the points R, M, and E must be lined up. This straight line is represented in Figure 5.3.

From Figure 5.3, one can see that if

$$x_{C,R} = \text{line RS or } \overline{RS}$$
$$y_{C,E} = \text{line EH or } \overline{EH}$$
$$x_{C,M} = \text{line MO or } \overline{MO},$$

then $$\frac{R}{E} = \frac{y_{C,E} - x_{C,M}}{x_{C,M} - x_{C,R}} = \frac{\overline{EH} - \overline{FH}}{\overline{MO} - \overline{RS}} = \frac{\overline{EF}}{\overline{MN}},$$

and by using a similar right angle triangle,

$$\frac{R}{E} = \frac{\overline{EF}}{\overline{MN}} = \frac{\overline{ME}}{\overline{RM}}. \tag{5.9}$$

5.1.3.2 Single-Stage Equilibrium Extraction

Consider the following example: 100 kg/h of vegetable oil with 10% (mass) of fatty acid and 100 kg/h of pure ethanol enter in a single equilibrium stage. The process is shown in Figure 5.4. The streams are mixed, and the exit streams R_1 and E_1 leave in equilibrium:

Global mass balance:

$$F + S_1 = E_1 + R_1 = M_1 = 200 \text{ kg/h}.$$

Apply lever-arm rule for overall composition:

$$\frac{\overline{FM_1}}{\overline{FS_1}} = \frac{S}{M} = \frac{100}{200} = 0.5,$$

Mass balance for component C:

$$x_{C,M_1} = \frac{x_{C,F}F + y_{C,S_1}S_1}{M_1},$$

Mass balance for component B:

$$x_{C,M_1} = \frac{x_{B,F}F + y_{B,S_1}S_1}{M_1},$$

Mass flows of extract and raffinate by lever-arm rule:

$$\frac{\overline{R_1M_1}}{\overline{E_1M_1}} = \frac{E_1}{R_1} = 0.9 \Rightarrow E_1 = 0.9R_1$$

$$E_1 = 94.74 \text{ kg/h}$$
$$R_1 = 105.26 \text{ kg/h}.$$

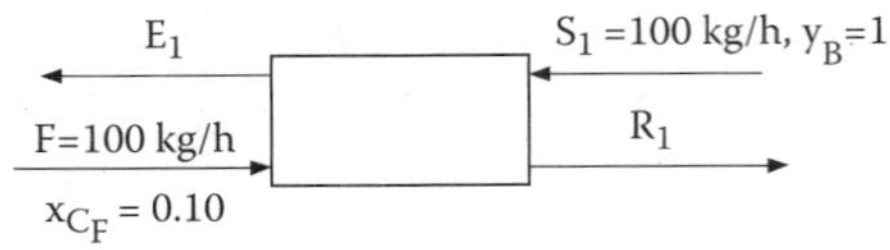

FIGURE 5.4 Single-stage extraction.

Composition of extract (E_1) and raffinate (R_1) stream from liquid–liquid diagram (Figure 5.5):

E_1	R_1
$y_{C,E1} = 0.052$	$x_{C,R1} = 0.048$
$y_{B,E1} = 0.925$	$x_{B,R1} = 0.120$
$y_{A,E1} = 1 - (y_{B,E1} + y_{C,E1}) =$	$x_{A,R1} = 1 - (x_{B,R1} + x_{C,R1}) =$
$= 1 - (0.052 + 0.925) = 0.023$	$1 - (0.078 + 0.120) = 0.832$

5.1.3.3 Multistage Crosscurrent Extraction

Consider the following example: 100 kg/h of vegetable oil with 10% (mass) of fatty acid and 100 kg/h of pure ethanol enter in the first stage of a multistage crosscurrent extractor. The process is shown in Figure 5.6. The streams that enter in each stage n are mixed and the exit streams R_n and E_n leave in equilibrium. The raffinate stream R is successively in contact with fresh solvent stream. In this case, we consider that $R_{n-1} = S_n$. The mass fraction of fatty acid in the final raffinate is 0.005.

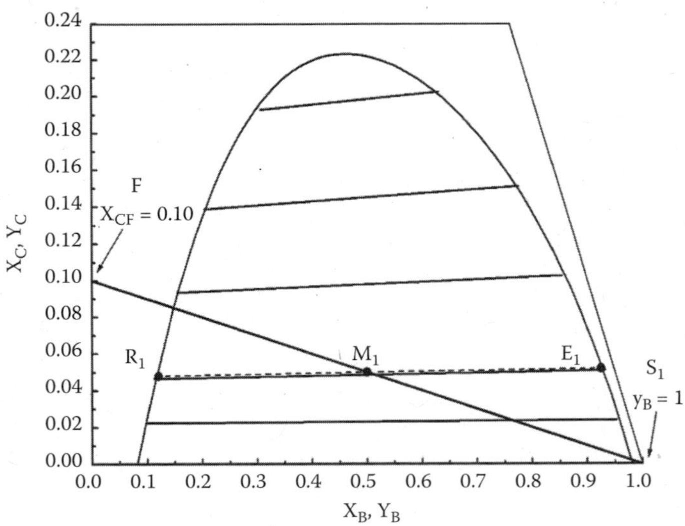

FIGURE 5.5 Phase diagram for single-stage extraction.

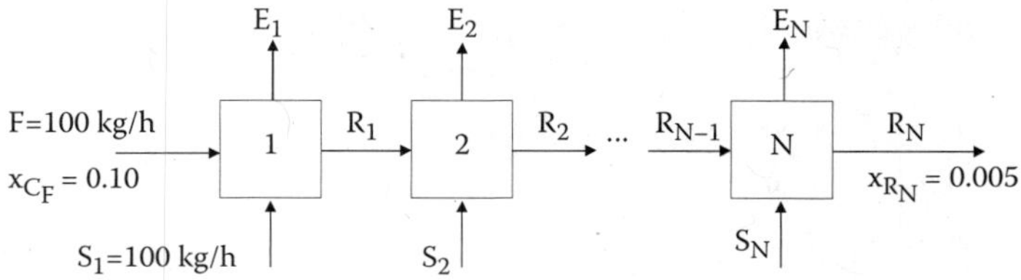

FIGURE 5.6 Flow sheet of crosscurrent extraction.

Mass balance in stage 1:

$$F + S_1 = M_1 = R_1 + E_1$$

Match with a line through the points F and S_1 (Figure 5.7)
Apply lever-arm rule to find point M_1:

$$\frac{\overline{FM_1}}{\overline{FS_1}} = \frac{S_1}{M_1} = \frac{100}{200} = 0.5.$$

$\overline{FS_1}$ is known, and then $\overline{FM_1}$ is found by lever-arm rule or by mass balance for components B and C (left column and right column, respectively):

$$x_{C,M} = \frac{x_{C,F}F + y_{C,S_1}S_1}{M_1} \qquad x_{C,M_1} = 0.5$$

$$x_{B,M} = \frac{x_{B,F}F + y_{B,S_1}S_1}{M_1} \qquad y_{C,M_1} = 0.5$$

If there is no tie line that passes in M_1 in the liquid–liquid diagram, it is necessary to interpolate a tie line to find E_1 and R_1 (Figure 5.7).

Mass balance for the next stage:

$$R_1 + S_2 = M_2 = R_2 + E_2.$$

Match the points R_1 and S_2, applying the lever-arm rule to find M_2 (Figure 5.7).
If $R_{i-1} = S_i$, then $R_1 = S_2$:

$$\frac{\overline{RM_2}}{\overline{RS_2}} = \frac{S_2}{M_2} = 0.5.$$

The segment $\overline{RS_2}$ is known, so $\overline{RM_2}$ is found.

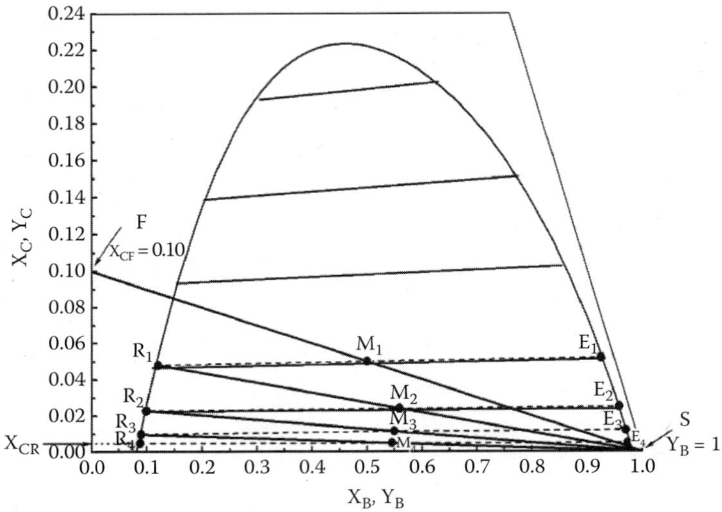

FIGURE 5.7 Phase diagram for crosscurrent extraction.

A new tie line passing through M_2 is traced, and the points E_2 and R_2 are found. This procedure must go on until $x_{C,R_N} \leq 0.005$. In this example, the extractor has four stages (Figure 5.7).

Stage 1:

$$S_1 = 100 \text{ kg/h}$$

$$R_1 + E_1 = M_1 = 200 \text{ kg/h}$$

$$\frac{E_1}{R_1} = \frac{\overline{R_1M_1}}{\overline{E_1M_1}} = 0.9 \Rightarrow E_1 = 0.9R_1$$

$$R_1 = 105.26 \text{ kg/h}$$
$$E_1 = 94.74 \text{ kg/h.}$$

Stage 2:

If $R_{i-1} = S_i$, then $R_1 = S_2$.

$$S_2 = 105.26 \text{ kg/h}$$

$$R_1 + S_2 = M_2 = R_2 + E_2$$

$$R_2 + E_2 = 210.52 \text{ kg/h}$$

$$\frac{E_2}{R_2} = \frac{\overline{R_2M_2}}{\overline{E_2M_2}} = 1.16 \Rightarrow E_2 = 1.16R_2$$

$$R_2 = 97.46 \text{ kg/h}$$
$$E_2 = 113.06 \text{ kg/h.}$$

Stage 3:
If $R_{i-1} = S_i$, then $R_2 = S_3$.

$$S_3 = 97.46 \text{ kg/h}$$

$$R_2 + S_3 = M_3 = R_3 + E_3$$

$$R_3 + E_3 = 194.92 \text{ kg/h}$$

$$\frac{E_3}{R_3} = \frac{\overline{R_3M_3}}{\overline{E_3M_3}} = 1.10 \Rightarrow E_3 = 1.10R_3$$

$$R_3 = 92.82 \text{ kg/h}$$
$$E_3 = 102.10 \text{ kg/h.}$$

Stage 4:
If $R_{i-1} = S_i$, then $R_3 = S_4$.

$$S_4 = 92.82 \text{ kg/h}$$

$$R_3 + S_4 = M_4 = R_4 + E_4$$

$$R_4 + E_4 = 185.64 \text{ kg/h}$$

$$\frac{E_4}{R_4} = \frac{\overline{R_4M_4}}{\overline{E_4M_{43}}} = 1.07 \Rightarrow E_3 = 1.07R_3$$

$$R_3 = 89.68 \text{ kg/h}$$
$$E_3 = 95.96 \text{ kg/h.}$$

The total mass flow of extract:

$$E = E_1 + E_2 + E_3 + E_4 = 405.86 \text{ kg/h.}$$

From the liquid–liquid diagram:

$$y_{C,E_1} = 0.052$$
$$y_{C,E_2} = 0.025$$
$$y_{C,E_3} = 0.012$$
$$y_{C,E_4} = 0.005$$

and

$$y_{C,E} = \frac{\sum_{i=1}^{4} E_i y_{C,E_i}}{E} = 0.023\cdot$$

5.1.3.4 Continuous Multistage Countercurrent Extractor

In this case, 100 kg/h of vegetable oil with 10% (mass) of fatty acid enters in the first stage and 300 kg/h of pure ethanol in the opposite side of the extractor. Extract and raffinate streams flow in a countercurrent arrangement. Figure 5.8 shows the flow sheet of the process. Each of the raffinate and extract streams that leave any of the stages are in equilibrium. In this case, the mass fraction of fatty acid in the final raffinate stream must be less than or equal to 0.005.

Global mass balance for the extractor:

$$F + S = M = R_N + S_1.$$

Mass balance for each stage:

Stage 1: $E_1 + R_1 = F + E_2 \Rightarrow E_1 - F = E_2 - R_1$

Stage 2: $E_2 + R_2 = R_1 + E_3 \Rightarrow E_2 - R_1 = E_3 - R_2$

...

Stage N: $E_N + R_N = R_{N-1} + S \Rightarrow E_N - R_{N-1} = S - R_N$

$$E_1 - F = E_2 - R_1 = E_3 - R_2 = \ldots = E_N - R_{N-1} = S - R_N = \Delta\cdot$$

Global mass balance for the extractor:

$$F + S = M = R_N + S_1 = 400 \text{ kg/h}.$$

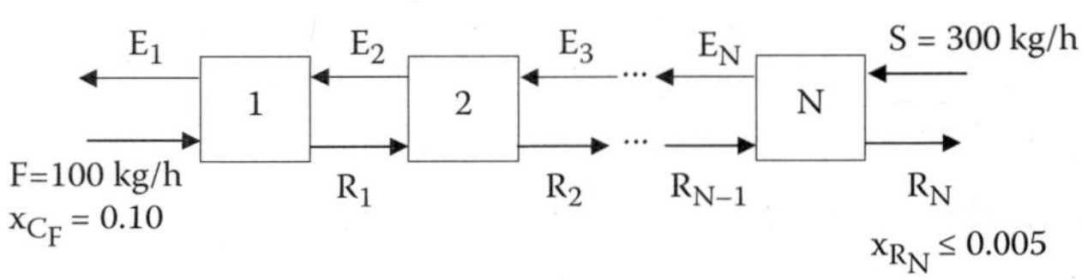

FIGURE 5.8 Flow sheet of countercurrent extraction.

Match the points F and S and applying the lever-arm rule (Figure 5.9):

$$\frac{\overline{FM}}{\overline{FS}} = \frac{S}{M} = \frac{300}{400} = \frac{3}{4}.$$

And from mass balance:

$$x_{C,M} = \frac{x_{C,F}F + y_{C,S}S}{M} \qquad x_{C,M} = 0.025$$

$$x_{B,M} = \frac{x_{B,F}F + y_{B,S}S}{M} \qquad x_{B,M} = 0.750.$$

Match the point R_N to M and find point E_1 in the binodal curve. The points R_N and E_1 are lined up by mass balance.

To find the point Δ, trace the lines $\overline{FE_1}$ and $\overline{R_NS}$; the interception of the two lines is the point Δ.

By mass balance the points F, E_1, and Δ and the points R_N, S, and Δ are lined up:

$$E_1 - F = S - R_N = \Delta$$

Match the point R_1 to Δ and find the point E_2 in the binodal curve:

$$E_2 - R_1 = \Delta.$$

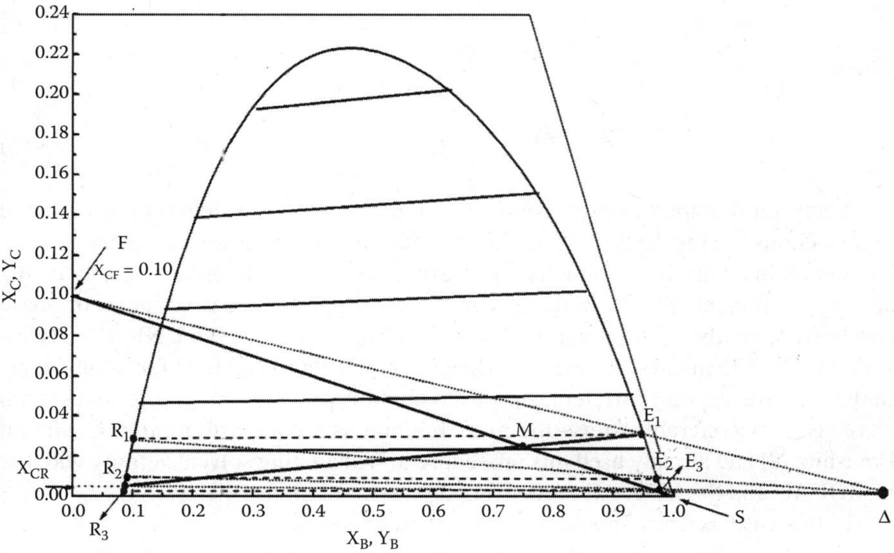

FIGURE 5.9 Phase diagram for countercurrent extraction.

Use this procedure until $x_{C,R_N} \leq 0.005$. In this example, three stages are necessary to reach this composition of component C in the raffinate stream.

The mass flows of raffinate and extract, the lever-arm rule is applied:

$$\frac{\overline{R_N M}}{\overline{E_1 M}} = \frac{E_1}{R_N} = 3.3 \Rightarrow E_1 = 3.3R_N$$

$$E_1 + R_N = 400 \text{ kg/h}$$
$$R_N = 93.02 \text{ kg/h}$$
$$E_1 = 306.98 \text{ kg/h}.$$

5.1.4 Thermodynamic: Phase Equilibrium

Design of chemical separation, such as liquid–liquid extraction, requires quantitative partial equilibrium properties of fluid mixture. When it is not possible to obtain all data for the desirable mixture in temperature and pressure conditions of interest, it is necessary to correlate the available experimental data to obtain the best interpolation.

The thermodynamic equilibrium condition for each component i in the mixture is given by the following:

$$f_i^I = f_i^{II}. \tag{5.10}$$

Using the definition of the activity coefficient we have

$$\gamma_i^I x_i^I f_i = \gamma_i^{II} x_i^{II} f_i, \tag{5.11}$$

where

$$\gamma_i^I x_i^I = a_i^I \quad \text{and} \quad \gamma_i^{II} x_i^{II} = a_i^{II}. \tag{5.12}$$

Many semi-empirical expressions have been proposed in literature to correlate excess Gibbs energy, mainly to the composition of the mixture. All these expressions contain adjustable parameters to fit experimental data in order to calculate the activity coefficient. The main molecular models suggested for description of phase equilibrium are the NRTL (Non-Random Two-Liquid) [6] and the UNIQUAC (Universal Quasi Chemical) [7] models. When the molecular weights of the components in the mixture are very different, such as in the fatty systems containing short-chain alcohols, it is preferable to use the mass fraction as a composition unit. Oishi and Prausnitz [8] had already used this procedure for calculating solvent activity with the UNIQUAC and the UNIFAC models in polymeric solutions.

In this case, activity should be rewritten as follows:

$$a_i = \gamma_i^x x_i = \gamma_i^w w_i, \tag{5.13}$$

where

$$\gamma_i^x = \gamma_i^w M_i \sum_j^n w_j / M_j. \tag{5.14}$$

In the NRTL model, the activity coefficient using composition expressed in mass fraction takes the following form:

$$\ln\gamma_i = \frac{\sum_j^C \frac{\tau_{ji} G_{ji} w_j}{M_j}}{\sum_j^C \frac{G_{ji} w_j}{M_j}} + \sum_{j=1}^C \left[\frac{w_j G_{ij}}{M_j \sum_k^n \frac{G_{kj} w_k}{M_k}} \left(\tau_{ij} - \frac{\sum_k^C \frac{\tau_{kj} G_{kj} w_k}{M_k}}{\sum_k^C \frac{G_{kj} w_k}{M_k}} \right) \right], \tag{5.15}$$

where

$$G_{ij} = \exp\left(-\alpha_{ij}\tau_{ij}\right) \tag{5.16}$$

$$\tau_{ij} = A_{ij}/T \tag{5.17}$$

$$\alpha_{ij} = \alpha_{ji}. \tag{5.18}$$

For the UNIQUAC model, it has the following form:

$$\ln\gamma_i = \ln\gamma_i^C + \ln\gamma_i^R \tag{5.19}$$

$$\ln\gamma_i^C = \ln\left(\frac{\phi_i'}{w_i/\zeta M_i}\right) + 1 - \frac{\zeta M_i \phi_i'}{w_i} + \frac{z}{2} M_i q_i' \ln\frac{\theta_i'}{\phi_i'} - \frac{z}{2} M_i q_i' \left(1 - \frac{\phi_i'}{\theta_i'}\right), \tag{5.20}$$

where

$$\zeta = \sum_j^C \frac{w_j}{M_j} \tag{5.21}$$

$$\theta_i' = \frac{q_i' w_i}{\sum_j^C q_j' w_j}; \; \phi_i' = \frac{r_i' w_i}{\sum_j^C r_j' w_j} \tag{5.22}$$

and

$$r_i' = \frac{1}{M_i} \sum_k^G v_k^{(i)} R_k; \; q_i' = \frac{1}{M_i} \sum_k^G v_k^{(i)} Q_k \tag{5.23}$$

$$\ln\gamma_i^R = M_i q_i' \left[1 - \ln\left(\sum_j^C \theta_j' \tau_{ji}\right) - \sum_j \left(\theta_i' \tau_{ij} \Big/ \sum_k^C \theta_k' \tau_{kj}\right)\right]. \quad (5.24)$$

The adjustable parameters τ_{ij} and τ_{ji} are defined as follows:

$$\tau_{ij} = \exp\left[-\left(\frac{u_{ij} - u_{jj}}{RT}\right)\right] = \exp\left[-\left(\frac{A_{ij}}{T}\right)\right] \quad (5.25)$$

$$\tau_{ji} = \exp\left[-\left(\frac{u_{ji} - u_{ii}}{RT}\right)\right] = \exp\left[-\left(\frac{A_{ji}}{T}\right)\right]. \quad (5.26)$$

Due to the similarity of the triacylglycerols, the vegetable oil can be represented by a single triacylglycerol having the average molecular weight of all triacylglycerols of the oil. The same reasoning can be extended to a mixture of fatty acids. Then the values of r_i' and q_i' for the UNIQUAC model can be calculated by Equation 5.23, which considers the composition of triacylglycerols and fatty acids of any vegetable oil and any mixture of fatty acids, respectively. The parameters R_k and Q_k can be taken from Magnussen et al. [9]:

$$r_i' = \frac{1}{\overline{M}_i} \sum_j^C x_j \sum_k^G v_k^{(i)} R_k; \; q_i' = \frac{1}{\overline{M}_i} \sum_j^C x_j \sum_k^G v_k^{(i)} Q_k, \quad (5.27)$$

where x_j is the molar fraction of the triacylglycerols of the vegetable oil or fatty acids of a mixture of fatty acids and $\overline{M}_i$ is the average molecular weight of the vegetable oil or a mixture of fatty acids.

There are many adjusted parameters of the NRTL and the UNIQUAC models that describe the liquid–liquid equilibrium of these fatty systems in the literature [10–19].

5.1.5 Group Contribution Models

In a group contribution method, the basic idea is that the number of functional groups is much smaller than the chemical compounds of interest in chemical technology. If the physical properties can be calculated by summing group contribution, it is possible to obtain a large number of these properties in terms of a much smaller number of parameters that characterize the contribution of functional groups in the mixture.

For calculating phase equilibrium in the simulation of deacidification of vegetable oils through liquid–liquid extraction, the group contribution models, the UNIFAC [20] and the ASOG [21], are more appropriate, because they avoid expanding the pseudo-ternary systems vegetable oil + fatty acids + short-chain alcohols in a multicomponent system with a small number of structural groups, and consequently, a small number of binary interaction parameters is required.

Both the UNIFAC and the ASOG models assume the following forms when compositions are expressed in mass fractions.

5.1.5.1 UNIFAC Model

$$\ln \gamma_i = \ln \gamma_i^C + \ln \gamma_i^R. \tag{5.28}$$

In this model, the combinatorial part is taken directly from the UNIQUAC model. The residual part is as follows:

$$\ln \gamma_i^R = \sum_k^C \nu_k^{(i)} \left[\ln \Gamma_k - \ln \Gamma_k^{(i)} \right], \tag{5.29}$$

where $\Gamma_k^{(i)}$ is the group activity coefficient of the group k in the reference solution containing only molecules of the same type i:

$$r_i' = \frac{1}{M_i} \sum_k^G \nu_k^{(i)} R_k; \quad q_i' = \frac{1}{M_i} \sum_k^G \nu_k^{(i)} Q_k \tag{5.30}$$

$$\theta_i' = \frac{q_i' w_i}{\sum_j^C q_j' w_j}; \quad \phi_i' = \frac{r_i' w_i}{\sum_j^C r_j' w_j} \tag{5.31}$$

$$\ln \Gamma_k = M_k Q_k' \left[1 - \ln \left(\sum_m^G \Theta_m' \Psi_{mk} \right) - \sum_m^G \left(\Theta_m' \Psi_{km} \Big/ \sum_n^G \Theta_n' \Psi_{nm} \right) \right] \tag{5.32}$$

$$\Theta_m = \frac{Q_m' W_m}{\sum_n^G Q_n' W_n}; \quad W_m = \frac{\sum_j^C \nu_m^{(j)} w_j}{\sum_j^C \sum_n^G \nu_n^{(j)} w_j} \tag{5.33}$$

$$\Psi_{mn} = \exp\left[-\left(\frac{U_{mn} - U_{nn}}{RT} \right) \right] = \exp\left[-\left(a_{mn}/T \right) \right]. \tag{5.34}$$

5.1.5.2 ASOG Model

$$\ln \gamma_i = \ln \gamma_i^{FH} + \ln \gamma_i^G \tag{5.35}$$

$$\ln \gamma_i^{FH} = \ln \left(\frac{\zeta \nu_i^{FH}}{\sum_j^C \frac{w_j}{M_j} \nu_j^{FH}} \right) + 1 - \frac{\zeta \nu_i^{FH}}{\sum_j^C \frac{w_j}{M_j} \nu_j^{FH}}, \tag{5.36}$$

where

$$\zeta = \sum_{j}^{C} \frac{w_j}{M_j} \tag{5.37}$$

$$\ln \gamma_i^G = \sum_{k}^{G} \nu_{ki} \left(\ln \Gamma_k - \ln \Gamma_k^{(i)} \right) \tag{5.38}$$

$$\ln \Gamma_k = -\ln \sum_{l}^{G} W_l a_{k,l} + 1 - \sum_{l}^{G} \left(W_l a_{l,k} \Big/ \sum_{m}^{G} W_m a_{l,m} \right), \tag{5.39}$$

where W is the mass fraction of the group, calculated from Equation 5.33:

$$a_{k,l} = \exp\left(m_{k,l} + \frac{n_{k,l}}{T} \right). \tag{5.40}$$

The functional groups of fatty systems in alcoholic solutions for the UNIFAC model are as follows: CH_3, CH_2, CH, CH_2COO, CH=CH, COOH, and OH and for the ASOG model are CH_2, COO, C=C, COOH, and OH. The UNIFAC parameters for LLE were published by Magnussen et al. [9] and the ASOG parameters by Tochigi et al. [22]. Batista et al. [23] adjusted some of the UNIFAC and the ASOG parameters for fatty systems, and the results in the prediction of the liquid–liquid equilibrium of these systems were better than those using original parameters.

5.1.5.3 Minor Component

Binary interaction parameters of the UNIQUAC or the NRTL models between minor component and any other component in the fatty system (triacylglycerols, free fatty acids, ethanol, water) can be determined, assuming that the minor component are at infinite (∞) dilution in the liquid–liquid equilibrium system. In this case, the distribution coefficient, calculated according to Equation 5.41 below, can be approached by the distribution coefficient at infinite dilution k_i^∞. Using the isoactivity criterion this distribution coefficient for minor component, k_i^∞, can be calculated by Equation 5.42:

$$k_i = w_i^{II} / w_i^{I} \tag{5.41}$$

$$k_i^\infty = \left(\gamma_i^{w,I}\right)^\infty / \left(\gamma_i^{w,II}\right)^\infty. \tag{5.42}$$

To calculate γ_i^∞, the compositions of both phases are required. Since the minor component is present in a very low composition, the phase compositions can be estimated taking in account only the major components (triacylglycerols, free fatty acids, ethanol, water). The binary interaction parameters between the major components are used to perform liquid–liquid flash calculations for the estimation of phase compositions on the basis of the overall experimental composition of the mixtures.

The infinite dilution activity coefficient (γ_i^∞) is obtained applying the limit in the UNIQUAC or the NRTL models, keeping constant the mass fractions of the other components of the mixture and making the minor component compositions tend to zero.

For the adjustment of interaction parameters between minor components and any other components, the estimation was based on the minimization of the distribution coefficient objective function, Equation 5.43 below, following the procedure developed by Pessôa Filho and described in Rodrigues et al. [13, 16] and Gonçalves [18]. In Equation 5.43, the additional term is a penalty function suggested by Kang and Sandler [24] and used to preclude interaction parameters with too large absolute values:

$$OF(k_i) = \left(\sum_{n=1}^{N} \left(k_i^{ex} - k_i^{calc} \right)^2 \Big/ N \right)^{1/2} + Q \sum_{l=1}^{L} (p_l^2) / L, \tag{5.43}$$

where n is the tie line index, N is the total number of tie lines, k_i is the minor compounds' distribution coefficient, *ex* and *calc* refer to experimental and calculated values, Q is a small value that does not alter significantly the function residue, l is the UNIQUAC or NRTL parameter index, L is the total number of adjustable parameters, and p_l is the UNIQUAC or NRTL parameter.

5.1.6 Simulation of a Liquid–Liquid Extraction Column

The schematic representation of a stagewise column is shown in Figure 5.10. The vegetable oil with free fatty acids stream (F) enters the column in stage 1 and the solvent stream (S) in the opposite side of the column. Extract and raffinate streams flow from stage to stage countercurrently and provide the formation of two product streams, the final extract ($E1$) and final raffinate (RN) streams. Extract (en) and raffinate (rn) streams leave stage n in equilibrium.

In the vegetable oil deacidification process, the final raffinate stream (RN) contains refined vegetable oil and a residual fraction of the solvent, and the final extract stream ($E1$) contains the solvent with the free fatty acids extracted and a residual fraction of vegetable oil.

The algorithm, suggested by Naphtali and Sandholm [25] and developed for simulation of distillation column, is suitable to simulate the liquid–liquid extraction with the modifications of mass balance and equilibrium equations.

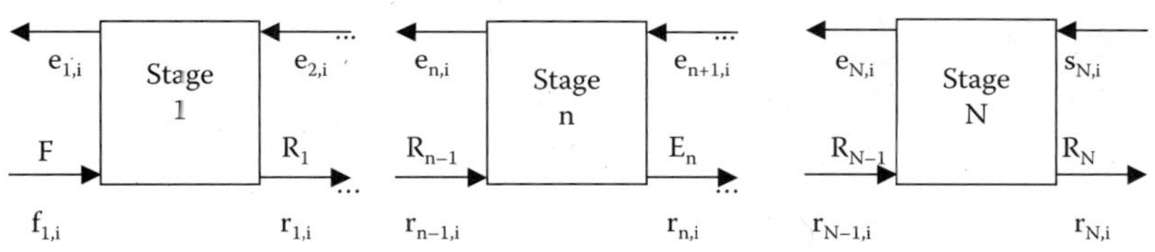

FIGURE 5.10 Schematic representation of a liquid–liquid extraction column.

The mass balance and liquid–liquid equilibrium equations are grouped for each component and each stage. The resultant group of equations has the structure of a tridiagonal block that permits a rapid solution with the Newton–Raphson method.

For each stage n, a set of dependent relationships (test functions $F_{k(n,i)}$) must be satisfied:

Mass balances of component i:

$$F_{1(n,1)} = r_{n,i} - r_{n-1,i} + e_n y_{n,i} - e_{n+1,i} \quad n = 2, 3, \ldots, N-1 \tag{5.44}$$

$$i = 1, 2, \ldots, C$$

$$F_{1(1,i)} = r_{1,i} - f_{1,i} + e_{1,i} - e_{2,i} \quad i = 1, 2, \ldots, C \tag{5.45}$$

$$F_{1(N,i)} = r_{N,i} - r_{N-1,i} + e_{N,i} - s_{N,i} \quad i = 1, 2, \ldots, C. \tag{5.46}$$

Equilibrium conditions:

$$F_{2(n,i)} = k_{n,i} E_n\, r_{n,i} / R_n - e_{n,i} \quad n = 2, 3, \ldots, N-1 \tag{5.47}$$

$$i = 1, 2, \ldots, C,$$

where

$$k_{n,i} = \gamma_{n,i}^{w,I} / \gamma_{n,i}^{w,II} = w_{n,i}^{II} / w_{n,i}^{I} \tag{5.48}$$

$$F_{2(1,i)} = k_{1,i} E_1\, r_{1,i} / R_1 - e_{1,i} \quad i = 1, 2, \ldots, C \tag{5.49}$$

$$F_{2(N,i)} = k_{N,i} E_N\, r_{N,i} / R_N - e_{N,i} \quad i = 1, 2, \ldots, C. \tag{5.50}$$

The above relationships comprise a vector of the test function:

$$F(x) = \begin{Bmatrix} F_1 \\ F_2 \end{Bmatrix} = 0 \tag{5.51}$$

which contains 2NC elements and which may be solved for equally many unknowns:

$$x = \begin{Bmatrix} e \\ r \end{Bmatrix}. \tag{5.52}$$

The iterative Newton–Raphson method solves Equation 5.51 using the prior set of values of the independent variables. In Newton–Raphson's interaction a new group of values, x_r, is generated from a previous estimation, x_{r-1}:

$$x_r = x_{r-1} - F_{r-1}(x_{r-1}) \big/ \left(\partial F/\partial x\right)\Big|_{x_{r-1}}. \tag{5.53}$$

When $|x_r - x_{r-1}|$ is small enough, the correct group of x is found and the iteration stops.

5.2 STATE OF THE ART—MINI-REVIEW OF THE LITERATURE

For the success of the commercial production of enzymes and proteins, there is a need for efficient downstream processing techniques. The downstream process for these biological materials requires purification techniques that are delicate enough to preserve the biological activity [26]. The purification protocols involve several steps, which increase the cost of the process and reduce the yield. The conventional procedures include ammonium sulphate precipitation, chromatography, dialysis, and filtration. Simpler and more efficient purification processes are needed.

Aqueous two-phase systems (ATPS) could be a good alternative to a first purification step because such systems allow removal of several contaminants by a simple and economic process. ATPS are formed by adding to water, either two structurally different hydrophilic polymers, such as dextran and polyethylene glycol (PEG) [27], or maltodextrin and PEG [28, 29], or a polymer and salt, such as PEG and potassium phosphate or PEG and sodium sulphate [30–32].

PEG + salt systems have been used in large-scale protein separation because of larger droplet sizes, a higher density difference between the phases, and lower viscosity, leading to a much faster separation than PEG + dextran systems. Industrial applications of the PEG + salt systems could be improved by the availability of commercial separators, which allow faster continuous protein separations [33–35].

The most common polymer + polymer system is composed of polyethylene glycol and dextran [36, 37]. Polypropylene glycol (PPG) is a polymer that is structurally closely related to PEG. PPGs of low molecular weight are soluble in water, whereas high molecular mass ones are only partially soluble [38]. Some recent purification techniques employing ATPS suggest the use of thermo-separating polymers, such as copolymers of ethylene oxide (EO) and propylene oxide (PO) units, to reduce the cost of polymer recovery [39, 40]. Dextran is a high-cost polymer that makes difficult the use of ATPS in large-scale processes. Maltodextrin (MD) can be used as a lower cost substitute for dextran [28, 30]. MD is a commercial polymer of D-glucose units linked primarily by $\alpha(1\rightarrow4)$ bonds. This polymer is obtained by acidic and enzymatic hydrolysis of starch. Low-molecular-mass saccharides, such as glucose, maltose, and sucrose, can also be used for dextran replacement, with the advantage that such compounds are of common occurrence in the food industry [41].

Phase equilibrium data for such systems are mainly found in the works of Albertsson [36] and Zaslavsky [37]. However, these data are not yet complete, particularly regarding the behavior of such systems at different experimental conditions, for example, temperature and pH.

Silva et al. [31] studied the effect of temperature, pH, and polymer molecular weight changes on the binodal curve and tie lines of the phase equilibrium diagrams for PEG + potassium phosphate + water systems. The equilibrium phase behavior of MD and PEG systems at 298.2 K and atmospheric pressure, under several conditions of concentrations and molecular weights of the polymers, was studied by Silva and Meirelles [28].

There are many reports in the literature concerning the partition of different enzymes and proteins in ATPS [26].

The behavior of the partition coefficients of bovine serum albumin (BSA), α-lactoalbumin (α-La), and β-lactoglobulin (β-Lg) in PEG/MD systems at 298.2 K, with several PEG/MD polymer concentrations and different polymer molecular weights, was published by Silva and Meirelles [29].

Alves et al. [42] performed an experimental study of the partitioning of different proteins, cheese whey α-La, β-Lg, and BSA, and porcine insulin in ATPS containing PEG (1500, 600, 1450, and 3350) and salt (potassium phosphate, and sodium citrate), and PEG (1450, 8000, and 10,000) and MD (2000 and 4000). The results showed the feasibility of α-LA and β-Lg purification. Partition coefficients of the BSA, α-LA, and β-Lg were also studied by Silva and Meirelles [30] in systems containing PPG 400 and MD at 25°C. Lima et al. [26] investigated the partitioning of four pectinolytic enzymes from a commercial pectinase preparation (Pectinex-3XL) in ATPS composed of PEG and potassium phosphate.

Another important application of liquid–liquid extraction is the organic acids purification such as citric, tartaric, lactic, and phosphoric acids. The recovery of carboxylic acids by liquid–liquid extraction with aliphatic tertiary amines dissolved in organic diluents has been studied by several authors [43–48].

The worldwide production of citric acid exceeds 500,000 ton/yr. In contrast with a lot of products that previously were obtained by microbiological methods and nowadays are obtained by synthetic methods, this acid continues to be manufactured, mainly by fermentation. Seventy percent of all citric acid produced is used by the food industry, and 18% is used by the pharmaceutical industry. Its use in the food industry represents 55%–65% of the total acidulants' market, in which 20%–25% corresponds to phosphoric acid and 5% to malic acid. The fermentation process technology for the industrial production of organic acids has been known for more than a century. Citric acid is one of the macro-fermentation processes of greater success within the bioproduct industries.

The classical method for recovering citric acid is based on the precipitation of calcium salts, by addition of calcium hydroxide in the fermentation broth. The solid is filtrated and treated with sulfuric acid (H_2SO_4) for the preferential precipitation of sulfate calcium. The free organic acid in the filtrate is purified using activated carbon or ion exchange and is concentrated by evaporation. The acid crystallizes with great difficulty and very low efficiency. Compared to the usual separation processes, liquid–liquid extraction seems to be a very promising alternative [49].

In relation to phosphoric acid, several publications deal with the modeling of the extraction of phosphoric acid from water by tri-*n*-butyl phosphate [50, 51]. In fact, phosphoric acid is an important raw material for fertilizer applications, as well as for products with higher purity standards [52].

The success of a liquid–liquid extraction process relies on solvent selection. Mixed solvents composed of tertiary amines and alcohol are suggested as appropriated solvents [43, 53]. The disadvantage of their use is their great toxicity and, consequently, higher purification costs. Welsh and Williams [54] studied several kinds of vegetable oils, as solvents to the recovery of organic compounds from aqueous solutions, such as corn oil, canola oil, olive oil, and others. The authors verified that short-chain alcohols and organic acids presented low recovery and small distribution coefficients, when the vegetable oils are used as single solvents. Therefore, there is great appeal to the search for new solvents, mainly combinations of solvents. The main difficulty is the analysis of mixed solvents because of the lack of equilibrium data.

Lintomen et al. [49] studied new solvents for the recovery of citric acid by liquid–liquid extraction using the following systems: water/citric acid/short-chain alcohol (2-butanol or 1-butanol) and water/citric acid/short-chain alcohol/tricaprylin.

Recently, Uslu [55] published a study of tartaric acid recovery from aqueous solutions using tertiary amine. Batch extraction experiments were performed with Alamine 336 dissolved in the diluents of various types—ketone (methyl isobutyl ketone), aromatic (toluene), different alkanes (hexane, cyclohexane), and alcohol (butan-1-ol).

Similar to that of citric acid, the interest toward lactic acid recovery from fermentation broth has been increased. This interest is caused by the increase in the demand for pure, naturally produced lactic acid, mainly for the food (as food additive and preservative) and pharmaceutical industries or for production of biodegradable polymers. Yankov et al. [56] investigated the lactic acid extraction from aqueous solutions and synthetic fermentation broth by means of a system composed of trioctylamine and an active (decanol) and an inactive (dodecane) diluent.

Essential volatile oils are vegetable products, which are basically a mixture of terpenic hydrocarbons and oxygenated derivatives such as aldehydes, alcohols, and esters. Citrus essential oil is used as a flavoring agent in pharmaceuticals as well as a fragrant ingredient in soaps, detergents, creams, lotions, and perfumes. From its components, oxygenated compounds are mainly responsible for the aroma and flavor, and their content has become a definitive parameter in establishing the price of the volatile oil and representing a reference of quality [57].

Citrus oils are obtained from the small balloon-shaped glands or vesicles located in the flavedo or colored portion of the citrus peel. The quality of these oils depends on factors such as soil, climate, extraction method of the oil, weather, maturity, and the variety of the fruit. Citrus oils are complex mixtures of over 200 chemical compounds, of which more than 100 have been identified. These include highly volatile components such as terpenes, sesquiterpenes, and oxygenated compounds and nonvolatile compounds such as pigments and waxes. The terpene fraction can constitute from 50% up to more than 95% of the oil. However, this fraction gives little contribution to the flavor and fragrance of the oil. Because terpenes are mostly unsaturated compounds, they are easily decomposed by heat, light, and oxygen to unpleasant off flavors and aromas. Therefore, it is common industrial practice to remove some of the terpenes and, as a consequence, to concentrate the oxygenated compounds, which are mainly responsible for the characteristic citrus flavor and fragrance. This procedure is known as "deterpenation" or "folding" and is carried out to improve oil stability, increase oil solubility, and reduce storage and transport costs [58–60].

Table 5.2 presents the main volatile compounds of citrus essential oils.

Deterpenation is currently done by distillation, solvent extraction, supercritical fluid extraction, or chromatographic separation [70–75]. The main drawbacks of these conventional processes are low yields, formation of thermally degraded undesirable by-products, and/or solvent contamination of the products [58, 73]. Solvent extraction is probably the most common process used by industry. The solvents most often used are hexane and chloroform, because of their intrinsic characteristics of selectivity related to terpenes and oxygenated compounds [76].

Alternative solvents have been suggested as substitutes of hexane and chloroform, such as acetonitrile, nitromethane, and dimethylformamide [77], diethylene glycol [78], 1,2-propanediol and 1,3-propanediol [79], aminoethanol [80], methanol [81], 2-butene-1,4-diol, ethylene glycol, and ionic liquids (1-ethyl-3-methylimidazolium methanesulfonate) [82].

In view of a possible future food, cosmetic, or pharmaceutical application of the extract, it is necessary to use solvents such as ethanol or water [57, 58, 83, 84].

The light components of the essential oil mixtures are completely soluble in ethanol but not completely soluble in water. The solution obtained by adding ethanol to water maintains the polar characteristics of water, but its polarity is lowered by the presence of the alcohol. Alcoholic extracts of citrus essential oils are particularly requested by the industry for the following reasons [83, 85, 86]:

1. They are highly soluble in aqueous solutions and can therefore be used to make drinks and perfumes;
2. They enhance the aromatic strength of the mixture; and
3. Oxidation reactions are reduced in the presence of alcohol [58].

Studies about essential oils deterpenation by liquid–liquid extraction are scarce in the literature. Massaldi and King [87] published an article concerning a simple technique for the determination of solubilities and activity coefficients of *d*-limonene, *n*-butylbenzene, and *n*-hexyl acetate in water and sucrose.

TABLE 5.2
Volatile Compounds Present in Essential Oils

Orange[a,b]	Mandarin[a,c]	Grapefruit[a,d,e]	Lemon[a,f,g,h]	Bergamot[f,i]
Etanal	Etanal	Etanal	Neral	Linalool
Octanal	Octanal	Decanal	Geranial	Linalyl acetate
Nonanal	Decanal	Ethyl acetate	β-Pinene	γ-Terpinene
Citral	α-Sinensal	*d*-Limonene	Geraniol	β-Pinene
d-Limonene	Thymol	Nootkatone	Geranyl acetate	*d*-Limonene
α-Pinene	γ-Terpinene		Neryl acetate	
	β-Pinene		Bergamoptene	

[a][61]; [b][62]; [c][63]; [d][64]; [e][65]; [f][66]; [g][67]; [h][68]; [i][69].

Ternary liquid–liquid equilibria for α-pinene + Δ_3-carene + polar compound (acetonitrile, nitromethane, and dimethylformamide) systems were determined by Antosik and Stryjek [77], at the temperature 298.2 K.

Thermodynamic behavior related to systems composed of essential oil compounds plus ethanol and water was firstly published by Gironi et al. [83]. The authors reported solubilities for the binary systems of water + limonene and water + citral at atmospheric pressure and at 293 K. Equilibrium data of ternary systems of water + ethanol + limonene and water + ethanol + citral were also determined.

Tamura and Li [81] tested methanol plus water as solvent for the deterpenation process. In this chapter, the authors measured the mutual solubilities of the terpenes dissolved in water or methanol and their multicomponent liquid–liquid equilibria.

Cháfer et al. studied the influence of the temperature on phase equilibrium of systems composed of limonene, ethanol, and water [88], and of linalool, ethanol, and water [58], respectively.

An ample study related to solvent choice for deterpenation of essential oils has been developed by Arce et al. [57, 78–80, 82, 84]. First, the authors evaluated the performance of diethylene glycol as solvent for systems containing limonene plus linalool at three different temperatures: 298.2, 308.2, and 318.2 K [78]. Subsequently, the following solvents were tested for the same oil systems: 1,2-propanediol e 1,3-propanediol [79], ethanol plus water [57, 84], 2-aminoethanol [80], 2-butene-1,4-diol, ethylene glycol, and 1-ethyl-3-methylimidazolium methanesulfonate [82].

Deacidification of vegetable oils can also be performed by liquid–liquid extraction. Oilseeds are the major source for the production of edible oils, which are regarded as an important component of the diet, being an important source of energy, of essential fatty acids (such as linoleic acid), and of fat-soluble vitamins (such as vitamins A and E). Crude vegetable oils are predominantly composed of triacylglycerols and free fatty acids, with mono- and diacylglycerols also present at lower levels. The refining of a vegetable oil consists of several steps, including its extraction from solid matrix by pressing and/or using organic solvents [89, 90], degumming, bleaching, deacidification, and deodorization [91, 92].

The removal of free fatty acids (deacidification) is the most difficult step of the oil purification process, mainly because it has the maximum economic impact on oil production. Deacidification of oils is performed industrially by chemical, physical, or miscella methods. However, for oils with high acidity, chemical refining causes high losses of neutral oil as a result of saponification and emulsification. Physical refining is also a feasible process for deacidification of highly acidic oils, because it results in lower losses of neutral oil than the traditional process, but more energy is consumed. Moreover, in some cases, the refined oil is subject to undesirable alterations in color and a reduction of stability with regard to resistance to oxidation [1].

New approaches for deacidification of vegetable oils have been proposed in the literature, such as biological deacidification, chemical reesterification, supercritical fluid extraction, membrane processing, and solvent (or liquid–liquid) extraction.

Liquid–liquid extraction is a separation process that takes advantage of the relative solubilities of solutes in immiscible solvents. A partial separation occurs when the components of the original mixture have different relative solubilities in the selected solvent phase [3]. The deacidification of oils by liquid–liquid extraction by means of an appropriate solvent is receiving attention because of its advantages in comparison to physical and chemical refining. As this process is normally carried out at room temperature and atmospheric pressure, less energy is consumed and the oil is submitted to softer treatments. Besides, liquid–liquid extraction has the advantages of avoiding the formation of waste products but still reducing the loss of neutral oil, and may preserve the nutraceutical compounds. Furthermore, solvent stripping from refined oil and solvent recovery from extract stream can be easily carried out because of the great difference between the boiling points of the solvent, fatty acids, and triacylglycerols. In fact, these operations can be accomplished by evaporation or distillation at relatively low temperatures, in most cases lower than 353 K [3, 93, 94].

The use of solvent extraction for deacidification of vegetable oils was first proposed by Bollmann [95]. In this patent the author suggests the use of methyl alcohol, ethyl alcohol, amyl alcohol, acetone or acetic ester not diluted or diluted with water. van Dijck [96] suggested a process combining liquid–liquid extraction and alkali refining. Free fatty acids from fats and oils were neutralized by adding a base, such as ammonia, and subsequently the soaps were removed by countercurrent extraction with a suitable solvent, such as ethanol.

Another study based on liquid–liquid extraction associated with alkali refining was patented by Nestlé Co. [97]. According to the inventors, free fatty acids are removed by controlled neutralization in an aqueous medium containing an alcohol or a polyol.

Swoboda [98] reports a process for refining palm oil and palm oil fractions, using as solvent mixtures of ethanol and water or isopropanol and water, preferably with a composition near the azeotropic one. According to the author, azeotropic mixtures are preferred because of the advantages of recycling the solvent.

Bhatacharyya et al. [99] and Shah and Venkatesan [100] studied the deacidification of rice bran and groundnut oils using aqueous 2-propanol as solvent. Kim et al. [101] and Kale et al. [102] tested methanol in the refining of rice bran oil (RBO). All these studies showed a decrease in the oil acidic value. Turkay and Civelekoglu [103] investigated the liquid–liquid extraction of sulfur olive oil miscella in hexane with aqueous ethanol solutions. Apelblat et al. [93] published an article that reports phase diagrams for soybean oil or jojoba oil plus oleic acid and several solvents (1,2-butanediol, dimethyl sulfoxide, *cis*-2-butene-1,4-diol, formamide, and *n*-methylformamide), at 298.2 K.

The extraction of free fatty acids from fatty materials using solvents has a long history, and several studies have already shown that this process is, in principle, feasible using short-chain alcohols, especially ethanol, as solvent [3, 93, 99, 100, 102, 104–111]. Ethanol has low toxicity, easy recovery in the process, good values of selectivity and of the distribution coefficient for free fatty acids [10, 11, 14, 15, 17, 106], and low losses of nutraceutical compounds [12, 13, 16, 18].

In the last years, equilibrium data for systems composed of several vegetable oils (canola, corn, palm, rice bran, Brazil nut, macadamia nut, grape seed, sesame seed, garlic, soybean, and cottonseed oils) plus saturated, monounsaturated, or diunsaturated free fatty acids, such as stearic, palmitic, oleic, and linoleic acids plus solvent (ethanol + water) have been published [10–19, 23, 112]. This set of works emphasizes that the mixture ethanol + water is more often recommended to be used as solvent for deacidification of vegetable oils. In fact, this new technique may produce vegetable oils with low acidic levels and simultaneously minimize the loss of neutral oil and nutraceutical compounds.

5.3 APPLICATIONS

5.3.1 Deacidification of Vegetable Oils

In this section we discuss some effects in the liquid–liquid equilibrium for fatty systems using short-chain alcohols. This information is useful in the choice of solvent or temperature for deacidification of vegetable oils by liquid–liquid extraction.

5.3.1.1 Effect of Temperature

The information about mutual solubility of the oil and solvent is contained in the base line of the liquid–liquid diagram (Figure 5.1). The mutual solubility for vegetable oil and short-chain alcohols increases with an increase in temperature, and above some temperatures, this binary mixture is totally soluble. The increase in mutual solubility with increasing temperatures affects the liquid–liquid equilibrium. The area underneath binodal decreases at higher temperatures, and the slopes of the tie line or distribution coefficients may change.

Batista et al. [10] presented the liquid–liquid equilibrium for the system containing refined canola oil + commercial oleic acid and short-chain alcohols at different temperatures. For systems with anhydrous methanol and anhydrous ethanol, the heterogeneous region decreases with the increasing in temperature from 293 to 303 K, and only a slight change in the distribution coefficient of oleic acid is observed. The increasing of mutual solubility of canola oil and anhydrous methanol or anhydrous ethanol with almost no impact on the slope of tie lines causes a decrease in the selectivity of the solvents with increasing temperatures.

Figure 5.11 shows the tie lines and binodal curves for the systems of refined canola oil + commercial oleic acid + methanol at 293 and 303 K.

5.3.1.2 Length Chain of Alcohols

Figure 5.12 represents the binodal curves for the system of refined canola oil + commercial oleic acid + anhydrous methanol or anhydrous ethanol. It can be seen that the heterogeneous region for the system with methanol is higher than for the system with ethanol, because the mutual solubility of refined canola oil with methanol is lower than that with ethanol, which can be explained by the higher polarity of the methanol chain in relation to that of ethanol.

The results proved that the distribution coefficient of oleic acid with anhydrous ethanol is somewhat larger than 1, whereas that for anhydrous methanol is somewhat smaller, which suggests that methanol has a somewhat lower capacity for extraction of fatty acids oil, thus presenting less selectivity than methanol.

As expected, the system of canola oil + oleic acid + anhydrous isopropanol at 293 K and canola oil + oleic acid + anhydrous *n*-propanol at 283 K formed only a minimum heterogeneous area.

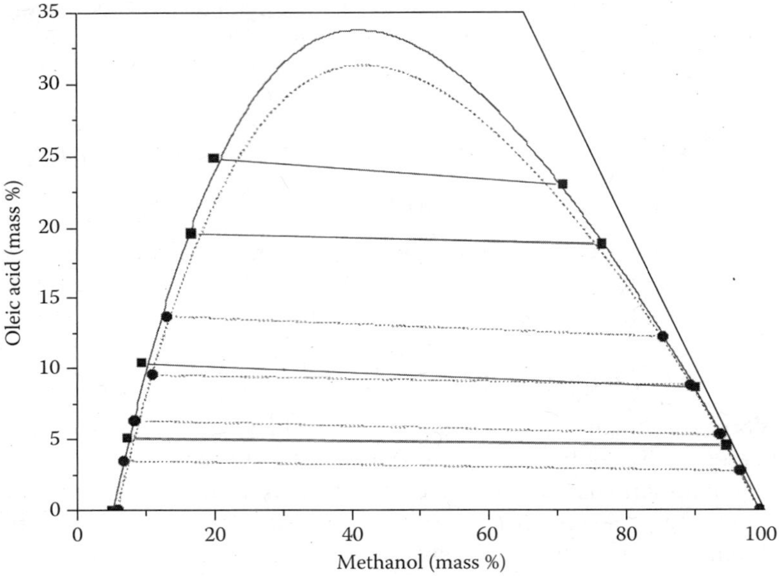

FIGURE 5.11 Experimental tie lines and binodal curves for the systems of refined canola oil + commercial oleic acid + anhydrous methanol at 293.2 K (—■—) and at 303.2 K (···●···).

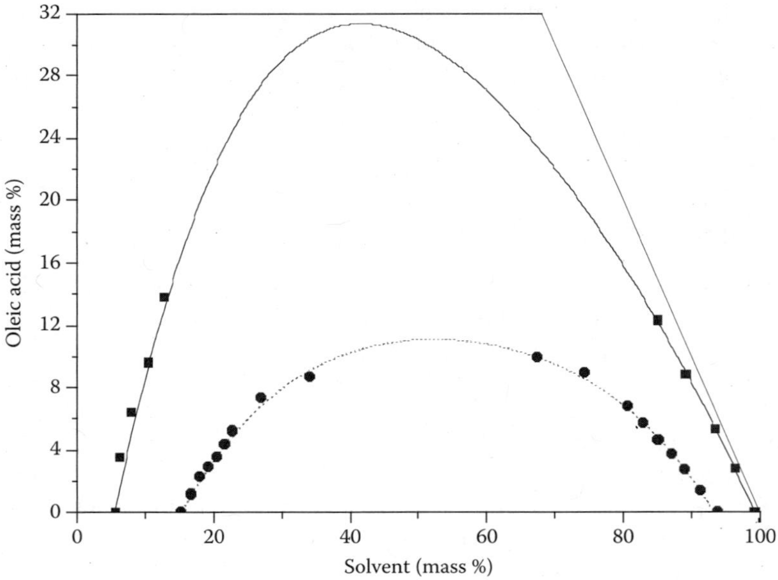

FIGURE 5.12 Binodal curves for the system refined canola oil + commercial oleic acid + solvents: anhydrous methanol (—■—) and anhydrous ethanol (···●···) at 303.2 K.

5.3.1.3 Addition of Water in the Solvent

The addition of water in ethanol increases its polarity and consequently decreases the mutual solubility of aqueous ethanol and vegetable oil. In Figure 5.13, one can see that the heterogeneous area at 303 K for the system of canola oil + oleic acid + anhydrous ethanol is lower than that for the fatty system with aqueous ethanol as solvent.

The addition of water in ethanol also decreases the distribution coefficient of the free fatty acid and in a stronger way the distribution coefficient of the vegetable oil. This effect represents that aqueous ethanol has lower capacity of extraction of free fatty acids, but the selectivity of the solvent increases and consequently reduces the loss of neutral oil in solvent extraction (see Figures 5.14 and 5.15).

Some articles [11, 12, 14] concluded that water content about 6% mass in the aqueous ethanol is appropriate for deacidification by solvent extraction, as it still provides distribution coefficients of the free fatty acid around unity and high selectivity of the solvent.

5.3.2 Deacidification of Vegetable Oils Retaining Bioactive Compounds

The majority of chemical compounds in human and animal organisms have clearly defined functions, and some of them are indispensable for maintaining the correct metabolism. Among these compounds there are polyunsaturated fatty acids, essential unsaturated fatty acids (EFAs) (linoleic, linolenic), and substances that protect them with antioxidant or other beneficial physiological properties—tocopherols, and tocotrienols belonging to the group of vitamin E, γ-oryzanol, and carotenoids [113].

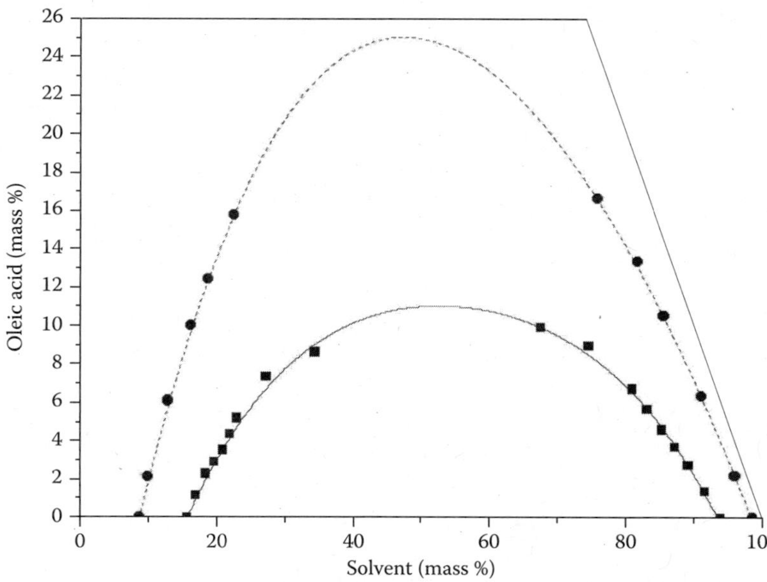

FIGURE 5.13 Binodal curves for the system refined canola oil + commercial oleic acid + solvents: anhydrous ethanol (—■—) and aqueous ethanol (···●···) at 303.2 K.

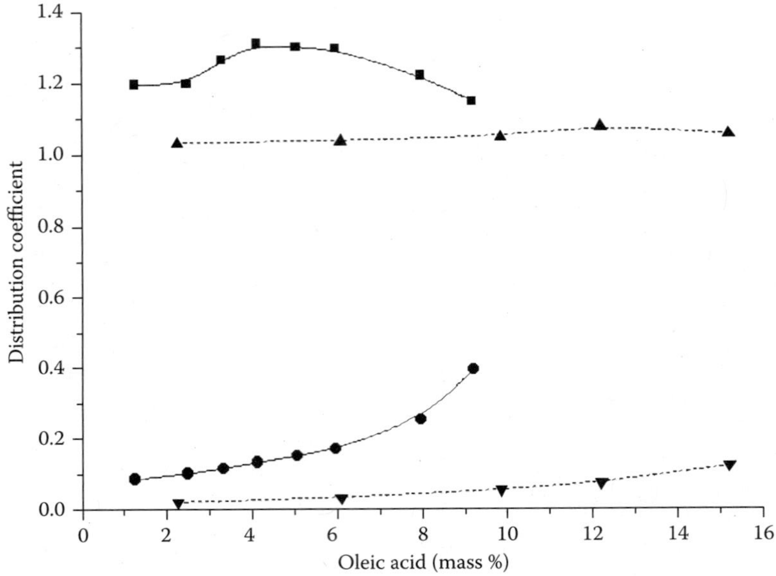

FIGURE 5.14 Distribution coefficient of: oleic acid (—■—) and canola oil (—●—) at 303.2 K in anhydrous ethanol, and oleic acid (···▼···) and canola oil (···▲···) at 303.2 K in aqueous ethanol.

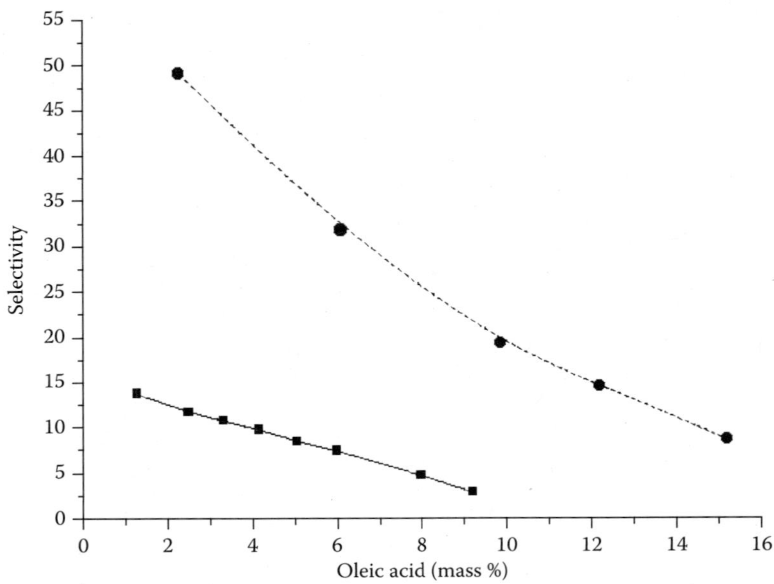

FIGURE 5.15 Selectivity of anhydrous ethanol (—■—) and aqueous ethanol (···●···) at 303.2 K.

These singular compounds are not synthesized by human or animal organisms, and so they have to be supplied in due time and in appropriate quantities [113]. Vitamin E and EFAs are substances of particular physiologic significance, and it is important to maintain their proper proportions [113–115].

Vitamin E (Figure 5.16) is a fat-soluble vitamin that comprises two major homologous series of compounds (tocochromanols), known as tocopherols and tocotrienols. The tocopherols are structurally characterized by a saturated side chain in the chroman ring, whereas the tocotrienols possess an unsaturated phytyl side chain. Four homologs of each type are known to exist in nature and have different degrees of antioxidant and vitamin E activity.

Gogolewski et al. [116] proposed a division of oils into three groups according to their nutritive value and contribution to the human organism's daily demand for fat, tocochromanols, and EFAs. The first group includes, e.g., the coconut, and olive oils; the quantity of EFAs and tocopherols in them is not sufficient for their protection from oxidation. The second group is formed by oils of which 100 g contains 30–32 g EFAs and 30–35 mg vitamin E. The third group is constituted of oils capable of supplementing the diet with vitamin E and the EFAs; among other oils there are those obtained from the wheat and maize germs with the highest content of EFAs and tocopherols and/or tocotrienols, such as rice bran, cottonseed, soybean, sunflower seed, and corn oils. Some authors suggest the optimum quantitative ratio of 0.5 mg of vitamin E equivalent to 1 g EFAs in the human organism [117–119].

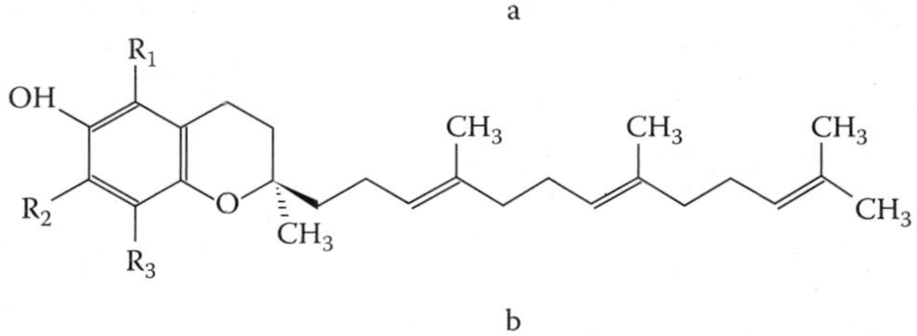

	R_1	R_2	R_3
α	methyl	methyl	methyl
β	methyl	hydrogen	methyl
γ	hydrogen	methyl	methyl
δ	hydrogen	hydrogen	methyl

FIGURE 5.16 Chemical structure of vitamin E (a: tocopherols; b: tocotrienols).

TABLE 5.3
Tocopherol Contents of Principal Edible Oils

Edible oil	Total tocopherols (mg/kg)
Palm oil	360–560
Rice bran oil	900
Cottonseed oil	830–900
Corn oil	870–2500
Olive oil	30–300
Soybean oil	900–1400
Peanut oil	330–480
Sunflower oil	630–700
Canola oil	690–695
Sesame seed oil	531–1000

In a general way, tocopherols and tocotrienols prevent formation of free radicals. They also take over the energy of the latter, inhibiting further metabolic transformations of polyunsaturated fatty acids during storage of oils, and after consumption, they participate in many physiologic processes in human organisms. In relation to the tocotrienol isomers, they present antioxidant and antitumor activities [120–124].

As can be seen in Table 5.3, vegetable oils are rich sources of tocopherols. Vitamin E has traditionally been extracted from the residues of the soybean refining industry. Tocotrienols, on the other hand, are predominantly found in palm oil and in cereal oils such as barley and RBOs. With the emergence of palm oil as the largest edible oil in the world markets [125], technological advances have been made enabling the extraction of tocotrienols from palm oil, which is currently available commercially.

Table 5.4 shows a typical tocols composition in crude palm and RBOs. Both vegetable oils present predominantly α-tocopherol and γ-tocotrienol.

TABLE 5.4
Tocols Composition in Crude Palm and Rice Bran Oils

Tocols	Crude palm oil (%)	Crude rice bran oil (%)
α-Tocopherols	21.5	23.2
β-Tocopherols	3.7	3.3
γ-Tocopherols	3.2	11.8
δ-Tocopherols	1.6	0.7
α-Tocotrienols	7.3	14.0
β-Tocotrienols	7.3	—
γ-Tocotrienols	43.7	44.3
δ-Tocotrienols	11.7	2.6

Refining of oils comprises several physical and chemical processes that aim at eliminating the unnecessary substances. During the refining process, substances with biological activity, such as tocopherols and tocotrienols, are also removed [126–129].

The contents of total and individual tocopherols and tocotrienols of vegetable oils at different stages of industrial chemical and physical refining processes gradually decrease until the end of the refining processes. The average losses of total tocopherol content in sunflower seed oil during the chemical and physical refining processes were found to be 30.2 and 35.5%, respectively [130].

The steam distillation (stripping) stage of the physical refining process causes greatest overall reduction (average 24.6%) in total tocopherol content in sunflower seed oil. In contrast to the physical refining process, the degumming–neutralizing stage in the chemical refining process causes greatest overall reduction (average 14.7%) in total tocopherol content. An additional average loss of 11.0 % occurs during deodorizing in the chemical refining process. In both chemical and physical refining, the bleaching stage causes similar effects. The physical refining process promotes a greater loss in the total and individual tocopherol contents when compared with the chemical refining process [117, 130, 131].

It has been reported that refined bleached deodorized (RBD) palm oil, palm olein, and palm stearin retain approximately 69, 72, and 76% of the original level of vitamin E in the crude oils, respectively. During the deodorization step refining process of RBO, a significant portion, about 25%, of vitamin E is stripped away with the distillate [132, 133].

Palm oil also plays an important role among the vegetable oils for being considered the world's richest source of natural plant carotenoids in term of retinal (pro-vitamin A) equivalent [134]. Figure 5.17 presents the chemical structure of the main carotenoid in palm oil (β-carotene). The typical composition of carotenoids in this oil is shown in Table 5.5.

Besides presenting vitamin A value, carotenoids reduce the risk of certain types of cancer and possess the ability of suppressing singlet oxygen [135]. Despite its nutritional value, carotenoids are removed in the physical refining process (generally used for oils with high acidity, such as palm oil) in order to obtain a clear color oil, which has better acceptance for industrial purposes [136]. Thus, some valuable characteristics of palm oil are lost during its processing, and the corresponding nutritional benefits remain available only in the crude oil [137].

In fact, the physical refining is responsible for great losses of nutraceutical compounds from palm oil. The carotenoid concentration (about 500–700 mg/kg in crude palm oil) is reduced by half during the bleached step of the physical

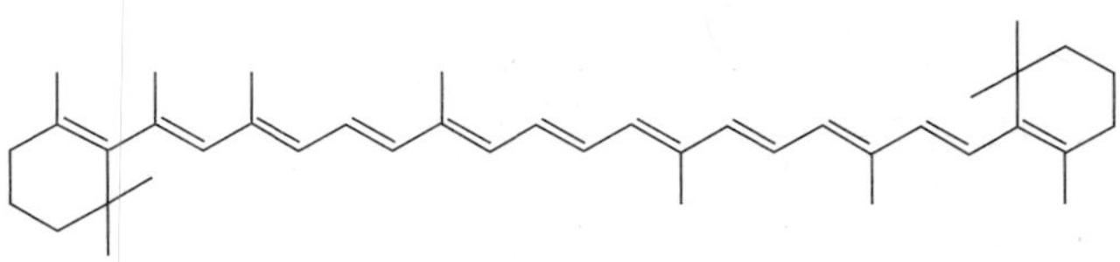

FIGURE 5.17 Chemical structure of β-carotene.

TABLE 5.5
Typical Carotenoid Composition of Palm Oil

Carotenoid	Percentage
β-Carotene	56.0
α-Carotene	35.2
cis–α-Carotene	2.5
Other carotenes (<2%)	6.3

refining process, because these components are completely destroyed during the high-temperature (240°C–260°C) and low-pressure (1–3 mmHg) deacidification–deodorization step.

In comparison with most vegetable oils, rice bran oil (RBO) has a qualitatively different composition of bioactive minor components, such as γ-oryzanol, tocotrienols, and phytosterols [132]. γ-Oryzanol derivatives, in particular, are found in only a very limited number of oils. γ-Oryzanol covers the whole group of ferulic acid esters of triterpene alcohols and phytosterols [138]. The four major components of γ-oryzanol in RBO have already been identified as 24-methylenecycloartanol

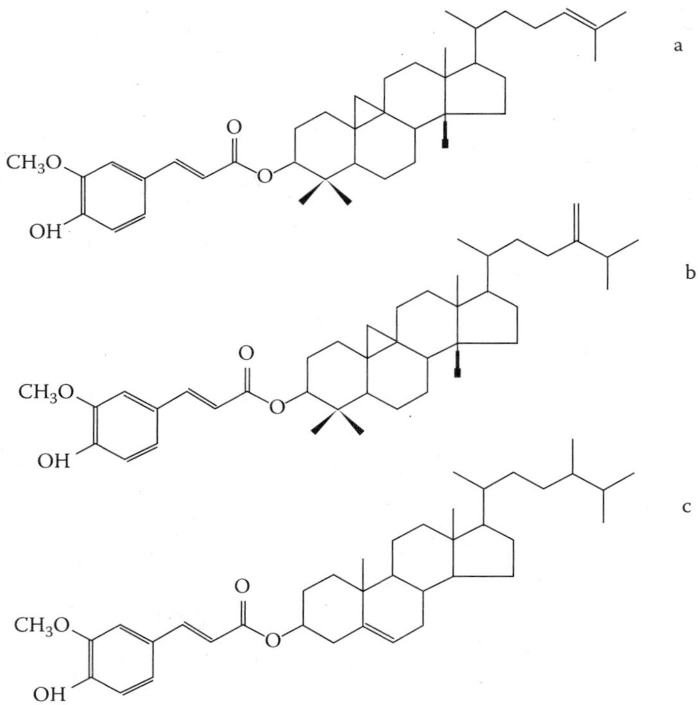

FIGURE 5.18 Chemical structure of γ-oryzanol [(a) cycloartenylferulate; (b) 24-methylencycloartanylferulate; (c) campesterylferulate].

TABLE 5.6
Components of γ-Oryzanol

Component	Molecular weight	Formula
Campesterylferulate	576.9	$C_{38}H_{56}O_4$
Campestanylferulate	578.9	$C_{38}H_{58}O_4$
b-Sitosterylferulate	590.9	$C_{39}H_{58}O_4$
Cycloartenylferulate	602.9	$C_{40}H_{58}O_4$
Cycloartanylferulate	604.9	$C_{40}H_{60}O_4$
24-Methylencycloartanylferulate	616.9	$C_{41}H_{60}O_4$

ferulate, campesterol ferulate, cycloartenol ferulate, and sitosterol ferulate [122, 124, 139]. Figure 5.18 shows the three major components in γ-oryzanol, and Table 5.6 shows the six main components in γ-oryzanol.

Medical studies indicate the hypocholesterolemic effect of RBO in humans and animals. The majority of such studies suggest that RBO is more effective in decreasing serum and liver cholesterol concentrations than oils with similar fatty acid composition, such as groundnut oil [120, 132, 140, 141]. The lowering of cholesterol levels by rice oil may be attributed to its high level of unsaponifiable matter [120, 132, 140].

Crude RBO may contain up to 5% of unsaponifiable matter. In fact, the majority of crude vegetable oils contain 1–5 g kg^{-1} of phytosterols, but RBO can contain up to 30 g kg^{-1} of phytosterols [133]. This level is reduced to values up to 1.5% in the refined RBO. In contrast, most refined vegetable oils contains only 0.3%–0.9% of unsaponifiable matter [142]. In addition to the hypocholesterolemic activity of these rice oil minor compounds, the isolated ingestion of γ-oryzanol may decrease early atherosclerosis [141] and may treat nerve imbalance disorders of menopause [143] and inflammatory processes [144].

Tocotrienols and γ-oryzanol are known as powerful antioxidants, which are associated with the prevention of cardiovascular diseases and some cancers [132, 145, 146]. Because of these beneficial effects, RBO has a high nutritional value and is therefore very appealing as a specialty oil in niche markets [132].

Refining processes have been optimized to obtain high-quality RBO for human consumption [132]. However, refining RBO is more complicated than refining other oils because of the difference in its composition of minor components [147]. The influence of refining processes on RBO has rarely been investigated. Yoon and Kim [148] briefly mentioned the effect of different chemical refining steps on the content of phosphorous, free fatty acids, total sterols, total tocopherols, and γ-oryzanol. That report mainly described the oxidative stability of RBO. Krishna et al. [149] studied the effect of refining on the retention of γ-oryzanol in chemically and physically refined oil.

van Hoed et al. [133] published an article that gives an overview of the effects of each individual step of the chemical refining process on the major and minor components of RBO. The total loss of γ-oryzanol in the whole process of refining is about

83%, being that 77% of the loss is related to the neutralization step [133, 141, 150]. In relation to physical refining, it is reported that most of the oryzanol (66%) can be retained in the refined oil [150].

As mentioned above, traditional methods of refining cause a significant decreasing of nutraceutical compound levels in edible oils. In this context, liquid–liquid extraction using appropriate solvents, such as short-chain alcohols, can be an alternative technique for refining nutritional oils.

Swoboda [98] reports a process for refining palm oil and palm oil fractions using as solvent mixtures ethanol and water or isopropanol and water, preferably with a composition near the azeotropic one. Crude palm oil subjected to solvent extraction may produce a raffinate containing a concentration of carotenoids similar to, or even larger than, the concentration of carotenoids in the original source.

With the purpose of obtaining RBO enriched with high levels of tocols—tocopherols and tocotrienols—and γ-oryzanol, Cherukuri et al. [151] suggested a liquid–liquid extraction process using lower aliphatic alcohols (C1 to C6, preferably methanol, ethanol, or isopropanol). The process involves mixing RBO and alcohol, separating the alcohol layer, and subsequently distilling this layer in order to recover enriched RBO.

A study of the process variable influence on the losses of γ-oryzanol and tocols during the deacidification process of RBO by liquid–liquid extraction was also reported by Rodrigues et al. [113]. The influence of process variables, such as acidity content in the oil, water content in the ethanolic solvent, and oil–solvent ratio, were analyzed using the response surface methodology. The results indicate that the increasing of the acidity level in the oil increases the loss of γ-oryzanol. The addition of water to the solvent reduces the solvent's capacity to extract this minor compound. In relation to tocols' losses the effect of the oil–solvent mass ratio is larger than the effect of water content in the solvent. The tocols' losses increase when the oil–solvent mass ratio is low.

Rodrigues et al. [12, 13] studied the partition coefficients of γ-oryzanol and tocopherols and tocotrienols in systems containing RBO, fatty acids, and aqueous ethanol. Their results show that most of the nutraceutical compounds from RBO can be kept on the refined oil after solvent extraction. These data were correlated by thermodynamic models, such as NRTL and UNIQUAC [13]. These models quantitatively described the systems.

In Rodrigues et al. [16], the equilibrium data for the systems containing cottonseed oil + commercial linoleic acid + ethanol +water + tocopherols were reported. The experimental data, obtained at 298.2 K, were correlated by the NRTL and UNIQUAC equations. These models quantitatively described the systems.

Recently, Gonçalves et al. [18] reported partition coefficients of carotenoids and tocopherols in systems containing palm oil + fatty acids + aqueous ethanol at 318.2 K and different water contents and oil–solvent mass ratios. The UNIQUAC model was used to correlate the partition coefficients of carotenoids and tocopherols.

Figures 5.19 and 5.20 show experimental and calculated data of nutraceutical compound partition coefficients commonly found in edible oils. The distribution coefficients are presented as a function of the water level in the ethanolic solvent [12, 13, 16, 18].

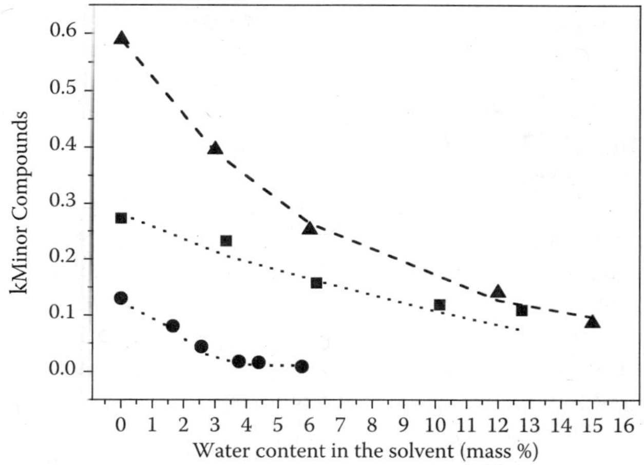

FIGURE 5.19 Minor compounds distribution coefficients (k) as a function of the water content in the solvent: (●) carotenoids in palm oil; (■) γ-oryzanol in rice bran oil; (▲) tocols in cottonseed oil; (·····) UNIQUAC model; (–––) NRTL model.

As can be seen in Figures 5.19 and 5.20, the addition of water in the solvent decreases nutraceutical compound distribution coefficients. This means that the larger the concentration of water, the smaller the solvent capacity for extracting the carotenoids, γ-oryzanol, and the tocols. It can also be observed that for all the aqueous solvents studied, the distribution coefficients of minor compounds were smaller than unity, indicating their preference for the oil phase. It is important to emphasize that this effect is desirable, once it demonstrates that most of such compounds remain in the oil refined by liquid–liquid extraction.

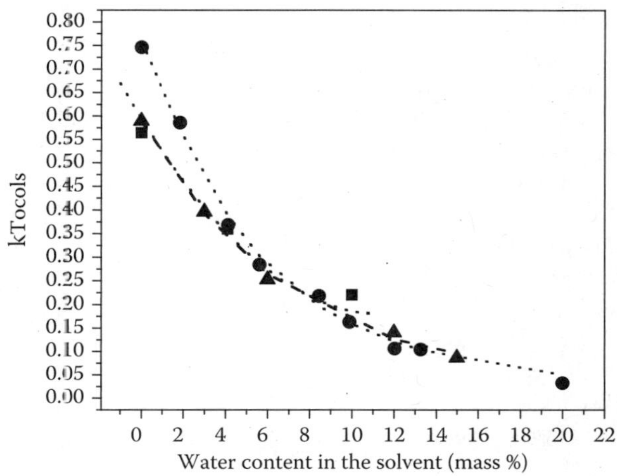

FIGURE 5.20 Tocols distribution coefficients (k) as a function of the water content in the solvent: (●) palm oil; (■) rice bran oil; (▲) cottonseed oil; (·····) UNIQUAC model; (–––) NRTL model.

Despite the same behavior, it can be observed that the tocols are transferred to the alcoholic phase in a major extension than γ-oryzanol. This can be attributed to structural differences between the molecules. Tocols are less hydrophobic than γ-oryzanol [152]. They are composed of smaller molecules that contain an unsaturated side chain in the tocotrienol series and a lower number of methyl substitutions than the oryzanol molecules.

It is also noticed that the tocols are transferred to the alcoholic phase in a larger quantity than the carotenoids. In fact, tocols and carotenoids are insoluble in water, because they have an apolar long chain (what makes them liposoluble). However, the OH group linked to the tocopherol aromatic ring enhances its solubility in ethanol.

In relation to the tocols' family, it can be seen in Figure 5.15 that the values of partition coefficients are independent of oil's chemical composition. It is possible to express the unsaturation level of fatty compounds by the iodine value. This can be calculated directly from fatty acid composition of oil according to method Cd 1c-85 AOCS [153]. Palm oil used by Gonçalves et al. [18] showed an iodine value of 55.0, whereas RBO and cottonseed oil studied by Rodrigues et al. [13, 16] presented values that equal 102.3 and 112.9, respectively.

The results showed that deacidification of vegetable oils by liquid–liquid extraction, using aqueous ethanol as solvent, allowed the retention of nutraceutical compounds in refined oil. For example, traditional physical refining usually provides a refined palm oil with approximately 0.03 mass % of tocopherols and exempt of carotenoids. In contrast, the solvent extraction process performed by using solvents containing about 6 mass % of water allows the maintenance of up to 99 mass % of carotenoids and about 80 mass % of tocopherols in refined palm oil.

5.4 NOMENCLATURE

Symbol	Description	Units in SI system	Dimension in M, N, L, T, and θ
a	UNIFAC or ASOG parameter	K	θ
A	NRTL or UNIQUAC parameter	K	θ
C	Number of components	—	—
D	Number of groups of data	—	—
E	Total mass flow of extract	$kg \cdot s^{-1}$	$M \cdot T^{-1}$
e	Mass flow of a component in extract	$kg \cdot s^{-1}$	$M \cdot T^{-1}$
F	Total mass flow of feed	$kg \cdot s^{-1}$	$M \cdot T^{-1}$
f	Mass flow of a component in feed	$kg \cdot s^{-1}$	$M \cdot T^{-1}$
G	Number of groups	—	—
G	NRTL parameter	—	—
k	Distribution coefficient	—	—
m	Group interaction parameter	—	—
M	Molecular weight	$kg \cdot kgmol^{-1}$	$M \cdot N^{-1}$
n	Group interaction parameter	—	—
N	Number of stage	—	—
Q	Group area parameter	—	—
q'	Area parameter	$kgmol \cdot kg^{-1}$	$N \cdot M^{-1}$
R	Group volume parameter	—	—
R	Total mass flow of raffinate	$kg \cdot s^{-1}$	$M \cdot T^{-1}$

Symbol	Description	Units in SI system	Dimension in M, N, L, T, and θ
R	Mass flow of a component in raffinate	$kg{\cdot}s^{-1}$	$M{\cdot}T^{-1}$
r'	Volume parameter	$kgmol{\cdot}kg^{-1}$	$N{\cdot}M^{-1}$
S	Total mass flow of solvent	$kg{\cdot}s^{-1}$	$M{\cdot}T^{-1}$
S	Mass flow of a component in solvent	$kg{\cdot}s^{-1}$	$M{\cdot}T^{-1}$
T	Temperature	K	θ
U	Interaction energy	$kg{\cdot}m^2{\cdot}s^{-2}{\cdot}kgmol^{-1}$	$M{\cdot}L^2{\cdot}T^{-2}{\cdot}N^{-1}$
u	Potential energy	$kg{\cdot}m^2{\cdot}s^{-2}{\cdot}kgmol^{-1}$	$M{\cdot}L^2{\cdot}T^{-2}{\cdot}N^{-1}$
w	Mass fraction	—	—
W	Group mass fraction	—	—
Superscript/Subscript			
C	Combinatorial part	—	—
$calc$	Calculated	—	—
ex	Experimental	—	—
FH	Size contribution	—	—
G	Group contribution	—	—
I	Oil phase	—	—
II	Alcoholic phase	—	—
i,j,k	Component	—	—
m,n,k,l	Group	—	—
R	Residual	—	—
Greek letter			
α	NRTL parameter	—	—
β	Selectivity	—	—
ϕ'	Volume fraction	—	—
γ	Activity coefficient	—	—
σ	Standard deviation	—	—
τ	NRTL or UNIQUAC parameter	—	—
θ'	Area fraction	—	—

5.5 REFERENCES

1. Antoniassi, R., W. Esteves, and A. J. A. Meirelles. 1998. Pretreatment of corn oil for physical refining. *Journal of the American Oil Chemists' Society* 75:1411–1415.
2. Bhosle, B. M., and R. Subramanian. 2005. New approaches in deacidification of edible oils—A review. *Journal of Food Engineering* 69:481–494.
3. Thomopoulos, C. 1971. Méthode de desacidification des huiles par solvant sélectif. Revve Française *Corps Gras*. 18:143–150.
4. Treybal, R. E. 1963. *Liquid extraction*. New York: McGraw-Hill.
5. Godfrey, J. C., and M. J. Slater. 1994. *Liquid-liquid extraction equipment*. Chichester: John Wiley & Sons.
6. Renon, H., and M. Prausnitz. 1968. Local compositions in thermodynamic excess functions for liquid mixtures. *AIChE Journal* 14:135–144.
7. Abrams, D. S., and J. M. Prausnitz. 1975. Statistical thermodynamics of liquid-mixtures—New expression for excess Gibbs energy of partly or completly miscible systems. *AIChE Journal* 21:116–128.

8. Oishi, T., and J. M. Prausnitz. 1978. Estimation of solvent activities in polymer solutions using a group-contribution method. *Industrial and Engineering Chemistry Process Design and Development* 17:333–339.
9. Magnussen, T., P. Rasmussen, and A. Fredenslund. 1981. UNIFAC parameter table for prediction of liquid-liquid equilibria. *Industrial and Engineering Chemistry Process Design and Development* 20:331–339.
10. Batista, E., S. Monnerat, K. Kato, L. Stragevitch, and A. J. A. Meirelles. 1999. Liquid-liquid equilibrium for systems of canola oil, oleic acid and short-chain alcohols. *Journal of Chemical Engineering Data* 44:1360–1364.
11. Gonçalves, C. B., E. Batista, and A. J. A. Meirelles. 2002. Liquid-liquid equilibrium data for the system corn oil + oleic acid + ethanol + water at 298.15K. *Journal of Chemical Engineering Data* 47:416–420.
12. Rodrigues, C. E. C., R. Antoniassi, and A. J. A. Meirelles. 2003. Equilibrium data for the system rice bran oil + fatty acids + ethanol + water at 298.2 K. *Journal of Chemical Engineering Data* 48:367–373.
13. Rodrigues, C. E. C., P. A. Pessôa Filho, and A. J. A. Meirelles. 2004. Phase equilibrium for the system rice bran oil + fatty acids + ethanol + water + γ-oryzanol + tocols. *Fluid Phase Equilibria* 216:271–283.
14. Gonçalves, C. B., and A. J. A. Meirelles. 2004. Liquid-liquid equilibrium data for the system palm oil + fatty acids + ethanol + water at 318.2K. *Fluid Phase Equilibria* 221:139–150.
15. Rodrigues, C. E. C., F. A. Silva, A. Marsaioli, Jr., and A. J. A. Meirelles. 2005. Deacidification of Brazil nut and macadamia nut oils by solvent extraction—Liquid-liquid equilibrium data at 298.2 K. *Journal of Chemical Engineering Data* 50:517–523.
16. Rodrigues, C. E. C., E. C. C. D. Reipert, A. F. Souza, P. A. Pessôa Filho, and A. J. A. Meirelles. 2005. Equilibrium data for systems composed by cottonseed oil + commercial linoleic acid + ethanol + water + tocopherols at 298.2 K. *Fluid Phase Equilibria* 238:193–203.
17. Rodrigues, C. E. C., A. Filipini, and A. J. A. Meirelles. 2006. Phase equilibrium for systems composed by high unsaturated vegetable oils + linoleic acid + ethanol + water at 298.2 K. *Journal of Chemical Engineering Data* 51:15–21.
18. Gonçalves, C. B., P. A. Pessôa Filho, and A. J. A. Meirelles. 2007. Partition of nutraceutical compounds in deacidification of palm oil by solvent extraction. *Journal of Food Engineering* 81:21–26.
19. Rodrigues, C. E. C., E. C. D. Peixoto, and A. J. A. Meirelles. 2007. Phase equilibrium for systems composed by refined soybean oil + commercial linoleic acid + ethanol + water, at 323.2 K. *Fluid Phase Equilibria* 238:193–203.
20. Fredenslund, A., J. Gmehling, and P. Rasmussen. 1977. *Vapor-liquid equilibrium using UNIFAC*. Amsterdam: Elsevier.
21. Kojima, K., and T. Tochigi. 1979. *Prediction of vapor-liquid equilibrium by the ASOG method*. Elsevier: Amsterdam.
22. Tochigi, K., D. Tiegs, J. Gmehling, and K. Kojima. 1990. Determination of new ASOG parameters. *Journal of Chemical Engineering of Japan* 23:453–463.
23. Batista, E., S. Monnerat, L. Stragevitch, C. G. Pina, C. B. Gonçalves, and A. J. A. Meirelles. 1999. Prediction of liquid-liquid equilibrium for systems of vegetable oils, fatty acids, and ethanol. *Journal of Chemical Engineering Data* 44:1365–1369.
24. Kang, C. H., and S. I. Sandler. 1987. Phase behavior of aqueous 2-polymer systems. *Fluid Phase Equilibria* 38:245–272.
25. Naphtali, L. M., and D. P. Sandholm. 1971. Multicomponent separation calculations by linearization. *AIChE Journal* 17:148–153.
26. Lima, A. S., R. M. Alegre, and A. J. A. Meirelles. 2002. Partitioning of pectinolytic enzymes in polyethylene glycol/potassium phosphate aqueous two-phase systems. *Carbohydrate Polymers* 50:63–68.

27. Marcos, J. C., L. P. Fonseca, M. T. Ramalho, and J. M. S. Cabral. 1999. Partial purification of penicillin acylase from *Escherichia coli* in poly(-ethylene glycol)-sodium citrate aqueous two-phase systems. *Journal of Chromatography B* 734: 15–22.
28. Silva, L. H. M., and A. J. A. Meirelles. 2000. Phase equilibrium in polyethylene glycol/maltodextrin aqueous two-phase systems. *Carbohydrate Polymers* 42:273–278.
29. Silva, L. H. M., and A. J. A. Meirelles. 2000. Bovine serum albumin, α-lactoalbumin and β-lactoglobulin partitioning in polyethylene glycol/maltodextrin aqueous two-phase systems. *Carbohydrate Polymers* 42:279–282.
30. Silva, L. H. M., and A. J. A. Meirelles. 2001. Phase equilibrium in aqueous mixtures of maltodextrin with polypropylene glycol. *Carbohydrate Polymers* 46:267–274.
31. Silva, L. H. M., J. R. Coimbra, and A. J. A. Meirelles. 1997. Equilibrium behavior of poly(ethylene glycol) + potassium phosphate + water two-phase systems at various pH and temperatures. *Journal of Chemical Engineering Data*. 42:398–401.
32. Snyder, S. M., K. D. Cole, and D. C. Szlag. 1992. Phase compositions viscosities and densities for aqueous two-phase systems composed of polyethylene glycol and various salts at 25 °C. *Journal of Chemical Engineering Data* 37:268–274.
33. Coimbra, J. R., J. Thömes, and M.-R. Kula. 1994. Continuous separation of whey proteins with aqueous two-phase systems in a Graesser contactor. *Journal of Chromatography A* 668:85–94.
34. Coimbra, J. R., J. Thömes, A. J. A. Meirelles, and M.-R. Kula. 1995. Performance of a Graesser contactor in the continuous extraction of whey proteins: Mixing, mass transfer and efficiency. *Bioseparation* 5:259–268.
35. Hustedt, H., K. H. Kroner, W. Stach, and M.-R. Kula. 1978. Procedure for simultaneous large-scale isolation of pullulanase and 1,4-alpha-glucan phosphorylase from *Klebsiella-Pneumoniae* involving liquid-liquid separations. *Biotechnology & Bioengineering* 20:1989–2005.
36. Albertsson, P. A. 1971. *Partition of cell particles and macromolecules.* New York: John Wiley.
37. Zaslavsky, B. Y. 1995. *Aqueous two-phase partitioning.* New York: Marcel Dekker.
38. Molyneaux, P. 1983. *Water soluble synthetic polymers: Properties and behavior.* Boca Raton: CRC.
39. Persson, J., A. Kaul, and F. Tjerneld. 2000. Polymer recycling in aqueous two-phase extractions using thermoseparating ethylene oxide-propylene oxide copolymers. *Journal of Chromatography B* 743:115–126.
40. Johansson, H.-O., G. Karlström, and F. Tjerneld. 1997. Temperature induced phase partitioning of peptides in water solutions of ethylene oxide and propylene oxide random copolymers. *Biochimica et Biophysica Acta* 1335:315–325.
41. Monteiro Filho, E. S., J. S. R. Coimbra, L. A. Minim, L. H. M. Silva, and A. J. A. Meirelles. 2002. Liquid-liquid equilibrium for ternary systems containing a sugar + a synthetic polymer + water. *Journal of Chemical Engineering Data* 47: 1346–1350.
42. Alves, J. G. L. F., L. D. A. Chumpitaz, L. H. M. Silva, T. T. Franco, and A. J. A. Meirelles. 2000. Partitioning of whey proteins, bovine serum albumin and porcine insulin in aqueous two-phase systems. *Journal of Chromatography B* 743:235–239.
43. Kertes, A. S., and C. J. King. 1986. Extraction chemistry of fermentation product carboxylic acids. *Biotechnology & Bioengineering* 28:269–82.
44. Tamada, J. A., A. S. Kertes, and C. J. King. 1990. Extraction of carboxylic acids with amine extractants. 1. Equilibria and law of mass action modeling. *Industrial & Engineering Chemistry Research* 29:1319–1326.
45. Bizek, V., J. Horacek, R. Rericha, and M. Kousova. 1992. Amine extraction of hydrocarboxylic acids. *Industrial & Engineering Chemistry Research* 31:1554–1562.

46. Prochazka, J., A. Heyberger, V. Bizek, M. Kousova, and E. Volaufova. 1994. Amine extraction of hydroxycarboxylic acids. 2. Comparison of equilibria for lactic, malic and citric acids. *Industrial & Engineering Chemistry Research* 33:1565–1573.
47. Juang, R. S., and R. H. Huang. 1997. Equilibrium studies on reactive extraction of lactic acid with an amine extractant. *Chemical Engineering Journal* 65:47–53.
48. Kirsch, T., and G. Maurer. 1997. Distribution of binary mixtures of citric, acetic and oxalic acids between water and organic solutions of tri-n-octylamine. *Fluid Phase Equilibria* 131:213–231.
49. Lintomen, L., R. T. P. Pinto, E. Batista, A. J. A. Meirelles., and M. R. W. Maciel. 2001. Liquid-liquid equilibrium of the water plus citric acid plus short chain alcohol plus tricaprylin system at 298.15 K. *Journal of Chemical Engineering Data* 46:546–550.
50. Zhang, Y., M. Valiente, and M. Muhammed. 1989. Extraction of nitric and phosphoric-acids with tributyl-phosphate. *Solvent Extraction and Ion Exchange* 7:173–200.
51. Ziat, K., B. Mesnaoui, T. Bounahmidi, R. Boussen, M. De la Guardia, and S. Garrigues. 2002. Modelling of the ternary system H_3PO_4/H_2O/TBP. *Fluid Phase Equilibria* 201:259–267.
52. Ziat, K., B. Mesnaoui, T. Bounahmidi, M. De la Guardia, and S. Garrigues. 2004. Liquid–liquid equilibria in the system H_3PO_4–KCl–H_2O–tri-n-butyl phosphate: Experiments and modeling. *Fluid Phase Equilibria* 224:39–46.
53. Mitchell, R. J., A. Arrowsmith, and N. Ashton. 1987. Mixed solvent systems for recovery of ethanol from dilute aqueous solution by liquid-liquid extraction. *Biotechnology & Bioengineering* 30:348–351.
54. Welsh, F. W., and R. E. Willians. 1989. The use of vegetable oils to recover compounds from aqueous solutions. *Journal of Chemical Technology & Biotechnology* 46:169–178.
55. Uslu, H. 2007. Liquid + liquid equilibria of the (water + tartaric acid + alamine 336 + organic solvents) at 298.15K. *Fluid Phase Equilibria* 253:12–8.
56. Yankov, D., J. Molinier, J. Albet, G. Malmaryb, and G. Kyuchoukov. 2004. Lactic acid extraction from aqueous solutions with tri-n-octylamine dissolved in decanol and dodecane. *Biochemical Engineering Journal* 21:63–71.
57. Arce, A., A. Marchiaro, J. M. Martínez-Ageitos, and A. Soto. 2005. Citrus essential oil deterpenation by liquid-liquid extraction. *Canadian Journal of Chemical Engineering* 83:366–370.
58. Cháfer, A., J. de la Torre, R. Muñoz, and M. C. Burguet, 2005. Liquid-liquid equilibria of the mixture linalool + ethanol + water at different temperatures. *Fluid Phase Equilibria* 238:72–76.
59. Wolford, R. W., J. W. Kesterson, and J. A. Attaway. 1971. Physicochemical properties of citrus essential oils from Florida. *Journal of Agricultural and Food Chemistry* 19: 1097–1105.
60. Barth, D., G. Couchi, E. Della Porta, M. Reverchon, and M. Perrut. 1994. Desorption of lemon peel oil by supercritical carbon dioxide: Deterpenation and psoralens elimination. *Journal of Supercritical Fluids* 7:177–183.
61. Fennema, Owen R. 1996. *Food chemistry*. New York: Marcel Dekker.
62. Haypek, E., L. H. M. Silva, E. Batista, D. S. Marques, M. A. A. Meireles, and A. J. A. Meirelles. 2000. Recovery of aroma compounds from orange essential oil. *Brazilian Journal of Chemical Engineering* 17:705–712.
63. Frizzo, C. D., D. Lorenzo, and E. Dellacassa. 2004. Composition and seasonal variation of the essential oils from two Mandarin cultivars of Southern Brazil. *Journal of Agricultural and Food Chemistry* 52:3036–3041.
64. Sun, D., and P. D. Petracek. 1999. Grapefruit gland oil composition is affected by wax application, storage temperature, and storage time. *Journal of Agricultural and Food Chemistry* 47:2067–2069.

65. Wilson, C. W., III, and P. E. Shaw. 1978. Quantitative composition of cold-pressed grapefruit oil. *Journal of Agricultural and Food Chemistry* 26:1432–1434.
66. Franceschi, E., M. B. Grings, C. D. Frizzo, J. V. Oliveira, and C. Dariva. 2004. Phase behavior of lemon and bergamot peel oils in supercritical CO_2. *Fluid Phase Equilibria* 226:1–8.
67. Benvenuti, F., F. Gironi, and L. Lamberti. 2001. Supercritical deterpenation of lemon essential oil, experimental data and simulation of the semicontinuous extraction process. *Journal of Supercritical Fluids* 20:29–44.
68. Verzera, A., A. Trozzi, M. Zappalaä, C. Condurso, and A. Cotroneo. 2005. Essential oil composition of Citrus meyerii Y. Tan. and Citrus medica L. cv. Diamante and their lemon hybrids. *Journal of Agricultural and Food Chemistry* 53:4890–4894.
69. Araújo, J. M. A., and A. P. F. S. Farias. 2003. Redução do teor de limoneno e bergapteno do óleo essencial de bergamota adsorvido em sílica gel pelo CO_2–Supercrítico. *Ciência e Tecnologia de Alimentos* 23:112–15 (Portuguese).
70. Owusu-Yam, J., R. F. Mathews, and P. F. West. 1986. Alcohol deterpenation of orange oil. *Journal of Food Science* 51:1180–1182.
71. Moyler, D. A., and M. A. Stephens. 1992. Counter current deterpenation of cold pressed sweet orange peel oil. *Perfumer & Flavorist* 17:37–38.
72. Meireles, M. A. A., and Z. L. Nikolov. 1994. Extraction and fractionation of essential oils with liquid carbon dioxide. In *Spices, herbs, and edible fungi*, ed. G. Charalambous, 171–199. Amsterdan: Elsevier Science.
73. Dugo, P., L. Mondello, K. D. Bartle, A. A. Clifford, D. G. P. A. Breen, and G. Dugo. 1995. Deterpenation of sweet orange and lemon essential oils with supercritical carbon dioxide using silica gel as an adsorbent. *Flavour and Fragrance Journal* 10:51–58.
74. Simões, P. C., H. A. Matos, P. J. Carmelo, E. Gomes de Azevedo, and M. Nunes da Ponte. 1995. Mass transfer in countercurrent packed columns: Application to supercritical CO_2 extraction of terpenes. *Industrial & Engineering Chemistry Research* 34:613–618.
75. Charara, Z. N., J. W. Williams, R. H. Schmidt, and M. R. Marshall. 1992. Orange flavor absorption into various polymeric packaging materials. *Journal of Food Science* 57:963–972.
76. Pangborn, R. M., and G. F. Russel. 1976. Part I. Food chemistry. In *Principles of food science*, ed. O. R. Fennema, 427–464. New York: Marcel Dekker.
77. Antosik, M., and R. Stryjek. 1992. Liquid-liquid equilibria in ternary α-pinene + Δ3-carene + polar compound systems. *Fluid Phase Equilibria* 71:321–331.
78. Arce, A., A. Marchiaro, O. Rodríguez, and A. Soto. 2002. Liquid-liquid equilibria of limonene + linalool + diethylene glycol system at different temperatures. *Chemical Engineering Journal* 89:223–227.
79. Arce, A., A. Marchiaro, and A. Soto. 2003. Propanediols for separation of citrus oil: liquid-liquid equilibria of limonene + linalool + (1,2-propanediol or 1,3-propanediol). *Fluid Phase Equilibria* 211:129–140.
80. Arce, A., A. Marchiaro, and A. Soto. 2004b. Phase stability of the system limonene + linalool + 2-aminoethanol. *Fluid Phase Equilibria* 226:121–127.
81. Tamura, K., and H. Li. 2005. Mutual solubilities of terpene in methanol and water and their multicomponent liquid-liquid equilibria. *Journal of Chemical Engineering Data* 50:2013–2018.
82. Arce, A., A. Marchiaro, O. Rodríguez, and A. Soto. 2006. Essential oil terpenless by extraction using organic solvents or ionic liquids. *AIChE Journal* 52:2089–2097.
83. Gironi, F., I. G. Farias, and L. Lamberti. 1995. Liquid-liquid equilibria for the water + ethanol + citral and water + ethanol + limonene systems at 293 K. *Journal of Chemical Engineering Data* 40:578–581.

84. Arce, A., A. Marchiaro, and A. Soto. 2004a. Liquid-liquid equilibria of linalool + ethanol + water, water + ethanol + limonene, and limonene + linalool + water systems. *Journal of Solution Chemistry* 33:561–569.
85. Babdyopadhyay, C. 1989. In *Frontiers of applied chemistry*, ed. A. Biswas, 206–221. New York: Springer-Verlag.
86. Broderick, J. 1995. *Manufacturing chemistry* 25:112–114.
87. Massaldi, H. A., and C. J. King. 1973. Simple technique to determine solubilities of sparingly soluble organics: Solubility and activity coefficients of d-limonene, n-butylbenzene, and n-hexyl acetate in water and sucrose solutions. *Journal of Chemical Engineering Data* 18:393–397.
88. Cháfer, A., R. Muñoz, M. C. Burguet, and A. Berna. 2004. The influence of the temperature on the liquid-liquid equilibria of the mixture linalool + ethanol + H_2O. *Fluid Phase Equilibria* 224:251–256.
89. Baryeh, E. A. 2000. Effects of palm oil processing parameters on yield. *Journal of Food Engineering* 48:1–6.
90. Resa, J. M., C. González, M. A. Fanega, S. Ortiz de Landaluce, and J. Lanz. 2002. Enthalpies of mixing, heat capacities, and viscosities of alcohol (C1–C4) + olive oil mixtures at 298.15 K. *Journal of Food Engineering* 51:113–118.
91. Leibovitz, Z., and C. Ruckenstein. 1983. Our experiences in processing maize (corn) germ oil. *Journal of the American Oil Chemists' Society* 60:347A–351A.
92. Cvengros, J. 1995. Physical refining of edible oils. *Journal of the American Oil Chemists' Society* 72:1193–1196.
93. Apelblat, A., T. Zaharoskin, J. Wisniak, and E. Korngold. 1996. Extraction of oleic acid from soybean oil and jojoba oil—Phase diagrams. *Journal of the American Oil Chemists' Society* 73:239–244.
94. Pina, C. G., and A. J. A. Meirelles. 2000. Deacidification of corn oil by solvent extraction in a perforated rotating disc column. *Journal of the American Oil Chemists' Society* 77:553–559.
95. Bollmann, H. 1921. Process for the removal of fatty acids, resins, bitter and mucilaginous substances from fats and oils. US patent, US19211371342.
96. van Dijck, W. J. G. 1942. US patent, US19422268786.
97. Bertholet, R. 2000. Process for refining fatty substances. WO20009637.
98. Swoboda, P. A. T. 1985. Refining of palm oils. GB19852144143.
99. Bhattacharyya, A. C., S. Majumdar, and D. K. Bhattacharyya. 1987. Refining of FFA rice bran oil by isopropanol extraction and alkali neutralization. *Oléagineaux* 42:431–433.
100. Shah, K. J., and T. K. Venkatesan. 1989. Aqueous isopropyl alcohol for extraction of free fatty acids from oils. *Journal of the American Oil Chemists' Society* 66:783–787.
101. Kim, S., C. Kim, H. Cheigh, and S. Yoon. 1985. Effect of caustic refining, solvent refining and steam refining on the deacidification and color of rice bran oil. *Journal of the American Oil Chemists' Society* 62:1492–1495.
102. Kale, V., S. P. R. Katikaneni, and M. Cheryan. 1999. Deacidifying rice bran oil by solvent extraction and membrane technology. *Journal of the American Oil Chemists' Society* 76:723–727.
103. Turkay, S., and H. Civelekoglu. 1991. Deacidification of sulfur olive oil. l. Single stage liquid-liquid extraction of miscella with ethyl alcohol. *Journal of the American Oil Chemists' Society* 68:83–86.
104. Fachini, S., and S. Samazzi. 1925. Behavior of alcohol in presence of olive oil which is acid. *Industria Olii e dei Grassi* 4:31–33.
105. Schlenker, E. 1931. Removal of fatty acids by means of alcohol. *Chemische Umschau—Fette, Oele, Wachse, und Harse* 38:108–110.
106. Sreenivasan, K., and D. S. Viswanath. 1973. Refining of cottonseed oil using solvents. *Indian Journal of Technology* 11:83–90.

107. Rius, A., and J. M Martínez-Moreno. 1947. Diagramas de solubilidad para la desacidificacion con disolventes del aceite de oliva. *Anales Fisica y Quimica* 123–148.
108. Rius, A., and Gutiérrez -Jodra. L. 1947. Diagramas de solubilidad para la eliminación de acidos grasos libres de los aceites e pescado por extracción con disolventes. *Anales de Fisica y Quimica* 245–68.
109. Rius, A., and M. A. Crespi. 1951. Desacidificacion de aceites vegetales por extraccion com disolventes. *Anales de la Real Socidad Espinolade Fisica y Quimica* 4:243–256.
110. Rigamonti R., C. Vaccarino, and A. Duzzi 1951. Sistemi ternari tra acido oleico, trioleina ed alcoli: Applicazione alla disacidazione degli oli vegetali. *Chimica e Industria* 10:619–623.
111. Rigamonti, R., and G. Botto. 1958. Équilibres de solubilité entre huile de coton, acétone et eau. *Oléagineux* 1:199–202.
112. Rodrigues, C. E. C., M. M. Onoyama, and A. J. A. Meirelles. 2006. Optimization of the rice bran oil deacidification process by liquid-liquid extraction. *Journal of Food Engineering* 73:370–378.
113. Elmadfa, I., and W. Bosse. 1985. *Vitamin E–eigenschaften, wirkungsweise und therapeutische bedeutung.* Stuttgart, Germany: Wissenschaftliche Verlagsgesellschaft mbH.
114. Nogala-Kalucka, M., M. Gogolewski, and E. Swiatkiewicz. 1993. Changes in the composition of tocopherols and fatty acids in post deodorisation condensates during refining various oils. *Fat Science Technology* 95:144–147.
115. Terao, J., and S. Matsushita. 1989. Fatty acid oxidation by singlet oxygen. *Journal of Food Processing and Preservation* 13:329–337.
116. Gogolewski, M., M. Nogala-Kałucka, and M. Szeliga. 2000. Changes of the tocopherol and fatty acid contents in rapeseed oil during refining. *European Journal of Lipid Science and Technology* 102:618–623.
117. Sawicki, J., and H. Niewiadomski. 1970. Study of the composition of the rapeseed oil deodorizer. *Scum. Zesz. Probl. Post. Nauk. Roln.* 91:237–277.
118. Harris, L. P., and N. D. Embree. 1963. Qualitative consideration of the effect of PUFA content in the diet upon requirement for vitamin E. *American Journal of Clinical Nutrition* 13:385–387.
119. Horwitt, M. K. 1962. Interaction between vitamin E and polyunsaturated fatty acids in adult men. *Vitamins and Hormones* 20:541–546.
120. Deckere, E. A. M., and O. Korver. 1996. Minor constituents of rice bran oil as functional foods. *Nutrition Review* 54:120S–126S.
121. Eitenmiller, R. R. 1997. Vitamin E content of fats and oils: Nutritional implications. *Food Technology* 51:78–81.
122. Kim, J., J. Godber, J. King, and W. Prinyawiwatkul. 2001. Inhibition of cholesterol autoxidation by the nonsaponifiable fraction in rice bran in an aqueous model system. *Journal of the American Oil Chemists' Society* 78:685–689.
123. Qureshi, A. A., B. A. Bradlow, W. A. Salser, and L. D. Brace. 1997. Novel tocotrienols of rice bran modulate cardiovascular disease risk parameters of hypercholesterolemic humans. *Journal of Nutritional Biochemistry* 8:290–298.
124. Xu, Z., and J. S. Godber. 1999. Purification and identification of components of γ-oryzanol in rice bran oil. *Journal of Agricultural and Food Chemistry* 47:2724–2728.
125. Soystats. 2007. http://www.soystats.com/2007/Default-frames.htm (accessed July 14. 2008).
126. Nogala-Kalucka, M., and M. Gogolewski. 1995. Qualitative und quantitative Änderungen von Tocopherolen und von Raps- und Sojaölen—roh und raffiniert. *Ernährung* 19:537–538.
127. Schuler, P. 1990. Natural antioxidants exploited commercially. In *Food antioxidants*, ed. B. F. J. Hudson. London: Elsevier.

128. Davidek, J., J. Valisek, and J. Pokorny. 1990. *Chemical changes during food processing*. Oxford: Elsevier.
129. Bermond, P. 1990. *Biological effects of food antioxidants*. In *Food antioxidants*, ed. B. F. J. Hudson. London: Elsevier.
130. Tasan, M., and M. Demirci. 2005. Total and individual tocopherol contents of sunflower oil at different steps of refining. *European Food Research and Technology* 220:251–254.
131. Karabulut, I., A. Topcu, A. Yorulmaz, A. Tekin, and D. S. Ozay. 2005. Effects of the industrial refining process on some properties of hazelnut oil. *European Journal of Lipid Science and Technology* 107:476–480.
132. MacCaskill, D. R., and F. Zhang. 1999. Use of rice bran oil in foods. *Food Technology* 53:50–52.
133. van Hoed, V., G. Depaemelaere, J. Vila Ayala, P. Santiwattana, R. Verhé, and W. De Greyt. 2006. Influence of chemical refining on the major and minor components of rice bran oil. *Journal of the American Oil Chemists' Society* 83:315–321.
134. Baharin, B. S., Y. B. Man, and R. A. Rahman. 2001. The effect of carotene extraction system on crude palm oil quality, carotene composition, and carotene stability during storage. *Journal of the American Oil Chemists' Society* 78:851–855.
135. Wrona, M., W. Korytowski, M. Roznowska, T. Sarna, and T. G. Truscott. 2003. Cooperation of antioxidants in protection against photosensitized oxidation. *Free Radical Biology & Medicine* 35:1319–1329.
136. Rossi, M., M. Gianazza, C. Alamprese, and F. Stanga. 2001. The effect of bleaching and physical refining on color and minor components of palm oil. *Journal of the American Oil Chemists' Society* 78:1051–1055.
137. Basiron, Y. 2005. Edible oils and fat products: Edible oils. Palm oil. In *Bailey's industrial oil and fat products,* ed. F. Shahidi, 333–429. Hoboken, NJ: John Wiley & Sons.
138. Piironen, V., D. G. Lindsay, T. A. Miettinen, J. Toivo, and A. Lampi. 2000. Review: Plant sterols: Biosynthesis, biological function and their importance to human nutrition. *Journal of the Science of Food and Agriculture* 80:939–966.
139. Shin, T. S., J. S. Godber, D. E. Martin, and J. H. Wells. 1997. Hydrolytic stability and change in E vitamers and oryzanol of extruded rice bran during storage. *Journal of Food Science* 62:704–708.
140. Rukmini, C. 1988. Chemical, nutritional and toxicological studies of rice bran oil. *Food Chemistry* 30:257–268.
141. Orthoefer, F. T. 1996. Rice bran oil: Healthy lipid source. *Food Technology* 50:62–64.
142. Rong, N., L. M. Ausman, and R. J. Nicolosi. 1997. Oryzanol decreases cholesterol absorption and aortic fatty streaks in hamsters. *Lipids* 32:303–309.
143. Nakayama, S., A. Manabe, J. Suzuki, K. Sakamoto, and T. Inagake. 1987. Comparative effects of two forms of ç-oryzanol in different sterol compositions on hyperlipidemia induced by cholesterol. *Japanese Journal of Pharmacology* 44:135–143.
144. Akihisa, T., K. Yasukawa, M. Yamaura, et al. 2000. Triterpene alcohol and sterol ferulates from rice bran and their antiinflammatory effects. *Journal of Agricultural and Food Chemistry* 48:2313–2319.
145. Xu, Z. M., N. Hua, and J. S. Godber. 2001. Antioxidant activity of tocopherols, tocotrienols, and g-oryzanol Components from rice bran against cholesterol oxidation accelerated by 2,2¢-azobis (2-methylpropionamidine) dihydrochloride. *Journal of Agricultural and Food Chemistry* 49:2077–2081.
146. Igbal, J., M. Minhajuddin, and Z. H. Beg. 2003. Suppression of 7,12-dimethylbenz [a]anthracene-induced carcinogenesis and hypercholesterolaemia in rats by tocotrienol-rich fraction isolated from rice bran oil. *European Journal of Cancer Prevention* 12:447–53.

147. Narayana, T., B. Kaimal, S. R. Vali, et al. 2002. Origin of problems encountered in rice bran oil processing. *European Journal of Lipid Science and Technology* 104:203–211.
148. Yoon, S H., and S. K. Kim. 1994. Oxidative stability of high-fatty acid rice bran oil at different stages of refining. *Journal of the American Oil Chemists' Society* 71:227–229.
149. Krishna, A. G. G., S. Khatoon, P. M. Sheila, C. V. Sarmandal, T. N. Indira, and A. Mishra. 2001. Effect of refining of crude rice bran oil on the retention of oryzanol in the refined oil. *Journal of the American Oil Chemists' Society* 78:127–131.
150. Orthoefer, F. T. 2005. Edible oils and fat products: Edible oils. Rice bran oil. In *Bailey's industrial oil and fat products,* ed. F. Shahidi, 465–489. Hoboken, NJ: John Wiley & Sons.
151. Cherukuri, R. S. V., R. Cheruvanky, I. Lynch, and D. L. McPeak. 1999. Process for obtaining micronutrient enriched rice bran oil. US19995985344.
152. Abidi, S. L., S. Thiam, and I. M. Warner. 2002. Elution behavior of unsaponifiable lipids with various capillary electrochromatographic stationary phases. *Journal of Chromatography A* 949:195–207.
153. American Oil Chemists' Society. 1988. *Official methods and recommended practices of the American Oil Chemists' Society*, vols. 1–2, 3rd ed., Champaign: AOCS.

6 Supercritical and Pressurized Fluid Extraction Applied to the Food Industry

Paulo T. V. Rosa, Juan Carlos Parajó, Herminia Domínguez, Andrés Moure, Beatriz Díaz-Reinoso, Richard L. Smith, Jr., Masaaki Toyomizu, Louw J. Florusse, Cor J. Peters, Motonobu Goto, Susana Lucas, and M. Angela A. Meireles

CONTENTS

6.1 Fundamentals of Supercritical Extraction from Solid Matrices 272
Paulo T. V. Rosa and M. Angela A. Meireles

6.1.1 Mass Transfer: Balance Equations and Kinetics 273
6.1.1.1 Mathematical Models to Describe the OECs 276
6.1.2 Thermodynamics: Equilibrium 281
6.1.3 Nomenclature 285
6.1.4 References 287

6.2 Obtaining Antioxidants by Supercritical Fluid Extraction 288
Juan Carlos Parajó, Herminia Domínguez, Andrés Moure, and Beatriz Díaz-Reinoso

6.2.1 Obtaining Antioxidants by Supercritical Fluid Extraction 288
6.2.1.1 Raw Materials and Their Conditioning 289
6.2.1.2 Operational Variables Affecting the SCF Extraction of Antioxidants 292
6.2.1.3 Processing Schemes Proposed for Antioxidant Extraction ... 293
6.2.1.4 Obtaining Antioxidants by SFE with Cosolvent 302
6.2.2 Obtaining Antioxidants by High-Pressure Water Extraction 306
6.2.2.1 Processing of LCM 306

6.2.2.2 Other Technologies Dealing with Hot Water Extraction of Vegetal Biomass 312
6.2.3 References 315
6.3 Obtaining Bioactive Compounds from Cashew Trees and Nuts 327
Richard L. Smith, Jr., Masaaki Toyomizu, Louw J. Florusse, and Cor J. Peters
6.3.1 Phenolic Lipids and Their Origin 328
6.3.2 Chemical Structures of Phenolic Lipids in Cashew 328
6.3.3 Bioactivity of Anacardic Acids and Uncoupling Effects 329
6.3.4 Cultivation and Production of Cashew 331
6.3.5 Cashew Trees and Processing of Cashew 332
6.3.6 Separation of Cashew with Supercritical CO_2 334
6.3.7 Phase Behavior 338
6.3.8 Measurements with a Synthetic Method 339
6.3.8.1 Procedure 340
6.3.8.2 Liquid–Liquid–Vapor Equilibria 340
6.3.9 Measurements with an Analytical Method 342
6.3.9.1 Procedure 343
6.3.9.2 Sampling 344
6.3.9.3 Vapor–Liquid Equilibria 344
6.3.10 Correlation of the Data 346
6.3.11 Separation Scheme for Cashew 347
6.3.12 Conclusions 347
6.3.13 Acknowledgments 348
6.3.14 References 348
6.4 Fractionation of Orange Volatile Oil 352
Motonobu Goto
6.4.1 Phase Equilibria for Citrus Oil Components 352
6.4.2 Liquid Material Processing 354
6.4.2.1 Countercurrent Extraction Process 354
6.4.2.2 Adsorption/Desorption Process 360
6.4.2.3 Pressure Swing Adsorption 362
6.4.3 Nomenclature 366
6.4.4 References 367
6.5 High-Pressure Adsorption/Desorption to Improve the Quality of Soluble Coffee Aroma 370
Susana Lucas
6.5.1 Introduction 370
6.5.1.1 Coffee Aroma 371
6.5.1.2 Conventional Techniques for Coffee Aroma Concentration 372
6.5.2 Supercritical Technology for Coffee Aroma Recovery 374
6.5.2.1 State of the Art 375

6.5.2.2 Process Description 376
6.5.2.3 Experimental Section 376
6.5.2.4 Influence of Process Operating Conditions 377
6.5.2.5 Results 379
6.5.3 Conclusions 383
6.5.4 Nomenclature 385
6.5.5 References 385
6.6 Cost of Manufacturing of Supercritical Fluid Extracts from Condimentary Plants 388
Paulo T. V. Rosa and M. Angela A. Meireles
6.6.1 Capital Cost 388
6.6.2 Cost of Manufacturing 390
6.6.2.1 Direct Manufacturing Cost 390
6.6.2.2 Fixed Manufacturing Cost 391
6.6.2.3 General Manufacturing Expenses 392
6.6.2.4 Estimation of the Cost of Manufacturing Condimentary Extracts 392
6.6.2.5 COM Extracts from Clove Buds 395
6.6.2.6 COM Extracts from Ginger 398
6.6.3 Nomenclature 400
6.6.4 References 400

Supercritical extraction and pressurized extraction find several applications in food and food-related industries. Thus, in Section 6.1 the fundamentals of supercritical fluid extraction (SFE) from solid matrices will be discussed. In Section 6.2 a review of the literature is focused on obtaining antioxidants by both SFE and hot water extraction (HWE). In Sections 6.3, 6.4, and 6.5 three applications of supercritical fluids will be discussed: (i) the processing of cashew nuts, (ii) the fractionation of orange volatile oil, and (iii) the improvement of the quality of soluble coffee aroma. Finally, in Section 6.6 the same methodology used in Chapters 2 and 4 to estimate the cost of manufacturing of extracts from aromatic, condimentary, and medicinal plants is presented and applied to obtaining extracts from clove buds and ginger. As in Chapters 2, 3, and 5, the mass transfer and the phase equilibria that govern the extraction and fractionation process are discussed. Section 6.3 gives an in-depth view of cashew processing as well as the phase equilibria involved in fractionating the valuable cashew products. Section 6.4 presents the fundamentals for the fractionation of orange volatile oil using adsorption/desorption; the phase equilibria involved in these processes are discussed. Section 6.5 gives a nice example of using adsorption/desorption at high pressure to improve the soluble coffee aroma. There are several other applications of supercritical and pressurized solvents in food and food-related industries; in *Supercritical Fluid Extraction of Nutraceuticals and Bioactive Compounds*, edited by J. L. Martínez (Boca Raton, FL: CRC Press, 2008), other applications of supercritical fluids in food and food-related industries can be found.

6.1 FUNDAMENTALS OF SUPERCRITICAL EXTRACTION FROM SOLID MATRICES

Paulo T. V. Rosa and M. Angela A. Meireles

Solid–fluid extraction is a unit operation that performs the separation of a solute or mixture of solutes present in a solid matrix by bringing it into contact with an adequate solvent. The solvent can be a regular liquid, as presented in Chapter 4, or a supercritical fluid. A fluid is considered to be in a supercritical state when the system pressure and temperature are above its critical point. At this region the fluid can be considered either as an expanded liquid or as a compressed gas.

Supercritical fluids display unique characteristics that enable then to be used as solvents. The density of these fluids is relatively high, and consequently they have high solvation power. Furthermore, the density can be easily changed by a small variation in the system pressure or temperature, mainly in the region near the critical point. This effect gives a certain degree of selectivity for these fluids and also allows an easy solvent–solute separation process. The separation can be achieved by decreasing the pressure or increasing the temperature of the mixture that leaves the extraction column. Another important characteristic is the relatively low viscosity and high diffusion coefficient that permits high extraction rates when these fluids are used.

The most used supercritical fluid is carbon dioxide, which has a critical point of 7.38 MPa and 304.2 K. This fluid has low critical temperature, which allows operations near room temperature, and mild critical pressure. It is nontoxic, nonflammable, nonexpensive, and readily available at good purity. The CO_2 is a good solvent for hydrophobic or slightly hydrophilic compounds. In general, when the system pressure is increased, some more hydrophilic compounds can be obtained. If even more hydrophilic compounds are aimed the solvent polarity can be changed by using more polar organic solvents such as ethyl acetate, ethanol, or methanol. These added organic solvents are known as cosolvents or modifiers.

A diagram of a supercritical extraction unit can be observed in Figure 6.1.1. The raw material that contains the desired solute is packed into the extraction column. To allow the solvent penetration into the raw material particles, it should be dried before the extraction. The supercritical solvent enters in one side of the extraction column and extracts the solutes as it flows into the system. At the exit of the extraction column the solvent–solute mixture goes to a separation vessel where, in general, the pressure is decreased to a value below the solvent's critical point. The solvation power of the gas is very low and the solute is precipitated. The solute is then collected in the separation vessel, and the gaseous solvent is recovered in a solvent cycle region. At the solvent cycle, the gaseous solvent is condensed by the decrease of temperature, the pressure is increased to a value above the critical point (but with temperature below this point) by a pump, and it is transformed in a supercritical fluid at the extraction temperature by flowing into a heat exchanger.

The main points that should be determined in the operation of a supercritical fluid extraction unit are the kinetic of the extraction process and the solubility of the solute in the supercritical fluid. These specific points will be discussed below in Sections 6.1.1 and 6.1.2.

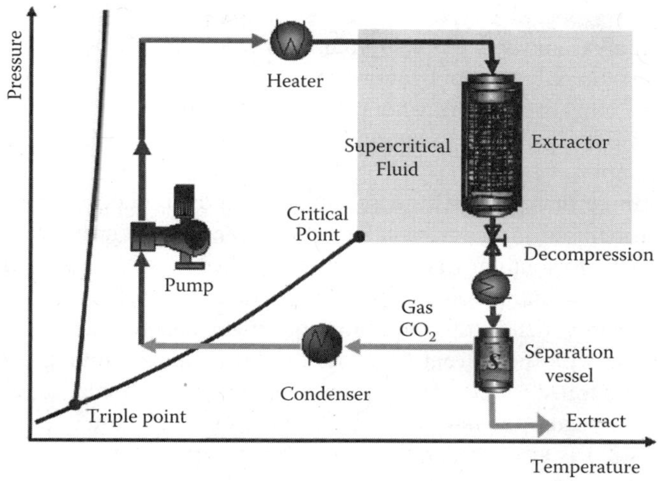

FIGURE 6.1.1 Diagram of a supercritical fluid extraction process.

6.1.1 Mass Transfer: Balance Equations and Kinetics

The variation in the supercritical solvent concentration into the extraction column can be determined by a mass balance performed in a differential volume of the particle bed as illustrated in Figure 6.1.2. In this figure the particles and the void space present in the bed were segregated into two distinct regions. This separation can only be true if the particle bed is homogeneous.

There are three mass transfer mechanisms presented in Figure 6.1.2: the mass transfer in the void region by convective transport, the mass transfer in the void region by the axial dispersion, and the mass transfer in the void–particle interface.

The convective term takes into account the mass that is transported with the flowing solvent. In the case of the homogeneous particle bed, the mean solvent velocity in the bed void space can be obtained by the following equation.

$$v = \frac{Q}{A_T \varepsilon}, \tag{6.1.1}$$

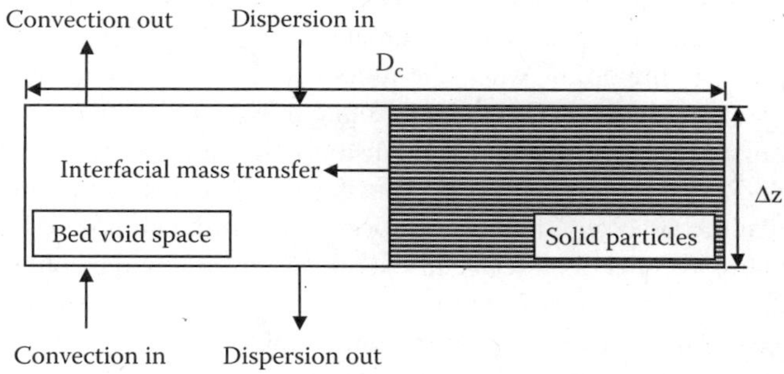

FIGURE 6.1.2 Variables representation for the differential mass balance.

where v is the solvent velocity, Q is the solvent volumetric flow rate, A_T is the column cross-section area, and ε is the void volume fraction or bed porosity. The ε value takes into account only the void volume of the interstitial region of the particle bed outside or, in other words, the void volume outside of the particles. The product of A_T and ε represents the cross-sectional area where the solvent can flow into the extraction column.

To determine the convective mass flux one needs to choose the concentration unit that will be used in the mass balance. The concentration in terms of the ratio of mass of solute to solution volume is generally used. But it is easier to express the concentration in the extraction column as the ratio between the solute mass and the solvent mass. Subsequently, this concentration will be denominated as Y. The Y value should increase with the solvent flow into the extraction column as a result of the interfacial mass transfer. Thus, the Y value should be a function of the axial position of the solvent. Because the interfacial mass flux should decrease as a function of the extraction time, Y is also a function of time. If the particle bed is homogeneous and the plug flow can be considered, the Y value should not be a function of the radial position.

Using the physical picture described above, the convective flux is given by the product of the solution density, solution velocity, and the solution concentration. Because the solution is usually diluted, the solvent density can be used instead of the solution density. Thus, the convective flux is given by

$$\textit{convective flux} = \rho v Y(z,t), \tag{6.1.2}$$

where ρ is the solvent density. The unit of the convective flux is mass of solute divided by area and time.

The axial dispersion occurs because of the differences in the Y value in the flow direction. Thus, there will be a molecular mass transfer in the opposite direction of the flow to decrease the mass gradient. This mass transfer is expressed in terms of the Fick's law. The axial dispersion flux is given by

$$\textit{axial dispersion flux} = \rho D_{az} \frac{\partial Y(z,t)}{\partial z}, \tag{6.1.3}$$

where D_{az} is the axial dispersion coefficient and z is the axial position.

The interfacial mass flux can occur by two mass transfer mechanisms: by the convection due to the solvent movement around the particles and by diffusion. The convection is important when the amount of solute present on the outside part of the particles is large. When there is solute only inside of the particles, the molecular mechanism of mass transfer, or diffusion, will be important for the process. In the intermediate situation, both mass transfer mechanisms can be important. This term is difficult to precisely define, and for now it will be considered as given by a function that represents the interfacial mass transfer rate by the column volume. We will identify this function as J, and it would have the solute concentration in the solvent and in the solid particles as independent variables, or $J = J(Y, X)$, where X is the mean solute concentration in the solid particles given in terms of the ratio between the mass of solute and solid particles free of solute.

Considering the bed void region presented in Figure 6.1.2 as the control surface, the mass balance equation can be described as follows:

$$A_T \varepsilon \rho v Y\big|_z + A_T \varepsilon \rho D_{az} \frac{\partial Y}{\partial z}\bigg|_{z+\Delta z} + J(Y,X) A_T \Delta z = A_T \varepsilon \rho v Y\big|_{z+\Delta z}$$

$$+ A_T \varepsilon \rho D_{az} \frac{\partial Y}{\partial z}\bigg|_z + \frac{\partial}{\partial t}\left(\rho A_T \Delta z Y\right). \tag{6.1.4}$$

The left-hand side of Equation 6.1.4 represents the mass of solute that enters into the control surface; the two first terms of the right-hand side correspond to the solute mass that leaves the control surface, and the last term symbolizes the solute mass accumulation. This mass balance represents that there is no chemical reaction during the extraction process. Each term of Equation 6.1.4 has unit of solute mass per time. The porosity term was used in the convective and dispersive terms to correct the real area that the solvent can flow into the particle bed.

Equation 6.1.4 can be rearranged to result in the following equation:

$$\frac{D_{az} \dfrac{\partial Y}{\partial z}\bigg|_{z+\Delta z} - D_{az} \dfrac{\partial Y}{\partial z}\bigg|_z}{\Delta z} + \frac{J(Y,X)}{\rho \varepsilon} = \frac{vY\big|_{z+\Delta z} - vY\big|_z}{\Delta z} + \frac{\partial Y}{\partial t}. \tag{6.1.5}$$

Applying the limit when Δz tends to zero in Equation 6.1.5, meaning that the region described in Figure 6.1.2 tends to one single plane into the column, we can determine the final expression for the differential mass balance in the extraction column. This expression is given by

$$\frac{\partial}{\partial z}\left(D_{az} \frac{\partial Y}{\partial z}\right) + \frac{J(Y,X)}{\rho \varepsilon} = v \frac{\partial Y}{\partial z} + \frac{\partial Y}{\partial t}. \tag{6.1.6}$$

Equation 6.1.6 represents the solute concentration variation in the supercritical phase as a function of the axial position and time. The variation in the solute concentration in the solid particles can be determined by a mass balance on the set of particles present in the differential volume presented in Figure 6.1.2. The mass balance equation on the differential volume is then given by

$$0 = J(Y,X) A \Delta z + \frac{\partial}{\partial t}\left[X \rho_s (1-\varepsilon) A \Delta z\right], \tag{6.1.7}$$

where ρ_s is the solid free of solute density. Considering that ρ_s is constant, the final mass balance equation in the solid phase is

$$\frac{\partial X}{\partial t} = -\frac{J(Y,X)}{\rho_s (1-\varepsilon)}. \tag{6.1.8}$$

Equations 6.1.6 and 6.1.8 are the starting point of the majority of the mathematical models present in the literature to describe the overall extraction curve.

This curve represents the variation with time of the solute concentration in the fluid phase at the exit of the extraction column.

To develop the mathematical models, the interfacial mass transfer term should be defined. For example, if the mass transfer is driven by the mass convection, the $J(Y, X)$ term is given by

$$J(Y,X) = \rho\varepsilon K_Y a(Y^* - Y), \tag{6.1.9}$$

where $K_Y a$ is the overall mass transfer coefficient (K_Y) multiplied by the interfacial area per column volume (a), and Y^* is the solute solubility in the solvent. The $\rho\varepsilon$ term is presented in Equation 6.1.9 to give the right unit for $J(Y, X)$. Substituting Equation 6.1.9 in Equations 6.1.6 and 6.1.8, we obtain

$$\frac{\partial}{\partial z}\left(D_{az}\frac{\partial Y}{\partial z}\right) + K_y a\left(Y^* - Y\right) = v\frac{\partial Y}{\partial z} + \frac{\partial Y}{\partial t} \tag{6.1.10}$$

$$\frac{\partial X}{\partial t} = -\frac{\rho\varepsilon K_y a\left(Y^* - Y\right)}{\rho_s(1-\varepsilon)}. \tag{6.1.11}$$

The solution of Equation 6.1.10 can provide the overall extraction curve (OEC) for the beginning of the extraction where the main mass transfer mechanism should be convection in the fluid phase. In this period, the variation of the solute concentration in the fluid does not depend on the solute concentration in the solid phase.

The resolution of the complete differential mass balance equation can be a cumbersome process, and some simplifications together with the definition of the interfacial mass transfer term are used to define the majority of the mathematical model presented in the literature. In the next section, some of these models will be presented.

6.1.1.1 Mathematical Models to Describe the OECs

The mathematical models used to describe the OECs can be divided into three main categories: empirical, those using similarity with heat transfer, and those obtained from the differential mass balance.

The first category of models relies on the hyperbolic shape of the OECs and then uses hyperbolic functions to fit it. Langmuir-like models were used by Naik et al. [1] and Esquievel et al. [2] to fit the experimental overall extraction curves. The general form of these models is given by

$$m_E = \frac{A_{1E}t}{B_{1E} + t}, \tag{6.1.12}$$

where m_E is the cumulative mass of solute obtained during the extraction, A_{1E} and B_{1E} are constants, and t is the extraction time.

The A_{1E} parameter represents the total amount of solute that can be obtained during the extraction, or the product of total mass of particles packed into the extraction column and the global yield. B_{1E} *is* related to the mass transfer in the extraction system. In spite of giving good fits in some cases, these models do not give practical information about the system and have limited application for scale-up of the extraction operation.

The model presented by Reverchon [3] uses the analogy proposed by Crank [4] between the mass diffusion and the heat conduction in porous media. All mass transfer is considered to happen by diffusion and an apparent diffusion coefficient can be determined. The model has the following equation:

$$m_E = \left(1 - \frac{6}{\pi^2}\sum_{n=1}^{\infty}\frac{-n^2\pi^2 D_{ap}t}{R_P^2}\right)x_0 m_t, \tag{6.1.13}$$

where D_{ap} is the apparent diffusion coefficient, R is the particle radius, x_0 is the global yield, and m_t is the total particle mass packed into the extraction column.

The application of this model is restricted to few systems because of the poor fit presented by the Crank model for the porous media. This behavior is expected because the apparent diffusion coefficient cannot describe properly the convective mass transport that dominates the beginning of the extraction process.

As previously stated, the differential mass balance is the starting point for the majority of the mathematical models used to describe the overall extraction curves. Here, only some of these models will be presented.

The Tan and Liou model [5] considers the variation of the solid phase concentration with time as a first-order kinetic equation and neglects the axial dispersion coefficient in the extraction column. With these restrictions, the differential mass balance equations for the fluid and solid phases are represented by

$$v\frac{\partial Y}{\partial z} + \frac{\partial Y}{\partial t} = -\frac{(1-\varepsilon)\rho_s}{\rho\varepsilon}\frac{\partial X}{\partial t} \tag{6.1.14}$$

$$\frac{\partial X}{\partial t} = -k_d X, \tag{6.1.15}$$

where k_d is the first-order constant also known as the desorption constant.

The variation with time of the solute concentration in the solid phase can be directly determined by integrating Equation 6.1.15, resulting in an exponential decay of the solute concentration with time. The initial condition for the solid phase was that the solid phase solute concentration at the beginning of the extraction is a constant value equal to the maximum attainable yield for a given condition of pressure and temperature. The expression for the solid phase concentration can be derived with respect to time and the obtained equation can be used in Equation 6.1.14. Thus, the variation in the fluid phase concentration with axial position and time can be determined by solving the resulting differential equation. The initial condition used for the fluid phase was that there is no solute in this phase at the beginning of the extraction and that the boundary condition was that the solvent enters pure into the extraction column. The expression obtained for the fluid phase at the exit of the column was

$$Y(H,t) = \frac{1-\varepsilon}{\varepsilon}x_0\frac{\rho}{\rho_s}\left[\exp\left(\frac{k_d H}{v}\right)-1\right]\exp(-k_d t). \tag{6.1.16}$$

The cumulative solute mass can be determined by integrating in time the solute mass rate produced at the exit of the column or

$$m_E = \int_0^t Y(H,t)Q\,dt. \tag{6.1.17}$$

For the Tan and Liou model [5], the final expression for the overall extraction curve is given by

$$m_E = \frac{Q(1-\varepsilon)x_0\rho_s}{\rho\varepsilon k_d}\left[1-\exp(k_d H / v)\right]\left[\exp(-k_d t)-1\right]. \tag{6.1.18}$$

The authors still tried to use an Arrhenius dependence of the k_d with temperature. In general, this model presents a poor fit of experimental overall extraction curves for several types of raw materials.

Goto et al. [6] presented a model used to describe the overall extraction curves from leaves of plants. The differential mass balance used a configuration different from the one presented in Figure 6.1.2. The mass balance in the solid phase was divided in two fractions: one that considered the solute present in the solid and the other the solute present in the fluid phase, which is located in the porous part of the solid matter. In the fluid phase, the convective term was considered to be practically constant so that this term could be changed by $Y/\varepsilon\tau$, where τ is the residence time of the solvent into the extraction column. The dispersion on the fluid phase was also neglected. The interfacial mass transfer term was considered to be the convective fluid phase, with driven force given by the difference of solute concentration in the bulk region of the fluid phase and in the entrance of the particle porous.

The OEC for this system was obtained by solving the differential mass balance in the fluid and solid phases, and the final equation obtained was

$$m_E = \frac{A_1\left[\beta / K + (1-\beta)\right]x_0\rho_s Q}{\rho}\left\{\frac{\tau}{a_1}\left[\exp(a_1\frac{t}{\tau})-1\right]-\frac{\tau}{a_2}\left[\exp(a_2\frac{t}{\tau})-1\right]\right\} \tag{6.1.19}$$

and

$$a_1 = \frac{1}{2}\left(-b+\sqrt{b^2-4c}\right) \quad ; \quad a_2 = \frac{1}{2}\left(-b-\sqrt{b^2-4c}\right)$$

$$b = \frac{\phi}{\beta+(1-\beta)K}+\frac{1}{\varepsilon}+\frac{\phi(1-\varepsilon)}{\varepsilon} \quad ; \quad c = \frac{\phi}{\left[\beta+(1-\beta)K\right]\varepsilon}, \tag{6.1.20}$$

$$A_1 = \frac{(1-\varepsilon)c}{a_1-a_2} \quad ; \quad \phi = \frac{3K_y a\tau}{3+B_i}$$

where K is the desorption equilibrium constant, β is the particle porosity, and B_i is the mass Biot number.

The Goto model can fit very well OECs that present a slow extraction curve at the beginning of the extraction followed by a constant extraction rate. In general, this behavior is observed in systems that have a huge accumulation term in the tubes after the extraction column and is not related to the real extraction kinetic.

Sovová [7] presented one interesting model where the solute was divided into two fractions: one present in broken cells and another in intact cells. The solute fraction present in broken cells (x_k) can be easily extracted. As the cells are broken during the milling process, the easily extracted solute should be located at the particle surface and it should be considered that it covers all surface area. So, the main mass transfer mechanism during the extraction of this fraction is the convection in the fluid phase. During this stage, the interfacial mass transfer term has a form similar to the one presented in Equation 6.1.9. The end of this constant extraction rate (CER) region is called t_{CER} in literature. The extraction rate of the easily accessible solute is characterized by a straight line that can be observed in the beginning of the extraction process.

As the extraction proceeds, there will be some places at the particle surface where the free solute finished and a slow decrease in the extraction rate can be observed. In this section both convection and diffusion will be important for the mass transfer in the system. The interfacial mass transfer term in this region is given by

$$J(Y,X) = K_x a X\left(1-\frac{Y}{Y^*}\right), \tag{6.1.21}$$

where $K_x a$ is the volumetric overall mass transfer coefficient in the solid phase.

The end of this transition region, identified as t_{FER}, is where the falling of the extraction rate can be observed, and thus is denoted in literature as the falling extraction rate (FER) period. When all easily extractable solute on the particle surface is finished, the extraction rate will be almost a straight line but with very low slope. The mass transfer will be limited by the solute effective diffusion from the particles. Equation 6.1.9 can still be used in this region because the Y value will be very small when compared with the solubility.

Additionally, Sovová [7] disregarded the dispersion term of the mass balance and neglected the variation of the fluid phase concentration with time because the residence time of the solute into the extraction column is relatively low to consider this variation. The transient behavior was considered to be noticed only in the solid phase. With these considerations, the final equations presented were

$$m_E = QY^*\left[1-\exp(-Z_1)\right]t \qquad \text{for } t \le t_{CER} \tag{6.1.22}$$

$$m_E = QY^*\left[t - t_{CER}\exp(Z_w - Z_1)\right] \qquad \text{for } t_{CER} \le t \le t_{FER} \tag{6.1.23}$$

$$m_E = m_{SI}\left\{x_0 - \frac{Y^*}{W}\ln\left[1+\left[\exp\left(\frac{Wx_0}{Y^*}\right)-1\right]C_1\right]\right\} \qquad \text{for } t > t_{FER}, \tag{6.1.24}$$

where

$$Z_1 = \frac{m_{IS}K_y a\rho}{Q(1-\varepsilon)\rho_s} \tag{6.1.25}$$

$$W = \frac{m_{IS}K_x a}{Q(1-\varepsilon)} \tag{6.1.26}$$

$$Z_W = \frac{Z_1 Y^*}{W x_0} \ln \left\{ \frac{x_0 \exp\left[\frac{WQ}{m_{SI}}(t_{CER} - t)\right] - x_k}{x_0 - x_k} \right\}, \tag{6.1.27}$$

where m_{IS} represents the mass of inert solid or the mass of solid particles minus the mass of extractable material.

The model of Sovová [7] in general can fit very well the extractions curve and can be used in the scale-up studies. In 2005, Sovová [8] proposed a model considering the fluid phase variation with time and changing the interfacial mass transfer term. The complexity of the model increases considerably.

The models presented so far consider the solute as one pseudocomponent and only the overall extraction curve can be obtained. Sometimes it is interesting to know the extraction of a family of compounds. Martínez et al. [9] proposed a mathematical model considering the interfacial mass transfer term as a summation of the several categories of compounds present in the solute. The mass transfer was considered to follow a logistic model for each category of compounds. Thus the interfacial mass transfer term was considered to be given by

$$J(Y,X) = \sum_{i=1}^{n} \frac{A_i b_i \exp\left[b_i (t_{mi} - t)\right]}{\left\{1 - \exp\left[b_i (t_{mi} - t)\right]\right\}^2}, \tag{6.1.28}$$

where A_i, t_{mi}, and b_i are the model parameters.

To integrate the fluid phase mass balance equation, the dispersion and transient terms were disregarded. With these assumptions, the cumulative mass of each fraction (m_{Ei}) was given by

$$m_{Ei} = \frac{QHA_i}{v\varepsilon} \left\{ \frac{1}{1 + \exp\left[b_i (t_{mi} - t)\right]} - \frac{1}{1 + \exp\left[b_i t_{mi}\right]} \right\}. \tag{6.1.29}$$

As for very long extraction times the cumulative mass tends to the total amount of that family of substance that is presented in the particle (m_{ti}), and Equation 6.1.29 can be written as follows:

$$m_{Ei} = \frac{m_{ti}}{\exp(b_i t_{mi})} \left\{ \frac{1 + \exp(b_i t_{mi})}{1 + \exp\left[b_i (t_{mi} - t)\right]} - 1 \right\}. \tag{6.1.30}$$

This model can also consider the mixture of solute as one pseudocomponent; in this case, the i index in Equation 6.1.30 can be dropped.

The applications of the various models for the system ginger/CO_2 are shown in Figure 6.1.3. Depending on the system, the fitting capacity of the models can change considerably, and no model can be elected as the best one for any situation.

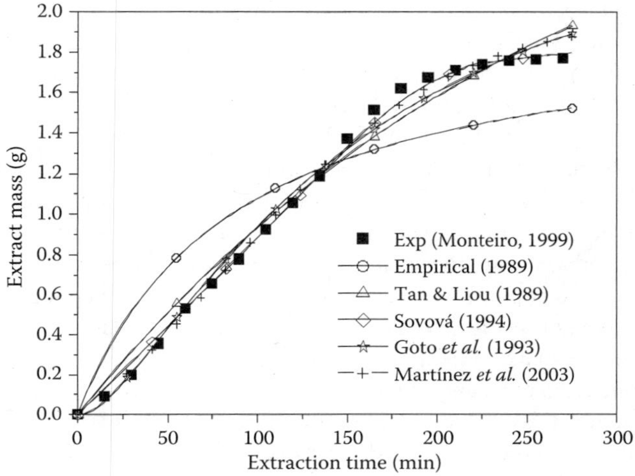

FIGURE 6.1.3 Comparison of experimental ginger oleoresin overall extraction curve with several mathematical models. Experimental condition: 15 MPa, 313.2 K, and 3.5 g/min of CO_2 mass flow rate.

6.1.2 Thermodynamics: Equilibrium

One of the most important pieces of information used to design the extraction column is the phase equilibrium between the supercritical fluid and the solutes that are extracted. The extraction system is quite complex, comprehending the supercritical solvent, a mixture of different compounds that forms the solute and a solid structure where the solute is distributed. The system can be simplified using different assumptions. The first one can consider only the equilibrium between the solvent and one pseudocomponent, with physical characteristics given by the main component of the solute or as a mean value of the mixture of compounds, calculated using, for instance, the Kay's rule [10]. The second considers the equilibrium between the solvent and the several components of the solute. In both cases a two-phase model is used to describe the system. The last one regards the equilibrium in a ternary system, including the influence of the solid matrix.

The experimental equilibrium data can be determined using several methodologies. The dynamic and static models can be used to do these measurements. In the dynamic model, the solvent is continuously admitted into an extraction column, at a given pressure and temperature, using a flow that assure its saturation at the exit of the column. Rodrigues et al. [11] used this method to determine the solubility of clove bud, ginger, and eucalyptus in supercritical CO_2. The authors used different extraction column configurations to validate the solubility measurement. It was observed that there is an optimum solvent flow rate that allows the solubility determination. For large flow rates, there is not enough contact time to saturate the solvent and for very low flow rates both the axial dispersion and the low interfacial mass transfer coefficient decreases the solute concentration. The optimum flow rate was a function of the used system, but the solubility values were the same in the different extraction column geometry, as expected.

The solubility of the binary system can be determined using the supercritical extracts of the raw material dispersed on the surface of a nonporous inert substratum. This dynamic method has the disadvantage of excluding the limitation of the solid matrix and using only a fraction of the solute, but it is easier to determine the solubility because in these systems there will be only a small influence of the mass transfer in the fluid phase, and the saturation can be readily attained.

In spite of the simplicity and high sensitivity of the dynamic methods, they are very sensitive to pressure fluctuations in the extraction column. Another factor that can have an influence on the solubility measurement is the possibility of solute accumulation in the system after the extraction column. Furthermore, these methods use a large amount of raw material to determine the solubility. In general, the binary solubility can be used to design the separation unit and the ternary solubility is used to design the extraction column. In Sections 6.3 and 6.4 the phase equilibria of cashew extracts and orange oil using CO_2 as solvents will be discussed.

In the static model a certain amount of extract or raw material is set into a vessel that is maintained at a constant temperature and pressure. After a long contact time, a sample of supercritical phase is withdrawn from the system and analyzed to give the equilibrium concentration in the supercritical phase. In general, the sensitivity of this method is quite low because only small samples of the supercritical phase can be taken without causing large disturbances in the system pressure. This method has been used for solutes that have high solubility in the supercritical phase. Another kind of static method for binary systems uses pressure cells containing a view port to observe the equilibrium. The most common system has a variable volume using an embolus. A certain amount of solute and supercritical solvent is admitted into the vessel, and the pressure is slowly increased by decreasing the system volume. The liquid solute is focused, and when the first droplets of solvent are observed, the pressure is annotated. This will be the bubble point of the system. The pressure is then increased until only one phase can be observed. After that, the pressure is slowly decreased, by increasing the vessel volume, until a cloud of small droplets can be observed. This will be the dew point of the binary system. Using this methodology the phase equilibria of systems of interest in food processing were measured: clove extract/CO_2 [12], fennel extract/CO_2 [13], and vetiver extract/CO_2 [14].

The solubility of compounds in supercritical fluids presented in isothermal systems increases as the pressure is increased. The solvent density increases with pressure and consequently the solvent power will be higher. The effect of temperature on the solute solubility is more complex to analyze. In general, the solute vapor pressure increases with temperature but the solvent density decreases. At pressures near to the critical point, the effect of temperature on the solvent density is stronger than on the solute vapor pressure. Thus, at these pressures the solute solubility decreases with temperature. For high pressures, the solvent density changes only slightly with temperature, and as a result the solute vapor pressure will be the main effect. Therefore, the solubility will increase with the temperature for high pressures. There will be an intermediate pressure where the solubility will not be a function of temperature. This pressure is known as the crossover point of the system. The value of this point will depend on the solute composition.

In the thermodynamic modeling of the system equilibrium, the equality of the fugacity of each component of the system in both phases is used. When a gas phase is considered, the fugacity of a component present in this phase is given by

$$\hat{f}_i^V = y_i \hat{\phi}_i^V P, \tag{6.1.31}$$

where $\hat{f}_i^V$ is the fugacity of the component i in the gas phase, y_i is its molar fraction, $\hat{\phi}_i^V$ is its fugacity coefficient, and P is the system pressure.

The solute, in general, can be considered as a mixture of liquids presented in the solid phase. For a liquid system, there are two ways to describe the fugacity of a component: using the activity coefficient and the fugacity coefficient. The expressions for the fugacity of liquids are represented by

$$\hat{f}_i^L = x_i \gamma_i P \tag{6.1.32}$$

$$\hat{f}_i^L = x_i \gamma_i P, \tag{6.1.33}$$

where $\hat{f}_i^L$ is the fugacity of the component i in the liquid phase, x_i is the molar fraction, γ_i is the activity coefficient of this component, and $\hat{\phi}_i^L$ is the fugacity coefficient of i in the mixture.

The supercritical fluid can be considered either as an expanded liquid or as a compressed gas. When the supercritical fluid is considered as an expanded liquid the activity coefficient should be calculated. In the majority of the cases, the supercritical fluid is considered as a compressed gas. The gas phase cannot be considered as an ideal gas because of the high pressures, and the fugacity coefficient is, generally, calculated using a cubic equation of state (EOS). The Peng–Robinson [15] and Soave–Redlich–Kwong [16] are the most used equations of state for supercritical fluids. The mathematical formula of these EOS can be observed in Table 6.1.1.

The fugacity coefficient for a component i present in a mixture of components can be obtained by

$$\ln \hat{\phi}_i = \frac{1}{RT} \int_{V \to \infty}^{V = ZRT/P} \left[\frac{RT}{V} - N \left(\frac{\partial P}{\partial N_i} \right)_{T,NV,N_{j \neq i}} \right] dV - \ln Z, \tag{6.1.34}$$

where Z is the compressibility coefficient, N is the total number of moles of the system, and N_i is the number of moles of i present in the system.

Equation 6.1.34 can be used for any phase, considering the compressibility coefficient of each phase. For instance, when the Peng–Robinson equation of state is used, the fugacity coefficient of the gas and liquid phases can be determined by

$$\ln \bar{\phi}_i^V = \frac{b_{Pi}}{b_P}\left(Z^V - 1\right) - \ln\left(Z^V - \frac{b_P P}{RT}\right)$$
$$- \frac{a_P}{2\sqrt{2}RT}\left(\frac{2\sum_j y_j a_{Pij}}{a_P} - \frac{b_{Pi}}{b_P}\right) \ln\left(\frac{Z^V + \left(1+\sqrt{2}\right)\frac{b_P P}{RT}}{Z^V + \left(1-\sqrt{2}\right)\frac{b_P P}{RT}}\right) \tag{6.1.35}$$

TABLE 6.1.1
Peng–Robinson and Soave–Redilich–Kwong Equations of State

Model	Equation	Parameter	
Soave–Redlich–Kwong	$P = \frac{RT}{V - b_P} - \frac{a_P(T)}{V(V + b_P)}$	$a_P = \frac{0.42747 R^2 T_c^2}{P_c} \alpha(T)$ $\sqrt{\alpha} = 1 + m\left(1 - \sqrt{\frac{T}{T_c}}\right)$	$b_P = 0.08664 \frac{RT_c}{P_c}$ $m = 0.48 + 1.574\omega$ $- 0.176\omega^2$
Peng–Robinson	$P = \frac{RT}{V - b_P} - \frac{a_P(T)}{V^2 + 2b_P V - b_P^2}$	$a_P = 0.45724 \frac{R^2 T_c^2}{P_c} \alpha(T)$ $\sqrt{\alpha} = 1 + K\left(1 - \sqrt{\frac{T}{T_c}}\right)$	$b_P = 0.0778046 \frac{RT_c}{P_c}$ $K = 0.375 + 1.542\omega$ $- 0.2699\omega^2$

R: universal gas constant; *V:* molar volume; *T*: system pressure; T_c: critical temperature; P_c: critical pressure; ω: accentric factor.

$$\ln \bar{\phi}_i^L = \frac{b_{Pi}}{b_P}\left(Z^L - 1\right) - \ln\left(Z^L - \frac{b_P P}{RT}\right) - \frac{a_P}{2\sqrt{2}RT}\left(\frac{2\sum_j y_j a_{Pij}}{a_P} - \frac{b_{Pi}}{b_P}\right)\ln\left(\frac{Z^L + \left(1+\sqrt{2}\right)\frac{b_P P}{RT}}{Z^L + \left(1-\sqrt{2}\right)\frac{b_P P}{RT}}\right) \quad (6.1.36)$$

where a_p and b_p are the Peng–Robinson parameters (Table 6.1.1), and Z^V and Z^L are the compressibility of the gas and liquid phases, b_{pi} is the Peng–Robinson parameter of component i, and a_{pij} is the "a_p" parameter for each pair of substance present in the mixture.

The a_p and b_p parameters from the Peng–Robinson or Soave–Redilich–Kwong equations can be determined using a mixing rule. The most used mixing rule was proposed by van der Waals, and is represented by

$$a_P = \sum_i \sum_j z_i z_j a_{Pij} \quad ; \quad a_{Pij} = \left(1 - k_{ij}\right)\sqrt{a_{Pi} a_{Pj}}$$

$$b_P = \sum_i \sum_j z_i z_j b_{Pij} \quad ; \quad b_{Pij} = \left(1 - l_{ij}\right)\frac{b_{Pi} + b_{Pj}}{2}, \quad (6.1.37)$$

where z_i and z_j are the molar fractions of i and j in one phase k_{ij} and l_{ij} and are adjustable parameters known as binary interaction parameters.

Thus, with Equations 6.1.35 through 6.1.37 it is possible to determine the equilibrium of the components distributed in the two phases if the ϕ–ϕ methodology is used to determine the phase equilibrium. When the γ–ϕ methodology is used, the activity coefficient should be used. To estimate the activity coefficient, the most used methodologies are the group contribution such as the UNIFAC (see Chapters 3 and 5). In some cases the fugacity of the liquid phase can be represented by Henry's law. Patel et al. [17] presented a comparison of several methodologies to estimate the phase equilibrium in supercritical fluids. Even Henry's law was able to represent the equilibrium when the system pressure was moderate (up to 10 MPa).

6.1.3 Nomenclature

Acronym	Description
CER	Constant extraction rate period
FER	Falling extraction rate period
Symbol	**Description**
$\hat{f}_i^V$	Fugacity of component i in the supercritical phase
$\hat{f}_i^L$	Fugacity of component i in the liquid phase

continued

Symbol	Description
a	Interfacial area per unit of column volume
A_1	Constant
a_1, a_2, b, c	Parameters of Goto's model
A_i, b_i, t_{mi}	Martínez's model parameters
A_{1E}, B_{1E}	Parameters of Equation 6.1.12
a_p, a_{pij}, b_p, b_{pi}	Peng–Robinson's equation parameters
A_T	Extraction column transversal section area
B_i	Biot number
D_{ap}	Apparent diffusion coefficient
D_{az}	Axial dispersion coefficient
H	Extraction column height
i	Component number or index
j	Component number or index
$J(X, Y)$	Interfacial mass transfer rate
k_d	First-order constant also known as the desorption constant
k_{ij}	Interaction parameter for "a_P" in the equation of state that is determined by fitting experimental data
K_x	Volumetric overall mass transfer coefficient in the solid phase
K_Y	Volumetric overall mass transfer coefficient in the supercritical phase
l_{ij}	Interaction parameter for "b" in equation of state mixing rule that is determined by fitting experimental data
m_E	Cumulative mass of extracted solute
m_{Ei}	Cumulative mass of fraction i
m_{IS}	Mass of inert solid
m_t	Total mass of particles packed into the extraction column
m_{ti}	Amount of a given class of substances present in the particle
M_w	Molecular mass
n	Integer number
N	Total number of moles
N_i	Number of moles of component i
P	Pressure
P_c	Critical pressure
Q	Solvent volumetric flow rate
R	Gas constant
R_P	Particle radius
T	Temperature
t	Time
T_c	Critical temperature
t_{CER}	Extension of constant extraction rate region
v	Solvent interstitial velocity
V	Molar volume
W, Z_1, Z_w	Sovová's model parameters
X	Solute mass ratio in the solid phase
x_k	Solute fraction presented in broken cells
x_o	Global yield
Y	Solute mass ratio in the supercritical phase
$Y(H, t)$	Solute mass fraction in the supercritical phase at the exit of the extraction column

Symbol	Description
Y^*	Solute solubility in supercritical solvent
y_i	Mole fraction of component i in the vapor or supercritical phase
z	Axial position
Z	Compressibility factor
Z^L	Compressibility factor of the liquid phase
Z^V	Compressibility factor of the supercritical phase
Greek letter	
β	Particle porosity
ε	Void volume fraction or bed porosity
ρ	Density
ρ_s	Solid free of solute density
ω	Acentric factor
$\hat{\phi}_i^V$	Fugacity coefficient of *i* in the mixture (vapor phase)
$\hat{\phi}_i^L$	Fugacity coefficient of *i* in the mixture (liquid phase)
γ_i	Activity coefficient of this component
τ	Residence time of the solvent

6.1.4 References

1. Naik, S. N., H. Lentz, and R. C. Maheshawari. 1989. Extraction of perfumes and flavours from plant materials with liquid carbon dioxide under liquid-vapor equilibrium. *Fluid Phase Equilibria* 49:115–126.
2. Esquível, M. M., M. G. Bernardo-Gil, and M. B. King. 1999. Mathematical models for supercritical extraction of olive husk oil. *Journal of Supercritical Fluids* 16:43–58.
3. Reverchon, E. 1997. Supercritical fluid extraction and fractionation of essential oils and related products. *Journal of Supercritical Fluids* 10:1–37.
4. Crank, J. 1975. *The mathematics of diffusion.* 2nd ed. Oxford: Claredon Press.
5. Tan, C., and D. Liou. 1989. Modeling of desorption at supercritical conditions. *AIChE Journal* 35:1029–1031.
6. Goto, M., M. Sato, and T. Hirose. 1993. Extraction of peppermint oil by supercritical carbon dioxide. *Journal of Chemical Engineering of Japan* 26:401–406.
7. Sovová, H. 1994. Rate of the vegetable oil extraction with supercritical CO_2. 1. Modeling of extraction curves. *Chemical Engineering Science* 49:409–414.
8. Sovová, H. 2005. Mathematical model for supercritical fluid extraction of natural products and extraction curve evaluation. *Journal of Supercritical Fluids* 33:35–52.
9. Martínez, J., A. R. Monteiro, P. T. V. Rosa, M. O. M. Marques, and M. A. A. Meireles. 2003. Multicomponent model to describe extraction of ginger oleoresin with supercritical carbon dioxide. *Industrial & Engineering Chemistry Research* 42:1057–1063.
10. Poling, B. E., J. M. Prausnitz, and J. P. O'Connel. 2001. *The properties of gases and liquids.* New York: McGraw-Hill.
11. Rodrigues, V. M., E. M. B. Sousa, A. R. Monteiro, O. Chiavone-Filho, M. O. M. Marques, and M. A. A. Meireles. 2002. Determination of the solubility of extracts from vegetable raw material in pressurized CO_2: A pseudo-ternary mixture formed by cellulosic structure + solute + solvent. *Journal of Supercritical Fluids* 22:21–36.
12. Souza, A. T., M. L. Corazza, L. Cardozo-Filho, R. Guirardello, and M. A. A. Meireles. 2004. Phase equilibrium measurements for the system clove (*Eugenia caryophyllus*) oil + CO_2. *Journal of Chemical Engineering Data* 49:352–356.

13. Moura, L. S., M. L. Corazza, L. Cardozo-Filho, and M. A A. Meireles. Phase equilibrium measurements for the system fennel (*Foeniculum vulgare*) extract + CO_2. *Journal of Chemical Engineering Data* 50:1657–1661.
14. Takeuchi, T. M., P. F. Leal, R. Favareto, L. Cardozo-Filho, M. L. Corazza, P. T. V. Rosa, and M. A. A. Meireles. 2008. Study of the phase equilibrium formed inside the flash tank used at the separation step of a supercritical fluid extraction unit. *Journal of Supercritical Fluids* 43:447–459.
15. Peng, D. Y., and D. B. Robinson. 1976. A new two-constant equation of state. *Industrial Engineering and Chemistry Fundamentals* 15:59–64.
16. Soave, G. 1972. Equilibrium constants from a modified Redilich-Kwong equation of state. *Chemical Engineering Science* 27:1192–1203.
17. Patel, N. C., V. Abovsky, and S. Watanasiri. 2001. Calculation of vapor–liquid equilibria for a 10-component system: Comparison of EOS, EOS–GE and GE–Henry's law models. *Fluid Phase Equilibria* 185:397–405.

6.2 OBTAINING ANTIOXIDANTS BY SUPERCRITICAL FLUID EXTRACTION

Juan Carlos Parajó, Herminia Domínguez, Andrés Moure, and Beatriz Díaz-Reinoso

6.2.1 Obtaining Antioxidants by Supercritical Fluid Extraction

The interest for cheap, renewable, and abundant sources of natural antioxidants has grown because of safety concerns, contradictory toxicological data about synthetic antioxidants, and consumer preferences for natural additives. Supercritical fluid extraction (SFE) can be more effective than conventional processing to selectively recover vegetal compounds with antioxidant action. SFE also shows advantages related to food regulations and environmental impact. Operation at reduced temperature prevents thermal degradation of labile compounds, and the absence of light and oxygen avoids oxidation reactions, a problem of major importance in antioxidant extraction. Carbon dioxide is the most suited solvent for SFE of thermolabile compounds because of its favorable properties (including nontoxic and nonflammable character, high availability at low cost, and high purity) and to its ability to produce isolates with optimal physicochemical, biological, and therapeutic properties. Extracts from SC-CO_2 processing are regarded as natural and have the GRAS status, because different microorganisms are inactivated and additional sterilization is not required. Propane, butane, and ethylene have also been proposed as solvents for SFE [1–3].

General aspects of SFE of antioxidants have been revised [4, 5], whereas other works emphasized the raw materials and antioxidant activities of the extracted products [6–8] or the operational conditions used for extraction and fractionation [2]. Depending on the raw materials and products considered, different process configurations have been proposed for extracting the major families of antioxidant compounds (phenolics, terpenoids, carotenoids, and tocopherols). Other types of compounds (such as proteins, oligosaccharides, and Maillard reaction products) also show antioxidant activity, but their SC-CO_2 solubility is low.

6.2.1.1 Raw Materials and Their Conditioning

A great effort is being devoted to the search for alternative, cheap sources of natural antioxidants, as well as to the development of efficient and selective extraction techniques. In most cases, solid materials have been considered as feedstocks, including traditional vegetal sources (plants, parts of plants, and trees), industrial processing wastes, and agricultural residues. Additionally, liquid streams from industrial processes or direct extracts from conventional solvent extraction (CSE) have been fractionated and/or purified by SC-CO_2 extraction.

6.2.1.1.1 Considerations on the Solid Raw Materials and Their Pretreatments Medicinal and aromatic plants are the most frequently used vegetal sources for SC-CO_2 processing. In this field, studies dealing with passion fruit [9], summer savory [10], sage [11], boldo [12], marjoram [13], rosemary [14–16], and lemon verbena and mango [17] have been reported. Leaves from trees have also been considered, including those from eucalyptus [18, 19], ginkgo [20], and tropical almond [21]. Tops, flowers, and stems have been used for antioxidant extraction from a variety of feedstocks, including medicinal herbs [22], sage [23], thyme [24], lemon balm [25], chamomile [26], curry plant leaves [27], and white lipia [28]. Both the epidermis and pulp gel from aloe [29] and the roots [30, 31] have also been considered. Studies on the extraction of a number of seeds have been published, including those from grape [32], coriander [33], black cumin [34], sesame [35], black pepper [36], or milk thistle [37]. Other antioxidant sources include fungal biomass such as micromycetes *Mortierella* sp [38], microalgae [39–41], and crustaceans (which have been extracted at an analytical scale) [42].

Agricultural and industrial wastes can be a profitable and reasonable choice to produce additives (antioxidants, flavors, colorants) with health-promoting activities. Mixed materials from residual origins have been assayed for this purpose, as is the case of pomace from the wine industry (a material composed of stems, seeds, and skins) [43], tamarind seed coat [44, 45], pistachio hulls [46], cacao hulls [47], rye bran [48, 49], palm fruit husks [3], potato waste [50], tomato waste [51–54], olive tree residues [55], and residues from the extraction of palm oil [56].

The content and extractability of bioactive compounds from a given raw material depend on crop-related factors (cultivar, maturity, edaphoclimatic conditions, etc.), structural features of the solid (leaves, roots, seeds, fruits, etc.), mechanical processing (cutting or milling), and thermal conditioning (drying). Conditioning operations are oriented to reduce the internal mass transfer resistance, because the solutes are frequently located in complex cellular structures or are linked to cell walls. On one hand, pretreatment operations control the particle size and modify the structure of the solid matrix, and therefore the kinetics and yield of the extraction. On the other hand, parameters such as porosity and apparent density of the fixed bed are also affected. Decisions on conditioning should be based on both techno-economic aspects and physicochemical and biological properties of the target compounds. Mechanical and thermal pretreatments, which are decisive to facilitate the extraction of intracellular solutes from natural matrices, are unnecessary when processing extracts coming from CSE.

Usually, the limiting step in solid–liquid extraction is the intraparticle solute diffusion, and small particle sizes lead to increased extraction rates and yields. Although fine grinding of the material is proposed at the lab scale or in characterization studies, other factors are influential at the industrial scale, because excessive grinding may result in losses by volatilization and degradation of active compounds, and too fine particles could limit the performance of fixed beds (owing to channelling, formation of dead zones, and compaction). Optimization of extraction kinetics on the basis of particle size has been frequently addressed based on grinding and sieving of the feedstocks. Different crushing degrees have been considered in the extraction of flavonoids from gingko [1] and carotenoids from microalgae [40, 57, 58], tomato wastes [59], apricot pomace [60], or carrot [61].

In bed extraction, the ground feedstock must be carefully packed to avoid channeling. This disposition is used for extracting natural materials, such as pepper [36], ginger [31], leaves [12, 21, 62], microalgae [63], and shiitake [64]. Extracts from CSE, commercial extracts, and oleoresins have also been processed [57, 65–67]. The bed can be covered on the bottom and top by glass wool [23], cotton wool [68], a porous plate [69], or a stainless steel frit [70] to ensure homogeneous solvent flow. The reported apparent densities of these beds were 117.4 kg clove basil/m^3 [71], 119.42 kg rosemary/m^3 [72], 350–400 kg/m^3 for ginger [31], lemon verbena and mango leaves [17], and 370 kg chamomile/m^3 [73]. Most studies were performed at lab scale, but a more frequent and effective approach at a higher scale is to improve the distribution of the solvent either with layers of inert materials or with homogeneous mixtures of inerts and samples. Glass beads have been used with grape seeds [67], ginger roots [74], leaves [68, 75, 76], tomato skins [53], and medicinal herbs [15]. Beds made up of rosemary and glass beads presented an apparent density of about 360 kg/m^3 [15], in comparison with 940 kg/m^3 for beds made up of *Spirulina maxima* and glass beads [39]. A nylon basket in combination with glass beads (to fill the dead space) has been used for extracting leaves [17]. Sea sand was used with medicinal herbs [16, 70, 77], glass wool with algae [57], silica gel with gingko biloba conventional solvent extracts [20], diatomaceous earth with eucalyptus leaves [18], and stainless steel beads with propolis [78]. Pelletized substrates have been proposed to increase the apparent density of beds, to avoid compaction, and to reduce the mass transfer resistance within the solid [79]. Enzyme treatment has been applied to disrupt cell walls, leading to improved conventional and SCF extraction from rosemary [66, 80].

Drying before SC-CO_2 extraction is necessary, as the presence of water can result in decreased effectiveness by either limiting the contact with apolar solutes or by acting as a cosolvent. Optimal drying of the feed material is essential for a suitable operation. Mild drying is required for conditioning aromatic plants in order to avoid decomposition and degradation of the target compounds, such as the pungent and natural flavors of ginger [74], phenolic diterpenes in fresh rosemary, and carotenoids. In the case of moisture-rich materials, such as fruits, mechanical pressing is preferred to thermal treatments in the initial drying stages, to protect thermolabile compounds [51]. The effect of drying on the extraction of antioxidant compounds was considered for different materials [54, 60, 81], and several technologies have been reported, including sun drying of origanum herbs [82], sun drying followed by vacuum drying of paprika and tomato [83, 84] and tomato wastes [59], spray drying of yeasts [85], vacuum drying or oven drying of sweet potato [86], and air drying of palm fruit [69] and tomato waste [54]. Freeze drying was selected for materials containing components sensitive to heat and oxygen, such as the antioxidants from aloe epidermis and pulp [29], and carotenoids from sweet potato [86], tomato wastes [51, 52], carrots [61], and algae [57]. Freeze drying causes little alteration in comparison with air and oven drying, but shows a limited ability to preserve bioactive compounds such as carotenoids, low-molecular-weight phenolics, and volatiles [87]. In addition, freeze drying is expensive, and other techniques could be more profitable at an industrial scale [61].

In the case of ginger rhizomes, freeze drying allowed higher yield than oven drying, but lower than the one obtained in an operation with the fresh material, because of the enhanced effective diffusivity within the moist particle [74]. However, the high capital cost associated with SFE is a deterrent of the utilization of moist solids, which requires the management of larger amounts of raw materials. For transport and storage, dried feedstocks are preferable to moist ones. These latter, when finely ground, can give operational problems (such as formation of a pulp or slurry, with reduction of the available interfacial area) [74].

6.2.1.1.2 Considerations on the Liquid Streams and Extraction Technologies

The liquid streams processed by SFE include fruit juices [88], vegetal oils [89–92] and their deodorizer distillates [90, 92–96], and streams generated during conventional solvent extraction [55, 97] or acid hydrolysis [98].

The oil deodorizer distillate (ODD) is the by-product of vegetable oil refining and contains valuable compounds such as tocopherols, tocotrienols, fatty acids, sterols, and squalene [94, 99, 100]. The by-product of physical refining of palm oil also contains provitamin A. Hydrolysis of both oil and distillates to free fatty acids and further conversion into ethyl or methyl esters has been proposed to increase their solubility in SC-CO_2, enabling the recovery of the target compounds in the raffinate [91].

When the desired compounds are present in a liquid stream, two operational methods can be used for extraction: batch mode or continuous countercurrent contact in a column. Alternatively, the solutes can be first adsorbed on a suitable solid material and then subjected to fractional desorption [101]. Liquid–liquid contact in SC-CO_2 extraction has been revised by Reverchon [101], Brunner [102], Gamse [103], and by Reverchon and De Marco [2].

Batch extraction of saponified and esterified soy deodorized distillate (SODD) has been carried out in a modified cell where the SC-CO_2 is bubbled through the liquid phase [95]. When the solute is in the liquid phase of a suspension, extraction in a packed bed could present operational problems derived from the aggregation of the solids on the packing elements. This type of feed can be processed by supercritical antisolvent extraction (SAE): the supercritical fluid and the liquid mixture are continuously fed to a pressurized vessel, where the liquid dissolves rapidly and the solid precipitates at the bottom [2]. This method has been applied to the concentration of flavonoids from a propolis ethanol tincture at the lab, pilot, and demonstration scales, to obtain a concentrated flavonoid fraction and a mixture of essential oil and ethanol [104]. Before entering the extraction vessel, propolis tincture was mixed with supercritical CO_2, which acted both as an antisolvent to precipitate high-molecular-weight components and as a solvent to extract the ethanol and soluble components of the propolis.

Semicontinuous and continuous processing of liquid feeds have been used, for example, in the extraction of sterols and tocopherols from olive oil [89], as well as for the extraction of ODD, enriching the top phase in squalene and the bottom phase in sterols [90]. Extraction of squalene from ODD has been carried out after converting the free fatty acids and the methyl and ethyl esters into their corresponding triglycerides [94] and from transesterified crude palm oil [91]. Countercurrent contact was also proposed for separating hydroxytyrosol from either olive oil-processing waters or their extracts (obtained with conventional solvents) [55].

Continuous processing may be performed using selected temperature profiles along the column for optimizing the composition of the mixtures at different levels [99]. Temperature gradients along the column induced an internal reflux, as a result of the change in solute solubility, and an optimal gradient can be established to maximize extraction yields and to improve the separation selectivity [92].

The solvent-to-feed (S/F) ratio affects the extraction efficiency. The ranges reported for the S/F ratio were 33–171 for ODD [90], 50–100 for the same case [92], and 5–40 for hexane extracts from olive leaves [105].

The packing material can be influential on the separation selectivity. Fenske rings were used for separating sterols and tocopherols from olive leaves selectively [105] and provided higher enrichment in the target compounds from olive oil than glass beads, Rasching rings, and Dixon rings [89, 97]. Sulzer rings and structured packing were selected for squalene and vitamin E recovery [90, 94] and hydroxytyrosol extraction [55]. Other types of packing materials used include Goodloe knitted-mesh packing for palm oil [106], Dixon packing for ODD [92], stainless steel filling [99], and glass beads impregnated in paprika oleoresin [83].

6.2.1.2 Operational Variables Affecting the SCF Extraction of Antioxidants

When the solute is in a solid matrix, both equilibrium and kinetics of the extraction are dependent on the experimental conditions and on the previous conditioning of the raw material. The major variables influencing the SFE of antioxidants (pressure, temperature, solvent flow rate, S/R, modifier type, and concentration) should be optimized before operation. Their effect on extraction yield and selectivity must be addressed for each particular case and have been previously reviewed [7, 8, 107].

Pressure and temperature affect both equilibrium and kinetics and control the solvent density and solvating power of CO_2. Solubilities of antioxidant compounds have been reported in the literature [5, 8, 12, 56, 101, 108–110].

Increased extraction pressure results in increased density and solvating power of the supercritical fluid, as well as in higher interaction between the fluid and the solid matrix. Pressures in the range 8–15 MPa are suited for essential oils [12, 101], whereas 15–40 MPa are the most usual ones for phenolic and terpenoids [8]. In antioxidant extraction, increased pressure can result in decreased selectivity as a result of the coextraction of compounds that reduce the purity and can confer color [20], as well as the prooxidant action to SCF extracts [111]. When the objective is to extract undesired components concentrating the antioxidants in the residue, increased pressure can be beneficial (because of the higher solubility and faster extraction), but coextraction of the target compounds could limit the selectivity of the separation.

The effect of temperature has to be considered on the basis of (i) the solvent power, (ii) the thermal stability of the solutes, (iii) the vapor pressure of the solute, and (iv) the properties of the matrix, which can make mass transfer difficult [31]. Mild extraction conditions (temperatures below 40–60°C) are frequently used to extract antioxidants from medicinal plants [26, 29], particularly phenolic acids [112], flavonoids and terpenoids [13, 15, 20], carotenoids [39], and tocopherols [113, 114].

For a given pressure, higher temperature leads to lower density and solvating power of SC-CO_2, but also to higher vapor pressure of the solute. Pressure also affects the SC-CO_2 density, which determines the solvating power of CO_2. The crossover effect of temperature and pressure has been observed in the extraction of antioxidant compounds.

6.2.1.3 Processing Schemes Proposed for Antioxidant Extraction

Different operational methods have been proposed for SFE of bioactive compounds from natural sources, the major dispositions being determined by the physical state of the feed [102]. Brunner [115] classified them in (i) single stage extraction of solids, (ii) multistage countercurrent extraction of liquid streams, and (iii) preparative chromatographic separations. In the first case, the solvent flows through a fixed bed of solids, and the process occurs in unsteady state in both solid and liquid phases. Batch or semibatch operations have been used at analytical and preparative scales. Fractionation of extracts can be achieved by supercritical preparative chromatography, whose major applications are related to analytical and preparative operations (for example, enantiomer separation or production of standards) and can be scaled up [115]. The most usual processes for extracting antioxidant compounds from a solid matrix are the following:

1. Single SFE stage and fractional separation in several vessels
2. Stagewise SFE at progressively increasing pressure
3. CSE and SFE processing of the extract to obtain the antioxidants either in the extract or in the residue
4. Processing by SFE and subsequent extraction of the solid residue with conventional solvents or by hydrothermal (HT) processing.

Alternatively, liquid–liquid extraction has been proposed for extraction, fractionation, and/or purification of antioxidant compounds present in liquid samples (including extracts coming from CSE).

6.2.1.3.1 Single SFE Stage and Fractionation in Several Separation Vessels

SFE is used to produce an extract that is further fractionated in separators (usually, one to three), according to the general principle shown in Figure 6.2.1. Two SFE stages have been used in studies dealing with scaling and continuous operation [17, 49, 116], together with a series of separators operated at controlled pressure and temperature. This disposition has also been used for analytical purposes and for preliminary SFE evaluation. Some examples performed at different scales are summarized in Table 6.2.1. The antioxidant potency of the extracts containing phenolics and terpenoids is expressed comparatively to standard antioxidants, and the yield and/or purity of tocopherols and carotenoids are listed.

SFE of antioxidants requires high pressure, conditions under which coextraction of other fractions (essential oils and waxes) can take place. Waxes are paraffinic compounds located on the surface of some vegetals and can be readily extracted in a process governed by solubility. Essential oils are inside the cell structure, and their extraction is controlled by internal mass transfer. Coextraction of waxes is undesirable, but some essential oils show antioxidant activity [10, 36, 81, 130]. Selective precipitation of cuticular waxes and fractions rich in essential oils has been reported

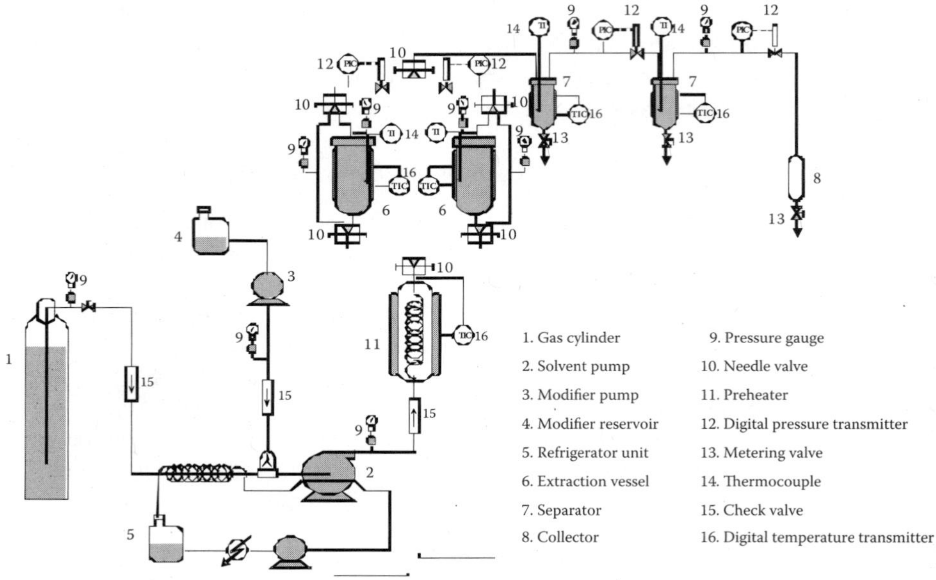

FIGURE 6.2.1 General flow diagram of a single stage SFE and fractionation in several vessels.

TABLE 6.2.1
Data Concerning Processes Based on Single-Stage SFE with Pure CO_2 and Fractional Separation

Phenolics and terpenoids

Feedstock	SFE: EV; P; T; nS[a]	Antioxidant activity	Reference
Aloe	1; 45; 323; 2	DPPH: T > CSE > BHT > SFE > αT	29
Black cumin	<0.1; 20; 313; —	βc: CSNA > SFE > CO	34
Black pepper	<0.1; 10; 333; —	GPO: SFE > CSO > CO	36
	<0.1; 28; 333; —	GPO: SFE > CE > CSE	
Boldo	0.1; 9; 323; 1	TEAC: CSE > SFE	119
Cacao husks	—; 15; 323; —	DPPH: T > CSE > SFE > BHA	47
Cape gooseberry	—; —; 40; 333	SFE> α-T > CSE	121
Clove basil	<0.1; 15; 313; 1	LA-βc: SFE > βc	71
Curry plant leaves	—; 26; 323; 1	$DPPH_{100\ \mu g/mL}$: BHA ≈ SFE βc: BHA > SFE	27
Chlorella	—; 40; 305; 2	DPPH: SFE > T > BHT > αT	120
Eucalyptus leaves	<0.1; 20; 323; 1	BHT > BHA > SFE > CSE	18
	—; 40; 343; 1	LAO: CSE ≈ BHT > SFE≈BHA	19
Lemon verbena	<0.1; 35; 318; 1	LA-βc: SFE > βc	17
Mango	<0.1; 25; 318; 1	LA-βc: SFE > βc	17
Marjoram	<0.1; 30; 313; 1	LA-βc: BHT > SFE	22
Oregano	4; 50; 368; 2	LO: BHA:BHT > SFE	116
	5; 45; 323; 1	CSE ≈ BHT > SFE	13
	<0.1; 30; 313; 1	LA-βc: BHT > SFE	22

	Phenolics and terpenoids		
Feedstock	**SFE: EV; P; T; nS[a]**	**Antioxidant activity**	**Reference**
Propolis	<0.1; 20.7; 323; 1	LDL: SFE > CSE > DHCA DPPH: FA > CSE > SFE	78
Propolis-ethanol	—; 20; 333; 3	AOE: SFR ≅ SFE ≅ S	122, 123
Rosemary	4; 50; 373; 2	LO: SFE > BHA:BHT	116
	0.3; 30; 313; 1	βc: SFE > βc	15
	0.28; 25; 313; 2	DPPH: AA > SFE	16
Sage	1; 25–35; 373; 3	SFE_3 > SFE_2 > SFE_1 > BHT	23
	<0.1; 30; 313; 1	LA-βc: BHT > SFE	22
Savory	4; 50; 368; 2	LO: SFE > BHA:BHT	116
	0.3: 12; 313; 4	SFR > CNA > SFE_3 > SFE_2 > EO_{SD}	10
Sesame	—; 20; 308; —	DPPH: T > BHA≈CSE > αT > SFE LA: BHA≈CSE > SFE > T > αT	35
Tamarind	<0.1; 10; 313; 1	TCN: CSE ≈ αT > SFE	44
	—; 30; 353; 1	LO: CSE > αT > SFE	45
Thyme	<0 1; 30; 313; 1	LA-βc: BHT > SFE	22
	5; 40; 333; —	SOO: BHT ≈ CSE ≈ SFE	24
Tropical almond leaves	—, 13; 313; 1	DETBA: αT > SFE	21
	<0.1; 20; 313;—	LAO: SFE > BHT PFO: BHT > SFE DPPH: BHT > SFE	62
		Carotenoids	
Feedstock	**SFE: EV; P; T; nS[a]**	**Compound (Yield, %; Purity, %)**	**Reference**
Buruti palm	1; 20; 313; 1	βc (80)	69
Chlorella	—; 35; 328; 1	TY (42; 66)	58
Chlorella	—; 30; 313; —	CSE (100) > SFE (69.1)	111
Dunaliella	<0.1; 44.8; 313; —	βc (0.7)	57
Haematococcus	<0.1; 55; 353; 1	TY (21.8) Ast (12.3)	125
Palm fruit husks	—; 25; 313; —	TY (0.77)	3
Palm oil processing residue	2; 50; 343; 1	c (0.455)	56
Paprika	0.85; 41.4; 313; 1	TY (11.6); βc (0.3)	118
Potato waste	0.5; 35; 313; 1	βc (91.5)	50
Spirulina	<0.1; 18; 303; —	TY (0.43)	39
Spirulina	0.285; 22; 328; 2	TY (0.70)	41
Stinging nettle	<0.1; 28; 313; 1	βc (0.024)	75
Tomato	10; 45; 340; 3	Lyc (35)	84
Tomato paste waste	—; 30; 338; 2	βc (40); Lyc (20)	214
Tomato skin	<0.1; 40; 373; 1	Lyc (0.12; 94)	53
Tomato skins	1; 30; 353; 2	Lyc (80); βc (88)	59, 117
Tomato waste	5; 46; 333; —	TY (15.05); Lyc (0.022; —)	54
		Tocopherols	
Feedstock	**SFE: EV; P; T; nS[a]**	**Compound (Yield, %; Purity, %)**	**Reference**
Lemon balm	0.4; 30; 323; —	TY (1.9); αT (0.3)	25

continued

TABLE 6.2.1 ***(continued)***

Feedstock	SFE: EV; P; T; nS[a]	Antioxidant activity	Reference
Olive leaves	0.075; 25; 313; —	TY (97.1)	127
Olive pomace	0.4; 35; 323; —	αT: 2084 ppm	113
Palm leaves	2; 30; 343; 1	αT (11.3)	56
Potato	0.5; 35; 313; 1	αT (76.8)	50
Sesame	1; 40; 308; 2 1; 30; 328; 2	TY (51.8; 20.44 μg/mL) DPPH: T > BHT > αT > SFE > CSE	124
Milk thistle	<0.1; 20; 213; 1	19.9%	37
Tomato waste	5; 30; 333; 1	SFE (5.9 mg/g) > CSE (3.7 mg/g)	54
Wheat germ	<0.1; 35.7; 316; 1	Total: SFE > CSE αT, γT, δT: SFE > CSE βT: CSE > SFE	114
		Tocopherols CC	
Feedstock	**SC-Extraction: B/CC; CH/EPH/EV; P; T; nS[b]**	**Compound (Yield, %; Purity, %)**	**Reference**
Crude palm oil	B: —; 14; 333; 1	βc (200-fold enriched) Sq (80-fold enriched)	91
Crude palm oil	CC;CH: —; 20; 340; —	To (—; 99.5)	90
Crude palm oil	B: <0.1; 20; 313; 1	TY (98.2)	128
Olive leaves	CC;CH: 0,15; 20; 308; 2	TY (84.4); Sq_{RCV} (69.76) αT (19.94); βc_{RCV} (63.74)	97
Olive oil	CC;CH: 0.18; 20; 313; 2	αT: Raf: 0.01% Ext. S_1:0.12%; Ext S_2:0.19%	89
OODD	CC;CH: 3; 15; 333; —	Sq (91; 90)	94
Palm oil	B: 0.3; 13.7; 323; 3; —	TY (90.2)	106
SODD	CC: —; 23; 353; —	Sq (—; 99)	90
SODD	B: 0.5; 18; 333; 1	TY (36)	95
SODD	CC: 0.17; 31; 343_B-363_T; 1	αT (84; 1.7 CF)	99
SuODD	CC; EPH: 180; 14; 338; 2	αT_{RCV} (98.0) St_{RCV} (97)	129, 130

[a] EV: Extractor volume (L); P: Extraction pressure (MPa); T: Extraction temperature (K); T: Top, B: Bottom; nS: Number of separation vessels.

[b] B/CC: Batch or countercurrent; CH/ EPH /EV: Column height (m), Effective packed height (cm), Extractor volume (L); SFR: Solvent-to-feed ratio.

CF: Concentration factor; RCV: Recovery; Ext: Extract in separators; Raf: Raffinate; SFE_n: Supercritical extract from the n separation stage; SFR: Supercritical residue; TY: Total extraction yield.

SODD: Soybean oil deodorizer distillate; SuODD: Sunflower oil deodorizer distillate.

AA: Ascorbic acid; Ast: Astaxanthin; BHA: Butylhydroxyanisol; BHT: Butylhydroxytoluene; DHCA: 3,5-diprenyl-4-hydroxycinnamic acid; βc: β-carotene; c: all carotene isomers.

CNA: Commercial natural antioxidant; CO: Commercial oil or oleoresin; CSE: Conventional solvent extract; CSNA: Commercial synthetic natural analogous (α-tocopherol); CSO: Conventional solvent oleoresin; EO_{SD}: Essential oil (steam distillation); Lyc: Lycopene; SCO: Supercritical oil; Sq: Squalene; T: Trolox; αT: α-Tocopherol, To: Tocochromanols; St: sterols.

AOE: Antioxidant enzymes; DETBA: Diethyl-2-thiobarbituric acid method; DPPH: 2,2-Diphenyl-1–picrylhydrazyl hydrazyl radical scavenging capacity; GPO: Ground pork oxidation; LA-βc: Linoleic acid-β-carotene; LAO: linoleic acid oxidation; LDL: Low-density lipoprotein oxidation; LO: Lard oxidation; PFO: Pork fat oxidation; PPOO: Pork patty oil oxidation; SOO: Sunflower oil oxidation; TEAC: Trolox equivalent antioxidant capacity; TCN: Thiocyanate.

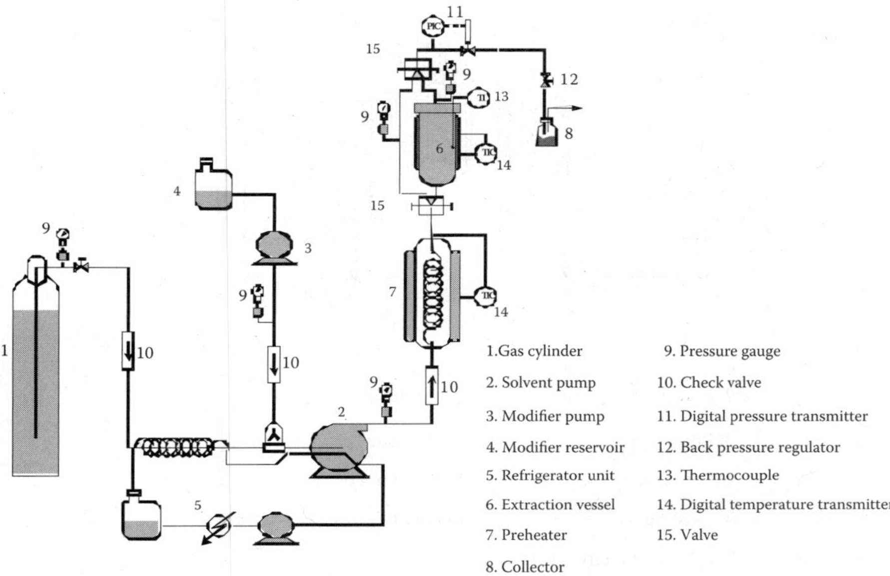

FIGURE 6.2.2 General flow diagram of a single stage SFE and stepwise collection.

[2, 101, 131]. When the coextraction of other compounds cannot be avoided, fractionation can be achieved either by using several separation vessels with independent control of pressure and temperature or (in systems with one separator) by withdrawing samples at different contact times. The first configuration has been called on-line fractionation [48], fractional separation [2, 10, 78], cascade fractionation [88], or cascade depressurization [78]. The second disposition (presented in Figure 6.2.2) has been named stepwise collection [132] or time fractionation [83], and fractions are collected at predetermined extraction periods [31, 64].

One or more separators (see Table 6.2.1) have been proposed for the recovery, fractionation, and purification of antioxidant extracts [8]. The separation of two different fractions has been used in the processing of medicinal herbs, enabling the recovery of antioxidant compounds in the first separator and essential oil in the second one [78, 115]. Fine tuning of the separation allowed the recovery of β-carotene isomers from the algae *Dunaliella bardawil*, based on their different solubility [57]. The pressures in the separators were selected to fractionate the desired products: below 10 MPa, the lycopene and most lipidic components are separated, whereas at 20 MPa only lycopene precipitates [117]. Separation of compounds with different activities (antioxidant and antimicrobial) from *Spirulina platensis* has been achieved using a related operational method [41].

In countercurrent supercritical extraction (CC-SFE), besides the fractions obtained in separators, the raffinate is collected at the bottom of the column (see Figure 6.2.3). The relative amounts of each fraction depend on the S/F ratio, as reported for the fractionation of orange juice. In this case, hesperidin, narirutin, naringin, and benzoic acid were found in almost all fractions, whereas flavanones were collected in the first separator, and sinensetin, nobiletin, and heptamethoxy flavone in the second

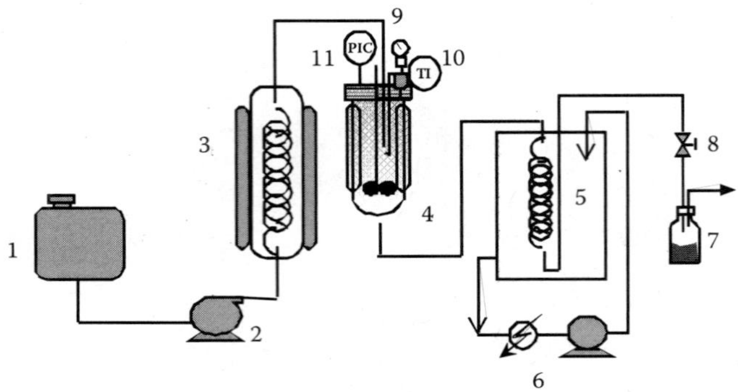

1. Feed
2. Pump
3. Preheater
4. Extraction vessel
5. Cold tank
6. Cold liquid circulator
7. Collector
8. Metering valve
9. Pressure gauge
10. Thermocouple
11. Digital pressure transmitter

FIGURE 6.2.3 Flow diagram of autohydrolysis process.

one [88]. The extraction pressure in CC-SFE controls the composition and yield of extracts in both separators and can be varied from those favorable to concentrate the compounds in the raffinate to others suitable for obtaining the target compounds in the separators [97].

6.2.1.3.2 Stagewise Extraction at Progressively Increasing Pressure

Stepwise increase of the extraction pressure was also named as a two-step process [82], discontinuous extraction [118], two-step presure gradient operation [86], multistep operation, and fractional extraction [2]. After a first stage at low pressure (<15 MPa) to extract non-polar compounds (such as volatile compounds, essential oil, and waxes), the solid residue is subjected to SFE. This operational mode limits the coextraction of some compounds in the first stage and allows the recovery of those scarcely soluble in SC-CO_2 in the second one. The solvent power of SC-CO_2 can be tuned by modifying pressure and/or temperature, enabling the extraction of more polar compounds. The same goal can be achieved using a modifier in the second stage [68, 116, 133].

Extraction of essential oil from raw materials (operating at 8–10 MPa and 313 K) and re-extraction of the more polar compounds from the residue under more severe conditions (18–40 MPa and 313–331 K) were proposed for processing rosemary [33, 80, 81], oregano [131], or paprika. In this latter case, lipids were obtained in the first stage and pigments in the second one [118]. Table 6.2.2 summarizes reported data concerning the stagewise extraction of antioxidant compounds and the antioxidant characteristics of the products obtained in the extract of the second stage.

Similar to the fractional separation, stepwise extraction allows the production of different products, and both dispositions provided the same overall yields for oregano oleoresin [131]. The second stage of stepwise extraction could be favored by the

TABLE 6.2.2
Reported Data Concerning Stagewise SFE of Solid Feedstocks at Progressively Increased Severity and Fractionation

Phenolics and terpenoids			
Feedstock	**SFE: n) EV; P; T; nS**	**Antioxidant activity**	**Reference**
Common balm	1) 0.4; 9; 323; — 2) 0.4; 30; 323; —	LAA: BHT > αT > SFE	25
Coriander	1) <0 1; 10; 313; — 2) <0.1; 18.8; 331; —	DPPH: Eu > SFE	33
Oregano	1) 4; 30; 313; 1 2) 4; 50; 313; 1	LO: BHA:BHT > SFE	116
Rosemary	1) 4; 30; 313; 1 2) 4; 50; 313; 1	LO: SFE > BHA:BHT	116
Rosemary	1) <0.01; 10; 313; — 2) <0.01; 40; 333; —	DPPH: $SFE_2 > SFE_1$	81, 80
Sage	1) 4; 30; 313; 1 2) 4; 50; 313; 1	LO: SFE > BHA:BHT	116

Carotenoids			
Feedstock	**SC-Extraction: n) EV; P; T; nS**	**Compound (Yield, %; Purity, %)**	**Reference**
Paprika	1) 0.85; 13.8; 313; 1 2) 0.85; 48.3; 313; 1	 TY (12.2); βc (6.67)	118
Paprika oleoresin	1) 2.6; 30; 333; 1 2) 2.6; 50; 353; 1	Tc (0,08; 48.9) TY (85.7; 18.5)	83
Yeasts	1) <0.1; 30; 333; 1 2) <0.1; 50; 333; 1	Ast (50; 4)	85

n): Extraction stage; EV: Extractor volume (L); P: Extraction pressure (MPa); T: Extraction temperature (K); nS: Number of separation vessels; SFE_n: Supercritical fluid extract produced in the *n* stage.
LO: Lard oxidation; LAA: Linoleic acid autoxidation; DPPH: 2,2-Diphenyl-1-picrylhydrazyl hydrazyl radical scavenging capacity.
αT: α-Tocopherol; Tc: Total carotenoid; BHT: Butylhydroxytoluene; Eu: Eugenol; BHA:BHT: mixture 1:1; Ast: Astaxanthin.

previous removal of waxes, but it is less effective than stagewise precipitation for fractionating the extracts into essential oil and pasty products. Stepwise extraction needs more than twice as much solvent as single extraction or fractional separation. An extraction profile with intermediate depressurization to 0.1 MPa resulted in lower CO_2 requirements and better extraction yields [112].

6.2.1.3.3 *Successive Extraction with Conventional Solvents and SC-CO_2 Processing of the Extract to Obtain the Antioxidants Either in the Extract or in the Residue*

Vegetal raw materials show low solute content and low bulk density. Both aspects limit the potential of SFE, because of their economic implications. In this context, SC-CO_2 extraction of commercial extracts or dried extracts from CSE presents operational and economic advantages, including operation under milder operational conditions, reduction in the volume of extraction vessels, and lower specific solvent consumption. This approach is recommended when the direct SFE of vegetal materials yields limited amounts of the target products, if their activities are low, or if the final product is dark colored, is difficult to manage, and/or shows low content of active compounds even under harsh extraction conditions or in the presence of a modifier [119]. CSE extracts can be processed by SFE either as a solution or after drying in order to recover the active compounds either in the extract (extraction) or in the raffinate (purification or dearomatization). Representative data reported on this approach are listed in Table 6.2.3.

Removal of apolar compounds by CSE followed by SCF extraction of large and/or polar molecules (such as polyphenols or terpenoids) can result in improved yields, with respect to the direct extraction of the raw material, particularly when the high-molecular-weight compounds are present at low concentrations and interact with the matrix [67]. The process proposed by Yang et al. [20], consisting of consecutive extractions with 70% ethanol and SC-CO_2, resulted in an extract from *G. biloba* leaves containing flavonoids and terpenoids with good color and solubility. CC-SFE of a liquid stream (wood hydrolyzates), whose primary objective was the removal of fermentation inhibitors (furan derivatives, phenolics, and aliphatic acids), gave a by-product stream containing antioxidants [99]. Vanillin, coniferyl aldehyde,

TABLE 6.2.3
Results Obtained in the Successive Extraction of Phenolics and Terpenoids with CSE and SFE of the Extract

Feedstock	CSE: S; T; t SFE: EV; P; T; nS	Antioxidant activity	Reference
Thoroughwax	Ethanol: —; 24; — CC: —; 20; 313; —	LP: $\alpha T > F_3 > EE > F_2 > R > F_1$ DPPH: $\alpha T > SFE_3 > SFE_2 > SFe_1 > EE > SFR$ SO: $SFE_3 > SFE_2 > R > EE > SFE_1$ H: $SFE_3 > SFE_2 > R > EE > SFE_1$	122, 123
Grape pomace	Ethyl acetate: —; —; — 0.4; 25; 318; 1	DPPH: SFE ≈ CSE ≈ CRE	43
Rosemary	Commercial: 0.5; 10; 308; 1	SFO: SFE > CE	14
Rosemary	2-propanol: —; —; — 0.005; 20; 333; —	DPPH: CSE > SFR	66

EV: Extractor volume (L); P: Extraction pressure (MPa); T: Extraction temperature (K); t: Extraction time (h).

nS: Number of separation vessels; CC: Countercurrent.

SFE_n: Supercritical extract from the *n* separation stage; SFR: Supercritical residue.

CE: Commercial extract; CRE: Commercial rosemary Extract; CSA: Commercial synthetic antioxidant; CSE: Conventional solvent extract; EO: Essential oil (steam distillation); SCO: Supercritical oil; CO: Commercial oil; EE: Ethanolic extract; αT: α-Tocopherol.

DPPH: 2,2-Diphenyl-1-picrylhydrazyl hydrazyl radical scavenging capacity; SO: Superoxide radical scavenging capacity; H: Hydroxyl radical scavenging capacity; SFO: Sunflower oil oxidation.

acetoguaiacone, and 4-hydroxybenzoic acid were quantitatively extracted from hydrolyzates, whereas the aliphatic fatty acids were only partially separated. Utilization of SFE as a concentration and purification step in combination with other technologies (membranes, adsorption in nonionic, polymeric resins) has been claimed in several patents [134–136].

Purification, deodorization, or dearomatization of the extracts from medicinal herbs (*Labiatae*) is required when the product shows undesired aroma or color. Extracts obtained with conventional polar solvents (ethyl acetate, acetone, methanol, ethanol, 1-propanol, 2-propanol, butanol, water, and/or mixtures) can be further treated by SC-CO_2 under mild conditions (10–15 MPa, 35–45°C) to remove the undesired compounds and to concentrate the target compounds in the residue, enhancing both its properties (activity, color, and odor) and antioxidant activity [14, 66].

Several processes have been proposed to remove residual aroma from aromatic herbs [66, 80, 137]. Extraction of pungent compounds from the red pigment of paprika oleoresins (produced by CSE or SFE) has been reported [65]. Although these processes are conceived to purify the antioxidants in the extract, CSE with ethanol was also applied with the aim of dehydrating orange peel before extracting β-cryptoxanthin by SC-CO_2 in the presence of a modifier [138].

If the conventional solvent extract contains several valuable compounds, some of them can be recovered in the extracts and others in the raffinate. This is the case of a raw extract of olive leaves in hexane, containing waxes, hydrocarbons, squalene, β-carotene, triglycerides, α-tocopherol, β-sitosterol, and alcohols. CC-SFE allowed the recovery of hydrocarbons in the separators, whereas waxes and α-tocopherol remained in the raffinate [105]. The extracts in hexane or in ethanol can be processed directly or concentrated to different degrees [105, 139]. The direct extraction of the solvent extract in countercurrent equipment was used to recover hydroxytyrosol, luteolin, caffeic acid, and *p*-coumaric acid [55].

6.2.1.3.4 *Processing by SFE and Subsequent Extraction of the Solid Residue with Conventional Solvents or by HT Processing*

Low pressure SFE has been proposed to remove volatile compounds and waxes from the solid substrate before extraction with conventional solvents (see Table 6.2.4). Fats can be removed from herbs by extraction with liquid or subcritical or supercritical CO_2. In a subsequent stage, the insoluble residue has been processed with alcohol to extract water-soluble antioxidants selectively [141]. Ribeiro et al. [140] observed that CSE of the solid residues obtained after the supercritical extraction of the oil from lemon balm leaves allowed higher yields of a more active extract than the direct CSE of the raw material. The extraction was faster from the supercritical solid residues than from the untouched plant because lipids and cuticular waxes, susceptible of hampering the extraction of polyphenols, had already been extracted. When the SFE is performed under high severity conditions, the product obtained in a subsequent CSE stage shows reduced activity, as reported for the ethanolic extracts from the residue of two SFE stages at 30 and 50 MPa [116]. If the antioxidants remaining in the solid residue after SFE are too polar or have high molecular weight, other solvent technologies (for example, hydrothermal processing) can be applied, as reported for the extraction of quinones and derivatives from SFE-treated bamboo [142].

TABLE 6.2.4
Data Reported on the Successive Extraction of Phenolics and Terpenoids with SFE and CSE of the Residue

Feedstock	SFE: EV; P; T; nS CSE: S; T; t	Antioxidant activity	Reference
Lemon balm	0.5; 10; 308; 2 W; 373; 1.5	PF: SFE > BHT	140
Oregano	4; 30; 313; 1 95% E; —; —	LO: BHA:BHT > SFE > SFE-CSE	116
Rosemary	4; 30; 313; 1 95% E; —; —	LO: SFE > BHA:BHT > SFE-CSE	116
Rosemary	—; 7.5; 305; 1 50% E; —; 1	RIM: BHT > SCF-CSE > αT	141
Sage	4; 30; 313; 1 95% E; —; —	LO: SFE > BHA:BHT > SFE-CSE	116
Thyme	4; 30; 313; 1 95% E; —; —	LO: BHA:BHT > SFE > SFE-CSE	116

EV: Extractor volume; P: Extraction pressure (MPa); T: Extraction temperature (K); t: Extraction time (h); nS: Number of separation vessels.

W: Water; E: Ethanol.

LO: Lard oxidation; RIM: Rodin iron method.

SFE-CSE: Conventional solvent extract from the supercritical residue; BHA:BHT: 1:1 mixture; αT: α-Tocopherol.

6.2.1.4 Obtaining Antioxidants by SFE with Cosolvent

Supercritical CO_2 is a good solvent for apolar solutes, but their solubility decreases with the molecular weight. Compounds of high molecular mass, such as flavonoids, are hardly soluble in pure CO_2. The solubility of polar organic compounds or their interaction with the matrix can be improved by either increasing pressure or adding a polar modifier. The extraction enhancement caused by a modifier may be related to different phenomena, including (i) change in polarity, density, and viscosity of the extraction fluid, (ii) miscibility of the modifier and solvent and the solute solubility, (iii) interaction between supercritical CO_2 and the matrix, and (iv) disruption of the bonding between solutes and the solid matrix. The effect of cosolvent results in changes in solubility, transport properties and intraparticle resistance in the matrix and can increase extraction yields and/or rates, depending on the pressure and temperature used. The solubility enhancement in the presence of cosolvents can be associated with intermolecular interactions between components, particularly hydrogen bonding [143].

Table 6.2.5 lists the most common modifiers used to extract antioxidants from different matrices. The modifier can be a pure compound or a mixture (for example, most alcohols are added as a water solution). Organic cosolvents present problems for industrial scale operation because of their cost, flammability, and disposal requirements. On the other hand, the process would not be solvent free, a major advantage of SFE. Ethanol and water are the more suited cosolvents for food-related applications. Ethanol is widely used to improve the extraction efficiency of phenolic acids, flavonoids,

TABLE 6.2.5
Modifiers Used in the Extraction and Fractionation of Antioxidant Compounds

Modifier	Matrix	Major solutes or target compounds	EYI[a]	Reference
Acetone	Pulp	Car	1.9	118
Acetonitrile	Seeds	Da, Ge	n.d.	144
Canola oil	Carrots	αc, βc	2.4	61
		Lu	5	
Chloroform	Leaves	Vi, Or, Ru	n.d.	9
	Tomatoes	Lyc	3.92	145
	Soy products	Da, Ge	1.27	146
DCM	Mushroom	PhC, Toc, βc	1.49	64
DMSO	Roots	Ggl	50	147
DMP	Pomace	βc	1.8	60
Ethanol and ethanol$_{aq}$	Seed coats	EC, DPA	2.6	44
		EC	2.8	45
	Leaves	Caf, EGC, EC, ECG, GA	1.3	132
		Q, KA, iR	7.3	20
		Lu, βc	n.d.	75
		Bo	3	12
		Ter	—	148
	Herbs	Sage extract	3.8	23
		Rosemary extract	S_1 (52) S_2 (8.4)	16
	Root	Ggl, Shg	7.1	30
			1.1	31
	Soy products	Da, Ge	1.8	146
	Caulomas and leaves	Lig, CA	—	76
	Tomato paste	Lyc βc	Lyc (2.2) βc (1.11)	126
	Yeasts	Ast	1.24	85
	Skins	Q	n.d.	149
	Bamboo	Etx	1.25	142
	Propolis	CA, F	—	104
	Pomace	α, β, γT	—	113
	Okara	SI	1.47	150
	Mushroom	PhC, Toc, βc	1.06	64
	Bran	Alk	4.3	48
	Microalgae	Ast	8	126
		Ast	1.25	151
		Phy	2.33	151
		Car, Xan, Phyt	n.d.	41
		Vit C, Vit E, ω3FA	1.5	120
	Hexane extract	Sq, βc, αT, βs	2.8	97
	Microalgae	Car	1.16	40

continued

TABLE 6.2.5 *(continued)*

Modifier	Matrix	Major solutes or target compounds	EYI[a]	Reference
Ethyl acetate	Leaves	Vi, Or, Ru	—	9
	Mushroom	PhC, Toc, βc	1.59	64
	Propolis	DHCA	3.7	78
Hazelnut oil	Tomato	Lyc	3	84
Methanol and methanol$_{aq}$	Leaves	Vi, Or, Ru	n.d.	9
	Flowers	Ap	2	26
		Ap-g	18	
	Pomace, seeds		0.74	32
	Hulls		7	46
	Roots	Ggl	70	147
	Soy products	Da, Ge	1.75	146
	Pulp		—	29
	Bran	Alk	4	48
	Grapes	Ant	—	152
	Soy beans	SI	—	153
2-propanol	Root	Ggl, Shg	1.02	31
Soybean oil	Microalgae	Car	1.10	111
Sunflower oil	Root	βc	—	61
THF	Peels	Bcx	—	138
Water	Leaves	Caf, EGC, EC, ECG, GA	—	132
	Grapes	Ant	—	152
	Seeds	Se	—	154

[a] EYI: Extraction yield increase, defined as number of times that the yield is increased; n.d., results that cannot be calculated because the solvent is not pure CO_2.

Alk: Alkylresorcinols; Ant: Anthocyanins; Ap: Apigenin; Ap-7-g: Apigenin-7-glucoside; Ast: Astaxanthin; Bo: Boldine; CA: Cinnamic acids; Caf: Caffeine; Car: Carotenes; αc: α-carotene; βc: β-carotene; βcx: β-cryptoxanthin; Da: Daidzein; DHCA: 3,5-diprenyl-4-hydroxycinnamic acid; DPA: 3,4-dihydroxyphenyl acetate; EC: (-) Epicatechin; ECG: Epicatechin gallate; EGC: Epigallocatechin; EGCG: Epigallocatechin gallate; Etx: Ethoxyquin; ω-3FA: ω-3 Fatty acids; F: Flavonoids; GA: Gallic acid; Ge: Genistein; Ggl: Gingerols; KA: Kaempferol; Lig: lignans; Lu: Lutein; Lyc: Lycopene; Or: Orientin; PhC: Phenolic compounds; Phy: Phycocyanine; Phyt: Phytopigments; Q: Quercetin; iR: Isorhamnetin; Ru: Rutin; Se: Sesamol; Shg: Shogaols; SI: Soy isoflavons; Sq: Squalene; Terp: Terpenoids; Toc: Tocopherols; α-, β-, γ-T: α-, β-, γ-Tocopherol; βs: β-sitosterol; Vi: Vitexin; Xan: Xanthophyll; DCM: Dichloromethane; DMP: 2,2-dimethoxypropane; DMS: Dimethylsulfoxide; THF: Tetrahydrofuran.

terpenoids, and carotenoids and can be easily removed from the final product by distillation. Processes using water as a cosolvent are clean, but some problems arise: (i) the formation of ice blockages during expansion, (ii) reduced solubility and extractability of ionizable compounds, (iii) hydrolysis of some components, and (iv) reduced shelf life of the product [82]. Water is used as a cosolvent in several industrial SC-CO_2 extraction processes (nicotine, caffeine, and vanillin), and has been proposed to extract phenolics [82, 130, 154] and to remove aroma compounds from conventional solvent extracts [66]. The utilization of water on SC-CO_2 has been revised by Balachandran et al. [74].

Water causes swelling of the solid, higher solute diffusivity, weakened interactions between the solute and the matrix due to the adsorption of water onto the polar sites, and the interactions of functional groups of the oxygenated compounds (charge-transfer complex formation, induced dipole, and hydrogen bonding) with water would result in increased yields. Opposite effects could occur at high pressure, as compression limits swelling and the increased polarity of CO_2 would be disadvantageous for extracting nonpolar components. The effects of moisture on the extraction yield depend on the considered solute. Neutral cosolvents such as vegetable oils are favorable for extracting high-molecular-weight compounds, such as β-carotene [61], an effect also observed in the extraction of carotenes from solid samples containing seeds [117].

When processing conventional solvent extracts by SFE in the presence of modifiers, the optimal cosolvent may depend on the solvent used in CSE, as it has been reported for olive leaves: with hexane extracts, ethanol was the cosolvent selected to concentrate β-sitosterol and terpenoids in the second separator and α-tocopherol in the raffinate, whereas with ethanol extracts, water was the selected modifier to concentrate eritrodiol and uvaol in the first separator and hydroxytyrosol in the raffinate [105].

The modifiers can be added to the SFE either mixed with CO_2 before being pumped to the extractor or mixed with the raw material. The addition of modifier to the CO_2 stream, also named as sequential [30], gradual [150], or continuous [61] cosolvent addition, is the most frequent choice. This operational procedure was used with yeast biomass [85], herbs [155], ginger [31], leaves [12, 76], rye bran [48, 49], carrots [61], or mushrooms [64]. For using water as a cosolvent, the CO_2 stream has been passed through an autoclave filled with moistened quartz sand [82].

Operation when the modifier is mixed with the feedstock has been referred to as batch, discontinuous, or individual addition [78]. This alternative was reported for diced onion skins [156], tomato [84], lyophilized aloe epidermis and pulp [29], seaweed [120], and propolis extract [78]. This strategy was used for processing conventional solvent extracts: the concentrates were dried and resuspended in ethanol, and the resulting dispersion was extracted with SC-CO_2 in the presence of ethanol as a modifier, to reduce the content of harmful compounds in the extracts (ginkgoic acid, bilobol, and ginkgol) and to increase the relative content of the active flavonoids [157]. The mode of cosolvent addition affects the extraction process. Leeke et al. [82] reported the largest increase from *Origanum vulgare* essential oils when water was added discontinuously (at a concentration of 80% w/w), whereas the continuous addition led to an increase of the coextracted waxy material. Previous mixing of modifier and the material to be extracted was also used in the antisolvent fractionation of propolis using an ethanol tincture [104].

Modifier concentrations in the range 5–15% are typically used for flavonoids and terpenoids, and 10% for carotenoids. Even though the total yield is favored with higher modifier concentrations, the selectivity in the extraction of target compounds can be maximal at intermediate values. Coextraction could be beneficial for the antioxidant activity, for example, in the joint recovery of carotenoids and xanthophylls from *Spirulina* [41], polyphenols and isoflavones from okara [150], and vitamin E and omega-3 fatty acids from *Chlorella pyrenoidosa* [120].

Increased cosolvent concentrations result in similar effects to those achieved by increasing pressure. This behavior enables the fractionation of solutes, extracting first the low polar compounds followed by the more polar ones [32]. Some cosolvents assayed at an analytical scale present difficulties for scaling up (toxicity, low miscibility with SC-CO_2) [26].

The physical properties of the extracts, affecting the overall product quality, can be influenced by the modifier. Variations in color were the most frequently reported [20, 41].

6.2.2 Obtaining Antioxidants by High-Pressure Water Extraction

Several technologies for biomass processing based on the utilization of aqueous media have been reported in literature. These studies deal with a wide variety of objectives, including chemical fractionation, structural alteration, and isolation of fractions with special properties. In this chapter, the attention is focused on aqueous treatments of lignocellulosic materials (LCM), leading to both the hydrolytic degradation of hemicelluloses and the solubilization of antioxidant compounds, as well as on other related technologies that have been applied to other types of vegetal biomass and/or with objectives different from hemicellulose decomposition to yield isolates with antioxidant activity.

Owing to the broad scope, other related methods are not included, such as those based on the utilization of chemicals different from water (for example, water–solvent mixtures, water–oxygen media such as those used in the wet oxidation technology, or media containing mineral acids such as those used, for example, in prehydrolysis treatments or in preimpregnation of substrates for catalyzed-steam processing). The general flow diagram is shown in Figure 6.2.4.

6.2.2.1 Processing of LCM

6.2.2.1.1 Hydrothermal Treatments

LCM, particularly those of residual origin coming from agroindustrial and forest activities, are promising sources of antioxidant compounds [158, 159]. Because LCM are heterogeneous and present a complex chemical nature, their integral benefit can be achieved by chemical fractionation, following the "biomass refining" philosophy [160], based on the selective separation of the main components to yield a variety of high added-value bioproducts.

Several studies on the fractionation of LCM by water or steam have been referred to in literature as autohydrolysis, hydrothermolysis, aqueous liquefaction or extraction, aquasolv, water prehydrolysis, hydrothermal pretreatment or treatment, and steam pretreatment or steam extraction [161]. All these studies are based on the same kind of reactions and are referred to here as hydrothermal or autohydrolysis treatments.

When LCM are contacted with water at temperatures in the range 413–493 K, a variety of effects are reached, including the following:

- Hydrolytic depolymerization of hemicellulose to give high-molecular-weight compounds (soluble fiber), oligosaccharides, monosaccharides, and sugar-degradation compounds (such as furfural and hydroxymethylfurfural).
- Extractive removal (including lipophylic compounds and low-molecular-weight phenolics).
- Acetic acid generation by cleavage of acetyl groups.
- Solubilization of acid-soluble lignin.
- Ash neutralization.

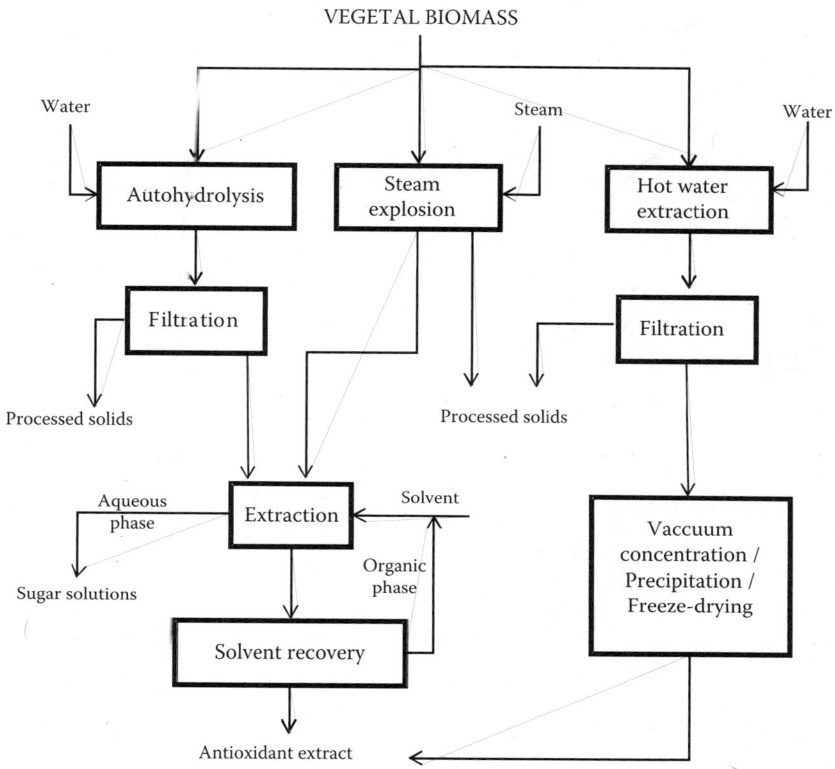

FIGURE 6.2.4 General flow diagram of an extraction process based on hot water extraction.

- Reactions involving proteins.
- Partial deetherification and depolymerization of lignin without causing significant cellulose damage [162]. The effects on lignin depend on the LCM feedstock: for example, softwood lignin (a typical guaiacyl lignin having methoxyl as the major functional group and lower amounts of other groups such as benzyl alcohol and phenolic hydroxyl) is less susceptible to hydrolytic decomposition than hardwood lignin [163], owing to differences in molecular weight and reactivity, which favor condensation over hydrolysis in the case of softwoods [164].

The effects of hydrothermal processing on the major fractions of vegetal biomass are shown in Figure 6.2.5. The most abundant hemicellulosic polymers are xylans, made up of xylose units. Xylans represent an immense resource of biopolymers for practical applications [165], accounting for 25%–35% of the dry biomass of woody tissues of dicots and lignified tissues of monocots, and occur up to 50% in some tissues of cereal grains. The structure of xylans depends on the source considered: the most common xylans are made up of a main backbone of xylose linked by β-1→4 bonds, where the structural units are often substituted at positions C2 or C3 with arabinofuranosyl, 4-*O*-methylglucuronic acid, and acetyl or phenolic substituents [166].

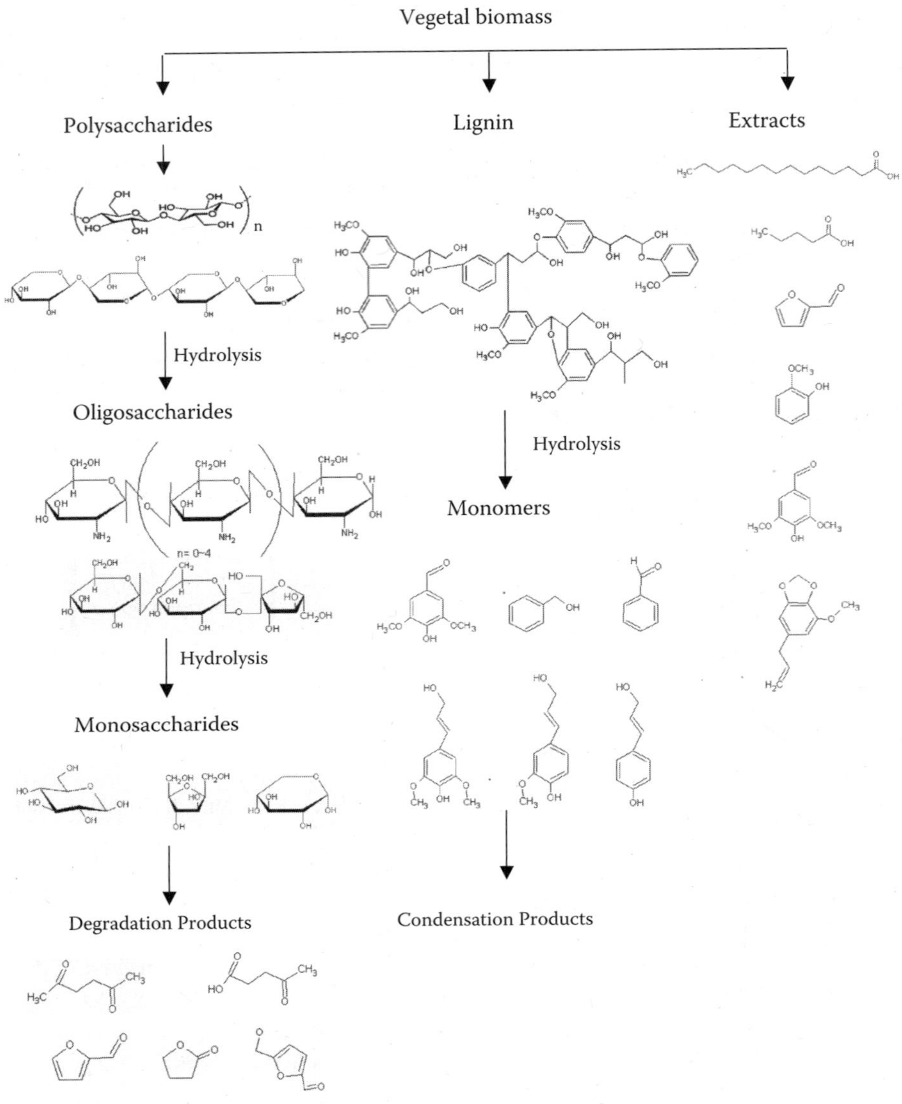

FIGURE 6.2.5 Effect of hydrothermal treatments on the major fractions of vegetal biomass.

When xylan-containing materials are used as feedstocks for hydrothermal treatments, the high-molecular-weight and oligomeric compounds derived from hemicelluloses are made up of xylose units (which can be substituted, for example, with acetyl groups, uronic acids, arabinose, or phenolic moieties). Several studies have been reported on the hydrothermal processing of a variety of xylan-containing feedstocks, such as crop residues (straws, corncobs), bamboo [142], hardwoods, softwoods, wine-making waste solids, wastes from olive oil production, and grain hulls

[163, 167–169]. Using flow-through reactors, the hemicellulose decomposition can be followed by a cellulose degradation stage by rising temperature above 230°C [163].

Hemicellulose-derived oligosaccharides have been proposed as prebiotic food ingredients based on their effect on the intestinal flora [167, 170–172], but obtaining food-grade products requires further purification to remove monosaccharides and nonsaccharide compounds. In this field, solvent extraction is useful for removing nonsaccharide components of hydrothermal liquors [173], yielding both a selectively refined aqueous phase and a solvent-soluble fraction mainly made up of phenolics and extractive-derived compounds.

The nonsaccharide compounds isolated from autohydrolysis liquors lack commercial value, and the development of practical applications for this fraction would be of scientific and economic interest. Based on the chemical nature of the compounds soluble in ethyl acetate and on their antioxidant activity, these compounds are potential candidates for commercial developments [174].

Typically, the nonsaccharide by-products present in autohydrolysis liquors include furans (furfural, hydroxymethylfural) from sugar dehydration, other compounds derived from sugars (ketones, lactones), terpenes, other lipophilic compounds, fatty acids, resin acids, nitrogen-containing compounds, and phenolics (monomeric phenols and lignin-related compounds).

Fatty acids (such as hexadecanoic acid and octadecanoic acid, which are present in barley husk autohydrolysis liquors) [175] or stearic acid, palmitic acid, oleic acid, 9–12 octadecanedienoic acid, and tetradecanoic acid, which are present in the autohydrolysis liquors of *Eucalyptus* [176], have been proposed for the manufacture of resins, as raw materials for the synthesis of other useful compounds such as industrial rubber, for applications in cosmetic industries, and as surfactants and components of soaps [177].

Phenolic compounds are the most important ones owing to their antioxidant activity. In this field, vanillin is usually one of the major phenolic components of autohydrolysis media (for example, in liquors from barley husks autohydrolysis liquors) [175]. Phenolic acids (such as ferulic acid, gallic acid, vanillic acid, isovanillic acid, homovanillic acid, 3-hydroxybenzoic acid, 3-methoxy-4-hydroxybenzoic acid, protocatechuic acid, syringic acid, *p*-coumaric acid, and cinnamic acid), aldehydes (such as benzaldehyde, benzeneacetaldehyde, syringaldehyde, sinapaldehyde, 4-hydroxy-2-methoxycinnamaldehyde, and 3,4 dihydroxybenzaldehyde), ketones (such as acetophenone, 2,5-dihydroxyacetophenone, acetovanillone, acetosyringone), and alcohols and other lignin-related compounds (such as benzyl alcohol, homovanillyl alcohol, 4-eugenol, isoeugenol, methoxyeugenol, guaiacol, 4-ethylguaiacol, 4-vinylguaiacol, and coniferyl alcohol) have been also identified in autohydrolysis liquors from *Eucalyptus* wood, corncobs, barley husks, wine-making waste solids, or rice husks [168, 175, 178].

Antioxidant properties have been reported for the ethyl acetate-soluble components of liquors from hydrolytic processing of biomass, a possible way for achieving an integrated benefit of the several fractions from autohydrolysis of LCM (oligosaccharide-containing aqueous phase from solvent extraction of liquors, antioxidant-containing organic phase from solvent extraction of liquors, and cellulose-enriched solid phase from autohydrolysis treatments) [167].

When pine wood was used as a feedstock for autohydrolysis, the yield obtained in the ethyl acetate extraction of liquors was more than five times higher than the ones reported for extractions with ethanol or methanol [179], but was slightly lower than the results reported for agricultural residues or for hardwoods. The experimental data suggest that some lignin depolymerization takes place under the operational conditions typical of autohydrolysis experiments. High severity promotes reactions involving guayacil units [180], which are the main constituents (85%–98%) of softwood lignin. Comparatively, the nonisothermal autohydrolysis of *Eucalyptus globulus* wood and corn cobs yielded 8.72 and 6.47 g ethyl acetate soluble solids/100 g, respectively [174], in comparison with 0.319 g/100 g oven-dried pine wood.

The antioxidant activity of ethyl acetate-soluble fractions extracted from autohydrolysis liquors depends on a variety of factors, including the type of LCM feedstock used in experiments, the operational conditions, and the possible implementation of refining treatments. It can be noted that the activity of crude extracts can be even higher than that of the purified fractions because of the presence of active compounds in small quantities and/or synergistic effects among various compounds [158]. In other situations, fractionation leads to concentrates with enhanced antioxidant activity [168]. In studies dealing with pine wood autohydrolysis, the antioxidant power of the aqueous hydrolyzate has been reported to be higher than that of the acetate-soluble fraction [181].

Garrote et al. [174] reported on the influence of the operational conditions (defined in terms of the severity factors) on the antioxidant properties of ethyl acetate-soluble phenolics from *Eucalyptus* wood and corncobs. The severity analysis included as dependent variables the yields in active fractions and their antioxidant activities. In the case of extracts from *Eucalyptus* wood, very active compounds (up to 60% more active than butylhydroxyanisol [BHA]) were obtained under mild autohydrolysis conditions (maximum temperature, 453 K), whereas harsher processing conditions resulted in improved yields, but also in decreased specific activity. Oppositely, the specific antioxidant activity of corncob extracts increased with the severity of treatments. Even though the specific activities of the fractions extracted from corncobs were lower than those of *Eucalyptus* for samples obtained under mild conditions, the specific activities of both wood- and corncob-derived fractions tended to reach a similar specific activity (about 60% of the specific BHA activity or 420% specific butylhydroxytoluene [BHT] activity) when the fractions were obtained under harsh treatment conditions [174].

Isolates with high specific antioxidant activity (up to 40 times more than BHT, 3.5 times more than BHA, three times more than gallic acid, eight times more than caffeic acid, and 25 times more than α-tocopherol) have been reported in experiments with pine wood autohydrolysis liquors [181].

As a summary, Table 6.2.6 lists the yields and comparative activities with respect to BHA and BHT of fractions isolated from autohydrolysis liquors of several raw materials. As an additional valuable feature, the antioxidants from ethyl acetate soluble-fraction of autohydrolysis liquors from red grape pomace after fermentation and distillation have been reported to show a better thermal stability than BHA or BHT, because limited weight loss was determined for the lignocellulose-derived antioxidants after prologed heating at 200°C (conditions under which the reference synthetic antioxidants were almost completely volatilized) [184].

TABLE 6.2.6
Yield and Antioxidant Activity of the Ethyl Acetate Extracts of Autohydrolysis Liquors from Selected Lignocellulosic Materials

Raw material	Conditions: hydrothermal treatment HTE[a] T; t; LSR	Yield (%)[b]	Comparative antioxidant activity (DPPH method)	Reference
Almonds shells	Isothermal 393; 1; 10	2.42	BHA > HTE	182
Chestnut burs	Isothermal 393; 3; 10	0.57	THE > BHA	183
Corn cobs	Non-isothermal 533; —; 8	6.47	BHA > HTE > BHT	174
Distilled grape pomace	Isothermal 373; 5; 8	1.10	HTE > BHA > BHT	168
Eucalyptus wood	Non-isothermal 533; —; 8	8.72	BHA > HTE > BHT	174
Pine wood	Non-isothermal 483; —; 8	3.50	THE > BHA > BHT	181

[a] HTE: Hydrothermal extract; T: Temperature (K); t: time (h); LSR: Liquid-to-solid ratio (g/g).
[b] As weight percent of the raw material.

Other related alternatives for antioxidant applications of hemicellulose-derived products explored in the literature are as follows:

- Utilization of high-molecular-weight compounds derived from hemicellulose fragmentation (soluble fiber) as antioxidant food ingredients [185, 186]
- Direct utilization of acidic xylooligosaccharides as antioxidants, based on their concentration-dependent, iron-reducing function [187]

6.2.2.1.2 Steam Explosion

Uncatalyzed steam explosion presents some features similar to autohydrolysis (utilization of water as sole reagent, fractionation effects on biomass mainly related to hemicellulose hydrolysis, extractive removal, extraction of acid-soluble lignin, hydrolytic effects on lignin), but in this case, the pressure is suddenly released to cause drastic structural alterations of the solid residue, yielding defibered materials suitable for dissolving pulp manufacture (prehydrolysis-kraft process) or fiberboard production or as substrates for the enzymatic hydrolysis of cellulose [188–196]. As the operational conditions are usually harsher than in the case of autohydrolysis, the amount of furans coming from sugar decomposition may become important, causing inhibition of further fermentation stages for utilization of pentoses and/or hexoses.

Low-molecular-weight phenolics have been cited as by-products of steam explosion. In some studies, the interest in these compounds was focused on their inhibitory activity, which can hinder further fermentation steps. For example, low-molecular-weight phenolics, related in structure to Hibberts ketones, have been identified as steam explosion products of the softwood *Pinus radiata* [164], whereas the inhibitory effects of aromatic monomers from steam-exploded poplar have been correlated with the functional groups attached to the benzene ring [197]. In this latter work, *p*-hydroxybenzoic acid, *m*-hydroxybenzoic acid, vanillic acid, syringic acid, *p*-hydroxybenzaldehyde, vanillin, syringaldehyde, cinnamic acid, cinnamaldehyde, and *p*-hydroxycinnamaldehyde were identified as reaction by-products. Vanillic acid, syringic acid, vanillin, and syringaldehyde have been found in the steam explosion of olive stones [198], as well as tyrosol and hydroxytyrosol, two simple phenolic compounds characteristic of olive fruit. Simple phenolics, including 4-hydroxy-3-methoxyhomovanillic acid, 4-hydroxybenzeneethanol, vanillyl alcohol, 4-allyl-2,6-dimethoxyphenol, syringaldehyde, 2,6-dimethoxyphenol, guaiacol, and benzaldehyde, have been identified in slurries of steam-exploded aspen [199], whereas guaiacol, catechol, vanillin, 4-propylguaiacol, 4-hydroxybenzoic acid, hydroxymethoxybenzoic acid, vanillic acid, syringic acid, and protocatechuic acid are present in steamed willow wood samples [200].

Even though the production of simple phenolics by uncatalyzed steam explosion is well established, scarce literature exists on their applications as antioxidants. In this field, the production of hydroxytyrosol by steaming of olive cake has been reported to yield up to 1.7 g/100 g of dry olive waste [194, 201].

6.2.2.2 Other Technologies Dealing with Hot Water Extraction of Vegetal Biomass

Water extraction of vegetal biomass different from LCM materials has been proposed to recover bioactive compounds, without focusing on substrate fractionation as a major objective. The advantages over CSE lie on chemical aspects (higher solubility, higher diffusion rates, and lower viscosity and surface tension) and environmental issues.

These operations have been proposed as emerging technologies providing alternatives to conventional extraction. Most studies have been performed in batch mode at small scale, and further studies to develop large-scale processes are needed because this technology is attractive for the extraction of plant material in a closed and inert environment, with reduced energy demands compared to steam distillation and reduced capital investment compared to SFE, although the need for special equipment to withstand with high presures and temperatures is required.

These operational methods have been refered to as high-pressure, high-temperature water extraction [202], pressurized liquid extraction or the trade name accelerated solvent extraction [7, 203], pressurized hot water extraction [11, 204], subcritical water extraction [205–214], hot water extraction [215–219], or simply water extraction.

Pressurized solvent extraction operates at high temperature and high pressure to keep the solvent as a liquid during operation. These conditions improve solute extraction and are of particular interest when the target compounds cannot be extracted at low temperatures.

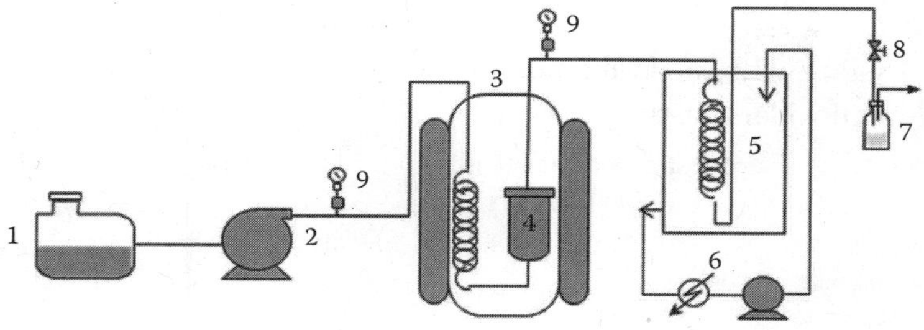

1. Feed
2. Pump
3. Oven
4. Extraction vessel
5. Cold tank
6. Cooling bath
7. Collector
8. Back pressure regulator
9. Pressure gauge

FIGURE 6.2.6 General flow diagram of subcritical water extraction.

The studied feedstocks include fruits or vegetables [202, 217, 218], wastes from industrial processing [214], seeds [205, 213, 219, 220], leaves [206, 208, 216, 221], peels or skins [202], plants or herbs [11, 209–212], roots [204, 215], skins [207], and algae [7, 203]. Some general reviews include the extraction of compounds with antioxidant activity [222].

Figure 6.2.6 shows the flow diagram of a subcritical water extraction process. Usually, the extraction system consists of a pump to provide a constant flow to the extraction cell. The water is purged with nitrogen to remove dissolved oxygen. The extraction cells are usually equipped with a frit at the inlet and at the outlet [213]. The extraction cell can be filled with sand [7, 11, 207, 223], glass beads [204], or with a cellulose filter at the bottom and top [214] or a frit [213]. The most common equipment is a packed column, including the commercial accelerated solvent extraction, but stirred vessels have also been proposed [12]. The fractions were collected in flasks along the extraction. Acidification of the media (with acetic acid, SO_2, or HCl) was proposed to enhance the extraction yields and/or improve the antioxidant activity [11, 207]. This option can provide higher extraction yields, probably caused by disruption of the cell walls, enhanced solubility, and improved diffusion and mass transfer [224]. Acid addition can favor the extraction of flavonols at lower temperatures and probably protects them from thermal degradation [202, 207].

Temperature has a marked effect on the extraction yield and selectivity of antioxidants. The dielectric constant of water decreases with temperature, enabling the extraction of nonpolar compounds. High temperature also enhances diffusivity of the solvent, improving extraction yields and facilitating the transport of solutes from the solid matrix. As a general trend, yields first increase with temperature and then decrease because of thermal degradation [12, 203, 214, 224, 225]. This behavior depends on the type of compounds considered: whereas the release of hydroxycinnamates from cell walls is favored at elevated temperature, anthocyanins can undergo degradation [207]. Increases in color caused by degradation of anthocyanins at increased extraction temperature have been reported [224]. Oxygen removal is required to minimize degradation.

TABLE 6.2.7
Studies Dealing with the Evaluation of Water-Extracted Products by Means of Multiple Antioxidant Tests

Feedstock	Extraction conditions: V; P; T; t[a]	Tests	Reference
Apple peels and pomace, grape marc, blueberry skins	2.5; 2.4; 498; —	FRAP	202
Black tea leaves	—; —; 483; —	CFAO: TE > Control	227
Boldo	—; —; 383, 3	ABTS: HPWE > SCE > CSE	12
Dunaliella	0.011; 10.3; 433; 0.5	TEAC: PLE_E > PLE_H > PLE_W	7
Grape	0.02;10.1; 433; —	ORAC: PLE > CSE	207
Noni root	10; 4; 473; —	DPPH: PHWE > CSE	204
Oregano	0.01; 10.3; 473; —	DPPH: SWE > SCE	206
Red grape	0.02;10.1; 433; —	ORAC: PLE > CSE	224
Rosemary	<0.01; 7; 373; —	DPPH: PWE ≈ HWE	208
Sage	0.34; 0.98; 373; —	DPPH: PHWE > CSE	11
Spirulina	0.011; 10.3; 388; 0.15	DPPH: PLE_H > PLE_W	229
Spirulina	0.011; 60; 388; 0.25 0.011; 60; 443; 0.15	βcB: BHT > PLE DPPH: AA > PLE	203
Taiwan yams	—; 2.08; 413; —	DPPH: HWE > CSE	230

[a] Extraction conditions: V: Extractor volume (L); P: Extraction pressure (MPa); T: Temperature (K), t: time (h).

PLE: Presurized liquid extraction using ethanol (E), hexane (H), and water (W) as solvents; HWE: Hot water extraction; PHWE: Pressurized hot water extraction; SWE: Subcritical water extraction; ABTS: 2,2′-azinobis (3-ethylbenzothiazoline 6-sulfonate); AA: Ascorbic acid; βcB: β-carotene bleaching; CFAO: Chicken fat accelerated oxidation (Rancimat); H: Hydroxyl radical scavenging activity; ORAC: Oxygen radical absorbance capacity.

A process with a stepwise pressure increase, consisting of a sequence of individual extractions, has been applied to black tea leaves [226] and rosemary leaves [208], as well as to the recovery of quercetin glycosides from onion waste [214], and to catechins and proanthocyanidins from winery by-products [213]. Oppositely to the extraction at a given temperature, sequential extraction allows the selective recovery of the most polar compounds at low temperatures and the less polar ones at higher temperatures.

Combined extraction procedures can have hot water extraction (HWE) as a first stage, followed by further processing of the extract by other extraction and/or modification methods. Examples of these kinds of technologies include the following:

- Thermal treatment of the extract at 130–190°C [227]
- CSE with a water-immiscible organic solvent [227]
- Incubation with tannase [226, 227], or with β-glucosidase [214]

Some experimental techniques can be assisted by ultrasound [46, 204], a novel method that shows potential for the extraction of nutraceuticals from solid plant matrices [228].

The above studies have been focused on a variety of targets, including the manufacture of extracted fractions with antioxidant activity [7, 11, 46, 202, 206, 208, 216–218], procyanidins and anthocyanins [207, 219], catechins and proanthocyanidins [213], anthraquinones [204, 215], quercetin glycosides [214], and oils [205, 209–212, 221].

In some of these studies, the antioxidant activity of the extracted products or fractions has been assessed. Table 6.2.7 summarizes representative data reported in this field.

6.2.3 References

1. Chiu, K. L., Y. C. Cheng, J. H.Chen, C. J. Chang, and P. W. Yang. 2002. Supercritical fluids extraction of Ginkgo ginkgolides and flavonoids. *Journal of Supercritical Fluids* 24:77–87.
2. Reverchon, E. and I. De Marco. 2006. Supercritical fluid extraction and fractionation of natural matter. *Journal of Supercritical Fluids* 38:146–166.
3. Antunes, O. A. C., D. A. G. Aranda, and L. Cardozo Filho. 2007. Process for extraction of carotenoids. World Patent WO2007045057.
4. del Valle, J. M. and J. M. Aguilera. 1999. Extracción con CO_2 a alta presión: Fundamentos y aplicaciones en la industria de alimentos. *Food Science and Technology International* 5:1–24.
5. Mukhopadhyay, M. 2000. *Natural extracts using supercritical carbon dioxide*. Boca Raton, FL: CRC Press.
6. Reglero, G., F. J. Señoráns and E. Ibáñez. 2005. Supercritical fluid extraction: An alternative to isolating natural food preservatives. *Food Science and Technology* 141:539–553.
7. Herrero, M., L. Jaime, P. J. Martín-Álvarez, A. Cifuentes, and E. Ibáñez. 2006. Optimization of the extraction of antioxidants from *Dunaliella salina* microalga by pressurized liquids. *Journal of Agricultural and Food Chemistry* 54:5597–5603.
8. Díaz-Reinoso, B., A. Moure, H. Domínguez, and J. C. Parajó. 2006. Supercritical CO_2 extraction and purification of compounds with antioxidant activity. *Journal of Agricultural and Food Chemistry* 54:2441–2469.
9. Moraes, M. L. L., J. H. Y. Vilegas, and F. M. Lanças. 1997. Supercritical fluid extraction of glycosilated flavonoids from *Pasiflora* leaves. *Phytochemical Analysis* 8:257–260.
10. Esquível, M. M., M. G. Bernardo-Gil, and M. B. King. 1999. Mathematical models for supercritical extraction of olive husk oil. *Journal of Supercritical Fluids* 16:43–58.
11. Ollanketo, M., A. Peltoketo, K. Hartonen, R. Hiltunen, and M. L. Riekkola. 2002. Extraction of sage (*Salvia officinalis* L.) by pressurized hot water and conventional methods: Antioxidant activity of the extracts. *European Food Research and Technology* 215:158–163.
12. del Valle, J. M., T. Rogalinski, C. Zetzl, and G. Brunner. 2005. Extraction of boldo (*Peumus boldus* M.) leaves with supercritical CO_2 and hot pressurized water. *Food Research International* 38:203–213.
13. Vági, E, E. Rapavi, M. Hadolin, et al. 2005. Phenolic and triterpenoid antioxidants from *Origanum majorana* L. herb and extracts obtained with different solvents. *Journal of Agricultural and Food Chemistry* 53:17–21.
14. Hadolin, M., M. Skerget, Z. Knez, and D. Bauman. 2004. Isolation and concentration of natural antioxidants with high-pressure extraction. *Innovative Food Science and Emerging Technologies* 5:245–248.
15. Carvalho, R. N., L. S. Moura, P. T. V. Rosa, and M. A. A. Meireles. 2005. Supercritical fluid extraction from rosemary (*Rosmarinus officinalis*): Kinetic data, extract's global yield, composition, and antioxidant activity. *Journal of Supercritical Fluids* 35:197–204.

16. Cavero, S., L. Jaime, P. J. Martín-Álvarez, F. J., Señoráns, G. Reglero, and E. Ibáñez. 2005. *In vitro* antioxidant analysis of supercritical fluid extracts from rosemary (*Rosmarinus officinalis* L). *European Food Research and Technology* 221:478–486.
17. Pereira, C. G., and M. A. A. Meireles. 2007. Evaluation of global yield, composition, antioxidant activity and cost of manufacturing of extracts from lemon verbena (*Aloysia triphylla* [L'hérit.] britton) and mango (*Mangifera indica* L.) leaves. *Journal of Food Process Engineering* 30:150–173.
18. Fadel, H., F. Marx, A. El-Sawy, and A. H. El-Ghorab. 1999. Effect of extraction techniques on the chemical composition and antioxidant activity of *Eucalyptus camaldulensis* var. *brevirostris* leaf oils. *Zeitschrift fur Lebensmittel-Untersuchung und -Forschung* 208:212–216.
19. El-Ghorab A. H., K. F. El-Massry, F. Marx, and H. M. Fadel. 2003. Antioxidant activity of Egyptian *Eucalyptus camaldulensis var. brevirostris* leaf extracts. *Nahrung* 47:41–45.
20. Yang, C., Y. R. Xu, and W. X. Yao. 2002. Extraction of pharmaceutical components from *Ginkgo biloba* leaves using supercritical carbon dioxide. *Journal of Agricultural and Food Chemistry* 50:846–849.
21. Mau, J. L., P. T. Ko, and C. C. Chyau. 2003. Aroma characterization and antioxidant activity of supercritical carbon dioxide extracts from *Terminalia catappa* leaves. *Food Research International* 36:97–104.
22. Dapkevicius, A., R. Venskutonis, T. A. Van Beek, and J. P. H. Linssen. 1998. Antioxidant activity of extracts obtained by different isolation procedures from some aromatic herbs grown in Lithuania. *Journal of Agricultural and Food Chemistry* 77:140–146.
23. Daukšas, E., P. R. Venskutonis, V. Povilaityte, and B. Sivik. 2001. Rapid screening of antioxidant activity of sage (*Salvia officinalis* L.) extracts obtained by supercritical carbon dioxide at different extraction conditions. *Nahrung* 45:338–341.
24. Simándi, B., V. Hajdu, K. Peredi, B. Czkur, A. Nobik-Kovacas, and A. Kery. 2001. Antioxidant activity of pilot-plant alcoholic and supercritical carbon dioxide extracts of thyme. *European Journal of Lipid Science and Technology* 103:355–358.
25. Marongiu, B., S. Porcedda, A. Piras, A. Rosa, M. Deiana, and M. A. Dessi. 2004. Antioxidant activity of supercritical extract of *Melissa officinalis* subsp. *officinalis* and *Melissa officinalis* subsp. *inodora*. *Phytotherapy Research* 18:789–792.
26. Scalia, S., L. Giuffreda, and P. Pallado. 1999. Analytical and preparative supercritical fluid extraction of chamomile flowers and its comparison with conventional methods. *Journal of Pharmaceutical and Biomedical Analysis* 21:549–558.
27. Poli, F., M. Muzzoli, G. Sacchetti, G. Tassinato, R. Lazzarin, and A. Bruni. 2003. Antioxidant activity of supercritical CO_2 extracts of *Helichrysum italicum*. *Pharmaceutical Biology* 41:379–383.
28. Stashenko, E. E., B. E. Jaramillo, and J. R. Martínez. 2004. Comparison of different extraction methods for the analysis of volatile secondary metabolites of *Lippia alba* (Mill.) N. E. Brown, grown in Colombia, and evaluation of its *in vitro* antioxidant activity. *Journal of Chromatography* 1025:93–103.
29. Hu, Q., Y. Hu, and J. Xu. 2005. Free radical-scavenging activity of *Aloe vera* (*Aloe barbadensis* Miller) extracts by supercritical carbon dioxide extraction. *Food Chemistry* 91:85–90.
30. Wang, H. C., C. R. Chen, and C. J. Chang. 2001. Carbon dioxide extraction of ginseng root hair oil and ginsenosides. *Food Chemistry* 72:505–509.
31. Zancan, K. C., M. O. M. Marques, A. J. Petenate, and M. A. A. Meireles. 2002. Extraction of ginger (*Zingiber officinale* Roscoe) oleoresin with CO_2 and co-solvents: A study of the antioxidant action of the extracts. *Journal of Supercritical Fluids* 24:57–76.
32. Palma, M., L. T. Taylor, R. M. Varela, S. J. Cutler, and H. G. Cutler. 1999. Fractional extraction of compounds from grape seeds by supercritical fluid extraction and analysis

for antimicrobial and agrochemical activities. *Journal of Agricultural and Food Chemistry* 47:5044–5048.
33. Yepez, B., M. Espinosa, S. López, and G. Bolaños. 2002. Producing antioxidant fractions from herbaceous matrices by supercritical fluid extraction. *Fluid Phase Equilibria* 194:879–884.
34. Machmudah, S., Y. Shiramizu, M. Goto, M. Sasaki, and T. Hirose. 2005. Extraction of *Nigella sativa L.* using supercritical CO_2: A study of antioxidant activity of the extract. *Separation Science and Technology* 40:1267–1275.
35. Xu, J., S. Chen, and Q. Hu. 2005. Antioxidant activity of brown pigment and extracts from black sesame seed (*Sesamum indicum* L.). *Food Chemistry* 91:79–83.
36. Tipsrisukond, N., L. N. Fernando, and A. D. Clarke. 1998. Antioxidant effects of essential oil and oleoresin of black pepper from supercritical carbon dioxide extractions in ground pork. *Journal of Agricultural and Food Chemistry* 46:4329–4333.
37. Hadolin, M., A. Rižner Hraš, D. Bauman, and Z. Knez. 2001. High pressure extraction of vitamin E-rich oil from *Silybum marianum*. *Food Chemistry* 74:355–364.
38. Vasil'eva, T. A., A. A. Zaporozhskii, O. I. Kvasenkov, and O. V. Kuznetsova. 2003. Food antioxidant from *Mortierella pulchella*. Russian Patent, RU2196489.
39. Canela, A. P. R. F., P. T. V. Rosa, M. O. M. Marques, and M. A. A. Meireles. 2002. Supercritical fluid extraction of fatty acids and carotenoids from the microalgae *Spirulina maxima*. *Industrial & Engineering Chemistry Research* 41:3012–3018.
40. Gouveia, L., B. P. Nobre, F. M. Marcelo, et al. 2007. Functional food oil coloured by pigments extracts from microalgae with supercritical CO_2. *Food Chemistry* 101:717–723.
41. Mendiola, J. A., L. Jaime, S. Santoyo, et al. 2007. Screening of functional compounds in supercritical fluid extracts from *Spirulina platensis*. *Food Chemistry* 102:1357–1367.
42. López, M., L. Arce, J. Garrido, A. Ríos, and M. Valcárcel. 2004. Selective extraction of astaxanthin from crustaceans by use of supercritical carbon dioxide. *Talanta* 64:726–731.
43. Louli, V., N. Ragoussis, and K. Magoulas. 2004. Recovery of phenolic antioxidants from wine industry by-products. *Bioresource Technology* 92:201–208.
44. Tsuda, T., K. Mizuno, K. Ohshima, S. Kawakishi, and T. Osawa. 1995. Supercritical carbon dioxide extraction of antioxidative components from tamarind (*Tamarindus indica L.*) seed coat. *Journal of Agricultural and Food Chemistry* 43:2803–2806.
45. Luengthanaphol, S., D. Mongkholkhajornsilp, S. Douglas, P. L. Douglas, L. I. Penqsopa, and S. Ponqamphai. 2004. Extraction of antioxidants from sweet Thai tamarind seed coat—preliminary experiments. *Journal of Food Engineering* 63:247–252.
46. Goli, A.H., M. Barzegar, and M. A. Sahari. 2005. Antioxidant activity and total phenolic compounds of pistachio (*Pistachia vera*) hull extracts. *Food Chemistry* 92:521–525.
47. Arlorio, M., J. D. Coïsson, F. Travaglia, et al. 2005. Antioxidant and biological activity of phenolic pigments from *Theobroma cacao* hulls extracted with supercritical CO_2. *Food Research International* 38:1009–1014.
48. Francisco, J. C., B. Danielsson, A. Kozubek, and E. S. Dey. 2005. Extraction of rye bran by supercritical carbon dioxide: Influence of temperature, CO_2, and cosolvent flow rates. *Journal of Agricultural and Food Chemistry* 53:7432–7437.
49. Francisco, J. C., B. Danielsson, A. Kozubek, and E. Szwajcer. 2005. Application of supercritical carbon dioxide for the extraction of alkylresorcinols from rye bran. *Journal of Supercritical Fluids* 35:220–226.
50. Okuno, S., M. Yoshinaga, and M. Nakatani. 2002. Extraction of antioxidants in sweetpotato waste powder with supercritical carbon dioxide. *Food Science and Technology Research* 8:154–157.
51. Gómez, M. S., M. L. Ruiz, J. G. Santa-María, G. P. Blanch, and M. Herraiz. 2003. Supercritical-fluid extraction and fractionation of natural-source carotenoids with a high lycopene content. World Patent, WO2003103645.

52. Ruiz, M. L., M. S. Gómez-Prieto, M. Herraiz, and G. Santa-María. 2003. Lipid composition in tomato skin supercritical fluid extracts with high lycopene content. *Journal of the American Oil Chemists' Society* 80:271–274.
53. Topal, U., M. Sasaki, M. Goto, and K. Hayakawa. 2006. Extraction of lycopene from tomato skin with supercritical carbon dioxide: Effect of operating conditions and solubility analysis. *Journal of Agricultural and Food Chemistry* 54:5604–5610.
54. Vági, E., B. Simándi, K. P. Vásárhelyiné, H. Daood, Á. Kéry, F. Doleschall, and B. Nagy. 2007. Supercritical carbon dioxide extraction of carotenoids, tocopherols and sitosterols from industrial tomato by-products. *Journal of Supercritical Fluids* 40:218–226.
55. De Magalhães, N. P. M. L., J. L. Cardador, A. A. Figueiredo, et al. 2007. Method of obtaining a natural hydroxytyrosol-rich concentrate from olive tree residues and subproducts using clean technologies. World Patent, WO013032.
56. Birtigh, A., M. Johannsen, G. Brunner, and N. Nair. 1995. Supercritical-fluid extraction of oil-palm components. *Journal of Supercritical Fluids* 8:46–50.
57. Gamlieli-Bonshtein, I., E. Korin, and S. Cohen. 2002. Selective separation of cis-trans geometrical isomers of β-carotene via CO_2 supercritical fluid extraction. *Biotechnology & Bioengineering* 80:169–174.
58. Mendes, R. L., B. P. Nobre, M. T. Cardoso, A. P. Pereira, and A. F. Palavra. 2003. Supercritical carbon dioxide extraction of compounds with pharmaceutical importance from microalgae. *Inorganica Chimica Acta* 356:328–334.
59. Sabio, E., M. Lozano, V. Montero de Espinosa, et al. 2003. Lycopene and β-carotene extraction from tomato processing waste using supercritical CO_2. *Industrial & Engineering Chemistry Research* 42:6641–6646.
60. Şanal, Í. S., A. Güvenç, U. Salgın, Ü. Mehmetoğlu, and A. Çalımlı. 2004. Recycling of apricot pomace by supercritical CO_2 extraction. *Journal of Supercritical Fluids* 32:221–230.
61. Sun, M. and F. Temelli. 2006. Supercritical carbon dioxide extraction of carotenoids from carrot using canola oil as a continuous co-solvent. *Journal of Supercritical Fluids* 37:397–408.
62. Ko, T. F., Y. M. Weng, and R. Y. Y. Chiou. 2002. Squalene content and antioxidant activity of *Terminalia catappa* leaves and seeds. *Journal of Agricultural and Food Chemistry* 50:5343–534.
63. Macías-Sánchez, M. D., C. Mantell, M. Rodríguez, E. Martínez de la Ossa, L. M. Lubián, and O. Montero. 2005. Supercritical fluid extraction of carotenoids and chlorophylla from *Nannochloropsis gaditana. Journal of Food Engineering* 66:245–251.
64. Kitzberger, C. S. G., Jr., A. Smânia, R. C. Pedrosa, and S. R. S. Ferreira. 2007. Antioxidant and antimicrobial activities of shiitake (*Lentinula edodes*) extracts obtained by organic solvents and supercritical fluids. *Journal of Food Engineering* 80:631–638.
65. Mohri, A., K. Morikawa, T. Matsuya, and S. Onaka. 1993. Materia colorante roja y método de elaborarla. Spanish Patent, ES 2038542.
66. López-Sebastian, S., E. Ramos, E. Ibáñez, et al. 1998. Dearomatization of antioxidant rosemary extracts by treatment with supercritical carbon dioxide. *Journal of Agricultural and Food Chemistry* 46:13–19.
67. Murga, R., R. Ruiz, S. Beltrán, and J. L. Cabezas. 2000. Extraction of natural complex phenols and tannins from grape seeds by using supercritical mixtures of carbon dioxide and alcohol. *Journal of Agricultural and Food Chemistry* 48:3408–3412.
68. Grigonis, D., P. R. Venskutonis, B. Sivik, M. Sandahal, and C. S. Eskilsson. 2005. Comparison of different extraction techniques for isolation of antioxidants from sweet grass (*Hierochloë odorata*). *Journal of Supercritical Fluids* 33:223–233.
69. França, L. F., G. Reber, M. A. A. Meireles, N. T. Machado, and G. Brunner. 1999. Supercritical extraction of carotenoids and lipids from buriti (*Mauritia flexuosa*), a fruit from the Amazon region. *Journal of Supercritical Fluids* 14:247–256.

70. Señoráns, F. J., E. Ibáñez, S. Cavero, J. Tabera, and G. Reglero. 2000. Liquid chromatographic-mass spectrometric analysis of supercritical-fluid extracts of rosemary plants. *Journal of Chromatography* 870:491–499.
71. Leal, P. F., F. C. M. Chaves, L. C. Ming, A. J. Petenate, and M. A. A. Meireles. 2006. Global yields, chemical compositions and antioxidant activities of clove basil (*Ocimum gratissimum* L.) extracts obtained by supercritical fluid extraction. *Journal of Food Process Engineering* 29:547–559.
72. Leal, P. F., M. E. M. Braga, D. N. Sato, J. E. Carvalho, M. O. M. Marqués, and M. A. A. Meireles. 2003. Functional properties of spice extracts obtained via supercritical fluid extraction. *Journal of Agricultural and Food Chemistry* 51:2520–2525.
73. Povh, N. P., M. O. M. Marques, and M. A. A. Meireles. 2001. Supercritical CO_2 extraction of essential oil and oleoresin from chamomile (*Chamomilla recutita* [L.] Rauschert). *Journal of Supercritical Fluids* 21:245–256.
74. Balachandran, S., S. E. Kentish, and R. Mawson. 2006. The effects of both preparation method and season on the supercritical extraction of ginger. *Separation and Purification Technology* 48:94–105.
75. Sovová, H., M. Sajfrtová, M. Bártlová, and L. Opletal. 2004. Near-critical extraction of pigments and oleoresin from stinging nettle leaves. *Journal of Supercritical Fluids* 30:213–224.
76. Sovová, H., L. Opletal, M. Bártlová, M. Sajfrtová, and M. Krenková. 2007. Supercritical fluid extraction of lignans and cinnamic acid from *Schisandra chinensis*. *Journal of Supercritical Fluids* 42:88–95.
77. Cavero, S., M. R. García-Risco, F. R. Marín, et al. 2006. Supercritical fluid extraction of antioxidant compounds from oregano: Chemical and functional characterization via LC–MS and in vitro assays. *Journal of Supercritical Fluids* 38:62–69.
78. Lee, Y. N., C. R. Chen, H. L. Yang, C. C. Lin, and C. M. J. Chang. 2007. Isolation and purification of 3,5-diprenyl-4-hydroxycinnamic acid (artepillin C) in Brazilian propolis by supercritical fluid extraction. *Separation and Purification Technology* 54:130–138.
79. Uquiche, E., J. M. del Valle, and J. Ortiz. 2004. Supercritical carbon dioxide extraction of red pepper (*Capsicum annuun* L.) oleoresin. *Journal of Food Engineering* 65:55–66.
80. Reglero, G., J. Tabera, E. Ibáñez, S. López-Sebastián, E. Ramos, and L. Ballester. 1999. Proceso de extracción con fluidos supercríticos para la producción de antioxidantes naturales y antioxidantes obtenidos mediante dicho proceso. Spanish Patent, ES2128996.
81. Ibáñez, E., A. Oca, G. De Murga, S. López-Sebastián, J. Tabera, and G. Reglero. 1999. Supercritical fluid extraction and fractionation of different preprocessed rosemary plants. *Journal of Agricultural and Food Chemistry* 47:1400–1404.
82. Leeke, G., F. Gaspar, and R. Santos. 2002. Influence of water on the extraction of essential oils from a model herb using supercritical carbon dioxide. *Industrial & Engineering Chemistry Research* 41:2033–2039.
83. Ambrogi, A., D. A. Cardarelli, and R. Eggers. 2002. Fractional extraction of paprika using supercritical carbon dioxide and on-line determination of carotenoids. *Journal of Food Science* 67:3236–3241.
84. Vasapollo, G., L. Longo, L. Rescio, and L. Ciurlia. 2004. Innovative supercritical CO_2 extraction of lycopene from tomato in the presence of vegetable oil as co-solvent. *Journal of Supercritical Fluids* 29:87–96.
85. Lim, G. B., S. Y. Lee, E. K. Lee, S. J. Haam, and W. S. Kim. 2002. Separation of astaxanthin from red yeast *Phaffia rhodozyma* by supercritical carbon dioxide extraction. *Biochemical Engineering Journal* 11:181–187.
86. Spanos, G. A., H. Chen, and S. J. Schwartz. 1993. Supercritical carbon dioxide extraction of β-carotene from sweet potatoes. *Journal of Food Science* 58:817–820.

87. Abascal, K., L. Ganora, and E. Yarnell. 2005. The effect of freeze-drying and its implications for botanical medicine: A review. *Phytotherapy Research* 19:655–660.
88. Señoráns, F. J., A. Ruiz-Rodríguez, S. Cavero, A. Cifuentes, E. Ibáñez, and G. Reglero. 2001. Isolation of antioxidant compounds from orange juice by using countercurrent supercritical fluid extraction (CC-SFE). *Journal of Agricultural and Food Chemistry* 49:6039–6044.
89. Ibáñez, E., A. M. Hurtado, F. J. Señoráns, and G. Reglero. 2002. Concentration of sterols and tocopherols from olive oil with supercritical carbon dioxide. *Journal of the American Oil Chemists' Society* 79:1255–1260.
90. Gast, K., M. Jungfer, C. Saure, and G. Brunner. 2005. Purification of tocochromanols from edible oil. *Journal of Supercritical Fluids* 34:17–25.
91. Chuang, M. H. and G. Brunner. 2006. Concentration of minor components in crude palm oil. *Journal of Supercritical Fluids* 37:151–156.
92. Fang, T., M. Goto, and X. Wang. 2007. Separation of natural tocopherols from soybean oil byproduct with supercritical carbon dioxide. *Journal of Supercritical Fluids* 40:50–58.
93. Brunner, G., T. Malchow, K. Stürken, and T. Gottschau. 1991. Separation of tocopherols from deodorizer condensates by countercurrent extraction with carbon dioxide. *Journal of Supercritical Fluids* 4:72–80.
94. Bondioli, P., C. Mariano, A. Lanzani, E. Fedeli, and A. Muller. 1993. Squalene recovery from olive oil deodorizer distillates. *Journal of the American Oil Chemists' Society* 70:763–766.
95. Nagesha, G. K., B. Manohar, and K. Udaya Sankar. 2003. Enrichment of tocopherols in modified soy deodorizer distillate using supercritical carbon dioxide extraction. *European Food Research and Technology* 217:427–433.
96. Vázquez, L., C. F. Torres, T. Fornari, F. J. Señoráns, and G. Reglero. 2007. Recovery of squalene from vegetable oil sources using countercurrent supercritical carbon dioxide extraction. *Journal of Supercritical Fluids* 40:59–66.
97. Tabera, J., A. Guinda, A. Ruiz-Rodríguez, et al. 2004. Countercurrent supercritical fluid extraction and fractionation of high-added-value compounds from a hexane extract of olive leaves. *Journal of Agricultural and Food Chemistry* 52:4774–4779.
98. Persson, P., S. Larsson, L. J. Jönsson, et al. 2002. Supercritical fluid extraction of a lignocellulosic hydrolysate of spruce for detoxification and to facilitate analysis of inhibitors. *Biotechnology & Bioengineering* 7:694–700.
99. Chang, C. M., Y. F. Chang, H. Z. Lee, J. Q. Lin, and P. W. Yang. 2000. Supercritical carbon dioxide extraction of high-value substances from soybean oil deodorizer distillate. *Industrial & Engineering Chemistry Research* 39:4521–4525.
100. Tan, Y. A., R. Sambanthamurthi, K. Sundram, and M. B. Wahid. 2007. Valorisation of palm by-products as functional components. *European Journal of Lipid Science and Technology* 109:380–393.
101. Reverchon, E. 1997. Supercritical fluid extraction and fractionation of essential oils and related products. *Journal of Supercritical Fluids* 10:1–37.
102. Brunner, G. 1998. Industrial process development: Countercurrent multistage gas extraction SFE processes. *Journal of Supercritical Fluids* 13:283–301.
103. Gamse, T. 2004. Supercritical Fluid extraction and fractionation of liquids. In *State of the art book on Supercritical Fluids,* ed. AINIA, 179–191. Valencia, Spain: AINIA.
104. Catchpole, O. J., J. B. Grey, K. A. Mitchell, and J. S. Lan. 2004. Supercritical antisolvent fractionation of propolis tincture. *Journal of Supercritical Fluids* 29:97–106.
105. Tabera, J., A. Ruíz, F. J. Señoráns, et al. 2005. Procedimiento para obtener compuestos de alto valor añadido a partir de hoja de olivo. Spanish Patent, ES 2 238 183.
106. Ooi, C. K., A. Bashkar, M. S. Yener, D. Q. Tuan, J. Hsu, and S. S. H. Rizvi. 1996. Continuous supercritical carbon dioxide processing of palm oil. *Journal of the American Oil Chemists' Society* 73:233–237.

107. Díaz-Reinoso, B., A. Moure, H. Domínguez, and J. C. Parajó. 2008. Antioxidant extraction by supercritical fluids. In *Supercritical fluid extraction of nutraceuticals and bioactive compounds*, ed. J. Martínez. Boca Raton: CRC Press.
108. Fornari, T., A. Chafer, R. P. Stateva, and G. Reglero. 2005. A new development in the application of the group contribution associating equation of state to model solid solubilities of phenolic compounds in SC-CO_2. *Industrial & Engineering Chemistry Research* 44:8147–8156.
109. Guçlü-Üstündağ, Ö, and F. Temelli. 2004. Correlating the solubility behavior of minor lipid components in supercritical carbon dioxide. *Journal of Supercritical Fluids* 31:235–253.
110. Su, C. S., and Y. P. Chen. 2007. Correlation for the solubilities of pharmaceutical compounds in supercritical carbon dioxide. *Fluid Phase Equilibria* 254:167–173.
111. Gouveia, A. F., C. Duarte, M. L. B. Costa, M. G. Bernardo-Gil, and M. Moldão-Martins. 2006. Oxidative stability of olive oil flavoured by *Capsicum frutescens* supercritical fluid extracts. *European Journal of Lipid Science and Technology* 108:421–428.
112. Smith, R. L., Jr., R. M. Malaluan, W. B. Setianto, et al. 2003. Separation of cashew (*Anacardium occidentale* L.) nut shell liquid with supercritical carbon dioxide. *Bioresource Technology* 88:1–7.
113. Ibáñez, E., J. Palacios, F. J. Señoráns, G. Santa-María, J. Tabera, and G. Reglero. 2000. Isolation and separation of tocopherols from olive by-products with supercritical fluids. *Journal of the American Oil Chemists' Society* 77:187–190.
114. Ge, Y., Y. Ni, H. Yan, Y. Chen, and T. Cai. 2002. Optimization of the supercritical fluid extraction of natural vitamin E from wheat germ using response surface methodology. *Journal of Food Science* 67:239–243.
115. Brunner, G. 2005. Supercritical fluids: Technology and application to food processing. *Journal of Food Engineering* 67:21–33.
116. Nguyen, U., G. Frakman, and D. A. Evans. 1991. Process for extracting antioxidants from *Labiatae* herbs. US Patent, US 5017397.
117. Sabio, E., A. Ramiro, J. F. González, et al. 2002. Procedimiento para la producción de un concentrado de licopeno libre de disolventes orgánicos, concentrado obtenido y composición que comprende dicho concentrado. Spanish Patent ES 2 172 442.
118. Jarén-Galán, M., U. Nienaber, and S. J. Schwartz. 1999. Paprika (*Capsicum annuum*) oleoresin extraction with supercritical carbon dioxide. *Journal of Agricultural and Food Chemistry* 47:3558–3564.
119. del Valle, J. M., C. Godoy, M. Asencio, and J. M. Aguilera. 2004. Recovery of antioxidants from boldo (*Peumus boldus* M.) by conventional and supercritical CO_2 extraction. *Food Research International* 37:695–702.
120. Hu, Q., B. Pan, J. Xu, J. Sheng, and Y. Shi. 2007. Effects of supercritical carbon dioxide extraction conditions on yields and antioxidant activity of *Chlorella pyrenoidosa* extracts. *Journal of Food Engineering* 80:997–1001.
121. Wu, S. J., J. Y. Tsai, and S. P. Chang, et al. 2006. Supercritical carbon dioxide extract exhibits enhanced antioxidant and anti-inflammatory activities of *Physalis peruvian*. *Journal of Ethnopharmacology* 108:407–413.
122. Wang, B. J., Y. H. Lien, and Z. R. Yu. 2004. Supercritical fluid extractive fractionation—Study of the antioxidant activities of propolis. *Food Chemistry* 86:237–243.
123. Wang, B. J., C.-T. Liu, C. Y. Tseng, Z. R. Yu. 2005. Antioxidant activity of *Bupleurum kaoi* Liu (Chao et Chuang) fractions fractionated by supercritical CO_2. *Lebensmittel-Wissenschatf & Technologie* 38:281–287.
124. Hu, Q., J. Xu, S. Chen, and F. Yang. 2004. Antioxidant activity of extracts of black sesame seed (*Sesaum indicum L.*) by supercritical carbon dioxide extraction. *Journal of Agricultural and Food Chemistry* 52:943–947.

125. Machmudah, S., A. Shotipruk, M. Goto, M. Sasaki, and T. Hirose. 2006. Extraction of astaxanthin from *Haematococcus pluvialis* using supercritical CO_2 and ethanol as entrainer. *Industrial & Engineering Chemistry Research* 45:3652–3657.
126. Baysal, T., S. Ersus, and D. A. J. Starmans. 2000. Supercritical CO_2 extraction of beta-carotene and lycopene from tomato paste waste. *Journal of Agricultural and Food Chemistry* 48:5507–5511.
127. de Lucas, A., E. Martinez de la Ossa, J. Rincón, M. A. Blanco, and I. Gracia. 2002. Supercritical fluid extraction of tocopherol concentrates from olive tree leaves. *Journal of Supercritical Fluids* 22:221–228.
128. Markom, M., H. Singh, and M. Hasan. 2001. Supercritical CO_2 fractionation of crude palm oil. *Journal of Supercritical Fluids* 20:45–53.
129. Vázquez, L., C. F. Torres, T. Fornari, N. Grigelmo, F. J. Señoráns, and G. Reglero. 2006. Supercritical fluid extraction of minor lipids from pretreated sunflower oil deodorizer distillates. *European Journal of Lipid Science and Technology* 108:659–665.
130. Mimica-Dukic, N., and B. Bozin. 2007. Essential oils from *Lamiaceae species* as promising antioxidant and antimicrobial agents. *Natural Products Communications* 2:445–452.
131. Simándi, B., M. Oszaqyan, E. Lemberkovics, et al. 1998. Supercritical carbon dioxide extraction and fractionation of oregano oleoresin. *Food Research International* 31:723–728.
132. Chang, C. J., K. L. Chiu, Y. L. Chen, and C.-Y. Chang. 2000. Separation of catechins from green tea using carbon dioxide extraction. *Food Chemistry* 68:109–113.
133. Ashraf-Khorassani, M., and L. T. Taylor. 2004. Sequential fractionation of grape seeds into oils, polyphenols, and procyanidins via a single system employing CO_2-based fluids. *Journal of Agricultural and Food Chemistry* 52:2440–244.
134. Im, S. H., M. S. Park, and S. Y. Ko. 2006. Method for extracting lycopene from tomato with higher extraction stability and yield for removing reactive oxygen species by using supercritical fluid extraction system with squalene, and anti-oxidative cosmetic composition containing the same. Korean Patent, KR2006070846.
135. Huang, M., S. Zhang, and S. Zhang. 2004. Tea polyphenols extracting process. Chinese Patent, CN1482126.
136. Mingsheng, C. 2000. Anti-oxidation process for natural bee glue. Chinese Patent, CN1274535.
137. Gerard, D., K. W. Quirin, and E. Schwarz. 1995. CO_2-extracts from rosemary and sage—Effective natural antioxidants. *Food Marketing Technology* 9:45–55.
138. Kyung, W. M., and L. S. Bin. 2004. Method for extracting β-cryptoxanthin from orange peel through one-step process using supercritical carbon dioxide. Korean Patent KR20040107902.
139. Wenbin, M., P. Baoling, and Q. Qianchua. 2003. Method of extracting rosemary as natural antioxidant. Chinese Patent, CN1281709C.
140. Ribeiro, M. A., M. G. Bernardo-Gil, and M. M. Esquível. 2001. *Melissa officinalis* L.: Study of antioxidant activity in supercritical residues. *Journal of Supercritical Fluids* 21:51–60.
141. Nakatsu, T. and A. Yamasaki. 2000. Water-soluble anti-oxidation agents. US Patent, US6123945.
142. Quitain, A. T., S. Katoh, and T. Moriyoshi. 2004. Isolation of antimicrobials and antioxidants from moso-bamboo (*Phyllostachys heterocycla*) by supercritical CO_2 extraction and subsequent hydrothermal treatment of the residues. *Industrial & Engineering Chemistry Research* 43:1056–1060.
143. Lucien, F., and N. R. Foster. 2000. Solubilities of solid mixtures in supercritical carbon dioxide: A review. *Journal of Supercritical Fluids* 17:111–134.
144. Araújo, J. M. A., M. V. Silva, and J. B. P. Chaves. 2007. Supercritical fluid extraction of daidzein and genistein isoflavones from soybean hypocotyl after hydrolysis with endogenous β-glucosidases. *Food Chemistry* 105:226–272.

145. Cadoni, E., M. Rita de Giorgi, E. Medda, and G. Poma. 1999. Supercritical CO_2 extraction of lycopene and β-carotene from ripe tomatoes. *Dyes and Pigments* 44:27–32.
146. Chandra, A., and M. Nair. 1996. Supercritical carbon dioxide extraction of daidzein and genistein from soybean products. *Phytochemical Analysis* 7:259–262.
147. Wood, J. A., M. A. Bernards, W. Wan, and P. A. Charpentier. 2006. Extraction of ginsenosides from North American ginseng using modified supercritical carbon dioxide. *Journal of Supercritical Fluids* 39:40–47.
148. Santoyo, S., R. Lloría, L. Jaime, E. Ibáñez, F. J. Señoráns, and G. Reglero. 2006. Supercritical fluid extraction of antioxidant and antimicrobial compounds from *Laurus nobilis* L. Chemical and functional characterization. *European Food Research and Technology* 222:565–571.
149. Martino, K. G., and D. Guyer. 2004. Supercritical fluid extraction of queracetin from onion skins. *Journal of Food Process Engineering* 27:17–28.
150. Quitain, A. T., K. Oro, S. Katoh, and T. Moriyoshi. 2006. Recovery of oil components of okara by ethanol-modified supercritical carbon dioxide extraction. *Bioresource Technology* 97:1509–1514.
151. Valderrama, J. O., M. Perrut, and W. Majewski. 2003. Extraction of astaxantine and phycocyanine from microalgae with supercritical carbon dioxide. *Journal of Chemical Engineering Data* 48:827–830.
152. Mantell, C., M. Rodríguez, and E. Martínez de la Ossa. 2003. A screening analysis of the high-pressure extraction of anthocyanins from red grape pomace with carbon dioxide and cosolvent. *Chemical Engineering Science* 3:38–42.
153. Rostagno, M. A., J. M. A. Araújo, and D. Sandi. 2002. Supercritical fluid extraction of isoflavones from soybean flour. *Food Chemistry* 78:111–117.
154. Byun, S. Y. 2004. Sesame oil having high content of sesamol and production thereof using sueprcritical carbon dioxide and ethanol. Korean Patent, KR20040074188.
155. Catchpole, O. J., N. B. Perry, B. M. T. da Silva, J. B. Grey, and B. M. Smallfield. 2002. Supercritical extraction of herbs I: Saw palmetto, St. John's wort, kava root, and echinacea. *Journal of Supercritical Fluids* 22:129–138.
156. Gorostiaga, K., and D. Guyer. 2004. Supercritical fluid extraction of quercetin from onion skins. *Journal of Food Process Engineering* 27:17–28.
157. Manabe, A., T. Yamashita, K. Harada, et al. 1989. Process for the supercritical extraction and separation of solid samples. US. Patent, US5178735.
158. Garrote, G., J. M. Cruz, A. Moure, H. Domínguez, and J. C. Parajó. 2004. Antioxidant activity of byproducts from the hydrolytic processing of selected lignocellulosic materials. *Trends in Food Science & Technology* 15:191–200.
159. Domínguez, H., J. L. Torres, and M. J. Núñez. 2001. Antioxidant phenolics as food additives from agricultural wastes. *Polyphenols Actualites* 21:26–30.
160. Myerly, R. C., M. D. Nicholson, R. Katzen, and J. M. Taylor. 1981. The forest refinery. *Chemical Technology* 11:186–92.
161. Garrote, G., H. Domínguez, and J. C. Parajó. 1999. Hydrothermal processing of lignocellulosic materials. *Holz als Roh- und Werkstoff* 57:191–202.
162. Garrote, G., H. Domínguez, and J. C. Parajó. 1999. Mild autohydrolysis: An environmentally friendly technology for xylooligosaccharide production from wood. *Journal of Chemical Technology & Biotechnology* 74:1101–1109.
163. Ando, S., T. Sakaki, T. Kokusho, M. Shibata, Y. Uemura, and Y. Hatate. 2000. Decomposition behaviour of plant biomass in hot compressed water. *Industrial & Engineering Chemistry Research* 39:3688–3693.
164. Clark, T. A., and K. L. Mackie. 1984. Fermentation inhibitors in wood hydrolysates derived from the softwood *Pinus radiata*. *Journal of Chemical Technology & Biotechnology* 34b:101–110.
165. Saha, B. C. 2003. Hemicellulose conversion. *Journal of Industrial Microbiology and Biotechnology* 30:279–291.

166. Ebringerová, A., and T. Heinze. 2000. Xylan and xylan derivatives—Biopolymers with valuable properties. 1. Naturally occurring xylans structures, isolation procedures and properties. *Macromolecular Rapid Comm*unications 21:542–556.
167. Moure, A., P. Gullón, H. Domínguez, and J. C. Parajó. 2006. Advances in the manufacture, purification and applications of xylo-oligosaccharides as food additives and nutraceuticals. *Process Biochemistry* 41:1913–1923.
168. Cruz, J. M., H. Domínguez, and J. C. Parajó. 2004. Assessment of the production of antioxidants from winemaking waste solids. *Journal of Agricultural and Food Chemistry* 52:5612–5620.
169. Ballesteros, M., M. J. Negro, P. Manzanares, I. Ballesteros, and J. M. Oliva. 2004. Recovery of phenolic compounds from a residual plant material by using a hydrothermal process. World Patent, WO2004009206.
170. Parajó, J. C., G. Garrote, J. M. Cruz, and H. Domínguez. 2004. Production of xylooligosaccharides by autohydrolysis of lignocellulosic materials. *Trends in Food Science & Technology* 15:115–120.
171. Vázquez, M. J., J. L. Alonso, H. Domínguez, and J. C. Parajó. 2001. Xylooligosaccharides. Manufacture and applications. *Trends in Food Science & Technology* 11:387–393.
172. Kabel, M. A., L. Kortenoeven, H. A. Schols, and A. G. J. Voragen. 2002. In vitro fermentability of differently substituted xylo-oligosaccharides. *Journal of Agricultural and Food Chemistry* 50:6205–621.
173. Vázquez, M. J., J. L. Alonso, H. Domínguez, and J. C. Parajó. 2006. Enhancing the potential of oligosaccharides from corncob autohydrolysis as prebiotic food ingredients. *Industrial Crops Products* 24:152–159.
174. Garrote, G., J. M. Cruz, H. Domínguez, and J. C. Parajó. 2003. Valorisation of waste fractions from autohydrolysis of selected lignocellulosic materials. *Journal of Chemical Technology & Biotechnology* 78:392–398.
175. Garrote, G., H. Domínguez, and J. C. Parajó. 2004. Production of substituted oligosaccharides by hydrolytic processing of barley husks. *Industrial & Engineering Chemistry Research* 43:1608–1614.
176. Vázquez, M. J., G. Garrote, J. L. Alonso, H. Domínguez, and J. C. Parajó. 2005. Refining of autohydrolysis liquors for manufacturing xylooligosaccharides: Evaluation of operational strategies. *Bioresource Technology* 96:889–896.
177. Quitain, A. T., N. Sato, H. Daimon, and K. Fujie. 2003. Qualitative investigation on hydrothermal treatment of hinoki (*Chamaecyparis obtusa*) bark for production of useful chemicals. *Journal of Agricultural and Food Chemistry* 51:7926–7929.
178. Garrote, G., E. Falqué, H. Domínguez, and J. C. Parajó. 2007. Autohydrolysis of agricultural residues: Study of reaction byproducts. *Bioresource Technology* 98:1951–1957.
179. Pinelo, M., M. Rubilar, J. Sineiro, and M. J. Núñez. 2004. Extraction of antioxidant phenolics from almond hulls (*Prunus amygdalus*) and pine sawdust (*Pinus pinaster*). *Food Chemistry* 85:267–273.
180. Canas, S., M. C. Leandro, I. Spranger, and P. Belchior. 1999. Low molecular weight organic compounds of chestnut wood (*Castanea sativa* L.) and corresponding aged brandies. *Journal of Agricultural and Food Chemistry* 47:5023–5030.
181. Moure, A., H. Domínguez, and J. C. Parajó. 2005. Antioxidant activity of liquors from aqueous treatments of *Pinus radiata* wood. *Wood Science Technology* 39:129–139.
182. Moure, A., M. Pazos, I. Medina, H. Domínguez, and J. C. Parajó. 2007. Antioxidant activity of extracts produced by solvent extraction of almond shells acid hydrolysates. *Food Chemistry* 101:193–201.
183. Ferrón, R., A. Moure, H. Domínguez, and J. C. Parajó. 2002. Extraction of soluble fiber and antioxidant compounds from chestnut hedgehog extracts. 9th Mediterranean Congress of Chemical Engineering. Barcelona, Spain, 26–29 November.

184. Cruz, J. M., E. Conde, H. Domínguez, and J. C. Parajó. 2007. Thermal stability of antioxidants obtained from wood and industrial wastes. *Food Chemistry* 100:1059–1064.
185. Saura-Calixto, F. 1998. Antioxidant dietary fiber product, a new concept and a potential food ingredient. *Journal of Agricultural and Food Chemistry* 46:4303–4306.
186. Ohta, T., S. Yamasaki, Y. Egashira, and H. Sanada. 1994. Antioxidative activity of corn bran hemicellulose fragments. *Journal of Agricultural and Food Chemistry* 42:653–656.
187. Yoshino, K., N. Higashi, and K. Koga. 2007. Antioxidant activities of acidic xylooligosaccharide. *Numazu Kogyo Koto Senmon Gakko Kenkyu Hokoku* 41:103–105.
188. Wallis A. F. A., and R. H. Wearne. 1985. Fractionation of the polymeric components of hardwoods by autohydrolysis–explosion–extraction. *Appita Journal* 38:432–437.
189. Kubikova, J., A. Zemann, P. Krkoska, and O. Bobleter. 1996. Hydrothermal pretreatment of wheat straw for the production of pulp and paper. *Tappi Journal* 79:163–169.
190. Saddler, J. N., H. H. Brownell, L. P. Clermont, and N. Levitin. 1982. Enzymatic hydrolysis of cellulose and various pretreated wood fractions. *Biotechnology & Bioengineering* 24:1389–1402.
191. Ropars, M., R. Marchal, J. Pourquié, and J. P. Vandecastelee. 1992. Large scale enzymatic hydrolysis of agricultural lignocellulosic biomass. Part 1: Pretreatment procedures. *Bioresource Technology* 42:197–204.
192. Glasser, W. G., and R. S. Wright. 1998. Steam-assisted biomass fractionation. Part II: Fractionation behavior of various biomass resources. *Biomass Bioenergy* 14:219–235.
193. Ibrahim, M. I., and W. G. Glasser. 1999. Steam-assisted biomass fractionation. Part III: A quantitative evaluation of the "clean fractionation" concept. *Bioresource Technology* 70:181–192.
194. Felizón, B., J. Fernández-Bolaños, R. Guillén, and A. Heredia. 2000. Steam-explosion pre-treatment of olive cake. *Journal of the American Oil Chemists' Society* 77:15–22.
195. Heitz, M., E. Capek-Ménard, P. G. Koeberle, et al. 1991. Fractionation of Populus tremuloides at the pilot plant scale: Optimization of steam pretreatment conditions using the STAKE II technology. *Bioresource Technology* 35:23–32.
196. Fernández-Bolaños, J., B. Felizón, A. Heredia, R. Rodríguez, R. Guillén, and A. Jiménez. 2001. Steam-explosion of olive stones: Hemicellulose solubilization and enhancement of enzymatic hydrolysis of cellulose. *Bioresource Technology* 79:53–61.
197. Ando S., I. Arai, K. Kiyoto, and S. Hanai. 1986. Identification of aromatic monomers in steam Pongnaravane exploded poplar and their influences on ethanol fermentation by *Saccharomyces cerevisiae*. *Journal of Fermentation Technology* 64:567–570.
198. Fernández-Bolaños, J., B. Felizón, M. Brenes, R. Guillén, and A. Heredia. 1998. Hydroxytyrosol and tyrosol as the main compounds found in the phenolic fraction of steam-exploded olive stones. *Journal of the American Oil Chemists' Society* 75:1643–1649.
199. De Bari, I., E. Viola, D. Barisano, et al. 2002. Ethanol production at flask and pilot scale from concentrated slurries of steam-exploded aspen. *Industrial & Engineering Chemistry Research* 41:1745–1753.
200. Jönsson, L. J., E. Palmqvist, N. O. Nilvebrant, and B. Hahn-Hägerdal. 1998. Detoxification of wood hydrolyzates with laccase and peroxidase from the white-rot fungus *Trametes versicolor*. *Applied Microbiology and Biotechnology* 49:691–697.
201. Fernández-Bolaños, J., G. Rodríguez, R. Rodríguez, A. Heredia, R. Guillén, and A. Jiménez. 2002. Production in large quantities of highly purified hydroxytyrosol from liquid-solid waste of two-phase olive ol processing or "alperujo." *Journal of Agricultural and Food Chemistry* 50:6804–6811.
202. Stanley, R. A. 2003. High-temperature and pressure extraction of phenolic antioxidants from fruits and vegetables. World Patent, WO 2003042133.

203. Santoyo, S., M. Herrero, F. J. Señoráns, A. Cifuentes, E. Ibáñez, and L. Jaime. 2006. Functional characterization of pressurized liquid extracts of *Spirulina platensis*. *European Food Research and Technology* 224:75–81.
204. Pongnaravane, B., M. Goto, M. Sasaki, T. Anekpankul, P. Pavasant, and A. Shotipruk. 2006. Extraction of anthraquinones from roots of *Morinda citrifolia* by pressurized hot water: Antioxidant activity of extracts. *Journal of Supercritical Fluids* 37:390–396.
205. Eikani, M. H., F. Golmohammad, and S. Rowshanzamir. 2007. Subcritical water extraction of essential oils from coriander seeds (*Coriandrum sativum* L.). *Journal of Food Engineering* 80:735–740.
206. Rodríguez-Meizoso, I., F. R. Marín, M. Herrero, et al. 2006. Subcritical water extraction of nutraceuticals with antioxidant activity from oregano. Chemical and functional characterization. *Journal of Pharmaceutical and Biomedical Analysis* 41:1560–1565.
207. Ju, Z. Y., and L. R. Howard. 2005. Subcritical water and sulfured water extraction of anthocyanins and other phenolics from dried red grape skin. *Journal of Food Science* 70:S270–S276.
208. Ibáñez, E., A. Kubátová, F. J. Señoráns, S. Cavero, G. Reglero, and S. B. Hawthorne. 2003. Subcritical water extraction of antioxidant compounds from rosemary plants. *Journal of Agricultural and Food Chemistry* 51:375–382.
209. Kubátová, A., B. Jansen, J. F. Vaudoisot, and S. B.Hawthorne. 2002. Thermodynamic and kinetic models for the extraction of essential oil from savory and polycyclic aromatic hydrocarbons from soil with hot (subcritical) water and supercritical CO_2. *Journal of Chromatography* 975:175–188.
210. Soto Ayala, R., and M. D. Luque de Castro. 2001. Continuous subcritical water extraction as a useful tool for isolation of edible essential oils. *Food Chemistry* 75:109–113.
211. Fernández-Pérez, V., M. M. Jiménez-Carmona, M. de Castro, and D. Luque. 2000. An approach to the static-dynamic subcritical water extraction of laurel essential oil: Comparison with conventional techniques. *Analyst* 125:481–485.
212. Jiménez-Carmona, M. M., J. L. Ubera, and M. D. Luque de Castro. 1999. Comparison of continuous subcritical water extraction and hydrodistillation of marjoram essential oil. *Journal of Chromatography* 855:25–632.
213. García-Marino, M., J. C. Rivas-Gonzalo, E. Ibáñez, and C. García-Moreno. 2006. Recovery of catechins and proanthocyanidins from winery by-products using subcritical water extraction. *Analytica Chimica Acta* 563:44–50.
214. Turner, C., P. Turner, G. Jacobson, et al. 2006. Subcritical water extraction and β-glucosidase-catalyzed hydrolysis of quercetin glycosides in onion waste. *Green Chemistry* 8:949–959.
215. Shotipruk, A., J. Kiatsongserm, P. Pavasant, M. Goto, and M. Sasaki. 2004. Pressurized hot water extraction of anthraquinones from the roots of *Morinda citrifolia*. *Biotechnology Progress* 20:1872–1875.
216. Farhoosh, R., G. A. Golmovahhed, and M. H. H. Khodaparast. 2007. Antioxidant activity of various extracts of old tea leaves and black tea wastes (*Camellia sinensis* L.). *Food Chemistry* 100:231–236.
217. Chu, C. Y., M. J. Lee, C. L. Liao, W. L. Lin, Y. F. Yin, and T. H. Tseng. 2003. Inhibitory effect of hot-water extract from dried fruit of *Crataegus pinnatifida* on low-density lipoprotein (LDL) oxidation in cell and cell-free systems. *Journal of Agricultural and Food Chemistry* 51:7583–7588.
218. Hsu, B., I. M. Coupar, and K. Ng. 2006. Antioxidant activity of hot water extract from the fruit of the Doum palm, *Hyphaene thebaica*. *Food Chemistry* 98:317–328.
219. Howard, L. R. and Z. Y. Ju. 2005. Pressurized water extraction of procyanidins from grape seeds. 23rd ACS National Meeting, Washington, DC.
220. Amin, I., and O. Mukhrizah. 2006. Antioxidant capacity of methanolic and water extracts prepared from food-processing by-products. *Journal of the Science of Food and Agriculture* 86:778–784.

221. Ozel, M. Z., F. Gogus, and A. C. Lewis. 2003. Subcritical water extraction of essential oils from *Thymbra spicata*. *Food Chemistry* 82:381–386.
222. Kaufmann, B., and P. Christen. 2002. Recent extraction techniques for natural products: Microwave-assisted extraction and pressurised solvent extraction. *Phytochemical Analysis* 13:105–113.
223. Bergeron, C., S. Gafner, E. Clausen, and D. J. Carrier. 2005. Comparison of the chemical composition of extracts from *Scutellaria lateriflora* using accelerated solvent extraction and supercritical fluid extraction versus standard hot water or 70% ethanol extraction. *Journal of Agricultural and Food Chemistry* 53:3076–3080.
224. Ju, Z. Y., and L. R. Howard. 2003. Effects of solvent and temperature on pressurized liquid extraction of anthocyanins and total phenolics from dried red grape skin. *Journal of Agricultural and Food Chemistry* 51:5207–5213.
225. Inoue, S., M. Asaga, T. Ogi, and Y. Yazaki. 1998. Extraction of polyflavanoids from radiata pine bark using hot compressed water at temperatures higher than 100°C. *Holzforschung* 52:139–145.
226. Mai, J., L. J. Chambers, and R. E. McDonald. 1989. Antioxidant compositions. US Patent, US4839187.
227. Chambers, L. J., J. Mai, and R. E. McDonald. 1984. Antioxidant compositions. European Patent, EP 0267 630 A2.
228. Wang, L., and C. L. Weller. 2006. Recent advances in extraction of nutraceuticals from plants. *Trends in Food Science & Technology* 16:300–312.
229. Herrero, M., E. Ibáñez, F. J. Señoráns, and A. Cifuentes. 2004. Pressurized liquid extracts from *Spirulina platensis* microalga: Determination of their antioxidant activity and preliminary analysis by micellar electrokinetic chromatography. *Journal of Chromatography* 47:195–203.
230. Chen, P. Y., Y. X. Tu, C. T. Wu, T. T. Jong, and C. M. J. Chang. 2004. Continuous hot pressurized solvent extraction of 1,1–diphenyl-2-picrylhydrazyl free radical scavenging compounds from Taiwan yams (*Dioscorea alata*). *Journal of Agricultural and Food Chemistry* 52:1945–1949.

6.3 OBTAINING BIOACTIVE COMPOUNDS FROM CASHEW TREES AND NUTS

Richard L. Smith, Jr., Masaaki Toyomizu, Louw J. Florusse, and Cor J. Peters

Biological features of the cashew tree and its fruit are summarized in this chapter section. The main bioactive compounds in cashew are phenolic lipids known as anacardic acids (AAs). The AAs have many bioactivities, but one notable one is that for uncoupling effects for mitochondria. This means that AAs have the possibility for controlling body fat in both animals and human beings. AAs occur in large concentrations in cashew nut shell liquid (CNSL), which can be considered as a natural protective agent for the edible cashew kernel. The removal of CNSL can be done simply without the use of organic solvents by the use of pressure swing and supercritical carbon dioxide. The phase behavior of CNSL and supercritical carbon dioxide is interesting and exhibits liquid–liquid–vapor equilibria at room temperature around the saturation pressure of CO_2. The phase behavior can be described quantitatively with cubic equations of state. Cashew has a bright future as an agrochemical crop and supercritical carbon dioxide can be used to maximize the quantity of bioactive compounds obtained from the nut and also to obtain bioactive compounds of the highest possible quality. More research is needed on processing the cashew tree and for developing new applications with the AAs, the cashew shell, gum leaves, and bark.

6.3.1 Phenolic Lipids and Their Origin

Phenolic lipids, which are primarily of plant origin, occur widely in the plant family Anacardiaceae, which includes poison ivy, poison sumac, mango, ginkgo, and cashew [1]. Phenolic lipids have a chemical structure that consists of a phenol group that is substituted with a hydroxy- or carboxyl- group and an alkyl or alkenyl chain that is generally from 3 to 27 carbons in length.

Phenolic lipids can also be described in terms of a catechol, a resorcinol, or a hydroquinone structure, which have a substituted alkyl chain with various degrees of unsaturation. The compounds are toxic and have high biological activities that are highlighted in a review by Kozubek and Tyman [1]. In general, the bioactivity of all phenolic lipids increases as the length of the alkyl chain increases and also as the degree of unsaturation increases. The reader is referred to the Web site of Kozubek (http://biochem.microb.uni.wroc.pl/liprez3.htm) for additional information both on the occurrence and on the structure of identified phenolic lipids.

6.3.2 Chemical Structures of Phenolic Lipids in Cashew

Chemical structures of the main phenolic lipids in cashew are shown in Figure 6.3.1, where it can be seen that AAs are distinguished from other phenolic liquids in cashew by the presence of the carboxylic acid group, which make them somewhat resemble salicylic acid in structure, where, instead of a hydrogen atom being attached at carbon 6, a 15-carbon alkyl group is present. The AAs in cashew (*Anacardium occidentale*) are recognized as being some of the most widely available natural bioactive compounds

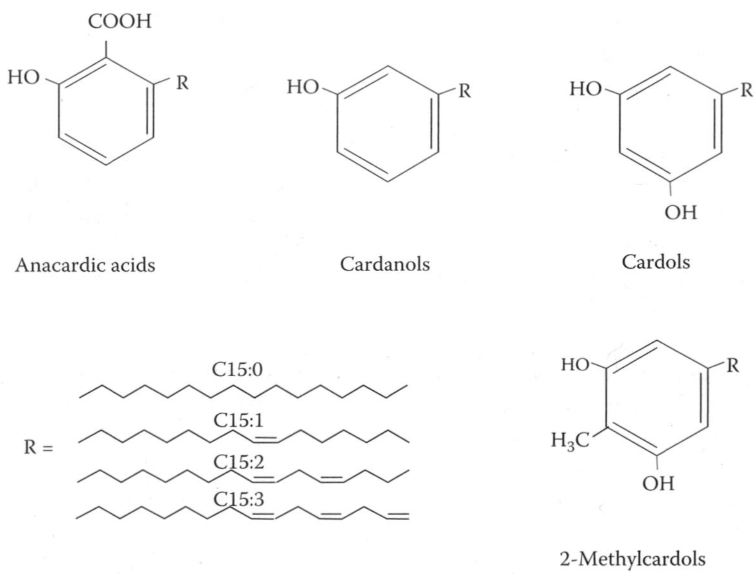

FIGURE 6.3.1 Phenolic lipids contained in cashew nut shell liquid.

[2–6]. Characterization of the alkyl phenols that occur in cashew show that they have antioxidant capacities [7]. In accordance with general bioactivity for phenolic lipids, AAs containing three double bonds in the alkyl side chain exhibit greater antioxidant and enzyme inhibition capacities than those having the other more saturated alkyl side chains. Most notably, AAs have been found to have uncoupling effects with energy transfer processes in mitochondria as described in the next section.

6.3.3 Bioactivity of Anacardic Acids and Uncoupling Effects

The bioactivity of AAs contained in cashew has been examined for its antitumor [8], antimicrobial [9], and potent molluscicidal effects [10]. However, one of the most interesting studies in bioactivity is that related to the uncoupling effect of AAs on oxidative phophorylation of mitochondria [5]. If a new type of uncoupler could be discovered from a natural source, for example, it could substantially contribute to controlling body fat in not only animals, but also in human beings.

Mitochondria, which are known as the powerhouses of cells, generate chemical energy in the form of adenosine triphosphate (ATP) that is used in metabolic processes in living organisms. Figure 6.3.2 shows a schematic based on an inner mitochondria membrane that contains four large enzyme complexes, I, II, III, and IV, which have functions related to the electron-transport chain. In the coupling situation of oxidative phosphorylation, which is a kind of metabolic pathway, high-energy electrons from molecules such as NADH and $FADH_2$ are transported down the electron-transport chain, and an electrochemical gradient is generated across the inner mitochondria membrane. As a result, both a pH gradient and an electrochemical gradient are

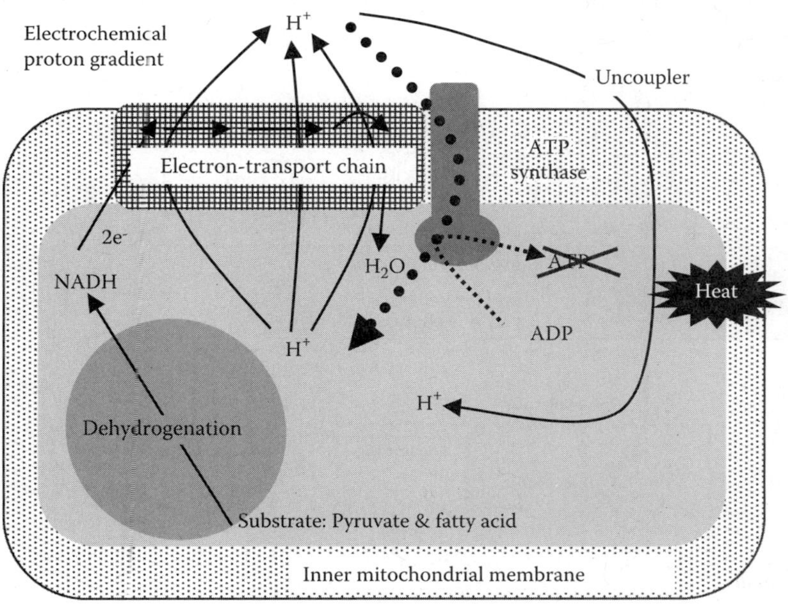

FIGURE 6.3.2 Oxidative phosphorylation in mitochondria showing the electron-transport chain and electrochemical proton gradient across the inner mitochondrial membrane. An uncoupler allows proton transport without driving ATP synthase and thus generates heat.

generated across the mitochondria membrane. Backflow of protons down this gradient drives ATP synthase to catalyze the conversion of adenosine diphosphate (ADP) to adenosine triphosphate (ATP). Uncouplers work to reduce these gradients by allowing protons to flow across the membrane to generate heat instead of ATP (Figure 6.3.2). Thus, the generation of heat instead of the ATP provides the basis for dietary control.

Figure 6.3.3 shows a schematic of a possible transport mechanism of AA and its interaction inside and outside a liposomal membrane. Anacardic acid diffuses across the mitochondrial membrane and forms anacardate, which induces inside-negative ΔpH. From the structural characteristics of anacardate, intramolecular hydrogen bonding is formed in anacardate, resulting in a stable six-member ring structure. This structure then permeates through the membrane according to the electrochemical gradient. Thus, a pH gradient, ΔpH, is generated and an electrochemical proton gradient, ΔΨ, is changed in liposomal membranes, and this implies that proton transport that would occur in mitochondria could do so without driving ATP synthase in the mitochondria. Detailed information of the process can be found in the literature [5, 11, 12], where it is shown that AA has an uncoupling effect on oxidative phosphorylation and that AA behaves both as an electrogenic (negative) charge carrier driven by ΔΨ and a proton carrier that dissipates proton gradients formed across liposomal membranes.

The reader is referred to a review by Skulachev [13] for detailed information on uncoupling and bioenergetics; it describes some of the main physiological functions of mitochondria. The main physiological functions of mitochondria, including those elucidated in other recent works, include (i) energy conservation, (ii) energy dissipation (heat), (iii) production of useful substances, (iv) decomposition

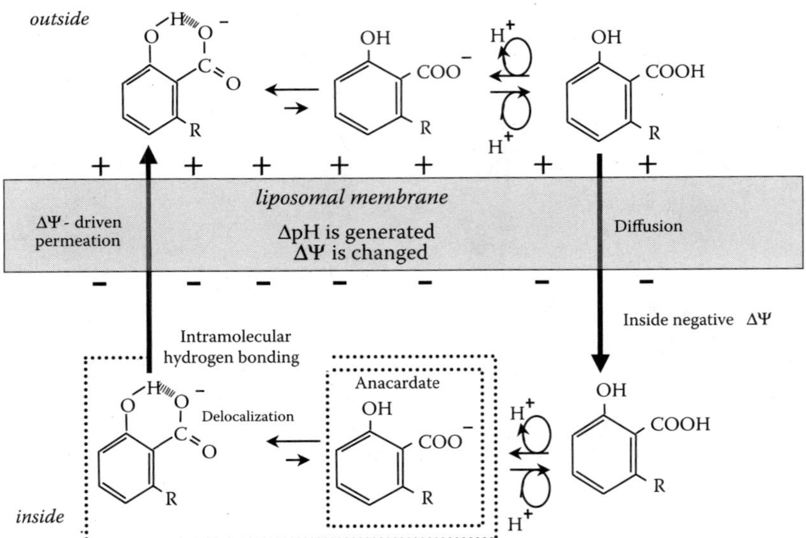

FIGURE 6.3.3 Transport mechanism of anacardic acid across a liposomal membrane showing diffusion and permeation processes.

of harmful substances, and (v) control of cellular processes, including reactive oxygen species (ROS). Some of their functions can be regulated by uncouplers. Therefore, sources of natural uncouplers and their function in food and diet are of great importance. In the next section, discussion will focus on one of the main sources of AAs that are available in large quantities contained in cashew.

6.3.4 Cultivation and Production of Cashew

Cashews (*A. occidentale*) are cultivated in tropical regions for their economic importance with regard to the edible nut and also as a source for resins, dyes, lacquers, oils, and waxes. The phenolic lipid content in the whole cashew fruit is very high, with the cashew nut shell liquid (CNSL) making up from 15 to 25% of the weight of the raw cashew nut-in-shell [14], but it can be as high as 32% [7]. Natural CNSL contains 80%–90% AAs, 10%–20% cardols (CDs), and small amounts of cardanols (CNs; 1%–2%) and methyl CDs (2%–3%) [14]. This makes cashew one of the largest renewable sources of phenolic lipids available in nature.

Some of the major countries producing cashew are shown in Table 6.3.1. From Table 6.3.1, it is clear that Vietnam was the top cashew producing country in 2005,

TABLE 6.3.1
Top 20 Cashew-Producing Countries in 2005

Rank	Country	Production (metric tons)
1	Vietnam	827,000
2	India	460,000
3	Brazil	251,268
4	Nigeria	213,000
5	Indonesia	122,000
6	United Republic of Tanzania	100,000
7	Côte d'Ivoire	90,000
8	Guinea-Bissau	81,000
9	Mozambique	58,000
10	Benin	40,000
11	Thailand	24,000
12	Malaysia	13,000
13	Kenya	10,000
14	Ghana	7,500
15	Philippines	7,000
16	Madagascar	6,500
17	Sri Lanka	6,200
18	Senegal	4,500
19	Burkina Faso	3,500
20	El Salvador	2,600
Total production		2,327,068

Source: From UN Food and Agricultural Organization (FAO), 2005. http://www.fao.org/es/ess/top/commodity.html?lang=en&item=217&year=2005 (accessed July 16, 2008).

with 827,000 metric tons of raw cashew being reported. However, some countries have developed extensive infrastructures for processing cashew, and thus, a number of these countries, including India and Vietnam, import raw cashew as a commodity product from producing countries. Of the processing countries, India presently has a highly developed cashew industry. According to estimates [15], Indonesia exports roughly half of the available cashew for this purpose. Most of the countries listed in Table 6.3.1, however, also process cashew on a small or local scale.

6.3.5 Cashew Trees and Processing of Cashew

Cashew trees have oval leaves and grow to heights of as much as 20 m with a diameter of about 1 m, under proper conditions [14]. However, many remarkable species exist. For example, the "Cashew Tree of Pirangi" (Cajueiro de Pirangi) in Brazil has a huge crown and occupies an area of almost 8400 m^2 and is the size of roughly 70 normal cashew trees. More common trees can also be found with large trunks of several meters. The raw nuts (nut-in-shell or NIS) provide the valuable cashew kernel and also contain the cashew nut shell liquid, which is used in many phenolic resin products. The cashew nut (fruit) grows off of a swollen root (peduncle) that is known as the cashew "apple."

Many parts of the cashew tree provide useful products. For example, the gum of the cashew tree has been suggested for use in protein extraction in two-phase aqueous systems [16–20], for use as polymeric agents or as thin films [21–24], for use as hydrogels [25, 26], or even for use as a flotation agent for phosphate mineral recovery [27]. In many countries, the cashew apple is used in making beverages and jams or fermented to make an alcoholic drink. In most countries, however, the main focus is on the cashew kernel.

The objective of most processing operations of cashew is to obtain the cashew kernel with as little damage as possible, while separating it from the highly vesicant cashew nut shell liquid, which surrounds the kernel within its testa and inner shell. Whole cashew that are light in color command a premium price. Of course, processing of cashew depends on the scale of the operation and the availability of infrastructure to provide markets for the by-products.

Figure 6.3.4 shows actual pictures of Indonesian cashew as donated to this research group by BPP Teknologi (Jakarta) and prepared at Tohoku University. As shown in Figure 6.3.4 (left), the cashew kernel and its tight fitting testa covering are contained within a double shell. The outmost shell or epicarp is light brown in color and is permeable to water and to some extent gases. The innermost shell or endocarp contains the cashew kernel (Figure 6.3.4, middle). In between the epicarp and endocarp is a kind of cellular matrix (Figure 6.3.4, right), that contains the CNSL that is made up of AAs and other compounds (Figure 6.3.1). The CNSL is bioactive, highly vesicant, and causes strong contact dermatitis as a result of the presence of the AAs. In processing, the edible kernels should not be allowed to come into contact with the cashew nut shell liquid, and if so, the kernels are considered to be spoiled. Thus, the processing problem becomes that of how to remove the cashew kernel from the shell without either contaminating the kernel with CNSL or breaking the kernel or changing its color, both of which affect the value of the product.

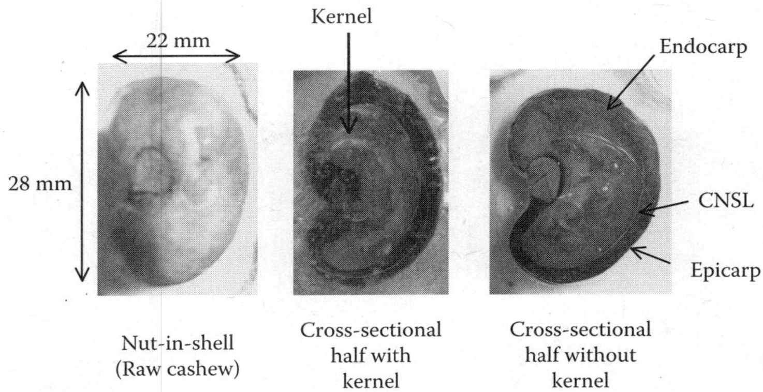

FIGURE 6.3.4 Photographs of Indonesian cashew showing the nut-in-shell (left), cross-sectional half with kernel (middle), and cross-sectional half without kernel (right). Samples are encased in resin for safety.

In the artisanal method of processing cashew, roasting of the raw cashew over a fire causes the AAs to decarboxylate (Figure 6.3.5) and releases CO_2 so that the CNSL foams and oozes from the shell and burns off with a pleasant aromatic odor, after which the embrittled shells can be removed, and the testa can be removed from the kernel before drying. Any remaining oil during shell removal, however, still has some activity and must be removed with care.

In the processing of cashew, two methods are common: the wet method and the dry method [28]. Local processing of cashew tends to use the wet method, because it does not require extensive equipment but does require experienced shellers. Both methods require considerable conditioning before and after kernel removal, which is discussed in detail in a Food and Agricultural Organization (FAO) of the United Nations report [29] and also in separate works [28].

In the wet method of processing, cashew nuts are sun-dried before peeling off the pericarp (epicarp, mesocarp, and endocarp) with a special tool. Figure 6.3.6 shows an example of this from a site in the Philippines. After the peeling process, the kernel in its testa is usually roasted to make it easier to remove the testa or is removed with a special tool (Figure 6.3.7). Local methods of processing tend to be highly labor-intensive and tend to produce only the kernel as product and the wasted shell that can be burned as by-product. The wet method also places considerable responsibility on workers for safety and health.

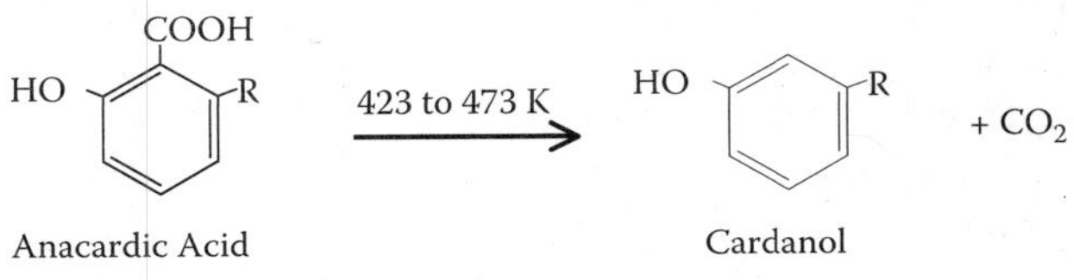

FIGURE 6.3.5 Range of decarboxylation temperatures of anacardic acids and reaction products. Polymerization (not shown) is also possible.

FIGURE 6.3.6 Cashew shelling in the Philippines. (Courtesy of Dr. Roberto Malaluan, Iligan Institute of Technology, Iligan City, Philippines.)

In the dry method of processing, hot (decarboxylated) CNSL is used to remove the raw CNSL from the shells and also to roast the shells. This can be performed in a batch dipping process or as a continuous process, where the nuts are allowed to move along a conveyor-belt type of system. Figure 6.3.8 shows an example of the dry method, with raw, preconditioned cashew being fed into an extraction chamber that contains hot, technical-grade CNSL. A belt conveyor allows the nuts to move through the extraction chamber for a given period of time that is generally within a couple of minutes. The hot CNSL serves to remove and decarboxylate the CNSL and causes it to foam and exude from the shell, and the heat causes the shell to become brittle. According to the FAO (1969), CNSL begins to decarboxylate and froth at 150°C and begins to polymerize at temperatures higher than 473 K. The ratio of the volume of CNSL to nuts is also important and must be maintained from 30:1 to 50:1 for good results, as described in the literature [29–31], although these recommended ratios seem to be based on an early work [31].

After extraction with the dry method, the nuts have to be cooled quickly to avoid scorching and color change. Then, the nuts are dried and shelled either by automatic or semi-automatic shelling machines, depending on the size of the operation and the grade of the cashew. Some countries may also use manual methods.

Other methods of processing cashew include steam processing at 543 K, quick roasting at 573 K, cold methods, and solvent extraction. The reader is referred to books on the subject [30–32] for discussion of some of these methods, including genetic modifications.

6.3.6 Separation of Cashew with Supercritical CO_2

In reviewing these methods, it is clear that many of the compounds contained within the raw cashew are damaged by heat. Thus, a nonthermal treatment method that

FIGURE 6.3.7 Manual removal of the cashew kernel at a local site in the Philippines. (Courtesy of Dr. Roberto Malaluan, Iligan Institute of Technology, Iligan City, Philippines.)

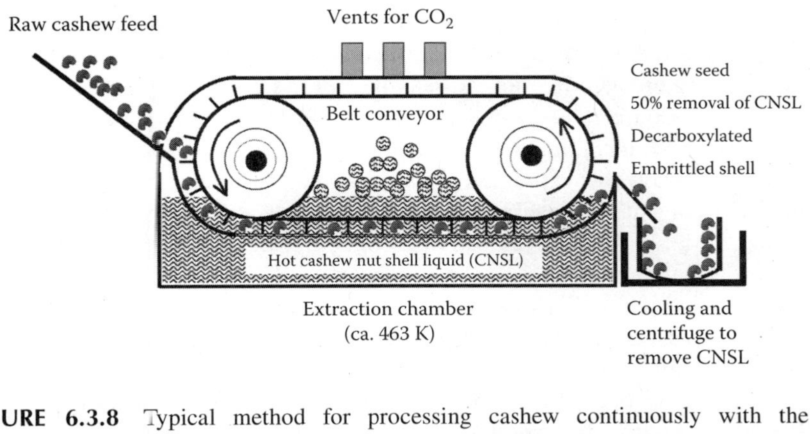

FIGURE 6.3.8 Typical method for processing cashew continuously with the dry method showing the feed, extraction chamber, and centrifuge. The dry process typically decarboxylates all of the CNSL in about 2 min and results in about 50% removal of the CNSL. The shell becomes brittle because of the heat, which makes its removal easier. Cooling and centrifugation are required to avoid color change of the kernel caused by scorching.
Source: Adapted from Budich, M., et al., *Journal of Supercritical Fluids,* 14:105–114, 1999.

TABLE 6.3.2
Constituents and Possible By-Products of Raw Cashew

Raw cashew constituent (wt %)[a]	Raw (metric tons)	Average	Low	High
Kernels (26%)	605,038			
Testa (2.5%)	57,401			
Tannin (25% of Testa)		14,350	14,350	14,350
Shells (71.5%)	1,664,629			
Cashew nut shell liquid (15%–30%)		374,542	249,694	499,389
Total (metric tons)	2,327,068	388,892	264,045	513,739

[a] Values are derived from averaging a wide range of six classes of raw cashew as reported by Ohler [30].

could remove the CNSL under dry conditions would be highly desirable to preserve possible by-products.

Table 6.3.2 shows the constituents and possible by-products of raw cashew based on averages of six sizes and grades of cashew [30] and using the currently available cashew production of the top 20 countries. As shown in Table 6.3.2, recovery of a huge quantity of raw CNSL, which consists mostly of AAs (80%–90%) and CDs (10%–20%) is possible. Further, the testa contains a high amount of tannins (25%), which can be used in leather tanning industries. Supercritical extraction of the CNSL from the cashew could be a good method to obtain the bioactive AAs from cashew if methods were developed.

The separation of the CNSL from the cashew shell material with supercritical extraction, however, has proven to be challenging. Early work [33] proposed a method to recover CNSL from cashew shells that used extraction with supercritical CO_2. The method provided phenolic lipids of high quality, but required the use of large amounts of CO_2 for a given quantity of cashew. The reason for this is that although the shells were ground or preprocessed, the solubility of the AAs is very low, as discussed in a later section. Researchers in India [34, 35] provide a detailed study on the economics of processing cashew using traditional supercritical fluid extraction with CO_2 including optimized conditions and yields for ground material (<8 mesh). They show that extraction pressure is a key parameter in the optimization and that CNSL of excellent quality could be obtained.

Arai et al. [36] proposed the use of alcohol entrainers with supercritical CO_2 for selective separation of CDs from CNs present in CNSL that gave good results. However, in that method [36], isopropanol was recommended as cosolvent, which means that the technique can only be used for postprocessing of CNSL from the shells after the edible cashew kernel is removed.

Smith et al. [37] proposed a method that uses supercritical CO_2 without any entrainers and that is discussed in this chapter. A typical apparatus is shown in Figure 6.3.9. It consists of a CO_2 gas or liquid (dip-tube) cylinder, a condenser-pump-heater unit (JASCO, SCF-Get) that delivers supercritical CO_2 at a given temperature and pressure, an

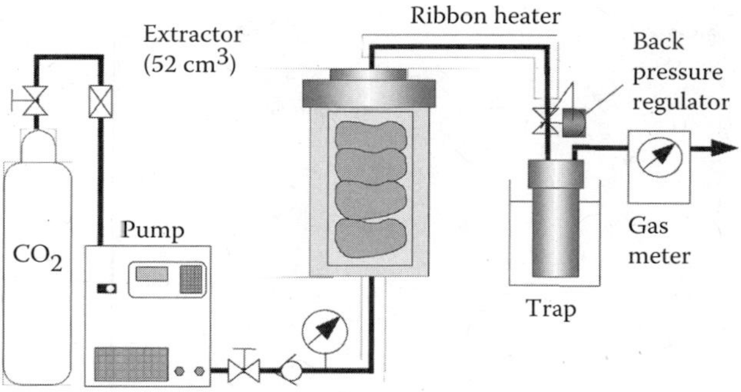

FIGURE 6.3.9 Typical extraction apparatus that can be used to study pressure-swing effects on the separations. Pressure swing and depressurization are controlled by an electronic back-pressure regulator. Flow can be oriented from bottom-to-top (as shown) or from top-to-bottom, depending on the density difference between the solute and that of CO_2 at the given conditions.

extractor, a back-pressure regulator, a trap, and a gas meter for flow rate measurement. In this system, cashews, either whole nuts-in-shells or cut-shells, are loaded into the extractor. For the case of cut-shells as described in this chapter, the cashew kernels are removed from the shell by hand to avoid any influence of cashew oils on the results. Pressure of the system is controlled by the automatic back-pressure regulator. If supercritical extraction is performed in the usual way, that is, by flowing supercritical CO_2 through the reactor and the product CNSL collected in the trap, the results are very disappointing as shown in Figure 6.3.10 by the line labeled as "Typical supercritical extraction method." If, on the other hand, pressure is cycled, even once, then the extraction yields greatly increase,

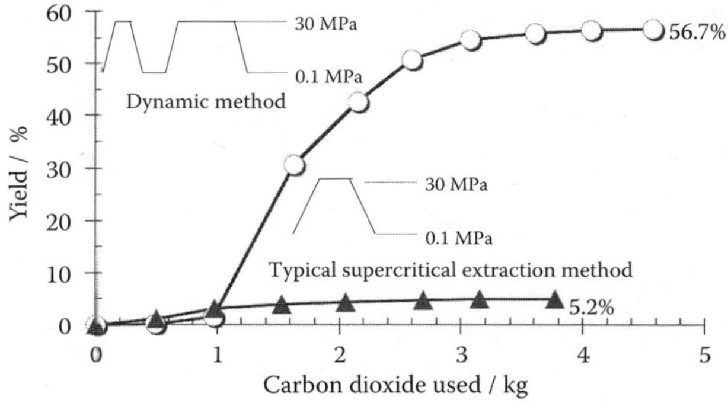

FIGURE 6.3.10 Yields of CNSL obtained from cut-shell cashew nuts using supercritical CO_2 showing (i) the typical method, and (ii) the dynamic method. Extraction conditions are 333 K and 30 MPa pressurization followed by 5 L/min flow of CO_2 at standard temperature and pressure (STP).

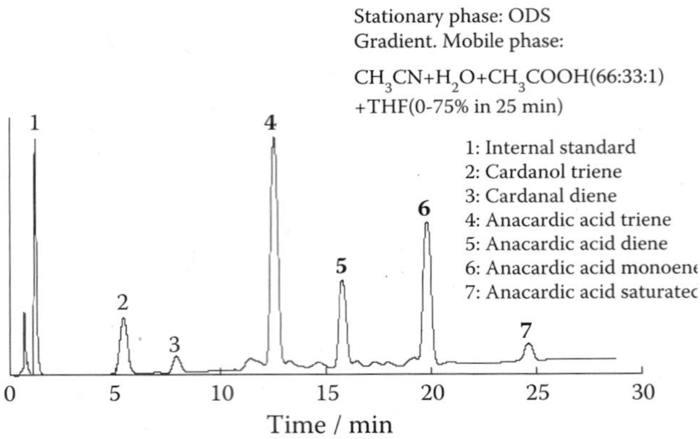

FIGURE 6.3.11 Chromatogram of CNSL from analysis with high-performance liquid chromatography (HPLC) using an acetonitrile–water–acetic acid mobile phase with THF gradient elution and *p*-tert-butyl phenol internal standard with the relative molar response (RMR) method of Tyman et al. [38].

as shown in Figure 6.3.10 by the line labeled "Dynamic method." As expected, the CNSL was of high quality, as shown by the HPLC chromatogram in Figure 6.3.11, and demonstrates that the method provides both high AA and high CN content with good selectivity. Remarkably, the increases are due to several phenomena as discussed later. When those authors used multiple pressure-swing (PS) steps, even higher yields could be obtained (>90%), which was similar to that which would be obtained if the de-kernelled shells were ground and loaded into the extractor as is. Results for multiple PS steps are shown in Figure 6.3.12. Each bend in the curve is associated with a PS step.

The effect of pressure on essential oil glandular trichomes has been discussed in the literature, and some detailed studies have been performed that use pressure as a mass separating agent [39–41]. In those studies, efficiency of the disruption process depended on many parameters including pre- and postexpansion pressures, exposure time, and decompression time. This seems to be true for cashew as well, with some contact time being necessary at a given pressure for the process to be effective. For cashew, the precontact time is greater than 5 min but less than 1 h. Some of the fundamental factors affecting separation of CNSL with CO_2 can be understood by examining the phase behavior of the system.

6.3.7 Phase Behavior

The phase behavior of CO_2 and phenolic lipids has not been well studied. In a review by Dohrn and Brunner [42], the closest related systems to AAs were measurements of alkyl benzenes with CO_2. In the review of Christov and Dohrn [43], the closest related systems to AAs that had been studied were those of Yamini [44], who reported measurements of dihydroxybenzene isomers, pyrocatechol, resorcinol, and hydroquinone in supercritical CO_2. In other works, Garcia-Gonzales et al. [45] reported solubility measurements of pyrocatechol in supercritical CO_2 and Francisco

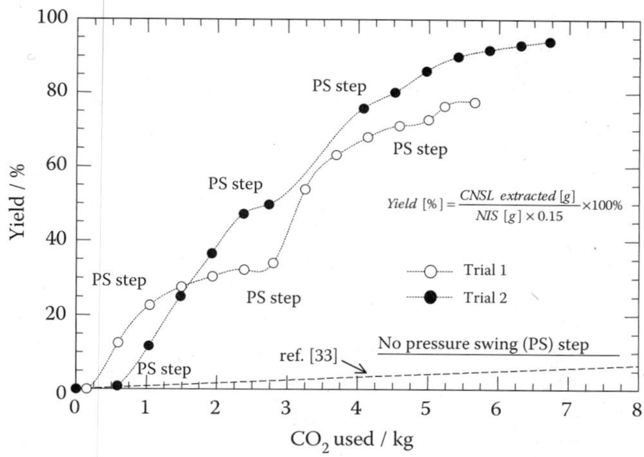

FIGURE 6.3.12 Yields of CNSL obtained from cut-shell cashew nuts using supercritical CO_2 showing the influence of pressure swing steps (dynamic method) on the yields. Extraction conditions are 333 K and pressurizations to either 9.8, 19.6, or 29.4 MPa followed by 5 L/min flow of CO_2 at standard temperature and pressure (STP). See Smith, R. L., Jr., et al. [37] for details.

et al. [46, 47] reported on extractions and isolation of alkylresorcinols related to rye bran. However, the data for AAs do not exist.

Because both the liquid and vapor phase behaviors are needed to understand the separation process and to discuss the mass transfer, we conducted some studies on both on the phase behavior of CNSL with CO_2 and the phase equilibria. Some results are shown using a synthetic method, in which composition of the system is fixed and the volume is varied and pressure is measured. This method can be used to study the pressure–temperature behavior of phase boundaries as shown by Peters and coworkers [48, 49]. Also, an analytical method was used to determine pressure–temperature–composition curves and equilibrium ratios. This method can be used to examine the trend of the equilibrium ratios ($K_i = y_i/x_i$) of the various components, AAs, CDs, and CNs, and the selectivities ($\alpha_{ij} = K_i/K_j$).

6.3.8 Measurements with a Synthetic Method

Measurements shown in this chapter were performed at Delft University with a Cailletet apparatus. The apparatus derives its name after Louis Paul Cailletet (1832–1913), who was a French physicist and the first scientist to liquefy a number of gases, including oxygen, in 1877. A Cailletet apparatus allows measurement of phase equilibria at fixed compositions for samples loaded into a capillary tube and has been described in the modern literature [49–53]. In the Cailletet apparatus, the sample is confined in a thermostatted capillary tube with a leg of mercury that transmits the pressure. The transmission of pressure by the mercury is controlled through a hydraulic oil system connected to a piston. At a given temperature, the pressure can be varied until a phase change is observed visually. A magnetic stainless steel ball within the capillary tube is used for mixing the various phases via an external magnet. Temperature can be increased or decreased as desired, and then the measurements can be repeated. This is known as the synthetic method, and it allows rapid and accurate phase boundaries to be determined for given compositions. A dead weight pressure gauge is used to measure the pressure of the oil transmission medium to within an accuracy of 0.03% of the reading. The temperature of the thermostat is controlled to better than a 0.01 K variation, and the sample temperature is measured to within an uncertainty of 0.02 K by a platinum resistance thermometer.

6.3.8.1 Procedure

The procedure for filling the capillary tube with CNSL and CO_2 is described next, because some details could be useful to the reader. Initially, a given amount of CNSL was injected into a sealed Cailletet tube with a micro-syringe. The amount injected was determined by mass difference. The sample inside the tube was frozen with liquid nitrogen, and the air was evacuated by connection to a high vacuum system (<0.00013 Pa). The sample was degassed by successive freeze-melt cycles under high vacuum. After the CNSL sample was prepared, a predetermined amount of CO_2 was added volumetrically and pressed into the tube using mercury. The Cailletet tube was then mounted into an autoclave and measurements were made. A detailed figure of the arrangement has been published [50].

6.3.8.2 Liquid–Liquid–Vapor Equilibria

It should be noted that although CNSL is a multicomponent mixture, the system can be treated as a pseudo-binary system, because the volatilities of CNSL components are relatively low compared with CO_2. In view of this, the phase equilibria were measured with a procedure described in the literature [54]. Briefly, after the loading of the samples into the capillary tube was completed, temperature of the sample mixture was set to a given value, and then pressure was varied until two phases were present. By adjusting the pressure until one phase disappeared, the phase boundary could be traced. The rise or fall of the meniscus with increasing pressure indicated whether the disappearing phase was vapor or liquid. Critical points were determined visually, and the reader is referred to the literature for other examples [50].

Figure 6.3.13 shows the trend of the three-phase boundary, where it can be seen that liquid–liquid–vapor (LLVE) equilibria occurs for the CNSL-CO_2 system over a range of temperatures up to those just above the critical temperature of CO_2 as might be expected. According to the measurements, the upper critical end point (UCEP) occurs at 304.28 K and 7.402 MPa, i.e., criticality was observed of the upper light liquid and vapor phase in presence of a heavy liquid phase. Because no binary CO_2-systems are known in the literature showing Type V behavior and the occurrence of Type IV is rarely met, this means that most likely the fluid phase behavior of this pseudo-binary system belongs to Type III, in terms of the van Konynenburg and Scott classification [55], which means that the critical line is discontinuous between the two components (CNSL-CO_2), that is, the pseudo pure component (CNSL) critical point and the pure CO_2 critical point.

The inset of Figure 6.3.13 shows the three phases—liquid CNSL, liquid CO_2, and vapor CO_2— in a larger-scale synthetic apparatus at Tohoku University. Conditions

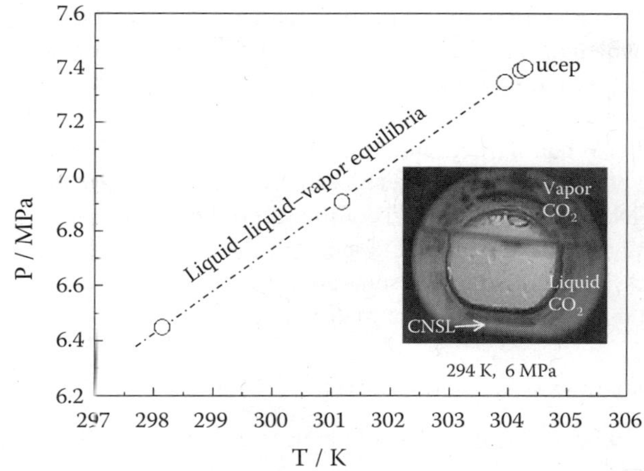

FIGURE 6.3.13 Temperature–pressure phase boundaries for the liquid–liquid–vapor (LLV) equilibrium lines for the CO_2 + cashew nut shell liquid (CNSL) system measured with a Cailletet apparatus of Delft University. Inset shows appearance of CNSL liquid saturated with CO_2, the CO_2 liquid phase, and the CO_2 vapor phase.

were changed in the apparatus and the behavior of CNSL was noted. As long as liquid was present, no great changes in the CNSL phase occurred when changing pressure. However, when only gas was present, reduction in pressure caused a large amount of foaming of the CNSL phase that appeared as if it were undergoing reaction or decarboxylation. Of course, no reaction was occurring, but it was clear that the liquid phase of CNSL contained a large amount of CO_2. Fundamental measurements allow one to understand some of the physical and chemical processes occurring in the larger scale separation experiments.

Figure 6.3.14 shows a possible extraction mechanism for the CNSL-CO_2 system. First, contact of the shells (cut or possibly whole) allows CO_2 to penetrate through

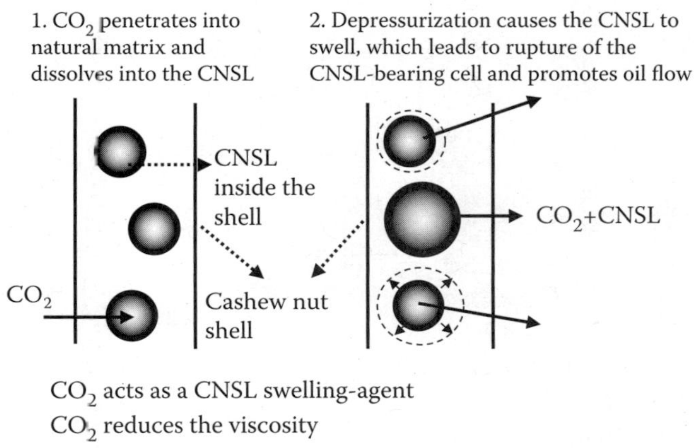

FIGURE 6.3.14 Separation mechanism of CO_2 in cashew nut shell.

the shell epicarp and dissolve into the oil. This causes a volume change in the CNSL-CO_2 mixture, which also reduces the viscosity. Depressurization then causes rupture of oil-bearing cells the CO_2 promotes oil flow. Foaming increases the separation. Multiple pressure swing steps probably help to build channels and to gradually rupture all cells. When we examined the treated cashew shell material, it was light and fluffy and could easily be scattered with one's breath, indicating that only trace quantities of CNSL were present. It is of interest to know the amount of CO_2 present in the CNSL phase and this can be done with the Cailletet apparatus. However, to explore the possibility of selective separation in the vapor phase of CO_2, analytical apparatus that use larger amounts of material are convenient, as described next.

6.3.9 Measurements with an Analytical Method

Measurements with recirculating static apparatus shown in this chapter were performed at Tohoku University. In this apparatus, a large sample is confined in a given volume, and after contacting the phases for an appropriate time by recirculation, phase equilibrium is established. Samples of both liquid and vapor phases are taken and analyzed with gas chromatography or liquid chromatography as appropriate. This is known as the analytical method, which is suitable for obtaining both the equilibrium ratios and component selectivities of binary and multicomponent mixtures. A schematic diagram of a recirculating static apparatus is shown in Figure 6.3.15 and is described here in some detail, because it may be of interest to readers making experimental measurements.

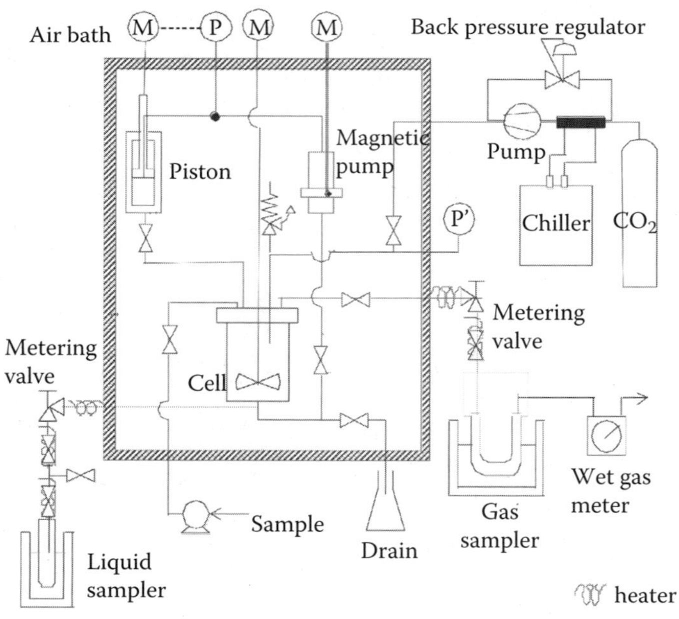

FIGURE 6.3.15 Recirculating static apparatus for measuring vapor–liquid equilibria.

The recirculating static apparatus shown in Figure 6.3.15 is custom-made (AKICO, Tokyo) and consists of a CO_2 loading system, an equilibrium cell with six diametrically opposed windows, an agitator, a gas-phase circulation system, and sampling loops for the liquid and vapor phases. Pressure in the cell is measured with an electronic gauge (Druck Japan, PTX 621) that has a full-scale (FS) range of 40 MPa and an accuracy of 0.05% F.S. The equilibrium cell, circulation system, and sample loops are inside an air bath that is controlled to a maximum deviation of ±0.5 K with a proportional-integral-derivative (PID) controller. The circulation system consists of a magnetic pump for circulating the gas phase and a pressure-controlled piston cylinder for allowing samples to be removed from the sample loops at constant pressure. The equilibrium cell that is shown has an internal volume of 500 cm^3, which is common for recirculating type apparatus. A special component of this apparatus is a piston cylinder that has a maximum internal volume of 200 cm^3 for the purpose of withdrawing up to 200 cm^3 of either vapor or liquid sample at constant pressure. Both equilibrium cell and piston are made of 316 stainless steel. The CO_2 loading system consists of a CO_2 compressed gas cylinder, a chiller (CA-111, EYELA), a diaphragm pump (MBS 3018, Orlita), and a back-pressure regulator (HPB-450, AKICO). The liquid loading system consists of a sample pump (LPS-12, GL Science). The gas sampling system consists of a metering valve, sampling cylinder (Whitey), and gas flow meter. The liquid sampling system consists of a metering valve, valves, and a sampling cylinder (Whitey). Lines including the metering valves are heat traced and controlled to the temperature of measurement.

In Figure 6.3.15, the special features of the apparatus are the (view) equilibrium cell, a magnetic pump for circulation of the vapor phase through the liquid phase, and metering valves for withdrawing samples from either the vapor or liquid phase. The apparatus has a PID-controlled piston that allows withdrawal of samples from the equilibrium cell at constant pressure. Although windows are present in the vessel, these can only be used for some operational checks.

6.3.9.1 Procedure

The general procedure for making measurements with the recirculating static apparatus are described first. Then, specific procedures of each system are discussed. The equilibrium cell and recirculation system (Figure 6.3.15) are controlled at the desired temperature with the air bath and evacuated with a vacuum pump through a liquid nitrogen trap (ca. 0.13 Pa). After this, approximately 300 to 400 cm^3 of liquid sample are fed into the cell either by pump or by syringe. Then, the system is pressurized with CO_2 and the vapor-phase magnetic pump and agitator are started. The position of the piston cylinder is such that the volume displacement is at a minimum. After recirculation for at least 6 h and close monitoring of the temperature and pressure, samples are taken of the liquid and vapor phases. In general, for this type apparatus, longer equilibration times are required for the first data point at a given temperature. In the general procedure, temperature is held constant and system pressure is changed. This means that the overall composition of the system also changes; however, this is inevitable for this type of experimental apparatus. Detailed investigation of multicomponent systems is highly time consuming, and for this reason, correlation equations are often developed to help extend and interpret the experimental trends. Correlation of the data is discussed in a section below.

6.3.9.2 Sampling

In making reliable measurements with the analytical method, it is important to obtain samples of the phases present that are representative of the given phase. In removing a sample from either a vapor or liquid phase, some considerations are (i) pressure change and (ii) volatilization of part of the liquid sample, condensation of part of the vapor sample, entrainment of droplets of liquid in the vapor phase, and entrainment of vapor phase bubbles in the liquid phase. In the literature, many tips are given on obtaining reliable samples. In the CNSL-CO_2 system, there are some fine points that should be mentioned.

In sampling the vapor phase, the sample is flowed through a trap that condenses any CNSL present. After the process, the sample can be both weighed and analyzed by chromatography. It is important that the trap be cooled to a low enough temperature. For the measurements shown, mixtures of acetone and dry ice are used, and some care has to be made in preparation of the cooling mixture safely. Briefly, crushed dry ice is slowly and gradually added into a dewar partially filled with acetone. Frothing of the cooling mixture occurs, and so care must be taken not to add the dry ice too fast, or otherwise, the froth can cause heat burns due to its cold (ca. 222 K) temperature.

In sampling of the liquid phase, normally the metering valve can be opened into a liquid sampling bomb that is evacuated and cooled to liquid nitrogen temperatures to collect a liquid sample. After a given period of time, the metering valve and shutoff valves are closed, and the sampling bomb can be weighed and its contents analyzed. The sampling lines can also be rinsed with solvent so that corrections can be applied. For the case of CNSL, however, this procedure did not work reliably, most probably because of the high viscosity of the liquid. To make a reliable procedure, the following method was devised. First, the sampling bomb for the liquid phase was not evacuated, but was pressurized with CO_2 gas at a pressure of about 5 MPa. The sample cylinder was maintained at room temperature until completion of the sampling operation. This prevented a drastic pressure reduction in the liquid sample pressures that were typically at 25 MPa and allowed reliable liquid samples to be obtained. Sampling lines still had to be washed with solvent to collect residual CNSL in the lines.

6.3.9.3 Vapor–Liquid Equilibria

Results for the experiments for the vapor–liquid equilibria (VLE) are shown in Figure 6.3.16 at temperatures from 323 to 343 K. The lines shown are discussed later. In terms of mole fraction, it can be seen that more than 60 mol % CO_2 dissolves into the CNSL phase and that the temperature does not play a large role in CO_2 saturation in the liquid phase. Further, the CNSL seems to saturate with CO_2 at about 15 MPa. In other words, above a certain pressure, increasing pressure does not lead to any

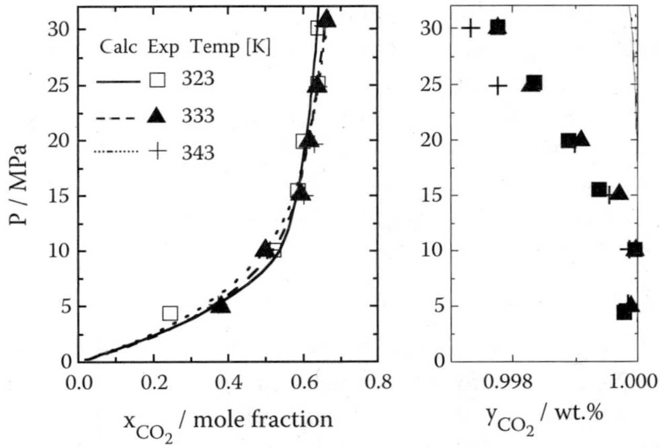

FIGURE 6.3.16 Vapor–liquid equilibria data and correlation of the CO_2–cashew nut shell liquid (CNSL) system.

further dissolution of CO_2 into the CNSL phase. The vapor phase, however, does show a variation with temperature and pressure. In interpreting these data in view of the separation results, it is important to remember that although the liquid phase compositions do not seem to change much with temperature, the physical properties of the liquid phase, such as viscosity or density, are strong functions of temperature and probably become pressure dependent.

It is of interest to examine the equilibrium ratios ($K_i = y_i/x_i$) of the various components, AAs, CDs, and CNs, and the selectivities ($\alpha_{ij} = K_i/K_j$). Figure 6.3.17 shows the equilibrium ratios plotted as a function of pressure at constant temperature. The equilibrium ratios initially show a downward trend and then increase with increasing pressure. The equilibrium ratios are very small, on the order of 10^{-4},

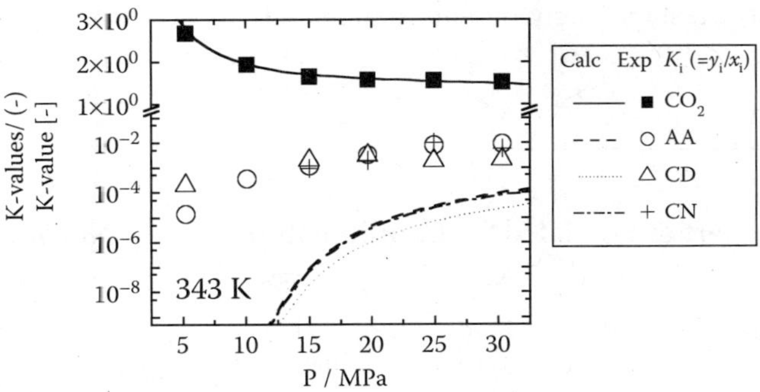

FIGURE 6.3.17 Separation ratios for the CO_2–cashew nut shell liquid (CNSL) system and estimation of the equilibrium ratios for the anacardic acids (AAs), cardols (CDs), and cardanols (CNs) in the mixture.

for all CNSL components. However, although the equilibrium ratios are similar for anacardic acid and CN, the ratio of that to CD is different by a factor of about 2. In other words, the selectivity is about 2 for anacardic acid versus CD or for CN versus CD, and this selectivity tends to increase as the temperature is decreased toward the critical point of CO_2. This trend for the solubility can be expected in view of the chemical structures shown in Figure 6.3.1, because addition of hydroxyl groups to an aromatic group tends to decrease CO_2-solute attraction.

6.3.10 Correlation of the Data

Correlation of the data is of interest for examining trends of the solubilities with temperature or pressure and for developing mass transfer models. Besides empirical relations, equations of state [56, 57] are frequently used. In applying an equation of state, physical properties of the components are needed and for CNSL these have to be estimated. Table 6.3.3 shows the physical properties estimated for some of the constituents in cashew nut shell liquid along with the pseudo-component CNSL and also for the solvent CO_2, in which experimental values for the pure component critical temperature, pressure, and acentric factor, ω, are available. Values in Table 6.3.3 were estimated with the Joback method [58], which uses the chemical structure of the molecule and experimental normal boiling point if available. Other methods, such as the group-contribution method [59] can also be applied to develop correlations considering isomeric factors.

In Figure 6.3.16, correlation of the liquid phase was done by treating CNSL as a pseudo-component so the system is a pseudo-binary of CNSL and CO_2. This procedure is frequently used for correlating data of complex mixtures. It can be seen in Figure 6.3.17 that correlation of the liquid phase was satisfactory; however, the vapor phase calculation was poor. The interaction parameters, k_{12} and l_{12}, obtained by minimizing the objective function (OF) of the absolute average differences of both vapor and liquid fractions, are shown in Table 6.3.4. These parameters are used with the properties in Table 6.3.3 to perform the calculations with the equations in Table 6.1.1 (see Section 6.1). From examination of the parameters, it is clear that they did not depend strongly on temperature and probably can be assumed to be constants.

TABLE 6.3.3
Physical Properties Estimated for Constituents of Cashew Nut Shell Liquid

Constituent	Mw	Tc (K)	Pc (MP)	ω
Anacardic acid	344.4	1187	1.4	1.198
Cardol	314.9	1088.4	1.64	1.168
Cardanol	299.2	998.1	1.45	0.973
CNSL	303.1	1164.1	1.44	1.184
CO_2	44.0	304.12	7.37	0.225

Mw: molecular weight; *Tc*: critical temperature; Pc: critical pressure; *ω*: acentric factor.

TABLE 6.3.4
Optimized Fitting Parameters for the CO_2 (1) and CNSL (2) Systems Determined for the Soave–Redlich–Kwong Equation of State

Temperature (K)	k_{12}	l_{12}	OF
323	0.0707	0.0010	0.0095
333	0.0777	0.0113	0.0046
343	0.0688	0.0105	0.0062

k_{12} *and* l_{12}: interaction parameters; OF: objective function.
Source: Based on Soave, G., *Fluid Phase Equilibria,* 82:345–359, 1993.

Figure 6.3.17 shows the calculation of the equilibrium ratios, K_i, for the case of fixing the k_{ij} values to those in Table 6.3.4 for CNSL components, AA, CD, and CN, and for performing the calculation with CO_2 for a given overall composition. The values for CO_2 are reproduced well, but those for the CNSL constituents are only qualitative. Both K_i values for anacardic acid and CN are close, meaning that the selectivity for AA versus CN ($\alpha = K_{AA}/K_{CN}$) is poor. However, it is interesting that the calculation also shows some differences in selectivities for AA versus CD ($\alpha_{AA,CD} = K_{AA}/K_{CD}$) and for CN versus CD ($\alpha_{CN,CD} = K_{CN}/K_{CD}$), which can be very useful, because CNSL consists mainly of AAs and CDs when processed without thermal treatment.

6.3.11 Separation Scheme for Cashew

In general, a processing scheme can be developed for cashew and for obtaining bioactive compounds from the cashew tree and nut as shown in Figure 6.3.18. In this processing scheme, water and CO_2 are used to extract or convert cashew into a multitude of products. From the point of view of food science, the cashew kernel and the cashew apple are the most important, with the cashew nut shell liquid, specifically, the AAs and CDs, being the most useful among cashew constituents for pharmaceuticals, drugs, biocides, and biopolymers. The initial processes in the scheme need to be mild in thermal nature to preserve the activity of the cashew constituents. Considering this, CO_2 can serve as an excellent solvent for promoting separation of CNSL from the cashew nut. Although not mentioned in this work, hot water and high-temperature water can also be used in processing the remainder of the cashew nut into liquid products or in processing other parts of the cashew tree. The cashew shell material has been shown by Smith et al. [60] to dissolve completely in high-temperature (ca. 600K) water, thus providing a source of liquid products. Still much research needs to be done in all of these areas including theoretical development.

6.3.12 Conclusions

AAs obtained from cashew have high bioactivity and are interesting as a class of phenolic lipids. AAs are best separated from cashew with supercritical carbon

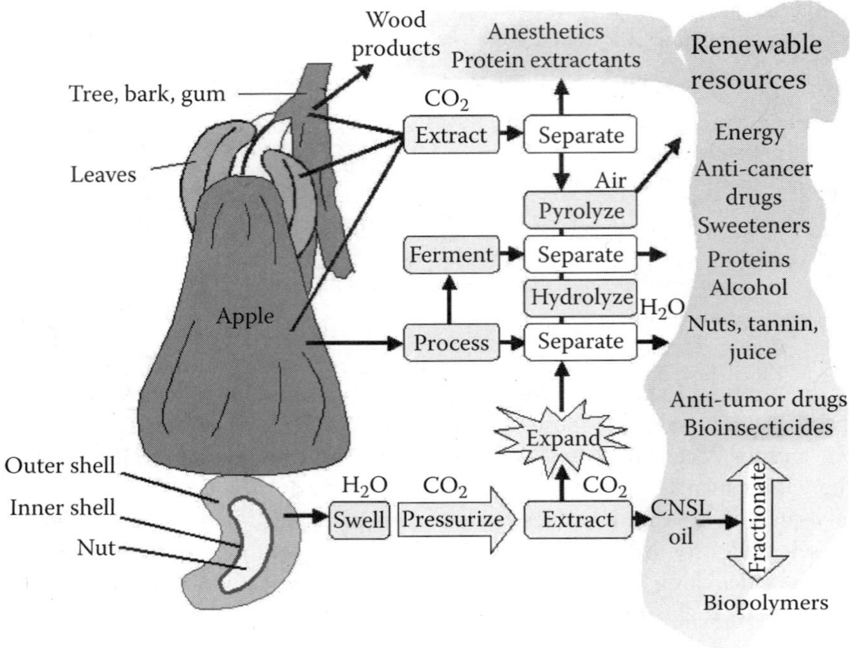

FIGURE 6.3.18 Scheme for processing cashew with CO_2 and water.

dioxide and using pressure-swing techniques. Cashew has a bright future as an agrochemical crop, and supercritical carbon dioxide can be used to maximize the quantity of bioactive compounds obtained from the nut and also to obtain bioactive compounds of the highest possible quality. More research is needed on processing the cashew tree and for developing new applications with the AAs, the cashew shell, gum leaves, and bark.

6.3.13 Acknowledgments

The authors acknowledge the Global Education Centers of Excellence program for partial support of this work. Also, the authors thank the students, Wahyu Setianto, Shouichiro Yoshikawa, Yuki Hanamura, Yusuke Sato, and Chisa Onuma, for diligence in their thesis work and for their great assistance in many of the figures. We also thank BPP Technologi (Jakarta) and specifically Mr. Priyo Atmaji for providing the wonderful overview of cashew processing in Indonesia and for supply the cashew nuts and Professor Roberto Malaluan for photographs of the actual cashew processing sites in the Philippines.

6.3.14 References

1. Kozubek, A., and J. H. P. Tyman. 1999. Resorcinolic lipids, the natural non-isoprenoid phenolic amphiphiles and their biological activity. *Chemical Reviews* 99:1–25.

2. Masuoka, N., and I. Kubo. 2004. Characterization of xanthine oxidase inhibition by anacardic acids. *Biochimica et Biophysica Acta—Molecular Basis of Disease* 1688:245–249.
3. Toyomizu, M., K. Okamoto, T. Nakatsu, and T. Konishi. 1994. A possible mechanism of uncoupling action of anacardic acid (AA) on oxidative-phosphorylation—Simultaneous determination of membrane potential-(Delta-Psi) and transmembrane pH difference (Delta-pH) in liposomes. *FASEB Journal* 8:A959–A959.
4. Toyomizu, M., S. Sugiyama, R. L. Jin, and T. Nakatsu. 1993. Alpha-glucosidase and aldose reductase inhibitors—Constituents of cashew, Anacardium-occidentale, nut shell liquids. *Phytotherapy Research* 7:252–254.
5. Toyomizu, M., K. Okamoto, T. Ishibashi, Z. Q. Chen, and T. Nakatsu. 2000. Uncoupling effect of anacardic acids from cashew nut shell oil on oxidative phosphorylation of rat liver mitochondria. *Life Sciences* 66:229–234.
6. Ha, T. J., and I. Kubo. 2005. Lipoxygenase inhibitory activity of anacardic acids. *Journal of Agricultural and Food Chemistry* 53:4350–4354.
7. Trevisan, M. T. S., B. Pfundstein, R. Haubner, G. Wurtele, B. Spiegelhalder, H. Bartsch, and R. W. Owen. 2006. Characterization of alkyl phenols in cashew (Anacardium occidentale) products and assay of their antioxidant capacity. *Food and Chemical Toxicology* 44:188–197.
8. Kubo, I. 1998. Molecular design of antioxidative and antimicrobial agents. Abstracts of Papers of the American Chemical Society 215:U7–U7.
9. Himejima, M., and I. Kubo. 1991. Antibacterial agents from the cashew Anacardium-occidentale (Anacardiaceae) nut shell oil. *Journal of Agricultural and Food Chemistry* 39:418–421.
10. Kubo, I., S. Komatsu, and M. Ochi. 1986. Molluscicides from the cashew Anacardium-occidentale and their large-scale isolation. *Journal of Agricultural and Food Chemistry* 34:970–973.
11. Toyomizu, M., K. Okamoto, Y. Akiba, T. Nakatsu, and T. Konishi. 2002. Anacardic acid-mediated changes in membrane potential and pH gradient across liposomal membranes. *Biochimica et Biophysica Acta—Biomembranes* 1558:54–62.
12. Toyomizu, M., K. Okamoto, T. Ishibashi, T. Nakatsu, and Y. Akiba. 2003. Reducing effect of dietary anacardic acid on body fat pads in rats. *Animal Science Journal* 74:499–504.
13. Skulachev, V. P. 1998. Uncoupling: New approaches to an old problem of bioenergetics. *Biochimica et Biophysica Acta—Bioenergetics* 1363:100–124.
14. Behrens, R. 1996. Cashew as an agroforestry crop: Prospects and potentials. Weikersheim, Germany: Margraf Publishers.
15. Food and Agricultural Organization. 2005. FAOSTAT—Detailed trade matrix. http://faostat.fao.org/ (accessed July 16, 2008).
16. Sarubbo, L. A., L. A. Oliveira, A. L. F. Porto, G. M. de Campos-Takaki, and E. B. Tambourgi. 2004. Partition of proteins in aqueous two-phase systems based on cashew-nut tree gum and poly(ethylene glycol). *Brazilian Archives of Biology and Technology* 47:685–691.
17. Sarubbo, L. A., L. A. Oliveira, A. L. F. Porto, H. S. Duarte, A. M. A. Carneiro-Leao, J. L. Lima, G. M. Campos-Takaki, and E. B. Tambourgi. 2000. New aqueous two-phase system based on cashew-nut tree gum and poly(ethylene glycol). *Journal of Chromatography B—Analytical Technologies in the Biomedical and Life Sciences* 743:79–84.
18. Sarubbo, L. A., L. A. Oliveira, A. L. F. Porto, J. L. Lima-Filho, G. M. Campos-Takaki, and E. B. Tambourgi. 2003. Performance of a perforated rotating disc contactor in the continuous extraction of a protein using the PEG-cashew-nut tree gum aqueous two-phase system. *Biochemical Engineering Journal* 16:221–227.

19. Oliveira, L. A., L. A. Sarubbo, A. L. F. Porto, G. M. Campos-Takaki, and E. B. Tambourgi. 2002. Partition of trypsin in aqueous two-phase systems of poly(ethylene glycol) and cashew-nut tree gum. *Process Biochemistry* 38:693–699.
20. Oliveira, L. A., L. A. Sarubbo, A. L. F. Porto, J. L. Lima, G. M. Campos-Takaki, and E. B. Tambourgi. 2002. Physical and rheological characterisation of polyethylene glycol-cashew-nut tree gum aqueous two-phase systems. *Journal of Chromatography B—Analytical Technologies in the Biomedical and Life Sciences* 766:27–36.
21. Silva, D. A., J. P. A. Feitosa, J. S. Maciel, H. C. B. Paula, and R. C. M. de Paula. 2006. Characterization of crosslinked cashew gum derivatives. *Carbohydrate Polymers* 66:16–26.
22. Silva, D. A., R. C. M. de Paula, J. P. A. Feitosa, A. C. F. de Brito, J. S. Maciel, and H. C. B. Paula. 2004. Carboxymethylation of cashew tree exudate polysaccharide. *Carbohydrate Polymers* 58:163–171.
23. Maciel, J. S., H. C. B. Paula, M. A. R. Miranda, J. M. Sasaki, and R. C. M. de Paula. 2006. Reacetylated chitosan/cashew gum gel: Preliminary study for potential utilization as drug release matrix. *Journal of Applied Polymer Science* 99:326–334.
24. Maciel, J. S., D. A. Silva, H. C. B. Paula, and R. C. M. de Paula. 2005. Chitosan/carboxymethyl cashew gum polyelectrolyte complex: Synthesis and thermal stability. *European Polymer Journal* 41:2726–2733.
25. Guilherme, M. R., G. M. Campese, E. Radovanovic, A. F. Rubira, J. P. A. Feitosa, and E. C. Muniz. 2005. Morphology and water affinity of superabsorbent hydrogels composed of methacrylated cashew gum and acrylamide with good mechanical properties. *Polymer* 46:7867–7873.
26. Guilherme, M. R., A. V. Reis, S. H. Takahashi, A. F. Rubira, J. P. A. Feitosa, and E. C. Muniz. 2005. Synthesis of a novel superabsorbent hydrogel by copolymerization of acrylamide and cashew gum modified with glycidyl methacrylate. *Carbohydrate Polymers* 61:464–471.
27. Ribeiro, R. C. C., J. C. G. Correia, M. B. M. Monte, P. R. Seidl, C. G. Mothe, and C. A. Lima. 2003. Cashew gum: A new depressor for limestone in the phosphate minerals flotation. *Minerals Engineering* 16:873–875.
28. Atmaji, P. 2003. Paper B-01 Cashew processing in Indonesia. International Mini-Symposium on Supercritical Fluid Extraction, 28–33. January 16–17, Sendai, Japan
29. Azam-Ali, S. H., and E. C. Judge. 2004. *Small-scale cashew nut processing.* Rome: FAO.
30. Ohler, J. G. 1979. *Cashew.* Amsterdam: Koninklijk Instituut voor de Tropen.
31. Russell, D. C. 1969. *Cashew nut processing.* Rome: Agricultural Services Bulletin (FAO).
32. Tyman, J. H. P. 1996. *Synthetic and natural phenols.* Amsterdam: Elsevier.
33. Shobha, S. V., and B. Ravindranath. 1991. Supercritical carbon-dioxide and solvent-extraction of the phenolic lipids of cashew nut (Anacardium-occidentale) shells. *Journal of Agricultural and Food Chemistry* 39:2214–2217.
34. Patel, R. N., S. Bandyopadhyay, and A. Ganesh. 2006. Economic appraisal of supercritical fluid extraction of refined cashew nut shell liquid. *Journal of Chromatography A* 1124:130–138.
35. Patel, R. N., S. Bandyopadhyay, and A. Ganesh. 2006. Extraction of cashew (Anacardium occidentale) nut shell liquid using supercritical carbon dioxide. *Bioresource Technology* 97:847–853.
36. Arai, K., M. Ajiri, S. Suzuki, and M. Nishimura. 1993. Japanese patent extraction of cardol and cardanol from cashew nutshell liquid. Japanese patent JP5000979.
37. Smith, R. L., Jr., R. M. Malaluan, W. B. Setianto, et al. 2003. Separation of cashew (*Anacardium occidentale* L.) nut shell liquid with supercritical carbon dioxide. *Bioresource Technology* 88:1–7.

38. Tyman, J. H. P., V. Tychopoulos, and P. Chan. 1984. Long-chain phenols, XXV: Quantitative-analysis of natural cashew nut-shell liquid (Anacardium-occidentale) by high-performance liquid-chromatography. *Journal of Chromatography* 303: 137–150.
39. Gaspar, F., T. J. Lu, R. Santos, and B. Al-Duri. 2003. Modelling the extraction of essential oils with compressed carbon dioxide. *Journal of Supercritical Fluids* 25:247–260.
40. Gaspar, F., T. J. Lu, R. Marriott, S. Mellor, C. Watkinson, B. Al-Duri, R. Santos, and J. Seville. 2003. Solubility of echium, borage, and lunaria seed oils in compressed CO_2. *Journal of Chemical and Engineering Data* 48:107–109.
41. Gaspar, F., R. Santos, and M. B. King. 2001. Disruption of glandular trichomes with compressed CO_2: Alternative matrix pre-treatment for CO_2 extraction of essential oils. *Journal of Supercritical Fluids* 21:11–22.
42. Dohrn, R., and G. Brunner. 1995. High-pressure fluid-phase equilibria—Experimental methods and systems investigated (1988–1993). *Fluid Phase Equilibria* 106:213–282.
43. Christov, M., and R. Dohrn. 2002. High-pressure fluid phase equilibria—Experimental methods and systems investigated (1994–1999). *Fluid Phase Equilibria* 202:153–218.
44. Yamini, Y., M. R. Fat'hi, N. Alizadeh, and M. Shamsipur. 1998. Solubility of dihydroxybenzene isomers in supercritical carbon dioxide. *Fluid Phase Equilibria* 152:299–305.
45. Garcia-Gonzalez, J., M. J. Molina, F. Rodriguez, and F. Mirada. 2001. Solubilities of phenol and pyrocatechol in supercritical carbon dioxide. *Journal of Chemical and Engineering Data* 46:918–921.
46. Francisco, J. C., B. Danielsson, A. Kozubek, and E. S. Dey. 2005. Extraction of rye bran by supercritical carbon dioxide: Influence of temperature, CO_2, and cosolvent flow rates. *Journal of Agricultural and Food Chemistry* 53:7432–7437.
47. Francisco, J. C., B. Danielsson, A. Kozubek, and E. Szwajcer. 2005. Application of supercritical carbon dioxide for the extraction of alkylresorcinols from rye bran. *Journal of Supercritical Fluids* 35:220–226.
48. Raeissi, S., J. C. Asensi, and C. J. Peters. 2002. Phase behavior of the binary system ethane plus linalool. *Journal of Supercritical Fluids* 24:111–121.
49. Raeissi, S. and C. J. Peters. 2005. Liquid-vapor and liquid-liquid-vapor equilibria in the ternary system ethane plus limonene plus linalool. *Journal of Supercritical Fluids* 33:201–208.
50. Raeissi, S. and C. J. Peters. 2001. Bubble-point pressures of the binary system carbon dioxide plus linalool. *Journal of Supercritical Fluids* 20:221–228.
51. Peters, C. J., J. D. Arons, J. M. H. L. Sengers, and J. S. Gallagher. 1988. Global phase-behavior of mixtures of short and long normal-alkanes. *Aiche Journal* 34:834–839.
52. Peters, C. J., and K. Gauter. 1999. Occurrence of holes in ternary fluid multiphase systems of near-critical carbon dioxide and certain solutes. *Chemical Reviews* 99:419–431.
53. Florusse, L. J., T. Fornari, S. B. Bottini, and C. J. Peters. 2002. Phase behavior of the binary system near-critical dimethylether and tripalmitin: Measurements and thermodynamic modeling. *Journal of Supercritical Fluids* 22:1–13.
54. Rovetto, L. J., C. J. Peters, and E. A. Brignole. 2005. Phase equilibrium behavior for hydrogenolysis components: Three-phase equilibria LLV and retrograde behavior. *Journal of Supercritical Fluids* 34:183–187.
55. Vankonynenburg, P. H., and R. L. Scott. 1980. Critical lines and phase-equilibria in binary vanderwaals mixtures. *Philosophical Transactions of the Royal Society of London Series A—Mathematical Physical and Engineering Sciences* 298:495–540.
56. Soave, G. 1972. Equilibrium constants from a modified Redlich-Kwong equation of state. *Chemical Engineering Science* 27:1192–203.
57. Soave, G. 1993. 20 Years of Redlich-Kwong equation of state. *Fluid Phase Equilibria* 82:345–359.

58. Joback, K. G., and R. C. Reid. 1987. Estimation of pure-component properties from group-contributions. *Chemical Engineering Communications* 57:233–243.
59. Constantinou, L., and R. Gani. 1994. New group-contribution method for estimating properties of pure compounds. *AIChE Journal* 40:1697–1710.
60. Smith, Jr., R. L., R. M. Malaluan, W. B. Setianto, H. Inomata, and K. Arai. 2002. Green processing of cashew nut (*Anacardium occidentale*) and cashew nut shell liquid with carbon dioxide and water. Asian Pacific Confederation of Chemical Engineering (APCChE) Paper 786.

6.4 FRACTIONATION OF ORANGE VOLATILE OIL

Motonobu Goto

In this section, separation processes for citrus oil using supercritical fluids are reviewed. The main objective for citrus oil processing is to remove terpenes from oxygenated aroma compounds. Because phase equilibria are the basis for the separation process, the literature for phase equilibria of citrus components is surveyed. For the separation of citrus oils, the extraction and adsorption/desorption processes are explained. Because citrus oils consist of a number of components having similar properties, a countercurrent extraction column is usually applied to obtain higher separation performance. Both semi-batch and continuous operations are used for the analysis of the separation process. The adsorption/desorption process is also used for citrus oil separation. Oxygenated aroma compounds are usually preferably adsorbed on silica gel. As a continuous operation process, the pressure swing adsorption process has been developed.

6.4.1 Phase Equilibria for Citrus Oil Components

In the supercritical extraction process either from natural solid feed or liquid feed, knowledge of the solubility of the components is one of the most important factors for design and analysis. The data have been generally interpolated by using equations of state. Solubility data of essential oil components were listed by Reverchon [1]. A main objective for the essential oil fractionation process is often deterpenation, where hydrocarbon terpenes are separated from oxygenated aroma compounds. Most of the data available in the literature are binary phase equilibria of pure components in supercritical CO_2. Solubility data for limonene have been measured by several researchers [2–8], because limonene is a major component in most essential oils and separation between limonene and linalool is one of the most difficult systems in essential oils. Figure 6.4.1 shows limonene + CO_2 phase equilibria [9]. Vapor–liquid equilibria were correlated by the Peng–Robinson equation of state (EOS) using conventional mixing rules with two interaction parameters. Equilibria for the linalool-CO_2 system, which is a major aroma component in orange oil, have been reported [6–8, 10]. In addition to these binary systems, equilibria for the ternary system, limonene–linalool–CO_2, have been reported [7, 11, 12]. To investigate the entrainer effect, phase equilibria including ethanol were studied [13, 14]. Instead of CO_2, the phase behaviors for the ethane system were reported by Raeissi and Peters [15, 16].

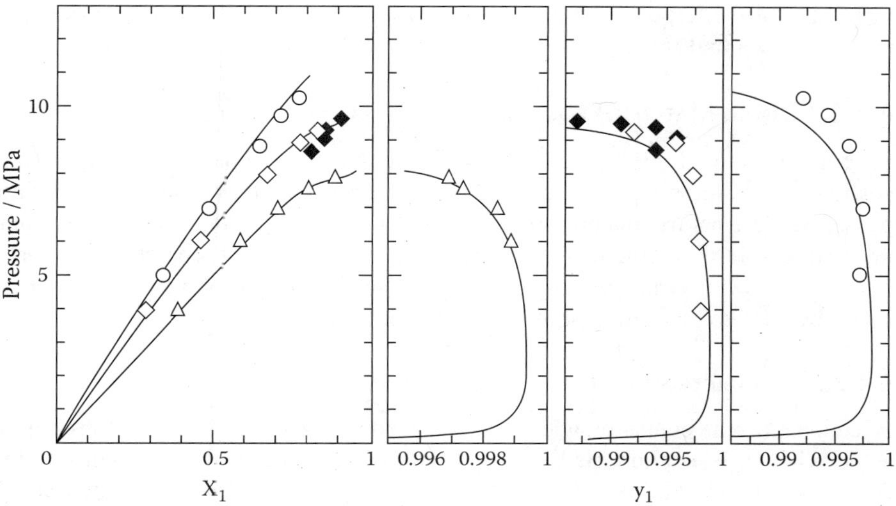

FIGURE 6.4.1 Vapcr–liquid equilibria for CO_2 (1) + limonene (2) and calculated results by Peng–Robinson equat:on of state (EOS).

Some authors studied phase equilibria of essential oils as a mixture (e.g., lemon oil [17, 18], orange oil [19–21]), and bergamot oil [18]. Figure 6.4.2 shows phase equilibria of orange oil and its fractions. Mutual solubilities reveal the extent of the two-phase region with respect to pressure, which is important for the design of a countercurrent separation process. Orange oil consists of terpenes (about 98 wt %) and aroma components. Mutual solubilities at isobaric and isothermal conditions

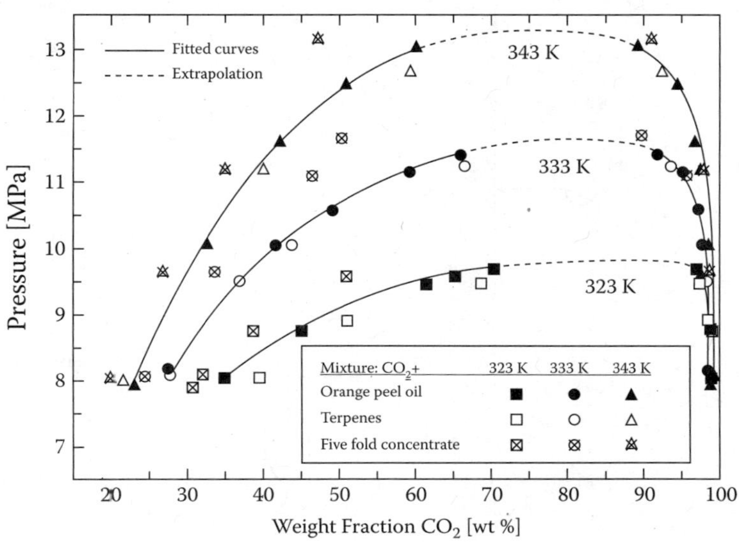

FIGURE 6.4.2 Phase equilibria of citrus oil fractions.

increased for the terpene fraction, whereas a fivefold aroma concentrate exhibits a much lower mutual solubility and increased critical points.

6.4.2 Liquid Material Processing

Separation of liquid mixture by supercritical fluid extraction is a process involving partioning and mass transfer between the supercritical fluid phase and the liquid phase. Because natural materials consist of a number of components having similar properties, a simple extraction process cannot achieve sufficient separation. In such a case, the following strategies may improve the separation: (a) countercurrent extraction process and (b) an adsorption/desorption process.

6.4.2.1 Countercurrent Extraction Process

A countercurrent contactor with a multistaged tower or packed bed tower may achieve higher separation. Because supercritical fluid density changes significantly from gas-like to liquid-like by changing the temperature or pressure, a supercritical fluid extraction tower can be regarded as a carrier gas distillation device or a countercurrent solvent extractor.

A schematic drawing of a countercurrent extractor is shown in Figure 6.4.3. In semi-batch operation mode, liquid feed is charged at the bottom and supercritical fluid flows from the bottom to the top. Composition of extracts changes with extraction

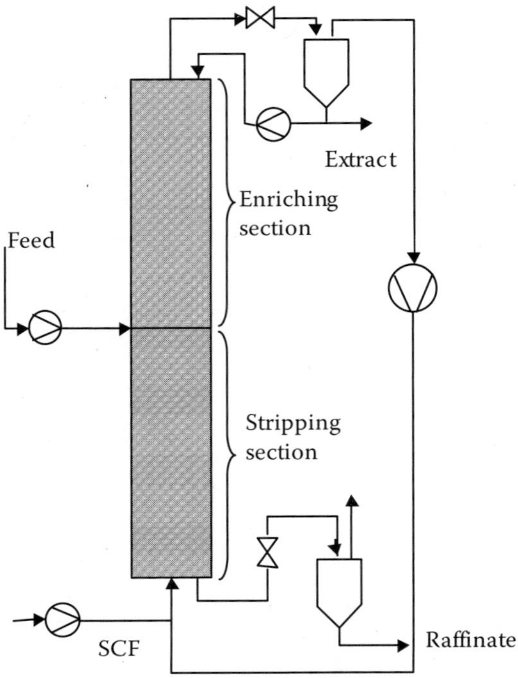

FIGURE 6.4.3 Process scheme of a countercurrent supercritical fluid extractor.

time, where components with higher solubility are extracted earlier and those with lower solubility are extracted later in time. More than two components can be separated into each fraction in series as a function of extraction time. The fractionation is enhanced by reflux, where a part of extracted solutes is liquefied and dropped down to give countercurrent contact with supercritical fluid within the column. The reflux can be achieved by internal reflux induced by a temperature gradient along the column or by external reflux from a separator set at lower pressure. In the continuous operation mode, liquid feed is supplied continuously at the middle of the column and supercritical fluid is fed at the bottom. The upper part of the extraction column serves as the enriching section and the lower part as the stripping section. Feed material is basically fractionated into two fractions of extract and raffinate. The fraction with higher solubility (light components) can be obtained from the top and the fraction with lower solubility (heavy components) at the bottom. For the separation of more than two components into each of the fractions, several extractor units have to be combined. For the separation of n components, n–1 extractors were necessary. The principle, design, and analysis of the countercurrent process are described by Brunner [19, 22].

The countercurrent extraction process has been applied to natural material processing such as citrus oils, unsaturated fatty acids, and squalene-tocopherol [19]. An important application is citrus oil processing, which is one of the most important subjects in the perfume and food industries. Citrus oil consists of terpenes, oxygenated aroma compounds, waxes, and pigments. A small amount of oxygenated compounds contributes to the specific flavor properties. Terpene content must be reduced to stabilize the products. Terpenes are conventionally removed by vacuum distillation or solvent extraction, which may involve thermal degradation and an organic solvent residue problem.

Simple extraction process does not achieve sufficient selectivity; mainly two methods, the countercurrent extraction and the adsorption/desorption processes, have been investigated. Stahl et al. [23] proposed a continuous countercurrent extraction process for orange oils. With an axial temperature profile 358 K in the middle, 348 K at the top, and 333 K at the bottom at a pressure of 8 MPa, they reduced the terpene content from 90 to 42% at the bottom. Perre et al. [24] described an pilot-scale apparatus.

Sato et al. [25–28] and Goto et al. [29, 30] have developed the countercurrent extraction process for orange oil processing. They have used a 20-mm wide, 2400-mm long column where the upper three-quarters was packed with 3-mm Dixon Packings. The extractor was used in both semi-batch mode and continuous mode of operation by using either cold-pressed orange oil or a model mixture composed of limonene, linalool, and citral (neral + geranial). In semi-batch mode, the operating condition used was 313–353 K with and without a temperature gradient at a pressure of 8.8 and 9.8 MPa. The temperature profile affected the separation behavior because of the internal reflux and countercurrent contact between the liquid phase and supercritical fluid phase. Figure 6.4.4 shows the results for the semi-batch extraction of a model mixture operated with temperature gradient 313–333 K at 8.8 MPa. Limonene, linalool, neral, and geranial were extracted sequentially. The separation selectivity increased by temperature gradient, and the selectivity was 2.87 at the optimal condition of 313–333 K. The separation selectivity between limonene and linalool

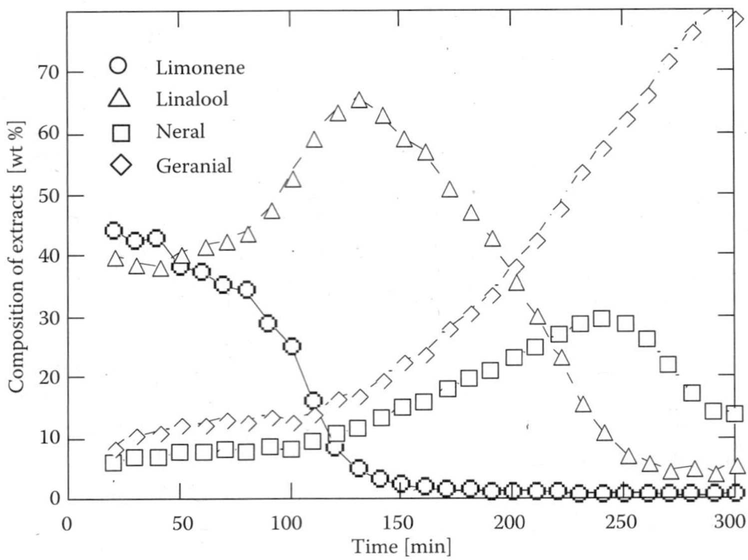

FIGURE 6.4.4 Change in composition of extracts for semi-batch extraction of a model citrus oil mixture at 8.8 MPa and 313–333 K.

was defined in terms of their mass ratio by $(Y_{\text{limonene}} / X_{\text{limonene}})/(Y_{\text{linalool}} / X_{\text{linalool}})$. The internal reflux ratio was calculated by the measurement of the extraction rates at the top and bottom of the column, and it was 7.6 at the above condition [26]. From the estimation of the physical properties such as solubility, density, and viscosity in the column, they found that these properties were remarkably changed in a temperature range of 313–333 K.

In the continuous operation, the terpene-rich fraction is recovered from the top of the column and oxygenated compounds are obtained from the bottom. Figure 6.4.5 shows the effect of the solvent-to-feed ratio (S/F) ratio on the extraction ratio of limonene at 333 K. The extraction ratio was defined by the ratio of weight of limonene in extract to that in feed. The extraction ratio increased with the increase in the S/F ratio and pressure. The extraction ratio was larger for the raw orange oil than the model mixture. Figure 6.4.6 shows the effect of the S/F ratio on the separation selectivity at 333 K. The selectivity also increased with the increase in the S/F ratio. When the model mixture was used as feed, high selectivity up to 705 was observed. The selectivity of the raw orange oil was lower than that of the model mixture because of the low stage efficiency induced by high terpene contents and the interaction among solutes.

For a system of the model mixture, the experimental HETS (height equivalent to a theoretical stage) was calculated by using the phase equilibria estimated by the Peng–Robinson EOS with binary interaction parameters and the Soave–Redlich–Kwong EOS with the interaction parameters set equal to zero. Figure 6.4.7 shows the effect of the S/F ratio on the calculated HETS for the model mixture. The HETS

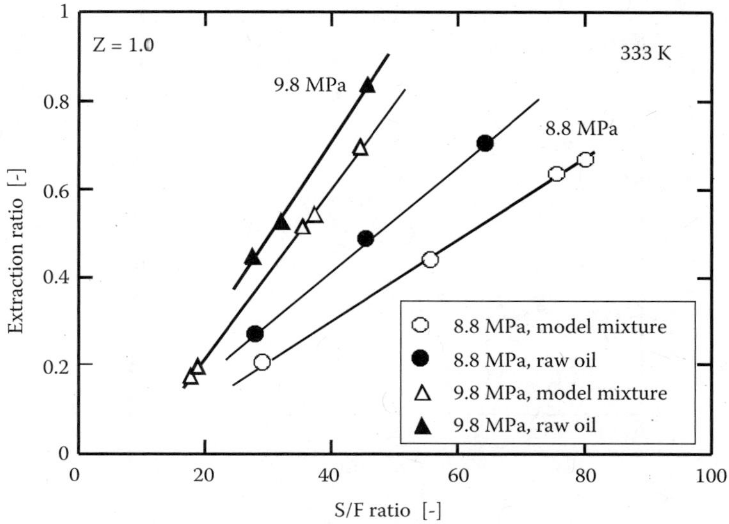

FIGURE 6.4.5 Effect of solvent-to-feed (S/F) flow ratio on the extraction ratio of limonene for a countercurrent extraction of orange oil.

decreased from 4.0 to 0.2 m as the S/F ratio increased from 30 to 80 at 333 K at 8.8 MPa or from 20 to 45 at 333 K at 9.8 MPa. For the raw orange oil processing, the HETS may be larger than that for the model oil processing.

For the raw orange oil processing, the effect of S/F ratio on the concentration factor, defined by the concentration divided by that in feed, was investigated for

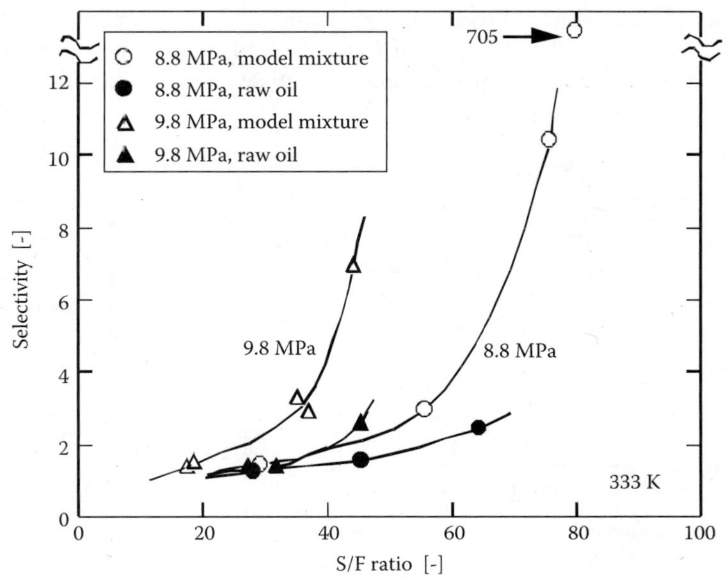

FIGURE 6.4.6 Selectivity between limonene and linalool for a countercurrent extraction of orange oil.

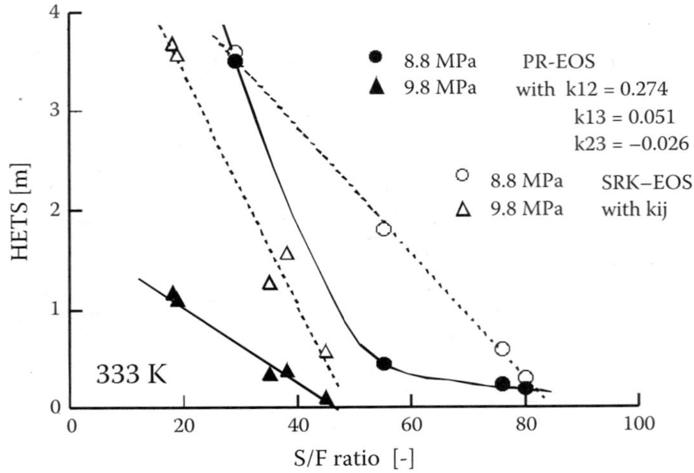

FIGURE 6.4.7 HETS for a countercurrent extraction of a model mixture.

major constituents. Figure 6.4.8 shows the variation in the concentration factor of each component and recovery of oxygenated compounds in raffinate at 333 K at 8.8 MPa. The solid symbols are terpenes, which are desired to be smaller than unity in raffinate, whereas the open symbols are oxygenated compounds, which are desired to be larger than unity in raffinate. Oxygenated compounds were concentrated with an increase in S/F ratio, without the decrease of the recovery yield. The deviations from unity for limonene and linalool were smaller than the other terpenes and oxygenated compounds. Therefore, when limonene and linalool were separated, the other components could be separated more selectively, that is, terpeneless oil could

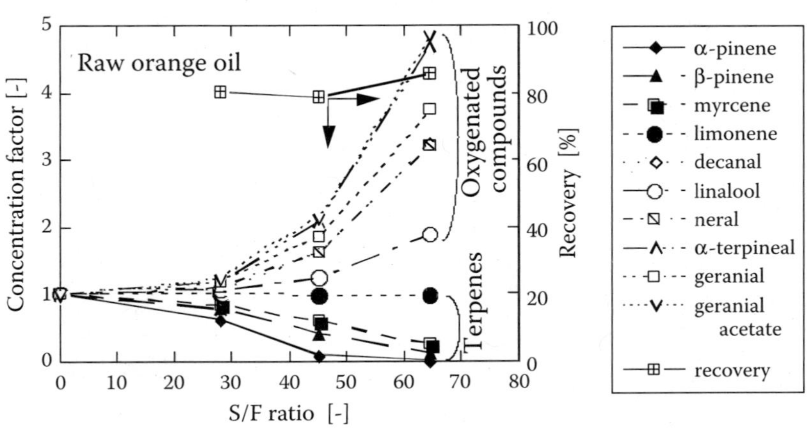

FIGURE 6.4.8 Concentration factor of each components and recovery of oxygenated compounds in the raffinate at 333 K and 8.8 MPa.

be obtained. The total concentration factor of oxygenated compounds in raffinate was 3.0 at the S/F ratio of 64 and at 333 K and 8.8 MPa.

For countercurrent operation, withdrawal of product from the side-cut stream at the intermediate of the column would be useful. Sato et al. [31] successfully fractionated orange oil into three fractions. Terpenes and oxygenated compounds were obtained from the stream at the top and the side-cut stream, respectively, as supercritical fluid phase stream. Waxes were recovered at the bottom of the column as a liquid-phase stream. Thus, terpenes, oxygenated compounds, and waxes in citrus oil could be simultaneously fractionated into extract, side-stream, and raffinate, owing to their solubility differences in supercritical CO_2.

Budich et al. [32] made a stage calculation based on the Jänecke diagram to evaluate the vapor–liquid equilibrium data of the system CO_2 + orange peel oil. Countercurrent column experiments were carried out to study the limit of separation. Flooding-point data were also determined to enable scale-up calculations [20]. They showed that 18 theoretical stages, equal to about 9 m of column height and a reflux ratio of 2.5, were required to produce a 20-fold concentrate containing 68.8 wt % of terpenes from a feed material of 98.25 wt % at 333 K and 10.7 MPa and a S/F ratio of 100. With a feed flow of 100 kg/h, the minimum inner diameter of the column determined from the flooding point data was 0.4 m.

Diaz et al. [33] applied a nonlinear programming model to formulate for the simultaneous determination of operating conditions and for the process and solvent cycle scheme, including investment and operating costs for deterpenation of orange peel oil. High solvent flow rate and reflux ratio together with a larger stripping section were necessary to obtain a fivefold product concentrate. Different solvent cycle schemes (pump and compressor) were included in the mathematical model. Net profit was maximized, taking into account capital and operating costs associated with the complete deterpenation process. The results showed that a compression cycle is the optimal solvent recovery system in all cases.

Lemon oil was fractionated by Kondo et al. [34]. Citral (geranial and neral) is the major component in the oxygenated component. The continuous operation with a linear temperature gradient from 313 to 333 K along the column at 8.8 MPa showed the highest selectivity. The selectivity increased with the increase in the S/F ratio, and oxygenated compounds were more concentrated with an increase in the S/F ratio without a decrease in the recovery.

Kondo et al. [35] applied the fractionation to bergamot oil. The composition of the bergamot oil was 40 wt % terpenes (25 wt % limonene) and 60 wt % oxygenated compounds (25 wt % linalyl acetate). Thus, the content of oxygenated compounds was much higher than that for orange oil or lemon oil. For the continuous countercurrent extraction at 333 K, the operation at 8.8 MPa gave higher selectivity than the operation at 9.8 MPa. At the S/F ratio of 63.2, the concentration of terpenes in the raffinate was reduced to less than 1 wt %, that is, terpeneless oil was obtained. For deterpenation of bergamot oil, a process simulator (Simsci Pro/II) was used to evaluate the separation behavior by Kondo et al. [36]. The effects of operating conditions on extraction ratio of limonene, separation selectivity, and recovery of linalyl acetate were observed as a function of the S/F ratio. Reflux of the top product was not a rewarding strategy for this system. The performance was improved with increase in the stage number at a higher S/F ratio.

6.4.2.2 Adsorption/Desorption Process

Separation in the countercurrent extraction process is achieved based on the phase equilibria. The operating condition is limited by the formation of homogeneous phase, because two phases, the liquid and supercritical fluid phases, must exist to achieve the separation. Because separation process by supercritical fluid extraction is based on phase equilibria and the difference in solubility, separation selectivity is often very small for natural components mixture. On the other hand, the adsorption process can be operated in the homogeneous phase because the separation occurs at the solid–fluid interface. Adsorption technology has been extensively used for both gas- and liquid-phase separation. When an adsorbent is used in a supercritical fluid, separation that would occur is based on adsorption equilibrium, which is often much more selective. Three methods have been used for essential oil fractionation by using an adsorbent in a supercritical fluid: preparative supercritical chromatography, desorption by supercritical fluid after adsorption in liquid phase, and adsorption/desorption in supercritical fluid.

Adsorbents were also used for citrus oil processing. Yamauchi and Saito [37] fractionated lemon peel oil with gradual increase in pressure (10–20 MPa at 313 K) by supercritical fluid chromatography. Four fractions were obtained and were mainly composed of terpenes, oxygenated terpenes, oxygenated terpenes removed by ethanol addition, and high-molecular-weight compounds. Barth et al. [38] and Chouchi et al. [39, 40] used supercritical CO_2 to desorb or extract the oxygenated aroma compounds after the equilibrium saturation under ambient conditions in a batch operation. They obtained a high-quality essential oil containing less terpenes and less nonvolatiles by supercritical CO_2 desorption with increasing pressure. However, high pressure or a cosolvent was required to regenerate the adsorber because the nonvolatiles such as waxes and pigments were more strongly adsorbed on silica gel than aroma compounds. These results suggested that aroma compounds could be more selectively adsorbed on silica gel than terpenes, relatively higher pressure is required to desorb them or regenerate the adsorbent, and the nonvolatiles must be removed to maintain the activity of adsorbent. Reverchon [41] studied selective desorption of limonene and linalool from silica gel by supercritical CO_2. The maximum selectivity was obtained by operation at 313 K and 0.1 kg/kg loading and in two successive pressure steps. The first step was performed at 7.5 MPa and produced the selective desorption of limonene; the second one was performed at 20 MPa and assured the fast desorption of linalool. Shen et al. [42] used silica gel and alumina as an adsorbent to concentrate flavor compounds from orange oil. Silica gel had a larger adsorption capacity than alumina. Orange oil was pumped to adsorbent column and then desorbed with supercritical CO_2. In the adsorption step, three-fourths of the terpene hydrocarbons were removed. Desorption at low temperatures and flow rates improved separation by supercritical CO_2. The oxygenated compounds were desorbed later than the terpene hydrocarbons. The ratio of aldehydes and alcohols to terpenes increased as desorption progressed. Decanal was concentrated to 20 times that of the feed oil using supercritical CO_2 at 13.1 MPa and 308 K.

The adsorption equilibrium constants of limonene and linalool on silica gel in supercritical CO_2 measured by an impulse response technique, are shown in Figure 6.4.9 [43]. Adsorption equilibrium constants were correlated linearly in log-log plot

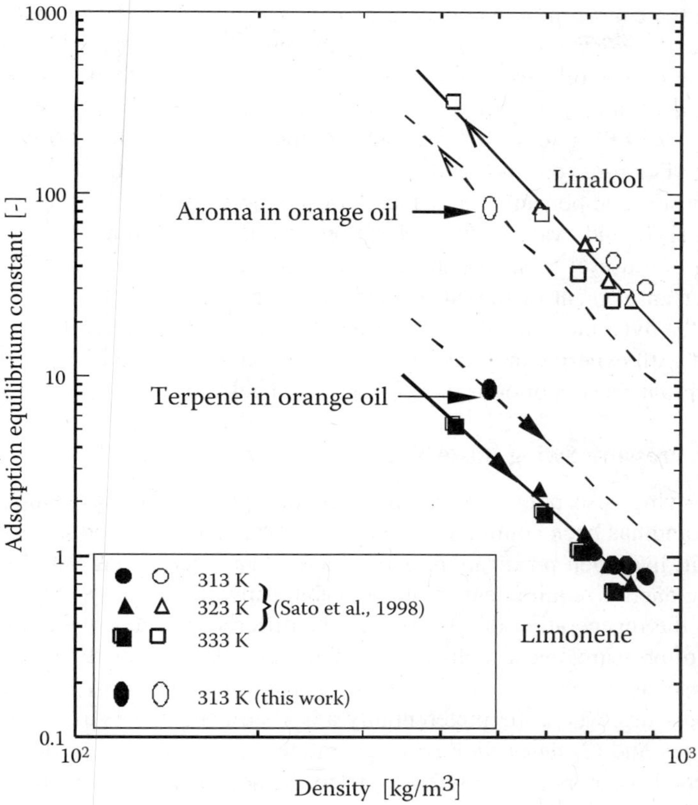

FIGURE 6.4.9 Adsorption equilibrium constant of orange oil components on silica gel in supercritical CO_2.

as a function of the density of supercritical CO_2 independent of pressure and temperature. Adsorbed amounts decreased with the increase in the solvent density for both limonene and linalool. These results suggest the possibility of a process where oxygenated compounds are selectively adsorbed on the adsorbent at a lower pressure and then desorbed at a higher pressure. Adsorption isotherm was also measured for aroma and terpene fractions of orange oil by a step response method and correlated with a multicomponent Langmuir equation:

$$q_i = \frac{q_s K_i C_i}{1 + \Sigma K_i C_i}, \tag{6.4.1}$$

where adsorption equilibrium constants were correlated as a function of CO_2 density: $K_1 = 4.237 \times 10^8 \rho^{-2.874}$ for terpene and $K_2 = 9.395 \times 10^{10} \rho^{-3.371}$ for aroma. The Langmuir equation was also applied by Reverchon [41] for limonene–linalool mixtures on silica gel.

From the equilibrium relation it is evident that adsorption is favored at a lower pressure and desorption is favored at a higher pressure. Adsorption and desorption behavior of orange oil was measured at 313 K [27]. After the feed, orange oil broke through the column at 8.8 MPa, and pure supercritical CO_2 was passed through the column at 19.4 MPa to desorb the solutes. Figure 6.4.10 shows the desorbed amounts and the variation in concentration of solutes desorbed divided by that in feed. Oxygenated aroma components were concentrated up to 50 times.

To analyze and evaluate the adsorption and desorption process, mathematical modeling is a useful tool. The differential mass balances for both fluid and solid phases for an element of the adsorbent column provide a mathematical model to describe the dynamic behavior of the process. The desorption process was modeled and fitted with experimental data by Reverchon [41]. Silva et al. [44] also modeled the desorption process of orange oil from a silica gel bed.

6.4.2.3 Pressure Swing Adsorption

Pressure swing adsorption (PSA) is an important process for the separation of gas mixtures and has been commercialized for air drying, oxygen and nitrogen separation of air, hydrogen purification, and various other separations [45, 46], because of its low energy requirements and low capital investment costs. PSA process is based on the regeneration of adsorber by the difference of adsorbed amounts as a function of pressure, because the adsorbed amounts decrease with decreasing pressure in gaseous systems. The process involves two fundamental steps: (1) adsorption at high pressure, where the preferentially adsorbed components are taken up from the feed gas and (2) desorption or regeneration step by pressure reduction, where the adsorbed components are removed from the adsorbent. Most commercial PSA processes are of multibed design to generate a near-continuous product stream. In

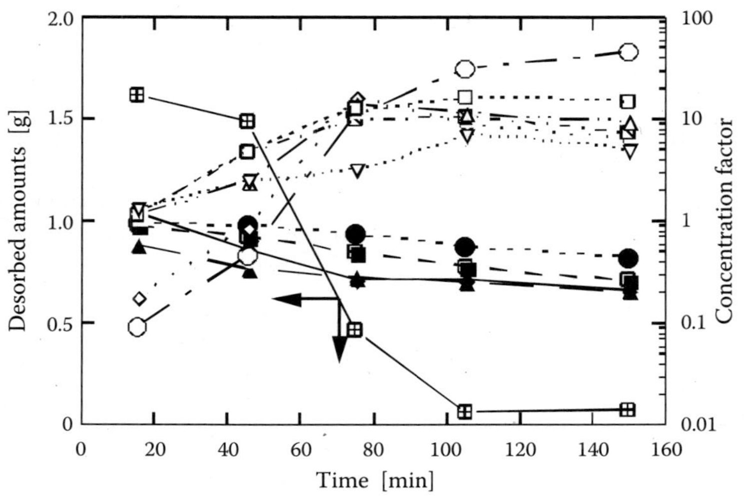

FIGURE 6.4.10 Desorption curves from silica gel saturated by orange oil. Definitions of symbols are the same as in Figure 6.4.8.

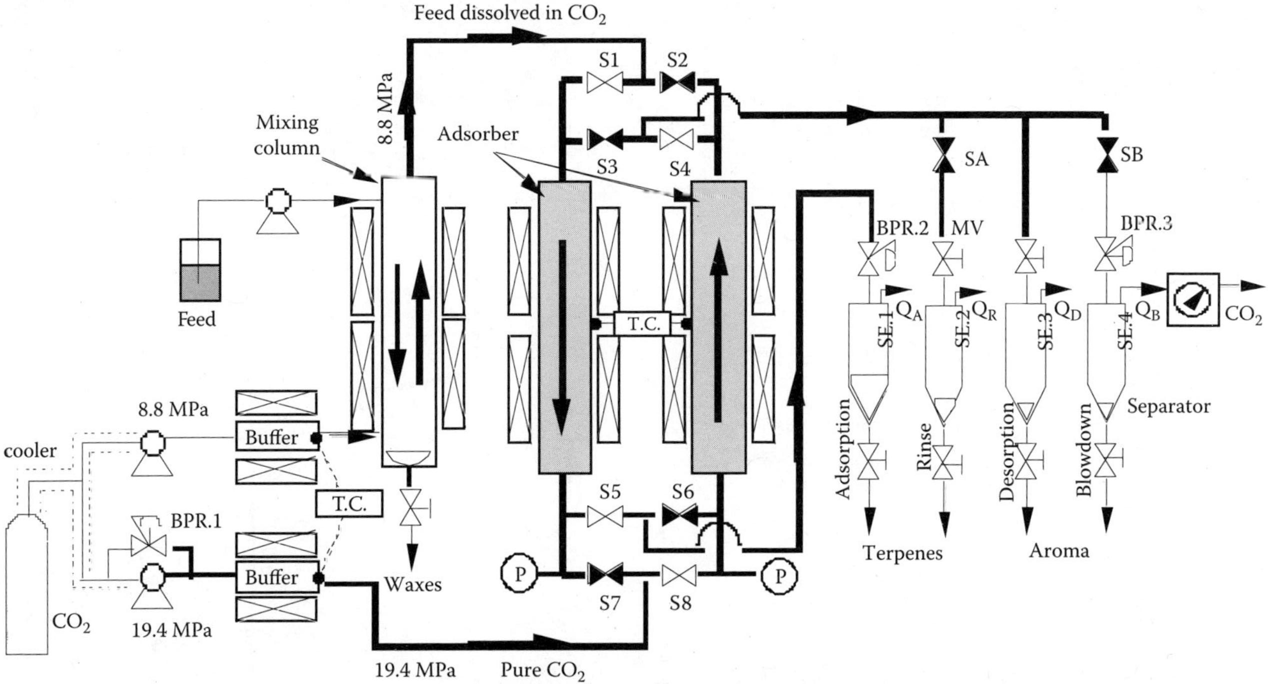

FIGURE 6.4.11 Experimental setup of supercritical pressure swing adsorption for citrus oil processing.

the case of a two-bed process, one bed is in the adsorption step, while the other one is in the desorption step.

Sato et al. [47] applied the pressure swing adsorption to supercritical fluid for the fractionation of citrus oil. Figure 6.4.11 shows the schematic diagram of the experimental setup. A continuous cyclic operation between the adsorption step, where a cold-pressed orange oil in supercritical CO_2 was continuously passed through the column at 8.8 MPa and 313 K, and the desorption step, where pure supercritical CO_2 was passed through the column at 19.4 MPa, including the rinse step, was demonstrated. All experiments were started with clean beds. In general, 10 half cycles were required to approach the cyclic steady state. The operation is shown in Figure 6.4.12.

The effect of desorption-to-adsorption CO_2 flow ratio Q_D / Q_A on the concentration factor and the recovery at a constant Q_A are shown in Figures 6.4.13 and 6.4.14, respectively. The concentration factor of 10 and the recovery of 65% were obtained at a Q_D / Q_A ratio of 2. An increase in the Q_D / Q_A ratio caused higher recovery in the desorption step. A mathematical model to simulate pressure swing adsorption process was developed. Model calculations agreed roughly with the experimental results as shown in Figures 6.4.13 and 6.4.14.

Figure 6.4.15 shows the gas chromatograms of the feed and the oil obtained from adsorption and desorption steps at a half cycle time of 120 min. The chromatogram for the adsorption step shows that aroma compounds in orange oil are adsorbed on the silica gel in the adsorption step; therefore, terpenes make up the major portion

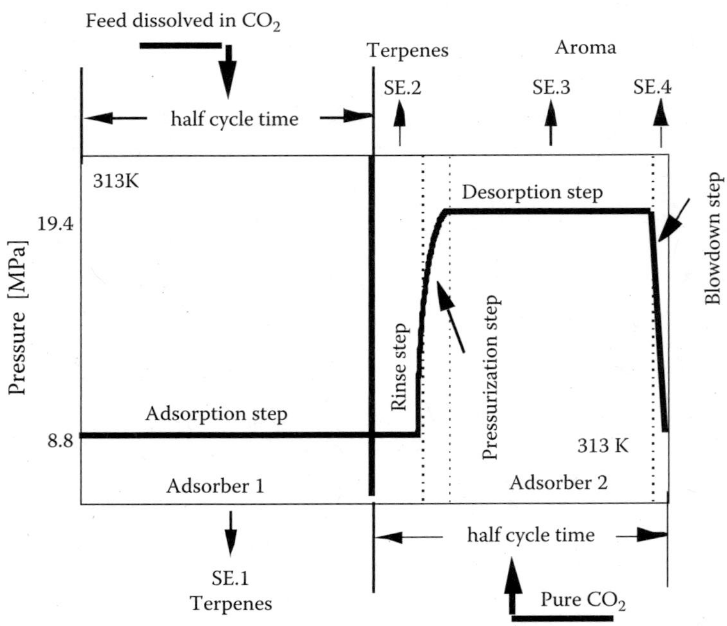

FIGURE 6.4.12 Configuration of a pressure swing adsorption for the citrus oil processing in supercritical CO_2.

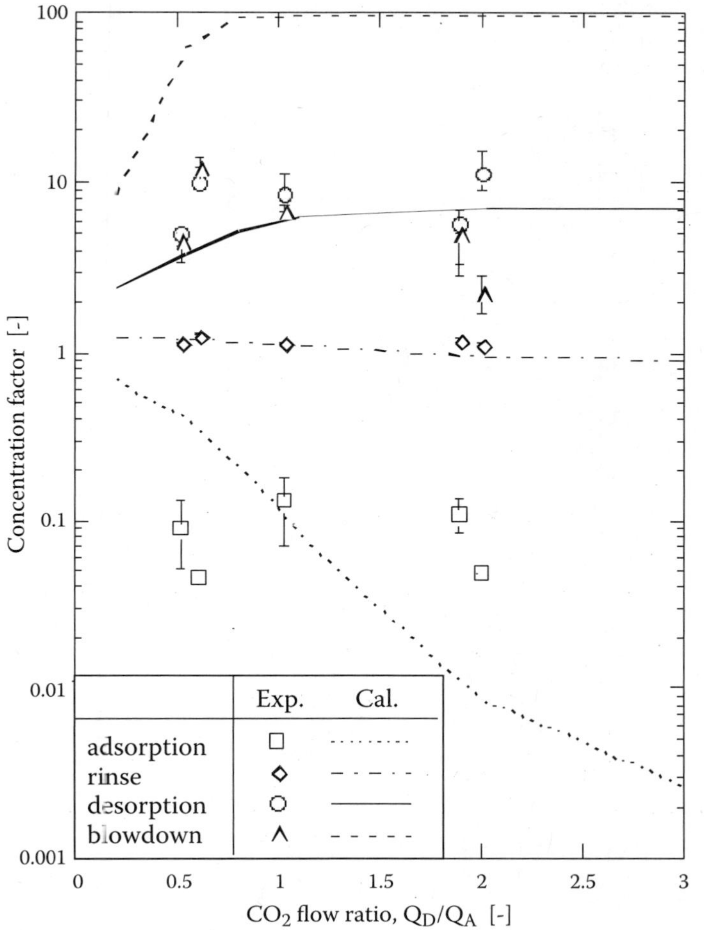

FIGURE 6.4.13 Concentration factor for the pressure swing adsorption of orange oil.

of the oil obtained in the adsorption step. On the other hand, the chromatogram for the desorption step shows that aroma compounds adsorbed on the silica gel in the adsorption step are desorbed, so that aroma compounds make up the major portion of the product oil obtained in the desorption step.

Pressure swing adsorption process was also applied to deterpenation of bergamot oil [48]. Because the content of oxygenated compounds in feed oil was considerably higher in comparison with orange oil, the separation performance was evaluated in terms of purity, defined by a fraction of oxygenated components in the product. The highest purity of 0.84 was obtained in the desorption step at a pressure ratio (desorption pressure/adsorption pressure) of 2.5 at a desorption pressure of 24.8 MPa. The recovery and yield increased with an increase in the pressure ratio.

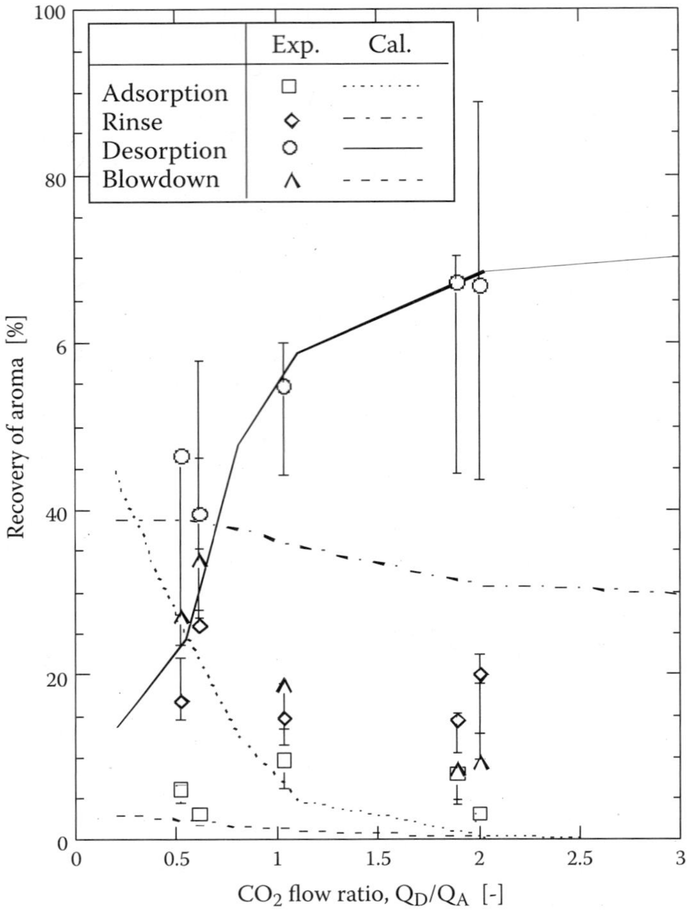

FIGURE 6.4.14 Recovery of aroma for the pressure swing adsorption of orange oil.

6.4.3 Nomenclature

Symbol	Description	Units in SI System	Dimensions in M, N, L, T, and θ
C_i	Concentration of component *i*	kg·m^{-3}	M·L^{-3}
K_i	Langmuir parameter	—	—
k_{ij}	Binary interaction parameter	—	—
Q_A	Flow rate in adsorption step	g·s^{-1}	M·T^{-1}
Q_D	Flow rate in desorption step	g·s^{-1}	M·T^{-1}
q_i	Adsorbed amounts of component *i*	kg·kg^{-1}	M·M^{-1}
q_c	Adsorbed amounts in equilibrium	kg·kg^{-1}	M·M^{-1}
X	Composition of solutes in liquid phase	—	—
Y	Composition of solutes in vapor phase	—	—
Z	Length of stripping section/length of rectification section	—	—
ρ	Density	kg·m^{-3}	M·L^{-3}

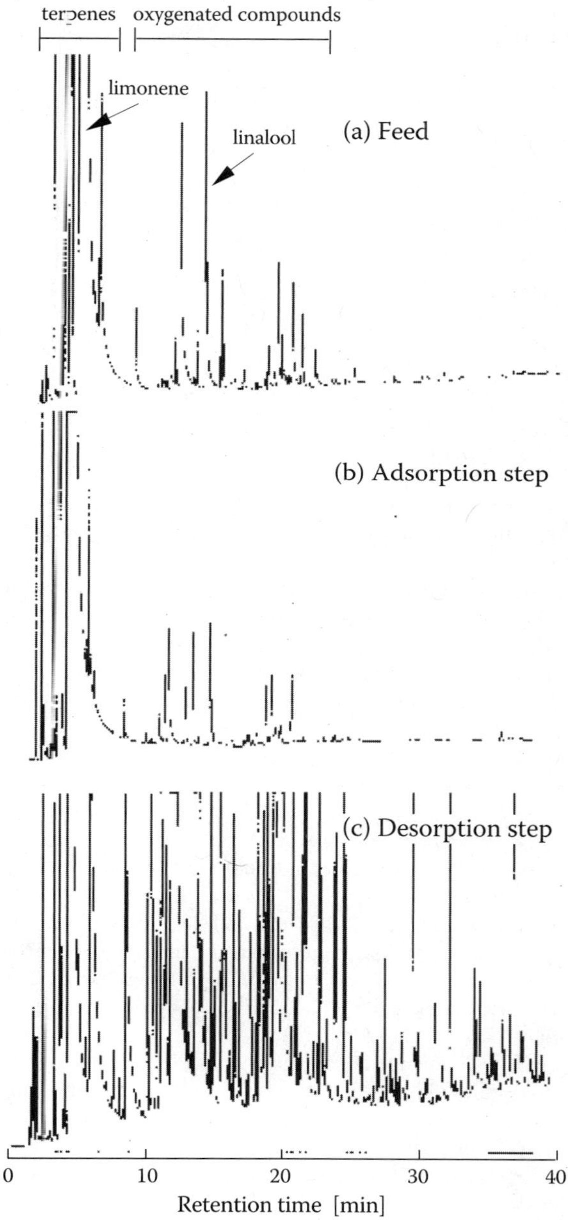

FIGURE 6.4.15 Gas chromatograms of orange oil for (a) feed, (b) adsorption step, and (c) desorption step in pressure swing adsorption processing.

6.4.4 References

1. Reverchon, E. 1997. Supercritical fluid extraction and fractionation of essential oils and related products. *Journal of Supercritical Fluids* 10 (1): 1–37.
2. Stahl, E., and D. Gerard. 1988. Solubility behavior and fractionation of essential oils in dense carbon dioxide. *Perfumer & Flavorist* 10:29–37.

3. Matos, H. A., E. G. D. Azevedo, P. C. Simoes, M. T. Carronde, and M. N. D. Ponte. 1989. Phase equilibria of natural flavors and supercritical solvents. *Fluid Phase Equilibria* 52:357–364.
4. Marteau, Ph., J. Obriot, and R. Tufeu. 1995. Experimental determination of vapor-liquid equilibria of CO_2 + limonene and CO_2 + citral mixtures. *Journal of Supercritical Fluids* 8 (1): 20–24.
5. Suzuki, J., and K. Nagahama. 1996. Measurement and correlation of solubility of limonene and linalool in high pressure carbon dioxide. *Kagakukougaku Ronbunshu* 22 (1): 199–200.
6. Temelli, F., J. P. O'Connell, C. S. Chen, and R. J. Braddock. 1990. Thermodynamic analysis of supercritical carbon dioxide extraction of terpenes from cold-pressed orange oil. *Industrial & Engineering Chemistry Research* 29:618–624.
7. Vieira de Melo, S. A. B., G. M. N. Costa, A. M. C. Uller, and F. L. P. Pessoa. 1999. Modeling high-pressure vapor-liquid equilibrium of limonene, linalool and carbon dioxide systems. *Journal of Supercritical Fluids* 1:107–117.
8. Berna, A., A. Chafer, and J. B. Monton. 2000. Solubilities of essential oil components of orange in supercritical carbon dioxide. *Journal of Chemical Engineering Data* 45:724–727.
9. Iwai, Y., N. Hosotani, T. Morotomi, Y. Koga, and Y. Arai. 1994. High pressure vapor-liquid equilibria for carbon dioxide + linalool. *Journal of Chemical Engineering Data* 39:900–902.
10. Iwai, Y., T. Morotomi, K. Sakamoto, Y. Koga, and Y. Arai. 1996. High pressure vapor-liquid equilibria for carbon dioxide + limonene. *Journal of Chemical Engineering Data* 41:951–952.
11. Raeissi, S., and C. J. Peters. 2005. Experimental determination of high-pressure phase equilibria of the ternary system carbon dioxide + limonene + linalool. *Journal of Supercritical Fluids* 35:10–17.
12. Chafer, A., A. Berna, J. B. Monton, and A. Mulet. 2001. High pressure solubility data of the system limonene + linalool + CO_2. *Journal of Chemical Engineering Data* 46:1145–1148.
13. Drescher, M., O. Seidel, and D. Geana. 2002. High pressure vapor-liquid equilibria in the ternary system orange peel oil (limonene) + ethanol + carbon dioxide. *Journal of Supercritical Fluids* 23:103–111.
14. Iwai, Y., M. Ichimoto, S. Takada, S. Okuda, and Y. Arai. 2005. Entrainer effect of ethanol on high-pressure vapor-liquid equilibria for supercritical carbon dioxide + limonene + linalool system. *Journal of Chemical Engineering Data* 50:1844–1847.
15. Raeissi, S., and C. J. Peters. 2002. Phase behaviour of the binary system ethane + limonene. *Journal of Supercritical Fluids* 22:93–102.
16. Raeissi, S., and C. J. Peters. 2005. Liquid-vapor and liquid-liquid-vapor equilibria in the ternary system ethane + limonene + linalool. *Journal of Supercritical Fluids* 33:201–208.
17. Kalra, H., S. Y.-K. Chung, and C. Chen. 1987. Phase equilibrium data for supercritical extraction of flavors and palm oils with carbon dioxide. *Fluid Phase Equilibria* 36:263–278.
18. Franceschi, E., M. B. Grings, C. D. Frizzo, J. V. Oliveira, and C. Dariva. 2004. Phase behavior of lemon and bergamot peel oils in supercritical CO_2. *Fluid Phase Equilibria* 226:1–8.
19. Brunner, G. 1998. Industrial process development: Countercurrent multistage gas extraction (SFE) processes. *Journal of Supercritical Fluids* 13:283–301.
20. Budich, M., and G. Brunner. 1999. Vapor-liquid equilibrium data and flooding point measurements of the mixture carbon dioxide + orange peel oil. *Fluid Phase Equilibria* 158–160:759–773.

21. Stuart, G. R., C. Dariva, and J. V. Oliveira. 2000. High-pressure vapor-liquid equilibrium data for CO_2-orange peel oil. *Brazilian Journal of Chemical Engineering* 17 (2): 181–189.
22. Brunner, G. 1997. *Gas extraction*. Darmstadt: Steinkopff and New York: Springer.
23. Stahl, E., K. W. Quirin, and D. Gerard. 1988. *Dense gases for extraction and refining*. New York: Springer-Verlag.
24. Perre, C., G. Delestre, L. Schrive, and M. Carles. 1994. Deterpenation process for citrus oils by supercritical CO_2 extraction in a packed column. *Proceedings of the 3rd International Symposium on Supercritical Fluids* 2:465–470.
25. Sato, M., M. Goto, and T. Hirose. 1995. Fractional extraction with supercritical carbon dioxide for the removal of terpenes from citrus oil. *Industrial & Engineering Chemistry Research* 34:3941–3946.
26. Sato, M., M. Goto, and T. Hirose. 1996. Supercritical fluid extraction on semibatch mode for the removal of terpene in citrus oil. *Industrial & Engineering Chemistry Research* 35:1906–1911.
27. Sato, M., M. Goto, A. Kodama, N. Tanoue, and T. Hirose. 1996. *High pressure chemical engineering*, ed. Ph. Rudolf von Rohr and Ch. Trepp, 303–308. Amsterdam, the Netherlands: Elsevier Science BV.
28. Sato, M., M. Goto, A. Kodama, and T. Hirose. 1997. Supercritical fluid extraction with reflux for citrus oil processing. In *Supercritical fluids*, ed. M. A. Abraham, and A. K. Sunol, 119–131. ACS Symposium Series 670. New York: Oxford University Press.
29. Goto, M., M. Sato, A. Kodama, and T. Hirose. 1997. Application of supercritical fluid technology to citrus oil processing. *Physica B* 239:167–170.
30. Goto, M., M. Kondo, M. Sato, A. Kodama, and T. Hirose. 1999. Supercritical fluid extraction process for the fractionation of citrus oil. *Recent Research Developments in Chemical Engineering* 3:9–20.
31. Sato, M., M. Goto, M. Kondo, A. Kodama, and T. Hirose. 1998. Fractionation of citrus oil by supercritical countercurrent extractor with side-stream withdrawal. *Journal of Supercritical Fluids* 13:311–317.
32. Budich, M., S. Heilig, T. Wesse, V. Leibküchler, and G. Brunner. 1999. Countercurrent deterpenation of citrus oils with supercritical CO_2. *Journal of Supercritical Fluids* 14:105–114.
33. Diaz, S., S. Espinosa, and E. A. Brignole. 2005. Citrus peel oil deterpenation with supercritical fluids. Optimal process and solvent cycle design. *Journal of Supercritical Fluids* 35:49–61.
34. Kondo, M., N. Akgun, M. Goto, A. Kodama, and T. Hirose. 2002. Semi-batch operation and countercurrent extraction by supercritical CO_2 for the fractionation of lemon oil. *Journal of Supercritical Fuids* 23:21–27.
35. Kondo, M., M. Goto, A. Kodama, and T. Hirose. 2000. Fractional extraction by supercritical carbon dioxide for the deterpenation of bergamot oil. *Industrial & Engineering Chemistry Research* 39:4745–4748.
36. Kondo, M., M. Goto, A. Kodama, and T. Hirose. 2002. Separation performance of supercritical carbon dioxide extraction column for the citrus oil processing: Observation using simulator. *Separation Science and Technology* 37 (15): 3391–3406.
37. Yamauchi, Y., and M. Saito. 1990. Fractionation of lemon-peel oil by semi-preparative supercritical fluid chromatography. *Journal of Chromatography* 505 (1): 237–246.
38. Barth, D., D. Chouchi, G. D. Porta, E. Reverchon, and M. J. Perrut. 1994. Desorption of lemon peel oil by supercritical carbon dioxide: Deterpenation and psoralens elimination. *Journal of Supercritical Fluids* 7:177–183.
39. Chouchi, D., D. Barth, E. Reverchon, and G. D. Porta. 1995. Desorption of bergamot peel oil. *Industrial & Engineering Chemistry Research* 34:4508–4513.

40. Chouchi, D., D. Barth, E. Reverchon, and G. D. Porta. 1996. Bigarade peel oil fractionation by supercritical CO_2 desorption. *Journal of Agricultural and Food Chemistry* 44:1110–1114.
41. Reverchon, E. 1997. Supercritical desorption of limonene and linalool from silica gel: Experiments and modeling. *Chemical Engineering Science* 52 (6): 1019–1027.
42. Shen, Z., V. Mishra, B. Imison, M. Palmer, and R. Fairclough. 2002. Use of adsorbent and supercritical carbon dioxide to concentrate flavor compounds from orange oil. *Journal of Agricultural and Food Chemistry* 50:154–160.
43. Sato, M., M. Goto, A. Kodama, and T. Hirose. 1998. Chromatographic analysis of limonene and linalool on silica gel in supercritical carbon dioxide. *Separation Science and Technology* 33 (9): 1283–1301.
44. Silva, E. A., L. Cardozo-Filho, F. Wolff, and M. A. A. Meireles. 2000. Modeling the supercritical desorption of orange essential oil from a silica-gel bed. *Brazilian Journal of Chemical Engineering* 17 (3): 1–16.
45. Ruthven, D. M., S. Farooq, and K. S. Knaebel. 1994. *Pressure swing adsorption*. New York: VCH Publishers.
46. Yang, R. T. 1987. *Gas separation by adsorption processes*. Boston: Butterworth.
47. Sato, M., M. Goto, A. Kodama, and T. Hirose. 1998. New fractionation process of citrus oil by pressure swing adsorption in supercritical carbon dioxide. *Chemical Engineering Science* 53 (24): 4095–4104.
48. Goto, M., G. Fukui, H. Wang, A. Kodama, and T. Hirose. 2002. Deterpenation of bergamot oil by pressure swing adsorption in supercritical carbon dioxide. *Journal of Chemical Engineering of Japan* 35 (4): 372–376.

6.5 HIGH-PRESSURE ADSORPTION/DESORPTION TO IMPROVE THE QUALITY OF SOLUBLE COFFEE AROMA

Susana Lucas

6.5.1 INTRODUCTION

The desirable smell in coffee is produced by a delicate balance in the composition of volatiles substances called aroma. It is important to recover coffee volatiles that are released during production of soluble coffee and to put them back into the liquid coffee extracts or dry products of the extract. This enhances the smell of coffee products and satisfies consumer preferences for such products.

In this section an overview of the main compounds responsible for coffee aroma and how they can be altered during coffee processing, the conventional techniques for volatile substances recovery, and the importance of supercritical technology as a selective extraction–concentration method are presented.

Supercritical extraction–adsorption processes have been demonstrated to be a powerful tool for aroma recovery studies but few efforts have been made in this field. In this section a method of recovering and returning the aromas to the coffee, based on an integrated process consisting of supercritical extraction (SFE) and separation by supercritical adsorption, is proposed. This study was performed in a two-step pilot plant comprising CO_2 supercritical extraction of volatile coffee compounds (the most valuable fraction) from roasted and milled coffee and a subsequent step of selective recovery of these flavor chemicals and removal of pungent volatiles by adsorption on activated carbon. The adsorbent is regenerated by heating and the concentrate stream of volatile coffee compounds is recovered by absorption with of coffee oil. The enriched coffee oil is sprayed on soluble coffee powders to improve the quality of the soluble coffee aroma before it is packed.

6.5.1.1 Coffee Aroma

Aromas, or more precisely volatile compounds, in coffee are normally various organic compounds present in low concentrations, typically ppm levels. Classes of organic compounds that can be regarded as coffee aroma are, for instance, alcohols, aldehydes, esters, lactones, carboxylic acids, hydrocarbons, phenols, ethers, and terpenes, and are mainly located within the coffee oil cells in the solid in the matrix (see Table 6.5.1). This fact is self-evident from the aromatic character of the oil derived by high-pressure mechanical expression or solvent extraction. Coffee oil is typically a vegetable glyceride oil that contains a high percentage of unsaponificable material.

The number of different volatile compounds in roasted coffee is now estimated to be more than 1000. The exact contribution, however, of many of these compounds to actual coffee flavor is not known. In fact, very many of these compounds may be present in insufficient quantity to be of significance for aroma. The discovery of the presence of new, and more, compounds has often tended to be a research objective itself. Maier [2] has indicated the progressive increase in the numbers identified, at about 50 in 1930, staying relatively constant until 1960, but then reaching some 350 in 1970. Maarse and Visscher [3] confirmed the complexity of coffee aroma involving more than 800 volatile compounds with a wide range of functional groups. In the last decade various studies have been focused on the most potent odorous constituents responsible for coffee aroma [4–6]. Grosch [7] found that 29 volatile compounds were mainly responsible for roast and ground coffee aroma. The sudden increase of

TABLE 6.5.1
Aromatic Compounds Identified in Roasted Jamaican Coffee (GC-MS)

Compound	Number
Furans + pyrazines	99+79
Ketones + pyrroles	70+67
Hydrocarbons + phenolic compounds	50+42
Esters + aldehydes	29+28
Thiazoles + oxazoles	28+27
Tiophenes + amines and N-compounds	26+24
Alcohols + acids	20+20
Sulphur compounds + pyridines	16+13
Nonclassified	17
Total	655

Source: From Lancashire, R. J., Jamaican coffee. The Department of Chemistry, University of the West Indies, Mona Campus, Jamaica, 2001. http://wwwchem.uwimona.edu.jm:1104/lectures/coffee.html (accessed July 16, 2008).

TABLE 6.5.2
Investigative Work for the Identification of Volatile Compounds in Coffee

Reference	Affiliation	Main reference dates
Blank et al. [4]	Gie[gerds]en University, Germany	1992
Gianturco et al. [10]	Coca Cola Co., Atlanta, GA	1969
Grosch [7]	Deutsch Forsch Anstalt Lebensmittelchem, Garsching, Germany	1998
Maarse and Visscher [3]	Food Analysis Institute, Zeist, The Netherlands	1996
Merritt et al. [11]	US Army Laboratories, Natick, MA	1969
Murkovic and Derler [12]	Graz University of Technology, Austria	2006
Nishimura and Mihara [13]	Ogawa and Company, Ltd., Tokyo, Japan	1990
Parliment et al. [14]	General Foods Corp., Tarrytown, NY	1973
Reymond et al. [15]	Nestlé Co., Vevey, Switzerland	1966
Sanz et al. [16]	Navarra University, Pamplona, Spain	2002
Shibamato [17]	University of California	1980
Silwar [8, 18, 19]	C. Melchers and Co. Produktions GmbH, Bremen, Germany	1986, 1988, 1993
Semmelroch and Grosch [20, 21]	Deutsch Forsch Anstalt Lebensmittelchem, Garsching, Germany	1995, 1996
Stephan et al. [9]	Hamburg University, Germany	2000
Stoll et al. [22]	Firminich et Cie, Geneva, Switzerland	1967
Stoffelsma et al. [23]	Polak Frutal Works, N.V., New York, NY	1968
Tressl et al. [6, 24–27]	Technical University of Berlin, Germany	1978, 1979, 1980, 1981, 1982
Viani et al. [28]	Nestlé Co., Vevey, Switzerland	1965
Vitzthum et al. [29–32]	Hag, Bremen, Germany	1974, 1975, 1978, 1979

numbers is, of course, attributed to the introduction of gas chromatographic (GC) techniques and mass spectral information developed in the past few decades. In this sense, Silwar [8] provided a review on instrumental measurement techniques used in coffee aroma analysis comprising sample preparation and identification techniques. More recently, Stephan et al. [9] gave a brief overview of the wide range of the aroma research (e.g., different isolation techniques, aroma characterization, quantification methods, and human perception). The main research in identifying volatile compounds in coffee is summarized in Table 6.5.2.

From an exhaustive revision of works related to coffee aroma composition and as previously mentioned, it can be concluded that certain aldehydes, esters, and ketones of low molecular weight and specific sulphur and phenolic compounds are responsible for the desirable and pleasant aroma of the coffee. However, polyphenolic compounds, melanoidines, and caffeine contribute to the bitter note of coffee aroma.

6.5.1.2 Conventional Techniques for Coffee Aroma Concentration

During coffee processing the aroma composition can be seriously altered as a result of chemical reactions or physical losses. After coffee roasting, two phenomena can be observed: the loss of the delicate fraction of volatile substances for exposure to ambient conditions and the spontaneous formation of undesirable substances by means of hydrolysis reactions, oxidations, Maillard reactions, or nonenzymatic reactions. A possible way of minimizing the changes is to use various separation techniques for aroma recovery and concentration. Techniques suitable for this task, both commercially available and being developed, are steam distillation, partial condensation, gas injection techniques, pervaporation, adsorption, organic solvent extraction, and supercritical fluid extraction (SFE) [33].

The following paragraphs show the most relevant works related to conventional techniques for coffee aroma recovery and concentration.

The first significant investigation was carried out by Staudinger and Reichstein [34]. They obtained the aroma oil by distilling the volatiles from roasted, ground, and prewetted coffee heated to 373.2 to 383.2 K at a pressure of 270 to 670 Pa and collecting the distillate on cold traps held at 253.2 to 93.2 K. Johnston and Frey [35] included variations on procedures formerly used, i.e., solvent extraction of ground coffee and distillation of ground coffee in a high vacuum in an inert atmosphere.

Reymond et al. [15] proposed a stripping method with helium with a subsequent condensation step at 193.2 K for the recovery of volatile compounds in a mixture of roasted coffee beans (10% wt.) dissolved in hot water at 353.2 K. Stoll et al. [22] and Gianturco et al. [10] studied the aroma preparations by high vacuum distillation of expelled coffee oil. Kroger [36] described in his patent a procedure for selective recovery of aroma from roasted and milled coffee using stripping with gas nitrogen and two consecutive condensation steps. Vitzthum and Werkoff [31] proposed a modified stripping method followed by a selective adsorption separation step with a suitable adsorbent (Tenax). This adsorbent is regenerated by desorption at high temperature.

Several authors analyzed the simultaneous distillation–extraction technique to isolate volatile compounds from ground coffee [23, 31]. In this sense, Tressl and Silwar [6] described a distillation–extraction process with distillated pentane ether as solvent. Recently, Nishimura and Satoru [13] proposed a steam distillation process for roasted coffee. The distillate is extracted with methylene chloride to obtain the volatile compounds (0.13% yield based on the original roasted coffee).

Sakano et al. [37] studied the adsorption process using A-type zeolites for effective removal of pungent smell components identified in the coffee aroma-containing gas evolved from roasted and ground coffee packed in a percolation vessel.

Morillo [38] mentioned and compared two additional techniques for coffee aroma recovery: gas and liquid CO_2 extraction. The extraction with CO_2 in gas phase was carried out at pressures of 500–4000 kPa and a subsequent step of condensation. Liquid CO_2 extraction allowed obtaining several oil components in extraction fraction compared to those obtained with CO_2 gas.

Sarrazin et al. [39] compared five different extraction methods: SFE with carbon dioxide, simultaneous distillation extraction, oil recovery under pressure, and vacuum steam-stripping with water or with organic solvent. Sensory testing of extracts revealed that vacuum steam-stripping with water provided the most representative aroma extract, for all three coffees tested (Arabica coffee: green, light-roasted, and medium roasted). In Figure 6.5.1 a summary of various extraction processes proposed in this work is shown.

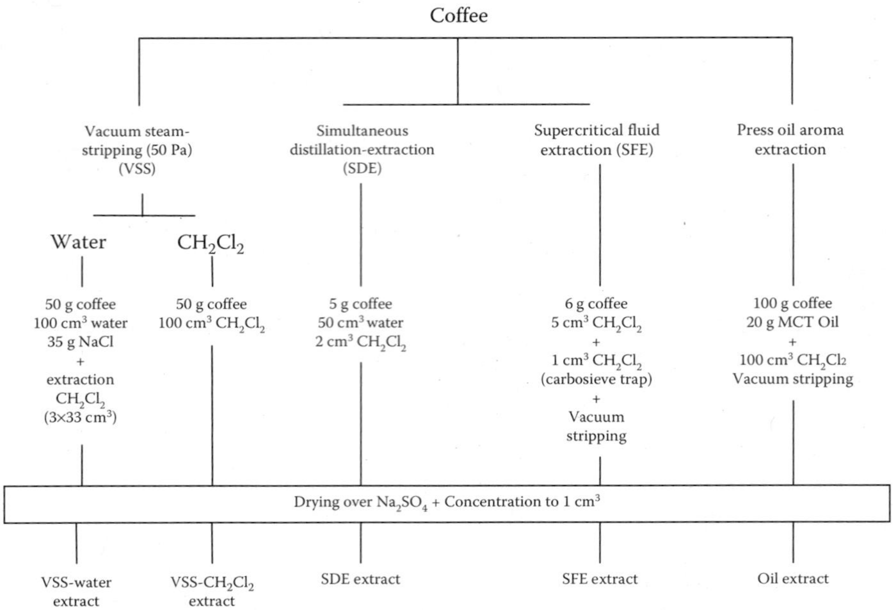

FIGURE 6.5.1 Extraction methods used for the isolation of coffee aroma. MCT oil [Delios, $C_{8:0}$ (60%) and $C_{10:0}$ (40%) triglycerides] was used as a neutral cosolvent for press oil aroma extraction. (Modified from Sarrazin, C., et al., *Food Chem.*, 70, 99–106, 2000.)

Mandralis et al. [40] proposed in their patent a process for the recovery of aroma components from a slurry of coffee grounds in an aqueous liquid by gas-stripping in countercurrent manner. The aroma components are then collected from the aromatized gas. The aroma components may be added to concentrated coffee extract prior to drying of the extract.

Common methods for recovery of coffee aroma, including steam distillation, hydrodiffusion, maceration, mechanical (cold) expression, and solvent extraction, present some drawbacks such as low efficiency of extraction, thermal and hydrophilic degradation of the product, loss of volatiles in the solvent separation, changes in composition by hydrolysis and oxidation reactions, and the wide demand for natural products free of solvents. In this sense, supercritical technology has a great potential to solve many of the mentioned problems: faster and more efficient extractions, extracts that preserve natural and fresh aroma without thermal degradation, the adjustable selectivity or solvent power, and a matrix free of solvent are the most relevant advantages connected with supercritical extraction technology.

6.5.2 Supercritical Technology for Coffee Aroma Recovery

The food industry has rapidly taken advantage of the possibility of using CO_2 as a nontoxic, environmentally safe, cheap, and selective extraction solvent [41, 42]. Carbon dioxide has a high selectivity for aroma recovery and adjustable solvent power ranging from gas to liquid depending on pressure and temperature; it is nonflammable, nonreactive, and is easy to separate from, and leaves no residue in, the raw material. Moreover, it has lower critical temperature (304.2 K) and moderate critical pressure (7.38 MPa), thus being an ideal solvent for compounds that may suffer thermal degradation.

6.5.2.1 State of the Art

Some research has been published on supercritical extraction of oil and volatile substances from coffee matrices [43–51] .

Ramos et al. [47] presented an optimized process based on SFE to obtain brewed coffee extracts with an aroma similar to the original brewed coffee. The composition of the extracts obtained at the optimal SFE conditions (T = 333.2 K, CO_2 density = 0.5 g/cm^3, CO_2 flow rate = 1.8 cm^3/min, extraction time = 1.4 min) was determined by using a purge-and-trap device coupled to a GC-mass spectrometry (GC-MS). For comparison, extracts obtained by using liquid–liquid extraction (with methylene chloride or pentane) and headspace solid-phase microextraction (SPME) were also obtained and analyzed by GC-MS. The study revealed that SC-CO_2 extraction of brewed coffee in the optimized conditions provided aroma extracts with high olfactory resemblance to the original brewed coffee. Moreover, the composition of the SC-CO_2 extract showed appreciable differences from those obtained by conventional techniques such as SPME and solvent extraction.

Sarrazin et al. [39] presented a supercritical CO_2 extraction process of ground coffee using ethanol as cosolvent. The extraction conditions were fixed at 20 MPa, 333.2 K, and 1% ethanol. The aromatic extract was collected in a test tube containing methylene chloride.

Lucas and Cocero [49] presented an optimized two-step integrated process consisting of CO_2 supercritical extraction of volatile coffee compounds from roasted and milled coffee and a subsequent step of selective removal of pungent volatiles by adsorption on activated carbon. Results showed that low adsorption pressures (12 MPa), low adsorption temperatures (308.2 K), and low CO_2 flow rates (3 kg/h) were suitable for removing the undesirably pungent and odorous components (e.g., furfural) and retaining the desirable aroma compounds (e.g. ethyl acetate).

Araújo and Sandi [51] optimized a supercritical CO_2 extraction process of green and roasted coffee to obtain the highest and lowest diterpene levels and the maximum coffee oil extraction. The operational temperatures (333.2–363.2 K) and pressures (23.5–38 MPa) were optimized for coffee oil extraction. Moreover, supercritical oil content levels and diterpene oil concentrations were compared to the results obtained with the extraction with Soxhlet apparatus, using hexane as the solvent. In general, an inverse correlation was observed between the amount of extracted oil and diterpene concentration levels. As a result, different oil contents with different diterpene concentrations could be obtained.

Lojkova et al. [50] studied the supercritical CO_2 extraction conditions of 4(5)-methylimidazole and 2-acetyl-4(5)-(1,2,3,4)-tetrahydroxybutyl-imidazole from ground coffee with high-performance liquid chromatographic-electrospray mass spectrometric quantification (HPLC/ESI-MS). The efficiency of the supercritical method was compared with that of solid-phase extraction.

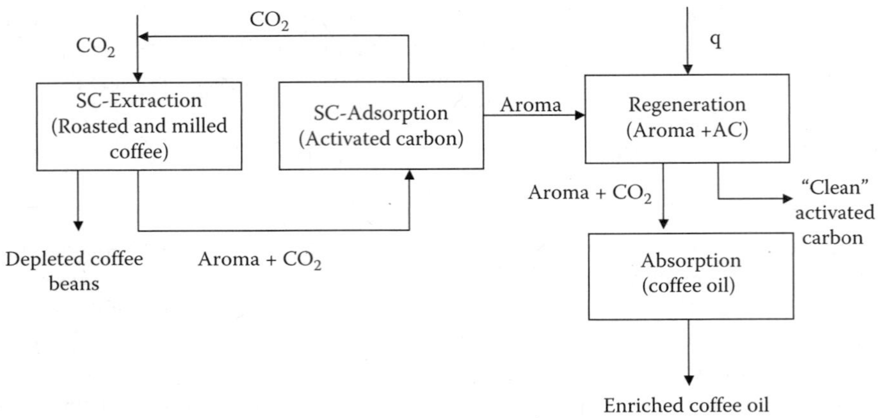

FIGURE 6.5.2 Integrated supercritical extraction–adsorption process for coffee aroma recovery.

6.5.2.2 Process Description

In this section an integrated method of recovery and put-back of the aromas of coffee based on a two-step process consisting of SCE and separation by adsorption is presented. The first step comprises CO_2 supercritical extraction of volatile coffee compounds (the most valuable fraction) from roasted and milled coffee, with a subsequent step of selective recovery of these flavor chemicals and removal of pungent volatiles by adsorption on activated carbon. The adsorbent is regenerated by heating and the concentrate stream of volatile coffee compounds is recovered by absorption within the coffee oil. The enriched coffee oil, analyzed by GC-MS, is sprayed on soluble coffee powders to improve the quality of the soluble coffee aroma before it is packed. A block diagram for the proposed concentration process is shown in Figure 6.5.2.

6.5.2.3 Experimental Section

6.5.2.3.1 Materials and Methods

Adsorbent materials

The granular activated carbon (CAL-Chemviron) evaluated in this research was obtained from Aguas de Levante S.A. (Barcelona, Spain).

Analysis of coffee aroma

A gas chromatograph (model Perkin Elmer Autosystem XL) with an MS detector (model Perkin Elmer QMASS 910) was used for measurement of the composition of aroma compounds. The separation was on a capillary column (SGL-20, 0.25 mm × 60 m). Oven temperature was raised from 313.2 to 453.2 K at 15 K/min, and 0.1 cm^3 of aroma gas was sampled with a gas-tight syringe and injected to the gas chromatograph. Each component in the aroma-containing gas was identified by comparison with standards.

Roasted coffee beans and coffee oil
Commercial coffee beans and coffee oil were used in this work.

6.5.2.3.2 Experimental Set-Up
A pilot plant for selective aroma recovering has been designed and built in the Chemical Engineering Department at the Valladolid University (Spain). It is a two-step integrated plant comprising CO_2 supercritical extraction and selective coffee aroma recovery by adsorption on activated carbon. The pilot plant was designed to operate at $P < 30$ MPa, $T < 353.2$ K, and a CO_2 mass flow rate of 1–20 kg/h and had a treatment capacity of 0.2 $kg_{coffe}/_{load}$. It consists of three pressurized vessels of 1000 cm^3 (inside diameter of 4 cm and length of 50 cm) that can operate as extractors or adsorbers depending on needs, a diaphragm pump to supply solvent and to recirculate CO_2 during operation (LEWA Herbert Leomberg type EH1), and auxiliary equipment such as heat exchangers, pressure, temperature, and flow meters, and valves and fittings suitable for high-pressure processes, together with the data acquisition system [52].

The pilot plant flow diagram is schematically presented in Figure 6.5.3. It is based on two consecutive integrated steps comprising CO_2 supercritical extraction and aroma recovery on the adsorbent. In the extraction, the supercritical CO_2 flows through a fixed bed of milled and roasted coffee beans and dissolves the extractable components of the solid. The loaded solvent is removed from the extractor and is fed to the adsorber where activated carbon is placed. The clean solvent evolved from the adsorber is recirculated to process operating the pilot plant under quasi-isobaric conditions (neglecting pressure drop). After 15 min, the pump is turned off and the adsorbent is regenerated by heating up to 338 K, and the concentrate stream of volatile coffee compounds is recovered by absorption with 15 cm^3 of coffee oil. The enriched coffee oil is then analysed by GC-MS.

6.5.2.4 Influence of Process Operating Conditions

In this section the influence of pressure, temperature, and solvent flow rate for the extraction and adsorption steps and the condition of the solid substrate on both processes are discussed.

6.5.2.4.1 Extraction Step
At process conditions of supercritical extraction from solid matrices, the solvent capacity in general increases with pressure at constant temperature. Therefore, the remaining content of solute in the solid substrate after a certain time of extraction will decrease with pressure and increase the corresponding solute concentration in the supercritical phase. A higher temperature often causes a higher extraction rate, if pressure is not low. One reason is the dependence of solvent power on temperature. At relatively low pressures, decrease of density and solvent power with increasing temperature prevails, whereas at relatively higher pressures, the increase in vapor pressure with temperature prevails. The other reason for a higher amount of extract per unit of time is increasing mass transfer rates with temperature. The solvent ratio is the most important parameter for supercritical extraction, once approximate values of pressure and temperature are selected. With increasing solvent ratio, the

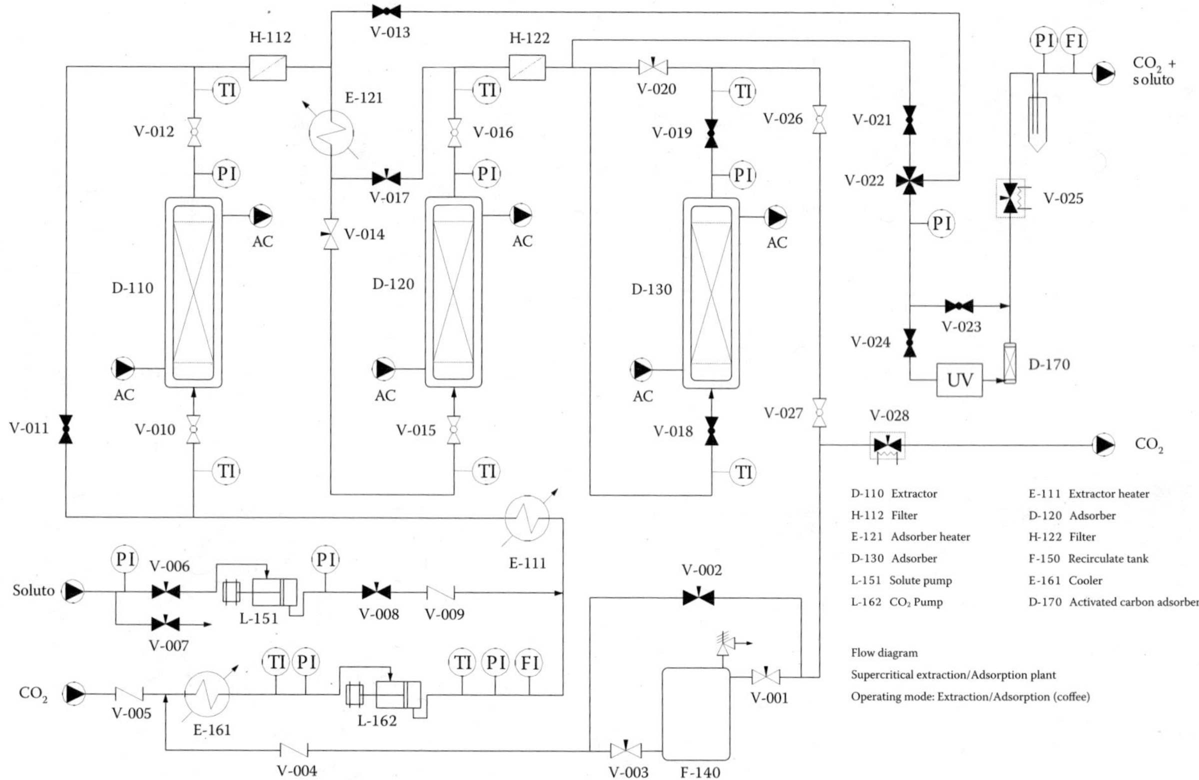

FIGURE 6.5.3 Flow diagram of the supercritical extraction–adsorption pilot plant. AC: activated carbon; V: valve; H: filters; D: extraction–adsorption columns.

extraction rate can be enhanced more than with changing process parameters within a relatively narrow limit. At low solvent ratios, the remaining amount of extract on the solid substrate is high after a certain time of extraction. In a medium range for the solvent ratio, its influence on the extraction result is the greatest. At very high solvent ratios, the remaining extract content seems to approach a lower limit. However, the influence of the solvent ratio cannot be discussed without considering economic consequences [41].

6.5.2.4.2 Adsorption Step

Previous works related to adsorption on solid matrices have revealed that operating at lower pressure, all mass transfer resistances decrease and the equilibrium is more favored [53–55]. This means that it is possible to get higher fractional bed utilization and shorter adsorption cycles. Moreover, the operating (pumping) and equipment costs are less at low pressure. The minimum operating pressure could be as low as the CO_2 critical pressure (7.1 MPa) and high enough to ensure a monophasic system.

At lower temperatures the equilibrium is improved and all mass transfer resistances increase, but internal mass transfer (the controlling step for the majority of these processes) remains constant. This means that similar fractional bed utilization can be achieved. However, the economic aspects suggest operating with lower temperature because the operating costs are smaller (heating). The same deduction is valid for the CO_2 flow rate. At a lower solvent flow rate, external and axial dispersion resistances increase but internal resistance remains constant. This means that similar fractional bed utilization is obtained. When the CO_2 flow rate is low, the operating costs (CO_2 and pumping) and fixed costs are less but on the other hand the adsorption cycles are longer, the treatment capacity per unit of time is lower, and hydrodynamic problems (channeling and compacting) can be observed. The optimum flow rate may be established with an economical viability study.

The complexity of the proposed supercritical extraction–adsorption process is shown by the high number of operating conditions connected with these separation processes along with the variability of raw materials (coffee beans and activated carbon). The pretreatment processes for both extraction and adsorption matrices will fix the extraction and adsorption rates. The size of solid particles and the humidity are other important process variables that it is necessary to take into account.

6.5.2.5 Results

Because of the high number of compounds responsible for coffee aroma and in order to study and simplify the overall process, several key compounds were selected. As an example, ethyl acetate and furfural were chosen as key components. Ethyl acetate is a desirable volatile compound responsible for the fruity and brandy component of coffee aroma, and it is the most common ester present in several kinds of fruit (e.g., apples, grapes). On the other hand, furfural is an undesirable volatile compound with a pungent or foul smell. Lucas et al. [54] reported adsorption equilibrium data for both compounds.

6.5.2.5.1 Key Compounds: Ethyl Acetate and Furfural

Pressure effect

Some supercritical adsorption experiments for ethyl acetate and furfural in the range of 12–17 MPa were performed in order to check the effect of operating pressure. The temperature was fixed at 310.2 K with a constant CO_2 flow rate of 3.5 kg/h.

The corresponding breakthrough curves were treated mathematically in order to obtain the characteristic adsorption parameters such as breakthrough and saturation times (t_b and t_s), breakthrough and saturation adsorptive capacities (q_b and q_s) and fractional bed utilization (FBU). From the results shown in Table 6.5.3 for both solutes, it can be deduced that at a low pressure (13 MPa) the adsorption cycle is faster (shorter breakthrough time), the capacity of the adsorbent (amount of solute adsorbed per kg of adsorbent) is higher, and utilization of the bed improves. This result suggests that at a low pressure the interaction forces between solute and activated carbon surface are higher than the corresponding solute–solvent binding forces [53]. Moreover at a low pressure all mass transfer resistances decrease, and it is possible to get a higher degree of fractional bed utilization [55].

The effect of temperature

The adsorption results for ethyl acetate and furfural obtained at temperatures of 308.2–323.2 K at a fixed pressure (14 MPa) and a constant CO_2 flow rate of 3.5 kg/h are shown in Table 6.5.4. Operating at lower temperatures (310.2 K) enables the obtainment of shorter adsorption cycles and higher adsorptive capacities, as can be deduced from analysis of Table 6.5.4. The fractional bed utilization decreases slightly with temperature. This affirmation is valid for both solutes and can be attributed to the increase in solvent power with temperature attributed to the increase in the vapor pressure. This means that at a lower temperature the solute-adsorbent interaction forces versus the corresponding solute–solvent attraction forces prevail.

TABLE 6.5.3
The Effect of Pressure: A Summary of SC Adsorption of Ethyl Acetate and Furfural

P (MPa)	t_b (min)	t_s (min)	q_b (g_{SOL}/g_{CA})	q_s (g_{SOL}/g_{CA})	FBU (%)	Removal ratio (%)[a]
			Ethyl acetate			
12.8	10.9	15.0	0.077	0.084	92.0	78.7
15.2	14.6	19.5	0.072	0.081	88.9	70.0
17.0	15.7	26.5	0.060	0.075	79.7	68.5
			Furfural			
13.0	12.9	15.9	0.089	0.098	90.8	80.4
15.6	14.8	18.0	0.084	0.094	89.1	77.5
17.2	13.1	18.4	0.081	0.092	87.5	75.0

[a] The removal ratio is the ratio of the amount of solute adsorbed to that fed into the adsorption column.

TABLE 6.5.4
The Effect of Temperature: A Summary of SC Adsorption of Ethyl Acetate and Furfural

T (°C)	t_b (min)	t_s (min)	q_b (g_{SOL}/g_{CA})	q_s (g_{SOL}/g_{CA})	FBU (%)	Removal ratio (%)[a]
			Ethyl acetate			
36.8	11.0	15.0	0.090	0.102	88.6	75.0
38.7	13.8	19.0	0.087	0.095	92.0	72.3
50.9	14.3	19.5	0.064	0.075	85.4	66.1
			Furfural			
36.6	10.8	14.3	0.092	0.107	86.4	80.3
38.5	10.8	14.7	0.090	0.103	87.7	79.4
50.9	14.8	18.4	0.087	0.098	88.3	75.2

[a] The removal ratio is the ratio of the amount of solute adsorbed to that fed into the adsorption column.

The effect of CO_2 flow rate

The adsorption results for ethyl acetate and furfural obtained with CO_2 flow rates of 3–5 kg/h at fixed pressure (14 MPa) and temperature (310.2 K) are shown in Table 6.5.5.

Operating at a low CO_2 flow rate produces longer adsorption cycles, although higher adsorptive capacities and higher fractional bed utilization are achieved (Table 6.5.5). The amount of solute adsorbed increases with the decrease in solvent flow rate because the solute–adsorbent contact time is shorter.

TABLE 6.5.5
CO_2 Flow Rate Effect: A Summary of SC Adsorption of Ethyl Acetate and Furfural

F_{CO_2} (kg/h)[a]	t_b (min)	t_s (min)	q_b (g_{SOL}/g_{CA})	q_s (g_{SOL}/g_{CA})	FBU (%)	Removal ratio (%)[b]
			Ethyl acetate			
3.0	15.5	23.5	0.085	0.098	86.3	78.7
4.4	13.7	19.0	0.072	0.084	85.6	75.6
5.2	12.1	16.8	0.062	0.073	85.4	69.1
			Furfural			
2.9	14.8	18.0	0.096	0.107	89.3	83.5
3.7	11.5	13.1	0.087	0.098	89.0	81.5
5.0	10.9	13.3	0.070	0.079	88.8	77.5

[a] F_{CO_2}: Solvent flow rate.

[b] The removal ratio is the ratio of the amount of solute adsorbed to that fed into the adsorption column.

From the adsorption point of view, similar adsorption curves with the same values of adsorptive capacities and fractional bed utilization were obtained for both solutes. The compounds have similar molecular masses (M_{EA} = 88.1 g/mol and M_{FF} = 96.1 g/mol) and molecular dimensions, which makes the selective adsorption of furfural (the undesirable component) more difficult than that of ethyl acetate. Nevertheless the furfural molecule has greater electronic mobility and reactivity associated with the carbonyl group–aromatic ring linkage. This phenomenon explains the stronger bonding forces between furfural and activated carbon and as a consequence, the higher values of the removal ratio for all the experiments. The higher adsorption heat of furfural (20–32 kJ/mol) compared to that of ethyl acetate adsorption heat (8–9 kJ/mol) corroborates this fact [54].

In this section, the operating parameters for the adsorption step of ethyl acetate (as a desirable coffee aroma compound) and furfural (as a pungent component) on activated carbon were optimized. Experiments were conducted at adsorption pressures of 12–17 MPa, adsorption temperatures of 308–323 K, and a solvent flow rate of 3–5 kg/h. In all cases, the solute concentration and the activated particle size were kept constant. Results show that low pressures (12 MPa), low temperatures (308.2 K), and low CO_2 flow rates (3 kg/h) are suitable for removing the undesirable pungent and smell components (e.g., furfural) and retaining the desirable aroma compounds (e.g., ethyl acetate).

6.5.2.5.2 Commercial Coffee

Some experiments were carried out with the commercial coffee in order to determine the optimal conditions for the extraction, adsorption, and regeneration steps involved in the overall process.

Extraction–adsorption pressure (Experiments 1–4)

$P_{ext\text{-}ads}$ = 6.5, 7.4, 8.5, 11.4 MPa; T_{ext} = 309.7 K; T_{ads} = 306.2 K; F_{CO_2} = 3.5 kg/h

From the results shown in Table 6.5.6, it can be seen that at a higher extraction–adsorption pressure (11.4 MPa) the amount of extractable compounds increased significantly in the final coffee oil. This effect of pressure may be due to the increase in density.

Extraction temperature (Experiments 5–7)

$P_{ext\text{-}ads}$ = 10.0 MPa; T_{ext} = 317.2, 323.5, 329.7 K; T_{ads} = 307 K; F_{CO_2} = 3.5 kg/h

When the extraction temperature was higher (329.7 K), the amount of the compounds extracted increased slightly. This behavior can be attributed to the increase in extraction rate with temperature (Table 6.5.6).

Adsorption temperature (Experiments 3 and 9)

$P_{ext\text{-}ads}$ = 8.5 MPa; T_{ext} = 310.2 K; T_{ads} = 307.2, 319.2 K; F_{CO_2} = 3.5 kg/h

At a lower adsorption temperature (307.2 K) the amount of extractable compounds fixed in the coffee oil increased meaningfully. This effect may be due to the

TABLE 6.5.6
GC-MS Analysis of Original and Final Coffee Oils Obtained by the Process of SC Extraction–Adsorption

		Experiments (% area)								
Compounds	**Oil**	**1**	**2**	**3**	**4**	**5**	**6**	**7**	**8**	**9**
2,4-Imidazolidindione	85.51	87.47	51.99	64.84	41.97	87.58	54.54	89.21	87.48	99.28
2-Aminopropanol	8.85	8.83	28.10	23.35	24.82	10.74	7.15	4.77	9.65	—
2-Acetoxi-propene	—	1.29	15.27	—	26.82	—	3.29	2.57	1.04	—
Ethyl acetate	—	0.77	2.88	4.12	2.37	0.82	1.12	1.17	0.57	—
Dichloromethane	—	—	—	—	1.82	0.47	11.71	0.47	—	0.37
Octametilcycletetraxyloxane	5.64	2.64	1.77	7.69	2.19	0.39	22.19	1.81	1.06	0.36
Aromatic profiles		B	R	F	R	B	B/C	B/C	B	E/M

B: Burnt; B/C: Buttery/Caramel; E/M: Earthy/Musty; F: Fruity; R: Roasty.

decrease in density with temperature versus the increase in vapor pressure at this operating pressure (Table 6.5.6).

CO_2 flow rate (Experiments 1 and 8)

$P_{ext\text{-}ads} = 6.6$ MPa; $T_{ext} = 305.2$ K; $T_{ads} = 306.2$ K; $F_{CO_2} = 3.5, 1.7$ kg/h

In the selected range (1.7–3.5 kg/h) no effect of flow rate can be observed in the final coffee oils. Similar concentrations of the main components detected by GC-MS were obtained as shown in Table 6.5.6 (Experiments 1 and 8).

Experiments with commercial coffee have demonstrated that low adsorption pressures (11.4 MPa), low adsorption temperatures (305.2 K), and relatively low CO_2 flow rates (1.7 kg/h) in the selected operating range were suitable for removing the undesirable pungent and odorous components and retaining the desirable aroma compounds in order to get a delicate balance in the composition of volatiles in the final coffee oil. This means that the operation with real roasted coffee corroborated the previous results obtained with the key compounds.

In Figure 6.5.4, a comparison of the original coffee oil chromatogram and that obtained under the optimal operating conditions is shown. This figure reveals that the original coffee oil is enriched in volatile fraction (aldehydes and ketones compounds such as methylbutanal, 2,3-butanedione, and 2,3-pentanedione), and its aromatic profile is improved from a burnt note to a buttery, caramel one.

6.5.3 Conclusions

The possibilities of supercritical technology for improving coffee aroma using an integrated and optimized process of supercritical CO_2 extraction with selective removal of pungent volatiles by adsorption on activated carbon have been demonstrated from experimental results. A relatively low extraction–adsorption pressure of 12 MPa (quasi-isobaric process), an extraction temperature of 329.2 K, an adsorption

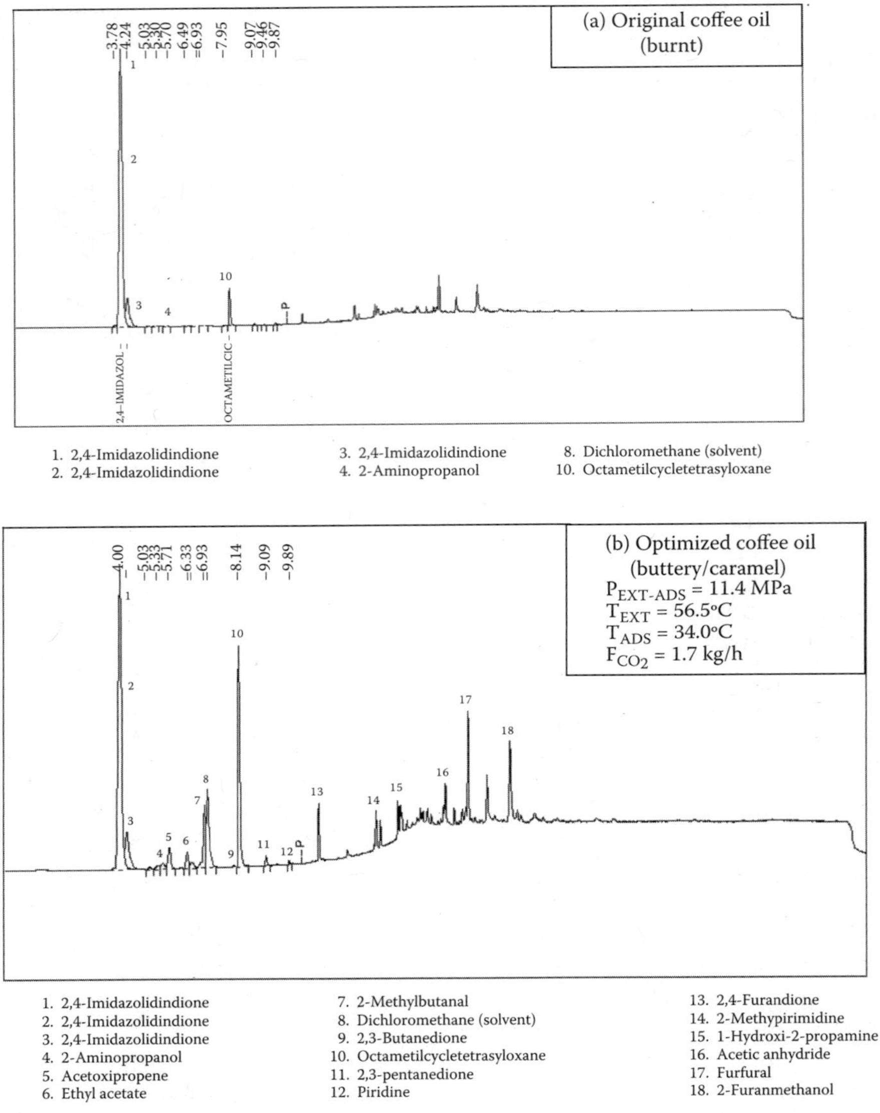

FIGURE 6.5.4 Chromatograms of original coffee oil (a) and enriched coffee oil obtained under the optimal operating conditions (b).

temperature of 308.2 K, and a CO_2 flow rate of 2 kg/h resulted in a delicate balance in the composition of volatiles in the final coffee oil.

Faster and more efficient extractions, extracts with natural and fresh aroma without thermal degradation, adjustable selectivity, and a matrix free of solvent are the most relevant advantages connected with supercritical technology for coffee aroma recovery and concentration.

6.5.4 Nomenclature

Symbol	Definition	Units in SI System	Dimensions in M, N, L, T, and θ
FBU	Fractional bed utilization	%	
F_{CO_2}	CO_2 flow rate	kgh^{-1}	MT^{-1}
M	Molecular weight	$g \cdot mol^{-1}$	
P	Input power per unit of fluid volume	$W \cdot m^{-3}$	$L^{-1}T^{-3}$
q_b	Adsorption capacity at breakthrough point	g_{SOLUTE}/g_{CARBON}	
q_s	Saturation adsorption capacity	g_{SOLUTE}/g_{CARBON}	
T	Absolute temperature	K	θ
t_b	Breakthrough time	min	θ
t_s	Saturation time	min	θ

6.5.5 References

1. Lancashire, R. J. 2001. Jamaican coffee. The Department of Chemistry, University of the West Indies, Mona Campus, Jamaica. http://wwwchem.uwimona.edu.jm:1104/lectures/coffee.html (accessed July 16, 2008).
2. Maier, H. G. 1981. *Kaffee*. Berlin and Hamburg, Germany: Paul Parey.
3. Maarse, H., and C. A. Visscher. 1996. Volatile compounds in foods. Quantitative and qualitative data. TNO Food Analysis Institute. The Netherlands: 7th Zeist.
4. Blank, I., A. Sen, and W. Grosch. 1992. Potent odorants of the roasted powder and brew of Arabica coffee. *Zeitschrift fuer Lebensmittel-Untersuchung und Forschung* 195:239–245.
5. Grosch, W. 1995. Instrumental and sensory analysis of coffee volatiles. 16th International Scientific Colloquium on Coffee, Kyoto, 147–155.
6. Tressl, R., and R. Silwar. 1981. Investigation of sulphur-containing components in roasted coffee. *Journal of Agricultural and Food Chemistry* 29:1078–1082.
7. Grosch, W 1998. Flavour of coffee. *Die Nahrung* 42 (6): 344–350.
8. Silwar, R. 1986. Analytical techniques for the investigation of coffee aroma. *Trends in Analytical Chemistry* 5 (3): 78–81.
9. Stephan, A., M. Bücking, and H. Steinhart. 2000. Novel analytical tools for food flavours. *Food Research International* 33 (3–4): 199–209.
10. Gianturco, M. A., J. J. Hilton, and R. E. Biggers. 1969. Differentiation between coffee arabica and robusta. *Journal of Chromatographic Science* 7:453–472.
11. Merritt, C., D. H. Robertson, and D. J. McAdoo. 1969. The relationship of volatile compounds in roasted coffee beans to their precursors. Proceedings of the 4th Colloquium on Coffee, ASIC, Amsterdam, 144–148.
12. Murkovic, M., and K. Derler. 2006. Analysis of amino acids and carbohydrates in green coffee. *Journal of Biochemical and Biophysical Methods* 69 (1–2): 25–32.
13. Nishimura, O., and M. Satoru. 1990. Investigation of 2-hydroxy-2-cyclopenten-1-ones in roasted coffee. *Journal of Agricultural and Food Chemistry* 38:1038–1041.
14. Parliment, T. H., W. Clinton, and R. Scarpellino. 1973. Trans-2-nonenal, coffee compound with novel organoleptic properties. *Journal of Agricultural and Food Chemistry* 21:485–487.
15. Reymond, D., F. Charan, and R. H. Egli. 1966. Changes in roasted coffee induced by salting. *Proceedings of the 1st International Congress of Food Science and Technology*. London: Gordon and Breach, 595–602.

16. Sanz, C., M. Czerny, C. Cis, and P. Schieberle. 2002. Comparison of potent odorants in a filtered coffee brew and in an instant coffee beverage by aroma extract dilution analysis (AEDA). *European Food Research and Technology* 214:299–302.
17. Shibamato, T. 1980. Application of HPLC for evaluation of coffee flavour quality. In *The quality of foods and beverages,* ed. E. Charamboulos. New York: Academic Press.
18. Silwar, R., and C. Lüllmann. 1988. The determination of mono- and disaccharides in green arabica and robusta coffees using high performance liquid chromatography. *Café Cacao Thé* 32:319–322.
19. Silwar, R., and R. Lüllmann. 1993. Investigation of aroma formation in Robusta coffee during roasting. *Café Cacao Thé* 35:145–152.
20. Semmelroch, P., and W. Grosch. 1995. Analysis of roasted coffee powders and brews by gas chromatography-olfactometry of headspace samples. *Lebensmittel-Wissenschatf & Technologie* 28:310–313.
21. Semmelroch, P., and W. Grosch. 1996. Studies of character impact odorants of coffee brews. *Journal of Agricultural and Food Chemistry* 44:537–543.
22. Stoll, M., M. Winter, F. Gautschi, I. Flament, and B. Withelm. 1967. Sur l'arome de café. Part I. *Helvetica Chimica Acta* 50:628–694.
23. Stoffelsma J., G. Sipma, D. K. Kettenes, and J. Pypker. 1968. New volatile compounds of roasted coffee. *Journal of Agricultural and Food Chemistry* 16:1000–1004.
24. Tressl, R., K. G. Grunewald, H. Koppler, and R. Silwar. 1978. Flüchtige phenole im röstkaffee. *Zeitschrift fuer Lebensmittel-Untersuchung und Forschung* 167:108–110.
25. Tressl, R., K. G. Grunewald, H. Kamperschroer, and R. Silwar. 1979. Verhalten einiger schwererfluctiger aromastoffe. *Chemie Mikrobiologie Technologie Lebensmittel* 6:52–57.
26. Tressl, R., K. G. Grunewald, H. Kamperschroer, and R. Silwar. 1980. Formation of pyrroles and aroma contributing sulphur compounds in malt and roasted coffee. *Progress in Food Nutrition Science* 4:1111–1129.
27. Tressl, R., M. Holzer, and H. Kamperschroer. 1982. Bildung von Aromastoffen in Röst-Kaffee in Abhangigkeit vom Gehalt an freien Aminosäuren und reduziertem Zucker. Proceedings of the 10th Colloquium on Coffee, ASIC, Salvador-Bahia, Brazil, 279–292.
28. Viani, R., F. Müggler-Cheven, D. Reymond, and R. H. Egli. 1965. Sur la composition de l'arome de café. *Helvetica Chimica Acta* 48: 1809–1815.
29. Vitzhum, O. G., and P. Werkhoff. 1974. Oxazoles and thiazoles in coffee aroma. *Journal of Food Science* 39:1210–1215.
30. Vilzthum, O. G. 1975. Chemie und bearbeitung des kaffees. In *Kaffee und caffein,* ed. O. Eichler, 3–77. Berlin: Springer-Verlag.
31. Vilzhum, O. G., and P. Werkhoff. 1978. Aroma analysis of coffee, tea and cocoa by headspace techniques. In *Analysis of foods and beverages,* ed. G. Charalambous, 115–133. New York: Academic Press.
32. Vilzhum, O. G., and P. Werkhoff. 1979. Messbare aromaveranderungen bei bohnenkaffee in sauerstoffdurchlassiger Verpackung. *Chemie Mikrobiologie Technologie Lebensmittel* 6:25–30.
33. Karlsson, H. O. E., and G. Trägardh. 1997. Aroma recovery during beverage processing. *Journal of Food Engineering* 34:159–178.
34. Staudinger H., and Reichstein T. 1928. Method of producing artificial coffee aroma. US Patent, US1696419.
35. Johnston, W. R., and C. N. Frey. 1938. The volatile constituents of roasted coffee. *Journal of the American Oil Chemists' Society* 60:1624–1630.

36. Kroger, C. 1970. Method of making aromatized oil. US Patent US3535118.
37. Sakano, T., K. Yamamura, H. Tamon, M. Miyahara, and M. Okazaki. 1996. Improvement of coffee aroma by removal of pungent volatiles using A-type zeolites. *Journal of Food Science* 61 (2): 473–476.
38. Morillo, M. 1999. Extracción de aromas de café con CO_2 supercrítico. Trabajo de Suficiencia Investigadora. Universidad de Valladolid, Spain.
39. Sarrazin, C., L. L. Le Quéré, C. Gretsch, and R. Liardon. 2000. Representativeness of coffee aroma extracts: A comparison of different extraction methods. *Food Chemistry* 70:99–106.
40. Mandralis, Z. I., S. Westfall, and K. Yunker. 2000. Aroma recovery process. US Patent, US6149957.
41. Brunner, G. 1994. Gas extraction: An introduction to fundamentals of supercritical fluids and the application to separation processes. Darmstadt: Steinkopff and New York: Springer.
42. McHugh, M., and V. Krukonis. 1986. *Supercritical fluid extraction: Principles and practice.* Burlington, MA: Butterworth-Heineman.
43. Aktiengesell-Schaft, H. 1973. Method of extracting coffee oil containing aroma constituents from roasted coffee. DT 1336276.
44. Roselius, W., O. Vitzthum, and P. Hubert. 1982. Method of extracting coffee oil containing aroma constituents from roasted coffee. US Patent US4328225.
45. Gopalakrishnan, N. 1990. Carbon dioxide extraction of coffee beans. *Indian Coffee* 54:7–10.
46. Brimmer, J. 1995. Process for extraction of coffee oil containing aroma constituents. German Federal Republic Patent Application (DE4335321A1).
47. Ramos, E., E. Valero, E. Ibañez, G. Reglero, and J. Tabera. 1998. Obtention of a brewed coffee aroma extract by an optimized supercritical CO_2-based process. *Journal of Agricultural and Food Chemistry* 46:4011–4016.
48. Lopez-Fontal, E. M., and J. J. Castano-Castrilon. 1999. Extracción de aceite a partir de subproductos de la trilla de café pergamino. *Cenicafe* 50:66–68.
49. Lucas, S., and M. J. Cocero. 2006. Improvement of soluble coffee aroma using an integrated process of supercritical CO_2 extraction with selective removal of the pungent volatiles by adsorption on activated carbon. *Brazilian Journal of Chemical Engineering* 23 (2): 197–203.
50. Lojkova, L., B. Klejdus, J. Moravcova, and V. Kuban. 2006. Supercritical fluid extraction (SFE) of 4(5)-methylimidazole (4-MeI) and 2-acetyl-4(5)-(1,2,3,4)-tetrahydroxybutyl-imidazole (THI) from ground coffee with high-performance liquid chromatographic-electrospray mass spectrometric quantification (HPLC/ESI-MS). *Food Additives and Contaminants* 23 (10): 963–973.
51. Araújo, J. M. A., and D. Sandi. 2006. Extraction of coffee diterpenes and coffee oil using supercritical carbon dioxide. *Food Chemistry* 101:1087–1094.
52. Cocero, M. J., E. Alonso, and S. Lucas. 2000. Pilot plant for soil remediation with supercritical CO_2 under quasi-isobaric conditions. *Industrial & Engineering Chemistry Research* 39:4597–4602.
53. Ryu, Y.-K., K.-L. Kim, and C.-H Lee. 2000. Adsorption and desorption of n-hexane, methyl ethyl ketone, and toluene on an activated carbon fiber from supercritical carbon dioxide. *Industrial & Engineering Chemistry Research* 39: 2510–2518.
54. Lucas, S., M. J. Cocero, G. Brunner, and C. Zetzl. 2004. Adsorption isotherms for ethyl acetate and furfural on activated carbon from supercritical carbon dioxide. *Fluid Phase Equilibria* 219: 171–179.
55. Lucas, S., M. P. Calvo, C. Palencia, and M. J. Cocero. 2004. Mathematical model of supercritical CO_2 adsorption on activated carbon. Effect of operating conditions and adsorption scale-up. *Journal of Supercritical Fluids* 32: 193–201.

6.6 COST OF MANUFACTURING OF SUPERCRITICAL FLUID EXTRACTS FROM CONDIMENTARY PLANTS

Paulo T. V. Rosa and M. Angela A. Meireles

Obtaining extracts from condimentary plants using supercritical fluids as solvents has been extensively proved to be technically viable for several systems. In spite of that, only a few supercritical extracts can be found on a commercial scale and with prices quite higher than the ones produced by conventional methods. In general, the supercritical fluid extraction (SFE) is not even considered as a feasible extraction method because of the cost consideration. In this section we will present a simple method to perform initial manufacturing cost estimation. In the following section the cost terms will be defined, and the methodology used to estimate the manufacturing cost will be discussed in two study cases.

6.6.1 Capital Cost

The capital cost is related to the expenses needed to construct the industrial installation. In other words, it is the investment required to transform the project into an operable production unit. The main components of the capital cost are the land where the factory will be constructed, the terrain preparation, the construction of the buildings, the construction and installation of equipment, and the first load of raw material that will be used in the process.

Some points that should be taken into account in the phase of choice of the industrial unit location are the following:

- Cost and availability of raw materials
- Logistics to get the raw material, CO_2, cosolvents, etc.
- Logistics to send the products to consumers
- Presence and cost of qualified and unqualified workers
- Cost to purchase the land

For example, in order to get extracts from plants native to remote places, sometimes it is better to export the raw material to a larger center than to construct the extraction unit close to the production region because of the lack of labor, logistics, and other supplies. On the other hand, the land cost can be higher in larger cities. Some countries have some safety restrictions for operating a high-pressure unit in regions with large populations. Some other points such as the availability of drinking water and a system to collect sewage should be important in the choice of an industrial location.

The terrain topography can have a strong influence in the cost to prepare the landscape for the building construction. The presence of swampy areas can increase the cost of this stage. The cost of the building construction will depend on the size of the extraction unit, presence of administrative areas, size of the raw material stock place, restrooms, and presence of a cafeteria, among others.

As one can notice, an estimation of the land, land preparation, and building construction costs will depend on several factors, including the locations and size of the extraction unit.

In the case of SFE, the equipment cost is, in general, the most important term of the capital cost. The unit is constructed in stainless steel and should support the high extraction pressure. The systems used to open and close the extractors should allow rapid and safe operation. Several safety systems should be present in order to allow proper operation of the extraction unit.

There are several methods to obtain an estimation of the extraction unit cost. The first one uses the information of the cost of a known-size unit. The cost of units with other sizes can, then, be obtained by the following equation:

$$C_v = C_u \cdot \left(\frac{v}{u}\right)^a \tag{6.6.1}$$

where C_u is the cost of the known extraction unit with extraction volume of u, v is the volume of the desired extraction unit, and C_v is its cost. a is a constant with a value about 0.6. This equation is known as the "six-tenths-factor rule." This equation can be used for the cost estimation of any chemical process.

Some correction for the inflation rate from the period when the known-size unit cost was determined to the present moment can be necessary to produce a better cost estimation. The main disadvantage of this procedure is the necessity to know the price of an industrial-scale unit and its extraction volume.

For the specific case of SFE units, Perrut [1] proposed a price index correlation that considers both the volume and flow rate variation. The final expression is given by

$$PI = A(10V_TQ)^{0.24} \tag{6.6.2}$$

where PI is a price index, V_T is the total volume of extraction columns and separators, Q is the maximum CO_2 flow rate, and A is a constant.

The constant A value can be obtained from a graphic of several extraction unit prices as a function of V_T and maximum CO_2 flow rate. Again, this procedure has the inconvenience of knowing the price of the industrial-scale units.

The third way to estimate the price of an industrial-scale unit is to determine the cost of the main parts of the process, namely extraction columns, separators, CO_2 reservoir, heat exchangers, and CO_2 pump. The extraction plant estimated cost can be obtained by multiplying the total cost of the plant parts by the Lang factor. This factor varies from 4 to 5, depending on the process, and incorporates the costs such as tubing, connections, insulation, instrumentation, safety items, installation, and painting.

The final way to create a cost estimation is to contact equipment suppliers and ask for a price quotation. There are several companies that produce SFE equipment. Among them are UHDE in Germany, Natex in Austria, Thar Technologies in the United States, Separex in France, and IIT in India. These companies can construct industrial-scale SFE plants with extractor volumes from a few liters to 17 m^3.

In general, the supercritical extracts from condimentary plants are obtained from columns from 10 to 1000 L. The industrial units have from one to four extractors and from one to four separators. Systems with more than one separator can produce fractions using sequential pressure reduction and the ones with more than one extractor can operate continuously. At this operation, one or more extractors are extracting while one is depressurized, unloaded, cleaned, loaded, pressurized, and reaches the extraction temperature. After these operations, the prepared column starts extraction and an exhausted column is removed from the system to be prepared for a new extraction cycle.

Another term that should be incorporated into the capital cost is the expense of the start-up process. Items such as CO_2, cosolvents, raw materials, cleaning process, lubrication, and thermal fluids for the heat exchangers, among others, are included in this category.

Finally, if the extraction plant is a stand-alone unit, some investment must be done in a steam generator and in a refrigeration system in order to supply utilities for the process. If the industry is located in a place where it can buy utilities, this term should be neglected. The cost of auxiliary equipment such as driers, knife mills, and sieving systems should also be considered.

6.6.2 Cost of Manufacturing

The manufacturing cost takes into account all terms that have some importance on the final production cost of the condimentary plant extract. It can be divided in three main categories: direct cost, fixed cost, and general expenses.

6.6.2.1 Direct Manufacturing Cost

The direct manufacturing costs (D_{MC}) are intrinsically related to the production scale, increasing when the production increases and decreasing when the fabrication is diminished. Some costs of this category are raw materials, operational labor, utilities, waste treatment, maintenance and repairs, operation supplies, laboratory analysis, and patents and royalties.

The main raw materials used in an extraction unit are the plants containing the extract, the carbon dioxide used as solvent, and the cosolvent if it is used in the process. The cost of the vegetable matter should take into account the price that is paid for the producers or distributors plus the costs to transport it to the extraction unit and to dry and to triturate it.

The operational labor is related to the workers who are responsible for the physical maintenance of the process. They are responsible to transport, dry, and triturate the raw material; to load and unload the extractors; to check and control the pressures, temperatures, and flow rates; and to remove and to pack the extract from the separators, among other attributions. There should be a minimum number of operators per shift who are able to accomplish all these tasks.

During the extraction process several kinds of utilities are used. Vapor is the main heating medium in medium- to large-size units. For small-scale units, sometimes electrical heating can be more feasible. Heating is used to increase the CO_2 temperature after the pump and in the extraction column. It is used in the separators to decrease the solute solubility after CO_2 expansion. The cold liquid, generally salt solutions, is used to condense the solvent after the separators and in the CO_2 reservoir. Electrical power is used in the CO_2 pump.

One very interesting point of the supercritical extraction is to use an environmentally safe solvent. The CO_2 is recycled into the extraction plant, decreasing the raw material cost. Thus, there is no gas residue from the extraction plant. The liquid effluent is the desired product. The solid disposal is the exhausted raw material that can be incorporated into the soil or burned in the steam-generating system, decreasing the utilities cost of the process. As a result of these described points, in general there is no waste treatment cost for the supercritical extraction process.

The maintenance of the supercritical extraction unit should be done carefully because of the high-pressure nature of the process. The inspection of relief valves, disrupting disks, pressure and temperature sensors, extraction column open and close systems, tubing, connections, columns, tanks, pumps, and heat exchangers should be done frequently, and any damaged part should be replaced or repaired as soon as possible. The repairs must be carried out by specialized professionals to assure the pressure limits of the installation. Periodical pressure tests, in general using liquids, should be scheduled to verify the resistance of the extraction plant. The replacement of the installation parts will depend on its utilization level. Different from the waste treatment cost, the maintenance has a considerable importance in the manufacturing cost.

The other terms of the direct cost are related to the replacement of lubrication and thermal fluids, filters, personal protection items, the chemical analysis used for quality control and solving processing problems, and the cost of licensing technology if it has any kind of intellectual protection.

6.6.2.2 Fixed Manufacturing Cost

The fixed manufacturing cost (F_{MC}) is the one that has less influence on the production level. It can be divided into three main categories: depreciation, local taxes and insurances, and plant expenses.

The depreciation considers the loss of value of the investment with the operation time. The linear depreciation rate is the more accepted one. The depreciation time will be characteristic of part of the extraction unit: the land is not depreciable, the building has a depreciation time of 25 yr, and the equipment from 10 to 15 yr. Depreciation value is used in both manufacturing cost and calculation of taxes. This term is related to the capital cost discussed previously.

The second term of the fixed cost considers the land and buildings taxes and the insurance that can be obtained to protect the business. In general, this term is calculated in terms of the investment and the operation risks. In spite of the high pressure used in the process, the solvent is nonflammable and the accident risk is not too high.

Plant expense is a term that comprehends the auxiliary operations that occur during the processing. Costs such as the payment and accountability, fire and safety protection programs, medical services, and the cafeteria operation are included in this category.

6.6.2.3 General Manufacturing Expenses

The cost of general manufacturing expenses (G_{ME}) takes into account the operations that are not related to the production process such as

- Administration
- Distribution and selling
- Research and development

Depending on the size of the extraction unit, these services can be contracted from specialized companies. The research and development can be supplied from consultants to develop specific applications.

6.6.2.4 Estimation of the Cost of Manufacturing Condimentary Extracts

There are several ways to estimate the manufacturing cost of a process. Some can take into account all parts of the cost and, thus, have a quite precise estimation of the cost of manufacturing (COM). In general, these procedures are difficult to carry out, and are a time-consuming process. Even using this approach, the estimated cost can differ from the real one because of the complexity of the real process.

As discussed in Chapters 2 and 4, Turton et al. [2] present a simplified model that can be used in the estimation of manufacturing cost considering only five major fractions of the cost: fixed capital investment (F_{CI}), operational labor (C_{OL}), waste treatment (C_{WT}), utilities (C_{UT}), and raw material (C_{RM}). All other costs can be estimated from these five basic costs. For instance, the maintenance and repairs can be obtained as a fraction of the fixed capital, the laboratory charges can be calculated as a portion of the cost of operational labor, and the distribution and selling cost can be estimated as a fraction of the COM. The multiplying factor used to estimate each cost from the five basic ones depends on the process. The expressions for the direct manufacturing cost, fixed manufacturing cost, and general manufacturing expenses, considering the most probable values of the multiplying factors, are given by

$$D_{MC} = C_{RM} + C_{WT} + C_{UT} + 1.33C_{OL} + 0.03COM + 0.069F_{CI} \tag{6.6.3}$$

$$F_{CM} = 0.708C_{OL} + 0.168F_{CI} \tag{6.6.4}$$

$$G_{ME} = 0.177C_{OL} + 0.009F_{CI} + 0.16COM. \tag{6.6.5}$$

The COM can be obtained by the addition of the three parts of the cost presented in Equations 6.6.3–6.6.5. Thus, the final expression to estimate COM is given by [2]

$$COM = 0.304 \times F_{CI} + 2.73 \times C_{OL} + 1.23 \times (C_{RW} + C_{WT} + C_{UT}) \cdot \tag{6.6.6}$$

In Equation 6.6.6, the depreciation factor was considered as 10% (10 yr of depreciation), which should be valid for the extraction equipment. The depreciation factor for buildings, in Brazil, is 4% (25 yr of depreciation), and the land should not be depreciated. In spite of this superestimation of the F_{CI} in the depreciation term represents only a small difference in the multiplying factor because the supercritical extraction unit cost is larger than the land and buildings costs.

To perform the COM estimation, some previous information such as the particle bed density; extraction conditions including temperature, pressure, and CO_2 mass flow rate; and the overall extraction curve of the raw material in a laboratory-scale extraction unit should be taken into account . In general, the overall extraction curve can be presented in one of the three forms that can be observed in Figure 6.6.1. The first form represents only the cumulative amount of extract that is obtained from the experiment. If one alters the amount of raw material packed into the column or changes the extraction column size, the curve will have a completely different shape. When the overall extraction curve is represented in terms of yield (mass of extract by mass of raw material ratio), it takes into account the total mass of raw material used during the experiment. The third form of the overall extraction curve gives the fraction of the total mass of extract that is obtained as a function of the extraction time. For instance, as indicated in Figure 6.6.1, after 30 min of extraction, 75% of the solute present in the particles was extracted.

The overall extraction curve has three distinct regions: a quite linear section with a high slope at the beginning of the experiment, a curved period for intermediate extraction times, and a low-slope linear portion for a long extraction time. The first part of the extraction curve is characterized by an extraction carried out by a convective mass transfer of the solute located at the surface of the particles. When

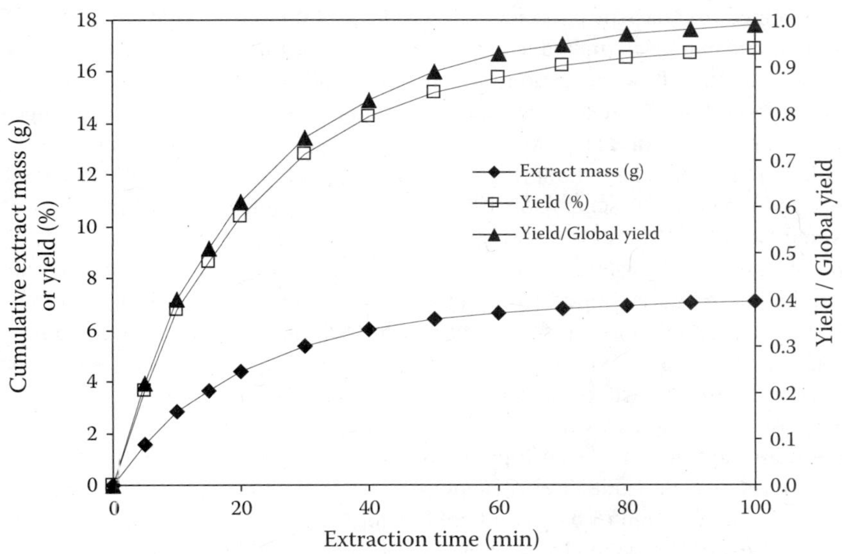

FIGURE 6.6.1 Clove bud overall extraction curve: 10 MPa, 308 K, and 9.0 ×10^{-5} kg/sec.

the solute present on the particle surface starts to finish, the mixed diffusive and convective mass transfer mechanisms are important. At the end of the extraction, there is only solute in the inner part of the particles, and the diffusive mass transport is responsible for the extraction. A larger fraction of the solute is obtained at the first region (at larger extraction rates), and as will be seen later, it is the operational region for a large-scale extraction unit.

The cost estimation presented in this chapter is similar to the one used by Rosa and Meireles [3]. The operational parameters used in the experimental extraction, such as particle size, bed density, and extraction pressure and temperature, are considered to be the same as in the large-scale unit. The performance of the industrial-scale extractor should be estimated in order to execute the cost estimation. This could be done using mathematical models that are used to fit the experimental overall extraction curve. There are several models that can be used such as VTII of Brunner [4], Naik et al. [5], Goto et al. [6], Sovová [7], and Martínez et al. [8]. These models are empirical or are obtained from the differential mass balance in the extraction column. In spite of the large number of models, there is none that can be used to predict the performance of the industrial-scale unit, mainly because of the lack of reliable mass transfer correlations for supercritical extraction systems.

One experimental scale-up procedure used by Rosa and Meireles [3] considers that the overall extraction curve, in term of yield variation with time, will have a similar shape in the large-scale unit if the mass of the raw material to the mass flow rate of solvent is kept equal to the laboratory unit. In general, this procedure is reasonably adequate if the column height-to-diameter ratio is also kept constant.

The calculation procedure is described next. Because the bed density is considered to be equal in the laboratory and the industrial columns, with the bed density used in the small-scale system and the volume of the large-scale column it is possible to calculate the mass of the particles used per extraction cycle in the industrial column. Once the mass of particles in the industrial column is calculated, the CO_2 mass flow rate can be estimated using the restriction of the constant mass of particle to the solvent mass flow rate ratio in the different scales units.

The amount of CO_2 lost during the processing is comprehended in the range of 1 to 3% (Thar Technologies, Pittsbutgh, PA, personal communication) of the total mass flow rate used during the processing. A loss of 2% was used in the cost calculation, and the cost was set as US$ 100.00/ton. The raw material cost in Brazil was obtained from the Brazilizn Institute of Geography and Statistics (IBGE) [9]. The final raw material cost was considered to be the market price plus US$ 30.00/ton for drying and grinding it (SuperPro Design Software). The number of operators was estimated using the tables presented by Ulrich [10], and the operational labor rate was assumed as US$ 3.00/hour. The waste treatment cost was neglected because there is no harmful waste produced in the extraction unit. Another restriction used in the cost calculation was that the utility cost can be estimated using the values of the pure carbon dioxide entropy-temperature diagram [4], because the solute solubility is low. The value of electric power was obtained for the Brazilian market and the saturated steam (5 barg) and cold water (278.2 K) cost was obtained from SuperPro Design Software (Demo Version). The costs of the extraction units used in this work were US$ 75,000.00, US$ 400,000.00, and US$ 2,000,000.00 for units with 2 × 5, 2 × 50, and 2 × 400 L, respectively. These costs were suggested from equipment supplier companies. Thus, using the considerations and values previously stated, the COM could be estimated.

In the next section, the results of the COM estimation of supercritical fluid extracts from clove buds and ginger will be presented. These two systems have different behaviors because clove buds have a large amount of extract and high solubility in supercritical CO_2. The ginger system has a lower amount of extract and lower solubility in supercritical carbon dioxide.

6.6.2.5 COM Extracts from Clove Buds

The overall extraction curve presented in Figure 6.6.1 was obtained at 10 MPa, 308 K, and the CO_2 flow rate of 9.0×10^{-5} kg/sec. This experiment was carried out with 42.26 g of clove bud particles in a 300-mL extraction column partially filled with glass beads at the CO_2 entrance. The bed density (ratio between mass of raw material and effective bed volume) was 750 kg/m^3. The cost of the clove buds used in the calculations was US$ 505.00/ton [9]. The manufacturing cost variation with extraction time can be seen in Figure 6.6.2 for three sizes of extractors.

For short extraction times the manufacturing cost is high because of the low use of the raw material. If one extraction unit could work with this short time, the raw material after extraction would still be rich in solute that can be easily extracted. As the extraction time increases, the extraction rate decreases and the impact of the fixed cost increases. One can also observe in Figure 6.6.2 that the increase of the extraction size has an important influence on the manufacturing cost. The calculation considered that the total annual operation time was 7920 h, which corresponded to working 330 days per yr, 24 h per day.

To maintain continuous operation, the extraction time should be sufficient to allow the depressurization, discharge of the exhausted raw material, cleaning of the

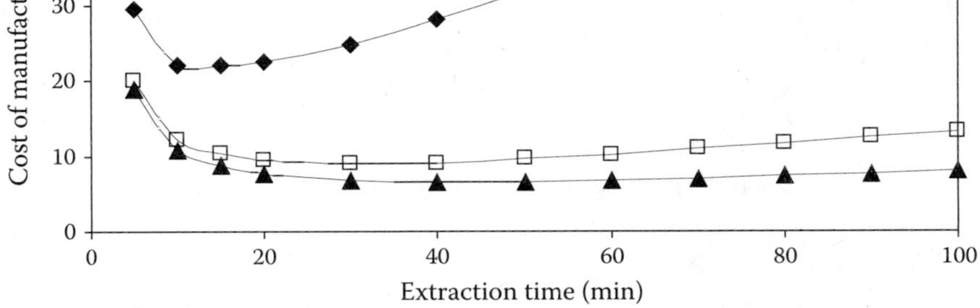

FIGURE 6.6.2 Cost of manufacturing clove bud extracts as a function of extraction time.

column, charge of the fresh raw material, pressurization of the column, and reaching the extraction temperature of the system. The use of extraction baskets can help to decrease the change and discharge times. The raw material is packed into the basket outside of the column, and the discharge and charge times are related only to the removal of or setting the basket into the extraction column.

Thus, for example, if the minimum extraction time is 60 min, the COM for clove bud extract are 6.80, 10.02, and 35.95 US$/kg, for extractions units of 2 × 400, 2 × 50, and 2 × 5 L, respectively. For this extraction time, the small-size unit can produce 4700 kg of extract per yr, the medium-size unit can produce 47,000 kg/yr, and the large-size unit can produce 376,000 kg/yr. These amounts of extract can be very high, and in general, the extraction unit will process several kinds of raw material.

The importance of each cost fraction in the manufacturing cost composition can be observed in Table 6.6.3. One can observe that as the extraction time increases, all cost fractions but the raw material increase. The decrease of raw material cost is a consequence of the better use of it, meaning that a large amount of the solute is recovered. As expected, the impact of operational labor for small-size units is larger than for large-scale units. In spite of the high-pressure operation, the utility cost represents only a small fraction of the manufacturing cost.

In general, the raw material cost can fluctuate considerably depending on the amount of production and the quality. One can observe in Figure 6.6.3 the influence of the raw material cost on the manufacturing cost for the 2×50L system. For short extraction times the influence of the raw material on the final cost is very high, and the percentage of the raw material increment is practically transferred for the COM. As the extraction time increases, the importance of the raw material cost decreases, and the cost fluctuation has a smaller impact on the COM. For instance, when the raw material cost increases from US$ 250.00/ton to US$ 2000.00/ton, or eight times larger, for 5 min of extraction time, the COM increases from US$ 11.56/kg to US$ 70.29/kg, or six times larger. For the same raw material cost fluctuation the COM for 60 min of extraction time increases from US$ 8.21/kg to US$ 21.87/kg, or 2.7 times larger. The effect of an increase in the raw material cost will be more important for large-scale units than for small-size units (Table 6.6.1). The extraction time that obtains the minimum COM increases as the raw material cost increases. For the raw material cost of US$ 250.00/ton, a minimum COM of US$ 6.39/kg is obtained

TABLE 6.6.1
Composition of the Manufacturing Cost of Clove Buds Extract

	% of the manufacturing cost of clove bud extract								
	2×5L			2×50L			2×400L		
	5 min	30 min	60 min	5 min	30 min	60 min	5 min	30 min	60 min
Investment	7.11	14.52	16.21	5.53	21.61	30.45	3.68	17.69	28.56
Raw material	61.27	20.97	11.78	89.27	58.50	41.52	95.14	76.64	62.30
Operational labor	31.24	63.76	71.17	4.56	17.79	25.07	0.61	2.91	4.70
Utilities	0.38	0.75	0.84	0.54	2.10	2.96	0.57	2.75	4.44

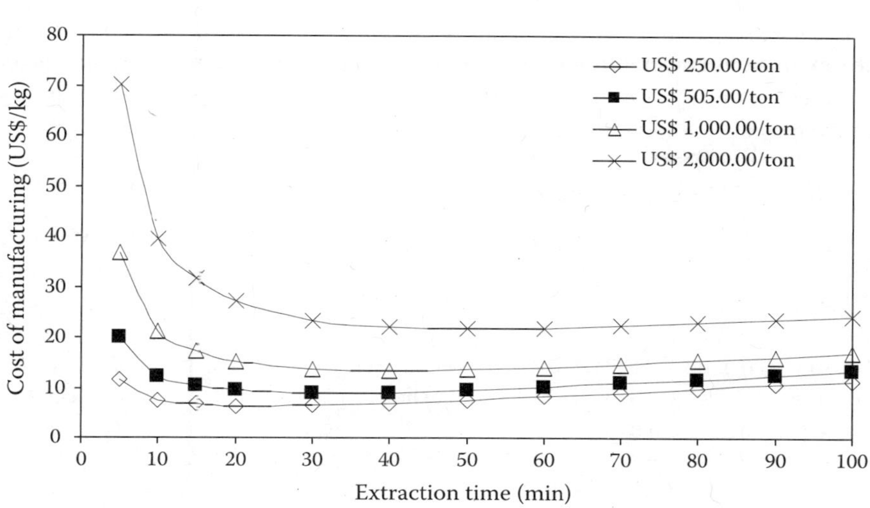

FIGURE 6.6.3 Influence of the raw material cost on the manufacturing cost of clove bud extract.

for 20 min of extraction time, and for the raw material cost of US$ 2,000.00/ton, a minimum cost of US$ 21.71/kg is obtained for 50 min of extraction time.

Another factor that can have influence on the COM is the number of shifts that the extraction unit works. In all calculations presented so far, the maximum operation time was considered. Table 6.6.2 presents some results of the impact on the number of operational shifts on the COM. For the small-scale unit there is only an increase of 32% in the COM whether the unit operates in three shifts or in one shift. This effect is more severe for the large-scale unit (57%). The difference is related to the large fixed cost of the large-scale unit, mainly in the depreciation factor. In spite of the relatively low increase of the COM, the amount of extract produced will decrease to one-third when the unit works in three shifts rather than that in one shift.

The cost of clove bud essential oil obtained by steam distillation varies from US$ 25.00/kg to US$ 88.00/kg, depending on the quality of the raw material, and the supercritical extract can be obtained by US$ 115.00/kg (Liberty Natural Products [11]). These costs are for purchase of relatively low amounts of extracts (4.54 kg), and

TABLE 6.6.2
Influence of the Number of Shifts on the Cost of Manufacturing Clove Buds Extract

	Cost of manufacturing clove bud extract (US$/kg)		
	2×5L	**2×50L**	**2×400L**
1 shift (2640h/year)	47.60	16.42	10.68
2 shifts (5280h/year)	38.86	11.76	7.77
3 shifts (7920h/year)	35.95	10.20	6.80

for larger amounts these values should be lower. To compare the extract cost estimated in the work to the commercial one, we should consider the production taxes and the extraction unit profits. It seems that the extraction unit with 2 × 50 L columns can produce extracts with manufacturing costs competitive with the market.

6.6.2.6 COM Extracts from Ginger

The ginger overall extraction curve used in this work was presented by Martinez et al. [8] and is illustrated in Figure 6.6.4. The experimental condition used to obtain the ginger overall extraction curve was 20 MPa, 313 K, and 5.6×10^{-5} kg/sec of CO_2 mass flow rate. The experiment was carried out in a 235-mL extraction column filled with 80 g of dried raw material. Thus, the bed density was 340 kg/m^3, which is less than half of the bed density of clove bud particles. The low bed density was a consequence of the ginger particle characteristics that do not allow the bed compression. The cost of the ginger used was US$ 495.00/ton [9], which is similar to the clove bud cost. One can observe that the ginger global yield is considerably lower than the one obtained for clove buds and that the extraction times are significantly larger than the ones observed for clove buds. The larger extractions times can be explained in terms of the lower solubility displayed by the ginger extract in the supercritical fluid [12] and the differences in the structures presented in the solid particles.

The lower global yield, lower bed density, and larger extractions times should give larger manufacturing costs than the ones obtained for clove buds. The estimated values of the COM ginger supercritical extracts are presented in Figure 6.6.5 for three extraction unit scales. The estimated values were at least one order of magnitude larger than ones observed for clove buds extracts. The shape of the

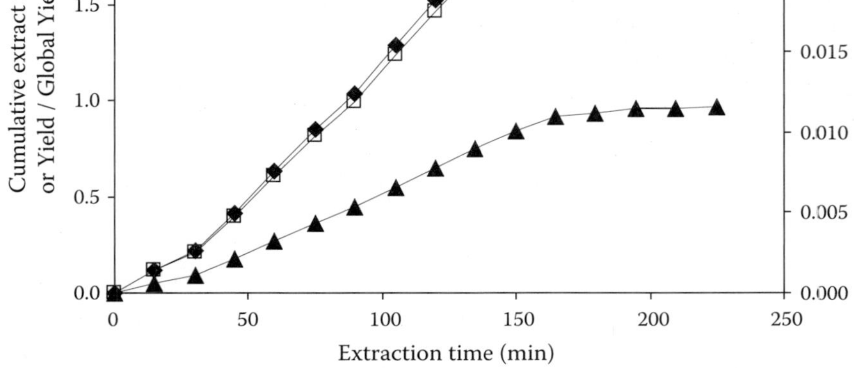

FIGURE 6.6.4 Ginger overall extraction curve: 20MPa, 313K, and 5.6×10^{-5} kg/sec.

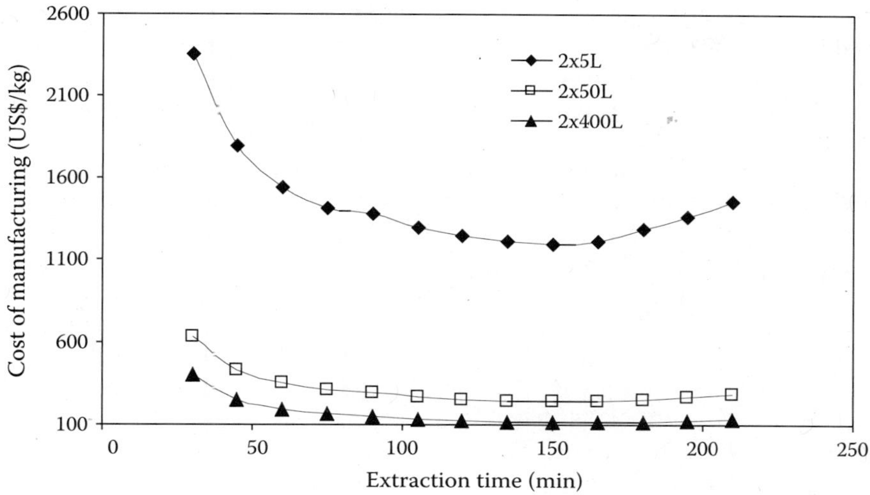

FIGURE 6.6.5 Cost of manufacturing ginger extracts as a function of extraction time.

curves is similar. There is a decrease in the COM when the extraction time is increased in the region where the extraction rate is high, and there is an increase when the extraction rate is too low. The COM decreases with the increase of the extraction unit scale. The minimum COM obtained occurs in the extraction times between about 150 and 165 min. The minimum costs obtained per kilogram of ginger extracts were US$ 1195.00, US$ 240.00, and US$ 113.00, from extraction units of 2 × 5, 2 × 50, and 2 × 400 L, respectively. The total amount of extract that can be obtained from these extraction units if the extraction time is set at 165 min was 125, 1250, and 10,000 kg/yr.

The importance of each cost fraction in the COM composition for three extraction unit sizes and three extraction times can be seen in Table 6.6.3. Again, all cost fractions but the raw material increase in importance when the extraction time is increased. In the case of the clove bud extraction in the medium- and large-size

TABLE 6.6.3
Composition of the Manufacturing Cost of Ginger Extract

	% of the manufacturing cost of ginger extract								
	2×5L			2×50L			2×400L		
	30 min	90 min	165 min	30 min	90 min	165 min	30 min	90 min	165 min
Investment	16.56	17.81	18.12	32.93	44.61	48.52	32.19	54.52	64.32
Raw Material	10.58	3.83	2.15	39.45	17.98	10.80	61.72	35.17	23.04
Operational Labor	72.72	78.22	79.58	27.12	36.73	39.95	5.30	8.98	10.66
Utilities	0.14	0.14	0.15	0.50	0.68	0.73	0.79	1.33	1.58

units, the cost with the major influence in the COM was the raw material. In the ginger extraction at these extraction units the most important cost was the investment. This different results are related to the large extraction times for the ginger. For the small-size unit, the operational labor cost is the most important fraction of the COM.

To decrease the extract cost, we should find a way to decrease the extraction time or to find a raw material with a larger amount of extract with the same commercialization price. The first can be obtained by optimizing the extraction condition by trying to find the experimental condition at which the ginger extract can have a larger solubility in the supercritical CO_2. This can be done by using a cosolvent, for instance. The second way to decease the extract cost is to improve the ginger cultivation conditions to increase the amount of extract present in the solid particles.

The commercial cost per kilogram of ginger essential oil is considered to be in the range of US$ 148.00 to US$ 240.00. For the ginger extract there is a considerable difference between the essential oil and the supercritical extract. The pungent fraction of ginger has a high boiling point and can be degraded during the extraction process. This fraction of ginger is the one responsible for its theraputic properties. The commercial cost of the supercritical fluid extract is in the range of US$ 300.00/kg to US$ 345.00/kg. This cost is not much higher than the one obtained in this work. We should stress that our estimated cost does not take into account the production and commercialization taxes.

6.6.3 Nomenclature

Symbol	Description
A	Constant in Equation 6.6.2
C_{OL}	Cost of operational labor
COM	Cost of manufacturing
C_{RM}	Cost of raw material
c_u	Known cost of an extraction unit with capacity u
C_{UT}	Cost of utilities
c_v	Cost of an extraction unit with known capacity v
C_{WT}	Cost of waste treatment
D_{MC}	Direct manufacturing cost
F_{CI}	Fixed capital investment
F_{MC}	Fixed manufacturing cost
PI	Price index
Q	Maximum CO_2 flow rate
u	Capacity of the reference extraction unit
V_T	Capacity of the desired extraction unit

6.6.4 References

1. Perrut, M. 2000. Supercritical fluid applications: Industrial developments and economic issues. *I&ECR* 39:4531–4535.
2. Turton, R., R. C. Bailie, W. B. Whiting, and J. A. Shaeiwitz. 1998. *Analysis, synthesis, and design of chemical process.* Upper Saddle River, NJ: Prentice Hall.

3. Rosa, P. T. V., and M. A. A. Meireles. 2005. Rapid estimation of the manufacturing cost of extracts obtained by supercritical fluid extraction. *Journal of Food Engineering* 67:235–240.
4. Brunner, G. 1994. *Gas extraction: An introduction to fundamentals of supercritical fluid and applications to separation processes*. Darmstadt: Steinkopff and New York: Springer.
5. Naik, S. N., H. Lentz, and R. C. Maheshawari. 1989. Extraction of perfumes and flavours from plant materials with liquid carbon dioxide under liquid-vapor equilibrium. *Fluid Phase Equilibria* 49:115–126.
6. Goto, M., M. Sato, and T. Hirose. 1993. Extraction of peppermint oil by supercritical carbon dioxide. *Journal of Chemical Engineering of Japan* 26:401–406.
7. Sovová, H. 1994. Rate of the vegetable oil extraction with supercritical CO_2. 1. Modeling of extraction curves. *Chemical Engineering Science* 49:409–414.
8. Martínez, J., A. R. Monteiro, P. T. V. Rosa, M. O. M. Marques, and M. A. A. Meireles. 2003. Multicomponent model to describe extraction of ginger oleoresin with supercritical carbon dioxide. *Industrial & Engineering Chemistry Research* 42:1057–1063.
9. Brazilizn Institute of Geography and Statistics (IBGE). 2003. www.sidra.ibge.gov.br/bda/agric/default.asp (accessed September 29, 2003).
10. Ulrich, G. D. (1984). *A guide to chemical engineering process design and economics*. New York: John Wiley & Sons.
11. Liberty Natural Products. 2003. Nature Source Botanical Ingredients. www.libertynatural.com/bulk/bulking.htm (accessed July 17, 2008).
12. Rodrigues, V. M., E. M. B. Sousa, A. R. Monteiro, O. Chiavone-Filho, M. O. M. Marques, and M. A. A. Meireles. 2002. Determination of the solubility of extracts from vegetable raw material in pressurized CO_2: A pseudo-ternary mixture formed by cellulosic structure + solute + solvent. *Journal of Supercritical Fluids* 22:21–36.

7 Concentration of Bioactive Compounds by Adsorption/Desorption

Lourdes Calvo and María José Cocero

CONTENTS

7.1 Fundamentals of Adsorption....................404
7.1.1 Introduction....................404
7.1.2 Fundamentals of Adsorption....................406
7.1.2.1 External Transport....................406
7.1.2.2 Internal Transport....................408
7.1.2.3 Equilibrium of Adsorption....................409
7.1.3 Types and Properties of the Adsorbents....................414
7.1.4 Adsorbent Regeneration....................417
7.1.5 Adsorption Processes....................418
7.1.5.1 Operation in Agitated Vessels....................419
7.1.5.2 Operation in Fixed Beds....................419
7.1.5.3 Operation in Moving Beds....................420
7.2 Applications of Adsorption in Food Processing....................422
7.2.1 Removal of Unwanted Natural and Harmful Anthropogenic Compounds from Edible Oils....................422
7.2.2 Purification of Drinking Water....................423
7.2.3 Removal of Color in Syrups....................424
7.2.4 Cane Sugar Refining....................425
7.2.5 Color and Taste Correction in Alcoholic Beverages....................426
7.2.6 Elimination of Color in Flavorings....................426
7.2.7 Purification of Carbon Dioxide for Use in Carbonated Drinks....................426
7.2.8 Decaffeination of Tea and Coffee....................427
7.2.9 Removal of Unwanted Odor or Color Compounds from Glycerin....................427
7.2.10 Purification of Fruit Juices....................427
7.2.11 Purification of Starch-Based Sweeteners....................428
7.2.12 Decolorization of Citric Acid....................428
7.2.13 Other Applications....................429

7.3 Nomenclature 432
7.4 Further Reading 434
7.5 References 435

The fundamentals of adsorption as a separation, concentration, and purification technique in the food industry are discussed in this chapter. The process is analyzed from the phenomenological point of view, describing the main stages for the adsorption of a solute on the inner surface of a porous material: external and internal transfer and equilibrium. The most commonly used models to describe these steps are presented and commented on in terms of the fluid-phase state: gas or liquid, and the possible applications. Then, the description of the properties and purposes of the most frequent adsorbents is given. The next section presents the different alternatives for the adsorbent regeneration after use because this step is crucial for the feasibility and cost of the whole process. They arise from diverse ways of altering the adsorption equilibrium such as increase in temperature, reduction of pressure, and the introduction of a purge or a desorbent. The different configurations and operation methods to carry out the process at industrial scale are illustrated with the aid of industrial examples. The batch operation in agitated tanks is described. The continuous operation is presented in various possibilities: by using series of fixed beds working with alternation of adsorption/desorption stages or by the utilization of simulated moving bed systems. The last section is devoted to briefly describing the main applications of adsorption in the food industry. Essentially, this operation is used to remove pigments, odors, and other kind of impurities. Finally a summary of novel separations and new procedures where adsorption is combined with other techniques is presented. A separate section is included at the end to discuss the use of adsorption in the selective recovery of bioactive compounds from crude solvent extracts. In most cases, the target compounds are polyphenols that are separated from plant wastes because of their potential health benefits.

7.1 FUNDAMENTALS OF ADSORPTION

7.1.1 Introduction

Adsorption is a technique for the separation of a substance (or various substances) from a fluid mixture by its retention, accompanied by its concentration, on the inner surface of a porous solid. The adsorbed solute is referred to as the *adsorbate,* whereas the solid material is the *adsorbent.* The phase formed by the molecules of adsorbate joined to the surface of the solid is called the *adsorbed phase.* The key factor is the existence of an affinity between the molecules of the adsorbate and the surface of the adsorbent, resulting in a higher concentration in the adsorbed phase than in the fluid after equilibrium is reached.

The affinity of the solid for the component to be separated results from the existence of adsorbate–absorbant interactions; these interactions can be of physical or chemical nature, depending on the intensity. In the physical adsorption, the forces are weak, such as van der Waals or hydrogen bonds. In the chemisorption, the forces are similar to those implied in chemical bonds. For physical adsorption, the amount of heat is the same order of magnitude as the heat of vaporization

(<80 kJ·mol^{-1}), whereas for chemisorption, it is close to the values of reaction enthalpies (80–400 kJ·mol^{-1}). The former is produced at temperatures close to room conditions, and it is a faster process with low activation energy, whereas the latter occurs at a wider interval of temperatures and requires surpassing much higher activation energy. For these reasons, a molecule that is physically adsorbed can be more easily desorbed by reducing the concentration or by increasing the temperature. On the contrary, the desorption of the chemically attached molecules is more difficult, requiring a greater increase in temperature. Chemisorption is restricted to a monolayer of molecules, and physical adsorption usually implies the formation of a multilayer of adsorbate. In industrial applications, it is more frequently the physical adsorption, although it is difficult to distinguish when chemisorption occurs because sometimes the first layer is chemically adsorbed and the successive layers are adsorbed physically.

The efficiency of the operation depends on the adsorbent capacity and selectivity. It is also important that the adsorbents are durable and mechanically strong. Adsorbents may be natural or artificial solids whose properties widely vary. New synthetic methods and adsorbents are highly investigated with the aim of delivering solids of better structures (higher surface) and properties promoting higher capacity, specificity, and easier regeneration.

On the other hand, the cost of the solid is usually high, and it is not environmentally acceptable to dispose of it without treatment, especially if the adsorbate is a contaminant compound; thus, when the adsorbent is exhausted, it is mandatory to recover it. This is accomplished in another stage that has to be considered as part of the whole process. Different methods for the regeneration can be used depending on the type of solid, the adsorbate, the fluid, and the mode of operation. All of them have in common the alteration of the adsorption equilibrium.

Adsorption is used as a separation operation for mixtures whose components present similar physical properties (volatility, solubility, etc.) or whose concentration is very low. The more mature operations such as distillation and extraction are not adequate to separate these mixtures because of the elevated energy requirements to boil (in distillation) or recover the solvent in the case of extraction. The heats of adsorption are usually much lower so the energetic costs are reduced compared to these conventional methods. This is the reason that adsorption is currently one of the most used alternatives for the elimination of contaminants in gas and liquid streams, the drying of air and organic liquids, or the purification of biochemicals. In the food industry, it is mainly used for the removal of colors and other impurities, as will be shown later. Nowadays, adsorption is also applied in the separation, concentration, and purification of bioactive compounds.

Depending on the application, an adsorption unit can be quite complex. The design has to take into account the type of fluid, the mode of solid–fluid contact, the way in which the regeneration of the solid is done, and the scale. The adsorption can be done in agitated tanks or in fluidized beds and continuously in moving and fixed beds. For example, the recovery of solvent vapors is conducted in fluidized beds. Fixed-bed adsorbers are used to separate air into N_2 and O_2. The removal of organic pollutants in wastewaters is usually conducted in agitated tanks.

All these aspects will be discussed in detail in the next section, specifying the peculiarities of this operation in the food industry with emphasis on the production of bioactive compounds. For this purpose, the chapter will be illustrated with examples and applications of interest for food technicians.

7.1.2 Fundamentals of Adsorption

The global rate of an adsorption process depends on the rate of each stage during the transportation and adsorption of the solute from the bulk medium onto the pores of the adsorbent. There are at least three resistances to be overcome:

1. External mass transfer from the bulk fluid to the entrance of the pore (surface of the solid).
2. Internal mass transfer from the entrance of the pore to the inner surface.
3. Adsorption of the solute onto the active sites of the internal pore surface. It is an equilibrium process that usually occurs at high rate.

During desorption, the process is reversed. Figure 7.1 shows that the external and internal mass transfer is coupled in series, whereas the adsorption may occur in parallel or in series to the inner transport. At the same time as molecules are adsorbed, the heat of adsorption is released, generating a temperature profile. The mathematical models to describe each stage are derived from the mass and heat transport fundamentals and the thermodynamic equilibrium, which are discussed in detail next.

7.1.2.1 External Transport

A simple way to describe the process is by the steady-state film theory [1]. According to this theory, the external mass transfer occurs by molecular diffusion through the boundary layer around the adsorbent particle because of a difference in concentration between the bulk fluid and the surface of the adsorbent. The adsorbate is then

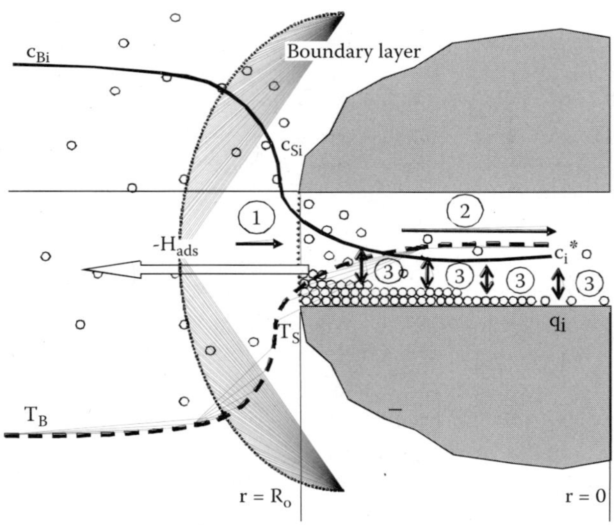

FIGURE 7.1 Scheme of the stages of an adsorption process.

transported at a rate directly proportional to the external mass transfer coefficient, k_g, and the concentration drop across the boundary film:

$$n_i = \frac{dN_i}{dt} = k_g a_p (c_{Bi} - c_{si}), \tag{7.1}$$

where a_p is the outer surface area of the particle, and c_{Bi}, c_{Si} are the concentrations of the adsorbate in the bulk fluid and at the surface of the particle, respectively.

The external mass transport coefficient, k_g, is calculated by fitting of experimental data or from mathematical correlations. A useful expression is the one developed by Wakao and Funazkri [2] from a wide set of gas- and liquid-phase mass transfer data as a function of the Reynolds (Re) and Schmidt (Sc) numbers:

$$Sh = 2 + 1,1\ Re^{0,6} Sc^{1/3}, \tag{7.2}$$

where $Sh = k_g d_p / D_m$; $Re = \rho v d_p / \mu$ and $Sc = \mu / \rho_{Dm}$. Based on this correlation, the mass transfer coefficient depends on the fluid properties (density, ρ, and viscosity, μ), the degree of turbulence in the fluid (e.g., v), the particle diameter (d_p), and the molecular diffusivity of the component in the mixture (D_m). The fluid properties should be introduced at the average temperature of the boundary layer. The expression is valid for the range $3 < Re < 10^4$, $0.6 < Sc < 70.600$, and $0.6 < d_p < 17.1$ mm.

When the particle is not spherical, d_p is replaced by d'_p:

$$d'_p = d_p \psi, \tag{7.3}$$

where sphericity, ψ, is a parameter that takes into account the difference between the surface areas of the particle and a sphere of the same volume. This parameter can be taken as 0.65 for most uses as discussed by Kunii and Levenspiel [3].

If the heat of adsorption is not negligible, as is the case of chemical adsorption and/or the heat does not disperse fast, a temperature profile between the temperature of the surface (T_S) and the temperature of the bulk fluid (T_B) across the boundary film is formed, provoking the transport of heat at a rate given by

$$q' = \frac{dQ}{dt} = h a_p (T_S - T_B). \tag{7.4}$$

Here again, the heat transfer coefficient, h, can be calculated from empirical correlations. The corresponding equation for this parameter is derived from Wakao and Funazkri [2]:

$$Nu = 2.0 + 1.1 Pr^{1/3} Re^{0.6}, \tag{7.5}$$

where $Nu = hd_p/k$ is the Nusselt number and $Pr = c_p \mu / k$ is the Prandlt number, which depends on c_p, the specific heat, and k, the thermal conductivity of the fluid.

7.1.2.2 Internal Transport

Once the molecule of adsorbate passes through the layer and reaches the surface of the pore, it diffuses through it. This internal transport can be described by different mechanisms:

1. Molecular diffusion due to concentration differences in the fluid that fills the pore.
2. Surface diffusion or surface migration in the adsorbed phase.
3. Knudsen diffusion, which occurs in micropores and at low pressures (only in gases).

The molecule travels within the pore, colliding with other molecules and/or with the pore walls. The diffusion in the fluid that fills the pores (gas or liquid) takes place when the collisions between the molecules of the adsorbate are the most frequent. It may be described with Fick's first law:

$$N_{im} = -D_m \frac{dc_i}{d\iota}, \tag{7.6}$$

where N_{im} is the molar rate of the component i through the fluid, ι is the distance in the pore, and D_m is the molecular diffusivity.

When the pore diameter is much smaller than the mean free path, the collisions of the molecules with the pore walls are more frequent than with themselves. This mechanism may become important for gases when the pore is very small and/or at low total pressure. The flux is now controlled by the Knudsen diffusion determined by the so-called coefficient, D_K [1]:

$$N_{iK} = -D_K \frac{dc_i}{d\iota}. \tag{7.7}$$

The surface diffusion, also known as surface migration, is important when the concentration of the adsorbed phase is high and the pores are small. The interaction degree with the adsorbent is not very strong and the molecules move from high- to low-concentration areas along the pore wall. This would explain why the surface diffusion coefficient, D_s, increases with coverage. The flux can be expressed in terms of the concentration of the adsorbed phase (q_i), as suggested by Schneider and Smith [4]:

$$N_{i_s} = -D_s \rho_p \frac{dq_i}{d\iota}, \tag{7.8}$$

where ρ_p is the particle density.

The surface diffusion is not easy to determine; thus, it is usually calculated after measurement of the total diffusion by subtraction of the theoretically predicted Knudsen and molecular diffusions.

All these mechanisms can occur at the same time because the adsorbents possess different pore sizes. To account for every form of diffusion, an effective diffusivity, D_e, may be used, which has to be evaluated empirically:

$$N_i = -D_e \frac{dc_i}{dt}. \quad (7.9)$$

The relative importance of the external and internal resistances in the global rate can be determined by the Biot number, *Bi*.

$$Bi = \frac{External\ resistance}{Internal\ resistance} = \frac{k_g R_o}{D_e}, \quad (7.10)$$

where R_o is the particle radius. If the *Bi* number is low, then the controlling step is the internal transport. This is the normal scenario for mass transfer in the adsorption processes. On the contrary, for heat transport, whose *Bi* number is defined as hR_o/k, the transfer across the boundary film usually restrains the global heat rate. The reason is that adsorbents exhibit high thermal conductivities, and temperature gradients within the particles are insignificant. Consequently, no heat transport evaluation inside the particle is generally needed.

7.1.2.3 Equilibrium of Adsorption

The majority of the separation processes by adsorption are based on the different capacities for equilibrium for each component by a determined adsorbent. Therefore, the knowledge of the equilibrium data of the system is fundamental for the design of the equipment.

In equilibrium, the concentration of the adsorbed phase, q (mol·kg^{-1} adsorbent or kg·kg^{-1} adsorbent) is related to the concentration of the adsorbate in the fluid c (mol·m^{-3} fluid or kg·m^{-3} fluid). For gases, partial pressure is used instead of concentration. This relation is specific for each adsorbate–adsorbent system and strongly depends on temperature. The plot of the q versus c data at constant temperature is called *adsorption isotherm* and limits the extent to which a solute is adsorbed on a determined adsorbent at the given operation conditions.

The isotherms are described with mathematical expressions. Many forms have been developed depending on whether only one component or more than one component (multicomponent) of the fluid mixture is adsorbed and depending on the type of fluid: gas or liquid. The most important isotherms are reviewed in the next section.

7.1.2.3.1 Gas Adsorption

The *linear isotherm of Henry* is the most simple equation to describe the equilibrium. It assumes that the adsorbed concentration is directly proportional to the fluid concentration:

$$q = Kp, \quad (7.11)$$

where p is the partial pressure of the component in the gas and K is the equilibrium constant, which usually follows an Arrhenius function of temperature:

$$K = K_0 exp\left(\frac{-\Delta H_{ads}}{RT}\right). \quad (7.12)$$

From this expression it is easy to predict the effect of temperature in the equilibrium. The term ΔH_{ads} is the variation of the enthalpy of a mole of adsorbate when it passes to the adsorbed phase; it is called heat of adsorption. Because the adsorption process is always exothermic, the variation of enthalpy is negative; thus, K decreases with temperature. This fact demonstrates that adsorbents could be regenerated by an increase in temperature, as will be shown in Section 7.1.4.

Henry's law describes the experimental data only when the partial pressure of the gas (or the concentration in the fluid) is very low. However, it is widely used in the theoretical models for adsorption because it allows a simple mathematical treatment. The other theoretical expressions have to tend to Henry's law for low concentrations.

Further expressions that are largely used in practical applications are the equations of Langmuir and Freundlich. The *Langmuir isotherm* was theoretically derived from kinetic considerations assuming that no adsorbate–adsorbate interactions occur in the adsorbed phase (e.g. only the adsorbate–adsorbant interactions are important) and the surface is energetically homogeneous, so the heat of adsorption is independent of the degree of the adsorbent covering. In partial pressure units, the following is the Langmuir equation:

$$q = \frac{q_{max} K p}{1 + K p}, \tag{7.13}$$

where q_{max} is the maximum capacity of adsorption corresponding to complete coverage of the surface by the gas, assuming a maximum of a monolayer coverage, and K is the adsorption–equilibrium constant, which is a measurement of the adsorbate–adsorbent affinity. The values of q_{max} and K can be obtained by fitting experimental data using the linearized form:

$$\frac{p}{q} = \frac{1}{q_{max} K} + \frac{p}{q_{max}}. \tag{7.14}$$

A graph of p/q versus p gives a straight line with a slope of $1/q_{max}$ and an intercept of $1/q_{max}K$. The Langmuir isotherm is generally applied to low concentrations.

The Freundlich isotherm is an empirical and nonlinear expression that assumes that the surface of the adsorbent presents a nonuniform distribution of the heat of adsorption:

$$q = k_F p^{1/n_F}, \tag{7.15}$$

where k_F and n_F are constants. The latter one increases with temperature and lies in the range of 1 to 5. Obviously, when $n_F = 1$, Equation 7.15 is equal to the Henry's law, but this only happens at high temperatures. Conversely to the Langmuir equation, the Freundlich isotherm does not predict a limit for q at high pressures.

The Freundlich isotherm can also be used in the linearized form to determine the empirical constants, k_F and n_F, by fitting the experimental data numerically or graphically:

$$\log q = \log k_F + (1/n_F) \log p. \tag{7.16}$$

When more than an adsorbate with affinity to the adsorbent surface is present in the mixture, the capacity of the adsorption for one of the components is affected by the presence of the other and its concentration. The influence can be positive, so the adsorption of one component increases the adsorption of the other, being negative if the contrary happens or causes no effect, depending on the interactions between the molecules of the adsorbents. It is complicated to accurately describe this situation, but to represent the equilibrium of such systems, it is necessary to use a multicomponent isotherm. A commonly used model to describe the competitive effect of the components is the extension of the Langmuir equation presented by Markham and Benton [5]. The interactions between the adsorbates are neglected; so the only effect is the rivalry between them for the vacant surface area:

$$q_i = \frac{q_{i,max} K_i p_i}{1 + \sum_j K_j p_j}. \tag{7.17}$$

This model presents the advantage that it uses the parameters corresponding to the pure component isotherms, so there is no need to get equilibrium data of the mixtures.

For many systems, the extended Langmuir equation has limited applicability especially for liquid mixtures. In those cases, empirical models are used. The fitting is better, but they cannot be extrapolated to concentrations over the interval at which the data was obtained. An example is the model developed by Yon and Turnock [6] by combination of the Freundlich and Langmuir equations:

$$q_i = \frac{q_{i,max} K_i p_i^{1/ni}}{1 + \sum_j K_j p_j^{1/nj}}. \tag{7.18}$$

It also uses the parameters of the pure component isotherms, although better results are obtained if the parameters are calculated from the experimental equilibrium data of the mixtures.

Similar equations can be applied to *liquid adsorption* using concentrations instead of partial pressures. The Langmuir isotherm for the adsorption of a single component would be as follows:

$$q = \frac{q_{max} Kc}{1 + Kc}. \tag{7.19}$$

The solvent is assumed not to be adsorbed so the change in the composition of the bulk liquid is only due to the adsorption of the solute in the porous solid. This situation is likely to occur on dilute solutions; however, at high concentrations, the solvent may compete and curious adsorption curves are obtained [7]. For these mixtures, it is convenient to use multicomponent isotherms such as those described for gases.

Anyway, the Freundlich equation is the preferable model for the adsorption of organic compounds from aqueous solutions unto activated carbon:

$$q = k_F c^{1/n_F}. \tag{7.20}$$

There are many other published empirical and theoretical equations to describe the adsorption isotherms of pure and multicomponent mixtures. However, it is important to find an equation that is adapted to the compromise of fitting the experimental data with the lowest possible mathematical complexity to elaborate the final model for the equipment design.

An important concept for food applications is the sorption equilibrium between the moisture content of food and the relative humidity of the storage atmosphere. Water in foods may be more or less "available," and so it is distinguished between free and bound water. The strength of the water attachment in food is measured by the water activity, a_w, defined as the decrease in the partial pressure of the water vapor:

$$a_w = \frac{P_w}{P_w^o}, \tag{7.21}$$

where P_w is the partial pressure of the water vapor in a food and P_w^o is the partial pressure of pure water vapor at the same temperature.

In the equilibrium, the water activity is related to the moisture content by the sorption isotherm, which is different depending on the physical structure of the food, the chemical composition, and the extent of water binding within the food. Figure 7.2 plots adsorption isotherms of several food products. All of them have the characteristic shape shown in Figure 7.3. The first part of the curve, to point A ($0 < a_w < 0.2$), corresponds to the adsorption of a monolayer of water. It cannot be removed by drying. The second part of the curve (AB) represents water adsorbed in multilayers and corresponds to weak-strongly bond water. The third portion, over B, is the free water condensed within the capillary structure.

The Brunauer–Emmet–Teller (BET) equation is the most applied theoretical model to describe such equilibrium because it takes into account that an infinite number of molecular layers can be adsorbed [9]:

$$\frac{a_w}{M(1-a_w)} = \frac{1}{M_1 K} + \frac{a_w(K-1)}{M_1 K}, \tag{7.22}$$

where M is the moisture as percentage dry weight, M_1 is the moisture corresponding to a monomolecular layer (dry weight basis), and K is the equilibrium constant ($= K_o \exp^{-\Delta H_{ads}/RT}$). From the slope and the intercept of the linear fitting to experimental data, the monolayer weight of water and the heat of adsorption may be calculated. The BET monolayer value represents the moisture content at which the food is most stable. At moistures below or above this level, food deterioration by chemical, enzymatic, or microbial activities is promoted. The BET monolayer is

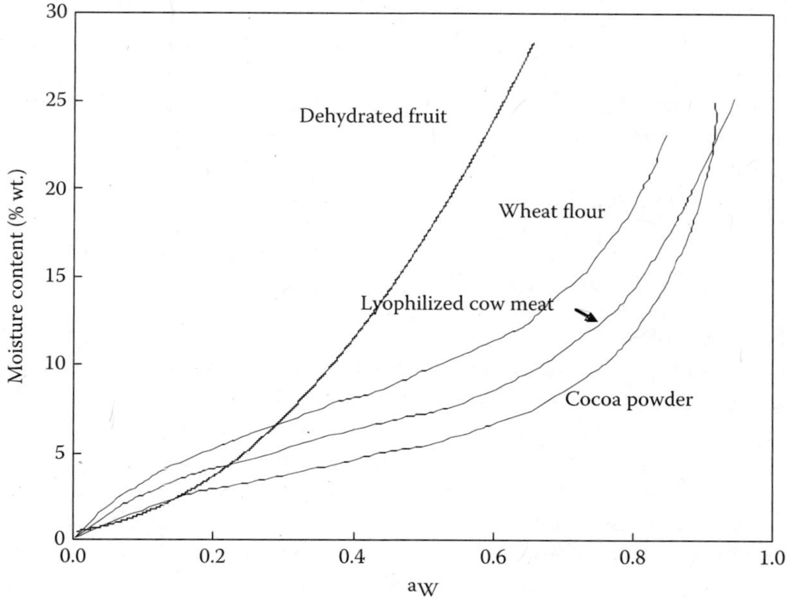

FIGURE 7.2 Water adsorption isotherms of some foods.

Source: Adapted from Brunauer, S., P. H. Emmett, and E. Teller, *Journal of the American Chemical Society* 60:309–319, 1938.

also used to indirectly determine the specific area of the adsorbent using nitrogen as reference adsorbate [10, 11].

The sorption isotherm differs according to whether the water is removed from the food (desorption) or added to dry food (adsorption; see Figure 7.3). This effect is termed as the *hysteresis loop,* and it also happens in other sorption processes.

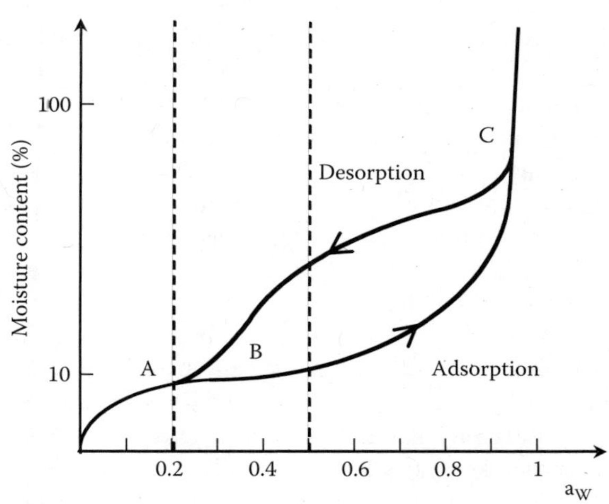

FIGURE 7.3 Typical sorption process in food products.

Several explanations have been given to this phenomenon [12]. It has been related to the capillary condensation that depends on the surface tension and the pore diameter. During adsorption, the contact angle between the water and the solid is higher than when water is removed (desorption). Besides, the diameter of the pore is wider when deepening. These two facts mean that the water vapor pressure needed to fill them is more elevated than that needed to empty them [8]. Hysteresis can also happen when strongly adsorbed impurities are present.

7.1.3 Types and Properties of the Adsorbents

Adsorbents are sold in granules, pellets, flakes, or powders whose size ranges from 50 μm to 1.2 cm. Because adsorption is a superficial phenomenon, the most important parameter is the surface area per unit of mass (specific area). Manufacturers have developed methods that give specific area up to 1500 $m^2 \cdot g^{-1}$ [13, 14]. This is possibly due to the high porosity of the particles. By the IUPAC, the materials could be classified according to the size of the pore into microporous (<20 Å), mesoporous (20–500 Å), and macroporous (>500 Å). The adsorbents may have up to 85% volume of micro- and mesoporous particles, and the pore size distribution may determine the selectivity of the adsorption because it allows the discrimination among the adsorbates as a function of its molecular size.

For commercial exploitation, the adsorbent should embody a series of characteristics: high selectivity and capacity to improve the efficiency of the separation, slow aging for maximum profit, and resistance against abrasion because agitation in the unit and transportation may cause solids rupture. Because of the high cost of the materials, to minimize the recharge and to save energy in the regeneration stage, the material should be easy to recover and clean.

The chemical nature of the material will determine the components to be adsorbed. One of the earliest and most used adsorbents is *activated carbon*. It is made by thermal decomposition of carbonaceous material (e.g., coal, wood, and bones), followed by activation by partial gasification with CO_2 or steam or by treatment with chemicals such as zinc chloride or phosphoric acid before carbonizing. The largest portion of the surface of the activated carbon is nonpolar, so it has a low affinity for water. Because of this, it has been used for the adsorption of organic components in aqueous solutions, e.g., in the purification of water or for the treatment of moist gases as in range hoods and other purification systems. Other applications include solvent recovery and the adsorption of gasoline vapors in automobiles. In the food industry, activated carbon is used in the removal of many kind of impurities including proteins, colorants, natural pigments, off-taste products, color precursors, mycotoxins (e.g., patulin), and trihalomethanes.

Activated carbons in commercial use are mainly in two forms: the powder form (particle size 1–150 μm) and the granular or pelletized form (particle size in the 0.5–4 mm range). Powdered activated carbon (PAC) is mostly used in processes where the liquids are purified batchwise. Granular activated carbon (GAC) is often used in continuous processes, especially in gas-phase adsorption, where the granules or pellets are set forming a bed. Spent GAC is usually regenerated by thermal treatment.

Other carbonaceous adsorbents are *carbon molecular sieves*. They are activated by special procedures to yield materials with a very narrow distribution of micropores ranging from 4 to 9 Å. Because of this characteristic, it is possible the discrimination of the adsorbates as a function of their molecular size. At the commercial level, this adsorbent was used to separate O_2 and N_2 from air.

Zeolites are composed of tetrahedra of silica and alumina arranged in various ways through shared oxygen atoms to form an open crystal lattice containing pores of molecular dimensions into which guest molecules can penetrate. The regular crystalline structure of these materials renders a uniform and known porous distribution where the pore size may vary between 3 and 10 Å. They can be found in nature or can be synthesized by hydrothermal reaction in autoclaves. Up to forty different types have been described under the general stoichiometric formula:

$$(M)_{x/n}\left[(AlO_2)_x(SiO_2)_y\right]zH_2O, \tag{7.23}$$

where M is the cation with valence n, z is the number of water molecules in each unit cell, and x and y are the integers such that $y/x \geq 1$. Depending on the type of cation and on the ratio between the silanol and aluminol groups, different structures are obtained, namely type A with the smallest pore diameter, types X and Y, mordenite, silicalite, and ZSM-5. The separation operations with zeolites can be based on size exclusion or on the difference in affinity for the components. Although the second option is the most frequently used, the narrow interval of pore size of the zeolites makes possible the separation of molecules with small size difference. The large majority of zeolites applications are purification processes in which the zeolite is used to remove an impurity such as water, CO_2,or SH_2 in gases. Other commercial separation processes include the separation of air into its components, the separation of linear and branched hydrocarbons, and the isolation of xylene isomers.

For their hydrophilic character, *silica gel* and *activated alumina* are used for drying gases and liquids. Silica gel is a partially dehydrated form of polymeric colloidal silicic acid whose chemical composition can be expressed as $SiO_2 \cdot nH_2O$. The water is chemically bound in the form of hydroxyl groups and amounts to about 5 wt %. Depending on the pore size, two types of silica gel (Types A and B) are used for commercial purposes. Type A has pores of 20–30 Å, whereas the pores of Type B are of 70 Å. This provides surface areas of 650 $m^2 \cdot g^{-1}$ (Type A) and 450 $m^2 \cdot g^{-1}$ (Type B). Although both are applied in the dehumidification of gases such as air of hydrocarbons, Type B is more suitable when the humidity is higher than 50%.

Active alumina is mainly γ-alumina, which is the porous form of aluminum oxide. Its specific area ranges between 150 to 500 $m^2 \cdot g^{-1}$, with a pore radius of 15 to 60 Å, depending on the production method. Apart from its use as a drying agent, active alumina is also used for the removal of polar gases from hydrocarbon streams.

Several natural silicates are used as adsorbents in the refining of food products. Diatomaceous earth is a naturally occurring, soft, chalk-like sedimentary rock that is easily crumbled into a fine white to off-white powder. This powder is very light, as a result of its high porosity. The typical chemical composition of diatomaceous earth is 86% silica, 5% sodium, 3% magnesium, and 2% iron. Bentonite and Fuller's earth are aluminum phyllosilicate, generally an impure clay consisting mostly of montmorillonite, $(Na,Ca)_{0.33}(Al,Mg)_2Si_4O_{10}(OH)_2\cdot(H_2O)_n$. The distinction between them is the dominant cation present in the clay, which gives them markedly different properties. Calcium is the principal cation in Fuller's earth, whereas sodium is in Bentonite. Bentonite has the interesting property of adsorbing relatively large amounts of protein molecules from aqueous solutions. Clay consists of a variety of phyllosilicate minerals rich in silicon and aluminum oxides and hydroxides, which include variable amounts of structural water. They are activated by disgregation in water, washing with sulfuric acid solutions, filtration, drying, and milling. Their acidic character may result in an increase on acidity in the treated products. Table 7.1 summarizes the most relevant properties of the above discussed adsorbents.

Apart from these adsorbents, nonionic polymeric and ion–exchange resins have being successfully employed in the selective recovery of bioactive compounds. Most typical resins are based on cross-linked polystyrene achieved by adding 0.5–25% of divinyl benzene to styrene at the polymerization process, producing materials that are fully ionized over the entire pH range. Weakly acid, cation exchangers are sometimes based on the copolymerization of acrylic acid and methacrylic acid. These two cross-linked copolymers swell in the presence of organic solvents and have no ion-exchange properties. Both types of polymeric adsorbents present highly porous structures whose internal surfaces can adsorb and then desorb a wide variety of different species, depending on the environment in which they are used.

TABLE 7.1
Physical Properties of the Most Frequently Used Adsorbents

Adsorbent	Particle porosity (ε_p)	Particle density (ρ_p 10^{-3} kg.m^{-3})	Bulk density (10^{-3} kg.m^{-3})	Average pore diameter (D_p, Å)	Surface area (S, 10^{-3} m^2.kg^{-1})	Sorptive capacity (drybasis) (kg.kg^{-1})
Activated alumina	0.50	1.25	0.70	10–75	320	0.20
Zeolites						
Type 3A	0.30		0.62–0.68	3	700	0.21–0.23
Type 4A	0.32		0.61–0.67	4	700	0.22–0.26
Type 5A	0.34		0.60–0.66	5	700	0.23–0.28
Type 13X	0.38		0.58–0.64	10	600	0.25–0.36
Silica gel	0.30–0.48	0.6–1.1	0.70–0.82	20–100	600–800	0.35–0.50
Clay, acid treated			0.85			
Diatomaceous earth			0.44–0.50		2	
Fuller's earth			0.6			0.18
Activated carbon						
Small pore	0.4–0.6	0.5–0.9	0.3–0.7	10–25	400–1200	<0.7
Large pore		0.6–0.8	0.3–0.7	>30	200–600	>0.3
Carbon molecular sieves	0.35–0.50	0.98		2–10	400	0.5–0.20

7.1.4 Adsorbent Regeneration

As adsorption is going on, the pores of the adsorbent are filling with adsorbate molecules until the capacity of the adsorbent is exceeded. The adsorbent is usually expensive, and for environmental concerns, the regeneration of the solid is currently legally obliged. Therefore, the adsorption operation really includes two stages: the adsorption and the regeneration, which is in fact a desorption process. In many occasions, the economy of the whole operation depends to a great extent on this later stage.

The regeneration of the exhausted adsorbent may be achieved by different procedures, all of them based on changes in the equilibrium:

1. *Thermal reactivation.* The desorption is accomplished by an increase in temperature, which leads to a decrease in the concentration of the adsorbed phase (q; Figure 7.4a). This option is indicated when the interaction adsorbate–adsorbent is high. The desorption temperature should be optimized because a too low temperature may lead to incomplete regeneration. On the contrary, if the temperature gets too high, the adsorbent may be damaged. In many occasions, the manufacturer recommends the most adequate regenerating temperature or at least an interval, but the optimum temperature also depends on the type of adsorbate.

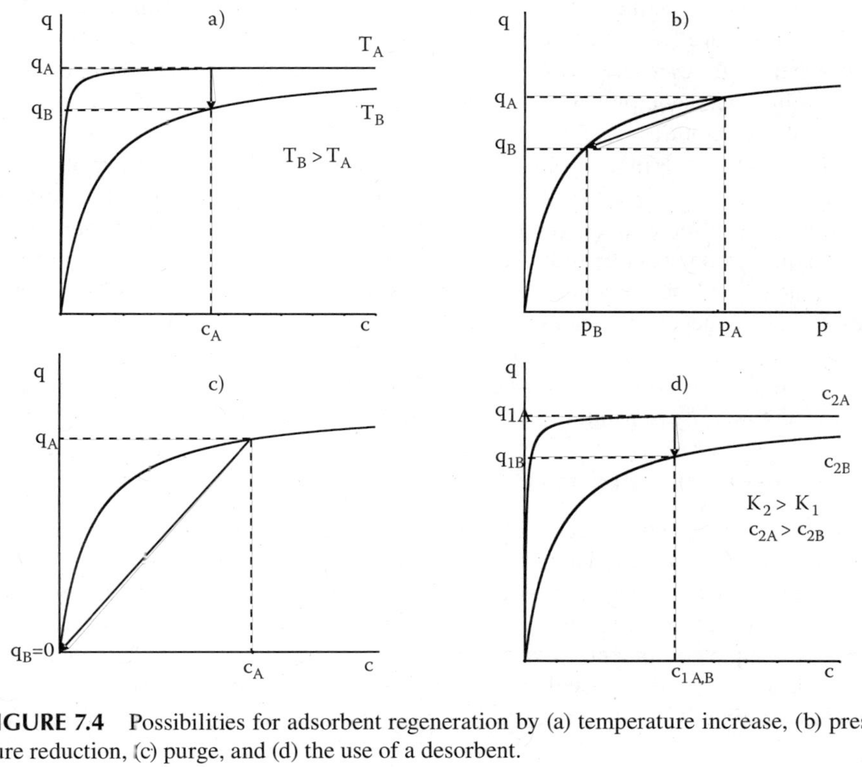

FIGURE 7.4 Possibilities for adsorbent regeneration by (a) temperature increase, (b) pressure reduction, (c) purge, and (d) the use of a desorbent.

2. *Regeneration by a decrease in pressure.* Obviously this possibility is only applicable with gases. In this case, pressure is decreased so the concentration in the adsorbed phase at equilibrium is less (Figure 7.4b). It is useful when the stream to be treated is needed at pressures above atmospheric. When the process is carried out at atmospheric pressure or below, the regeneration can be done at a vacuum. It is indicated for weak interactions, allowing quick adsorption–desorption cycles.
3. *Regeneration by purge.* It is achieved by feeding an inert, nonadsorbing purge that reduces the adsorbate concentration and thus its degree of adsorption (Figure 7.4c). Sometimes the purge is hot, e.g. steam, favoring the thermal regeneration at the same time.
4. *Regeneration by displacement* with an adsorbate of higher affinity, called the desorbent. This type of operation is adequate when the adsorbate has a strong interaction with the adsorbent and no thermal regeneration is possible (Figure 7.4d). The process is more complex because an extra step is needed: the separation of the adsorbate and the desorbent after regeneration is completed. In gas applications, the separation can be done by partial condensation. In liquid systems, the desorbent is usually a solvent that is recovered by distillation.

An example of thermal reactivation is the GAC recovery after use. The process can be done either on- or off-site. For larger volumes on-site reactivation is more economical. For small quantities, replacement or off-site reactivation is more profitable. In this latter case, the spent GAC is delivered to specialized reactivation centers where it is segregated and reactivated. Reactivation involves treating the spent carbon in special equipment (e.g., multiple-hearth furnaces, fluidized beds, or rotary kilns) at temperatures of 850°C and above.

During this thermal treatment four steps occur: drying, desorption of volatile compounds, carbonization/calcinations/pyrolysis of nonvolatile compounds, and finally, gasification of the carbonaceous residue. In this way, the undesirable organics on the carbon are fully destroyed. Residence time in the kiln must be optimized. Too long a residence time and the carbon is overreactivated and loses its hardness, resulting in higher attrition rates. A short residence time will not permit the reactivation to be completed. Typically the residence time ranges from about 30 to 45 min. Once the reactivation procedure is finished, the customer's original carbon is recycled and returned with only the addition of fresh material as make-up.

7.1.5 Adsorption Processes

A wide variety of configurations and operation methods are used for commercial adsorption applications. The batch operation can be conducted in agitated tanks or fluidized beds, whereas continuous flow may be achieved in fixed and moving beds. At industrial scale, fixed beds are mainly used for an efficient adsorbent use and simple equipment. Nevertheless, in liquid-phase processes, agitated vessel adsorbers are frequently used. This section focused on the qualitative explanation of these configurations whose operation will be illustrated with the aid of industrial examples.

7.1.5.1 Operation in Agitated Vessels

In the batch mode, the adsorbent is added as powder to form a slurry in the tank with the liquid (see Figure 7.5). The agitation is connected so the solute is adsorbed and its concentration in the liquid is reduced with time. The operation is stopped when concentration of the liquid reaches a prespecified value. Then, the slurry is discharged from the vessel and filtrated to remove the solids from the liquid. Finally, the adsorbent is regenerated, usually by thermal treatment.

A less frequent mode of operation in agitated vessels is the continuous mode, in which both the liquid and adsorbent are continuously added to and removed from the tank. In certain cases, the adsorbent is loaded at the beginning of the operation, while the liquid is continuously fed. The modelling of these systems is explained in Suzuki [10] and Seader and Henley [12].

7.1.5.2 Operation in Fixed Beds

In the operation with fixed beds, also known as *percolation*, the fluid is fed by the bottom part and is collected free from adsorbate by the upper part (if gas). The contrary happens if liquid. It is then a semicontinuous process (continuous with respect to the fluid but discontinuous with respect to the adsorbent). When the bed is saturated, the adsorbate is detected in the exit stream, which is necessary to proceed with its regeneration. This is the reason to normally operate with two or more fixed beds connected in parallel, so while some of them are in the adsorption stage, the rest are in regeneration.

Figure 7.6 plots a simple scheme for the possible separation of oxygen and nitrogen from air with two fixed beds that operate with alternation of adsorption–desorption stages and pressure changes, known as pressure swing adsorption (PSA) [15]. Air is fed to the bed on the left at high pressure. N_2 is adsorbed, while the exiting product gas is mainly O_2. Part of the produced oxygen is used as purge to regenerate the other bed at atmospheric pressure. When the bed on the left reaches saturation, the position of the valves is changed and the operation is repeated introducing the mixture by the bed on the right. Now the bed on the left is in the desorption stage

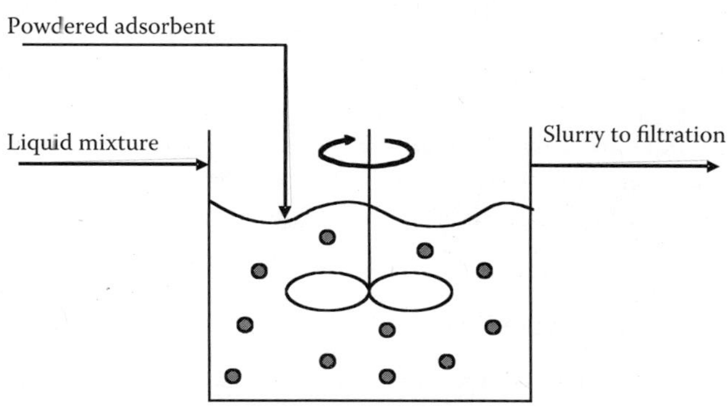

FIGURE 7.5 Batch adsorption in an agitated tank.

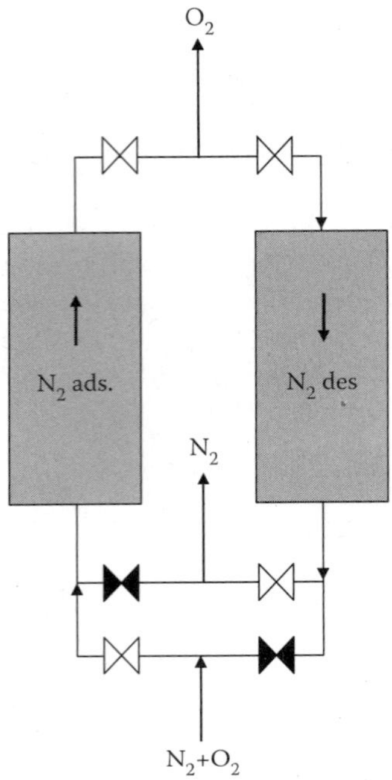

FIGURE 7.6 A two-bed pressure swing unit for the separation of air into oxygen and nitrogen.

at low pressure. The synchronization of the flow rates, pressure swings, and stream inlets makes possible short cycles, resulting in a steady-state operation. Major uses for PSA processes include gas purification (air dehumidification) as well as applications where contaminants are present at high concentrations (bulk separation). When adsorption is carried out at atmospheric pressure and desorption occurs at vacuum, the operation is referred to as vacuum swing adsorption (VSA) [16].

A similar mode of operation is carried out in the thermal (temperature)-swing adsorption (TSA). The cycles are now based on changes in the bed temperature. While one bed is adsorbing the solute at near-ambient temperature, the other bed is regenerated by desorption at a higher temperature. This latter step is usually accompanied by the introduction of a purge to avoid the readsorption of the solute when the bed gets cooled. The purge can be a portion of the feed or another fluid. Because the changes in temperature cannot be done quickly, the cycles in TSA operations may take hours or even days. TSA is applied to the removal of contaminants at low concentrations in gases and liquids. A deep discussion of this technology may be found elsewhere [15].

7.1.5.3 Operation in Moving Beds

The moving bed units put in contact the adsorbent and the fluid in countercurrent so that the maximum capacity for the adsorbent is achieved. The exhausted solid is extracted and regenerated continuously, normally via thermal treatment, returning

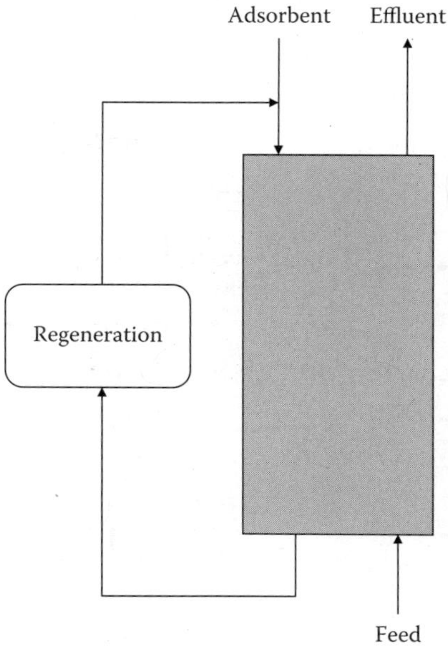

FIGURE 7.7 Continuous countercurrent adsorption in a moving bed apparatus.

it to the adsorber afterward (see Figure 7.7). However, this configuration has the disadvantage that the solid needs to be circulated as a moving bed, with the corresponding problems of mechanical abrasion and the crumbling of the solid particles. An application of these systems is the recovery of diluted solvents in air with activated carbon in petroleum refineries and in sugar manufacturing to remove the color.

A successful alternative is the simulated moving bed system, known generally as the Sorbex process, whose scheme is presented in Figure 7.8. In this case, the adsorbent is held stationary in one column that is equipped with numerous entries and lateral exits controlled by a valve of multiple vias [17]. A desorbent (D) is used for regeneration. The benefit of the countercurrent contact is achieved by moving the positions of the feed inlets and product exits, so that in some zones the adsorption of the component of higher affinity occurs (A), whereas in others the component of lesser affinity is desorbed (B). The mixtures of A + D and B + D are further separated in two adjacent distillation columns. Sorbex-like processes have been developed for a number of industrially important separations in the petrochemical industry [11, 18].

In the food industry, an application of the Sorbex process is the Sarex process for the separation of fructose from a feed mixture, such as an invert sugar solution or corn syrup. The adsorbent is either a cation exchange resin or a zeolite (X or Y) containing sodium cations at the exchangeable cationic sites. The separation is based on the uniquely adsorptive selectivity of these materials for a ketose with respect to an aldose, particularly fructose with respect to glucose. Further details can be found in Neuzil and Jensen [19]. If an X zeolite containing potassium cations is used, then glucose is adsorbed while the other compounds are eluted [20].

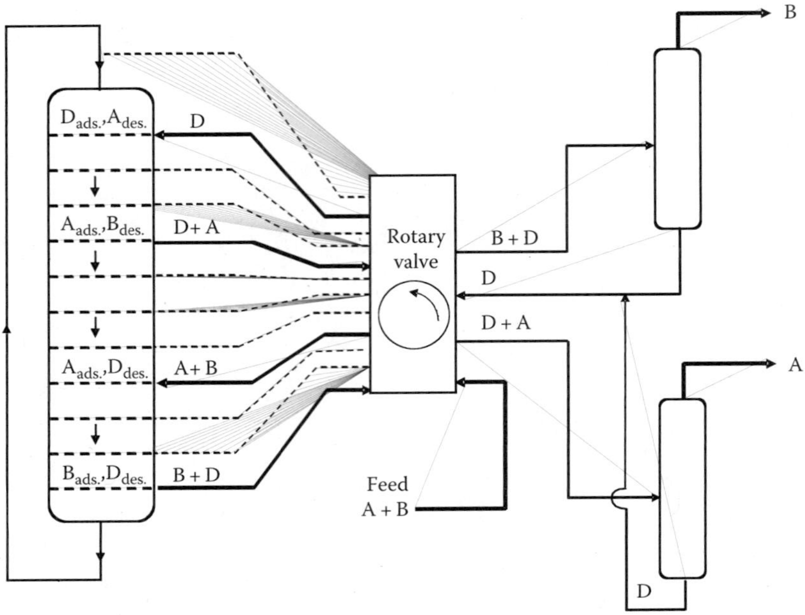

FIGURE 7.8 Sorbex simulated moving bed process. A: more strongly adsorbed component; B: less strongly adsorbed component; D: desorbent.

7.2 APPLICATIONS OF ADSORPTION IN FOOD PROCESSING

There are some important applications of adsorption in the food and beverage industries related to the removal of impurities from liquid mixtures. Activated carbon is the adsorbent in most cases. The commercially exploited ones are reviewed in the next section. Also, the trends in the investigation of the use of this technique are briefly summarized. At the end of the section, a literature review on the recovery and concentration of bioactive compounds by adsorption is summarized and discussed.

7.2.1 Removal of Unwanted Natural and Harmful Anthropogenic Compounds from Edible Oils

Adsorption is a relevant operation in the refining procedure of oils and greases. The objective of this operation is the elimination of undesirable pigments (e.g., carotenoides and chlorophylls) as well the rest of the soap, heavy metal traces, autooxidation products, and residual amounts of phosphorous substances [21]. The most frequent adsorbents are acid-treated clays [22] or activated carbon. The latter is very efficient in removing the red color, but because of its higher price, it is a common practice to use it in a mixture with 90–95 wt % clays. However, to be labeled as "ecologically" refined oil, only activated carbon can be used because it is authorized for practice in ecological agriculture by European Community regulation 2092/91 [23]. The concentration of the adsorbent may vary between 0.2 to 0.6 wt %.

The classical equipment for decoloration operates in batch mode. However, the most updated installations have introduced continuous systems such as the one schematized in Figure 7.9 [24]. The greasy substance previously dried and heated to 60°C–70°C enters in the mixer C where it is put into contact and intimately blended with the adsorbent coming from the continuous dozer B and the homogenizer A. The slurry generated in C goes to the decolorator D. The contact time between the adsorbent and the grease in the equipment is about 30 min. A special pump impels the exiting slurry to the filtration stage. The operation is conducted under vacuum conditions (6.7–9.3 kPa). An installation of 10-m^3 capacity is capable of treating 200 tons per day, and it is more economically profitable than the discontinuous equipment.

7.2.2 Purification of Drinking Water

Apart from the treatment of municipal water, many other processes include adsorption steps for the purification of water in the food industry, for example, in the production of ice cream, juices, soft drinks, and beer. Each type of water presents different characteristics (e.g., organic material, metals, nitrates, and hardness) and must be treated to achieve a constant yearlong quality. The objectives of the treatment are the

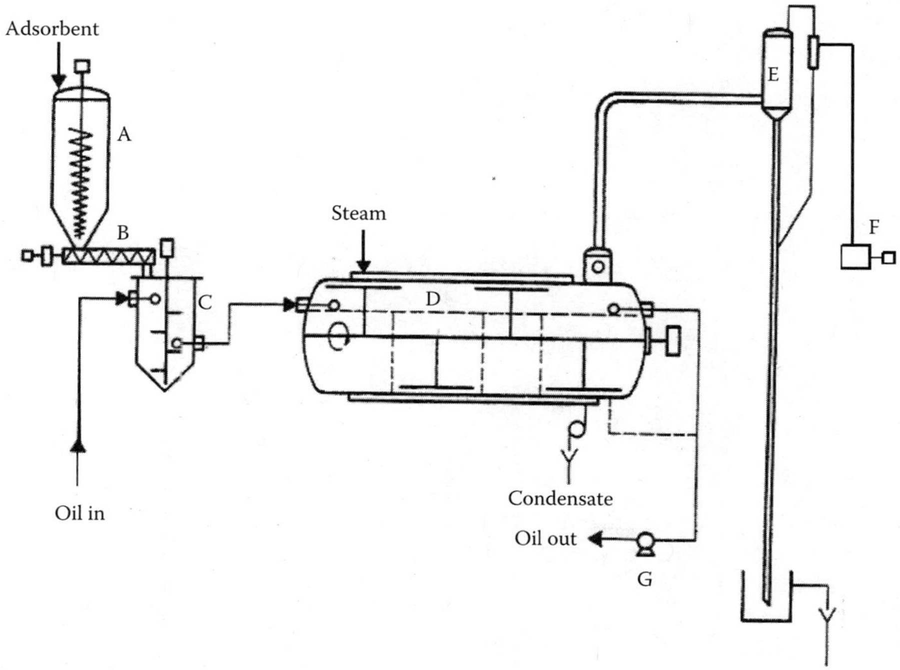

FIGURE 7.9 Scheme of a continuous decolorization unit operating at vacuum conditions. A: Homogenizator; B: Dozer; C: Mixer; D: decolorator; E: barometric condensator; F: Vacuum pump; G: Extraction pump.

Source: Adapted from Bernardini, E., *The New Oil and Fat Technology,* 2nd ed., Tecnologie SRL, Rome, 1973.

elimination of colloids and materials in suspension; the removal of color, unpleasant odors, and flavors; the reduction of alkalinity; and sterilization.

The core of a water treatment plant is the flocculation tank where a coagulant (aluminium or iron sulphate), polyelectrolytes, and lime are added [25]. Sodium hypochlorite or more frequently chlorine gas is also put in as a bactericide. A gelatinous precipitate that coagulates, forming flocs with the organic matter, is produced. In new installations the purification is done by reverse osmosis or by ionic exchange resins. Next, the water is filtered in a sand filter, followed by percolation over an activated-carbon bed (see Figure 7.10) to remove the chlorine excess as well as possible reaction products such as trihalomethanes (THM) and eventually other organic contaminants. Carbon's dechlorinating capability results from its ability to act as a reducing agent. Sometimes the water is finally radiated with ultraviolet rays to ensure the disinfection.

7.2.3 Removal of Color in Syrups

The other ingredient used to produce soft drinks is the syrup, which is elaborated from sucrose, glucose, or fructose syrups or granulated sugar. In this latter option,

FIGURE 7.10 Image of an opened activated carbon filter for drinking water purification (courtesy of Aguas de Valladolid, Spain). The bed dimensions are 16 m long, 3 m wide per channel, and 1.50 m deep. Residence time of the water is about 13 min.

the sugar and water are continuously fed to a mixer. A pump delivers the mixture to a heat exchanger to be pasteurized at 348–361 K. The syrup then goes to a filter to remove the solid impurities. When the liquid sugar is still hot, it is treated with activated carbon for decolorization and improvement of the sensory characteristics. With this aim, a suspension of this adsorbent is dosed, and then the mixture goes to a tank where the slurry remains for a certain period of time until the demanded degree of decolorization is achieved [26]. GAC decolorization in continuous mode is done similarly as in sugar refining, which is explained next.

7.2.4 Cane Sugar Refining

Traditionally, sugar cane has been processed in two stages: extraction from freshly harvested sugar cane and purification to produce refined white sugar (mainly sucrose) [27]. After the extraction, the juice is screened and heated to its boiling point. The remaining fibrous solids, called bagasse, are burned for fuel. Then, the suspended solids and colloidal materials in the juice are precipitated with lime, and the clarified juice is concentrated in a multiple-effect evaporator to make a syrup about 60%–65% by weight in sucrose. This syrup is further concentrated under vacuum until it becomes supersaturated and then is seeded with crystalline sugar to produce the sugar crystals in a three-stage crystallization process. A centrifuge is used to separate the sugar from the remaining liquid, molasses.

The raw sugar is then transported to the refinery, where it is dissolved with heavy syrup and centrifuged using hot water wash. This process is called "affination"; its purpose is to wash away the outer coating of the raw sugar crystals, which is less pure than the crystal interior. After centrifugation, the washed raw sugar is melted in high-purity sweetwater with low-pressure steam and or/vapor. The affination syrup is adjusted with lime slurry to pH 7.

This liquor has a yellow-to-brown color as a result of the presence of phenolic, polyphenolic, and flavonoid compounds that are originally attached to plant cell walls and to factory-formed colorants such as melanoidins (from Maillard reactions of glucose and fructose) and caramels formed by thermal degradation of sugar and other carbohydrates. Therefore, in sugar refining, the sugar solution must be further purified.

Clarification is conducted by the addition of carbon dioxide and calcium hydroxide to produce a calcium carbonate precipitate that entraps wax, gum, polysaccharides, colorants, and ash, mostly sulfate. An alternative option is to add phosphoric acid and calcium hydroxide, which combine to precipitate calcium phosphate. Carbonate cake is removed by filtration, and the press filter liquor is pumped to a supply tank. An additional color removal step is needed to ensure that the white sugar meets the product color specification. This additional color removal process is almost always adsorbent based, using GAC or ion-exchange resins. GAC is used in both fixed- and moving bed installations.

The purified syrup is then concentrated to supersaturation by evaporation and is repeatedly crystallized under vacuum, to produce white refined sugar. As in the sugar mill, the sugar crystals are separated from the molasses by centrifuging. Drying is accomplished first in a hot rotary dryer and then by blowing cool air through it for several days.

An alternative option has been recently developed to incorporate the refining process in the mill [28]. After the first evaporation effect at 20%–25% weight solids, the syrup is ultrafiltered to remove high-molecular-weight material. Then it is cooled and subjected to a subsequent ion-exchange separation. Under acidic conditions, sucrose breaks into fructose and glucose. The heart of the process is the combination of a continuous ion-exchange, demineralization-simulated moving bed followed by a decolorization bed adsorber charged with an industry standard strong-base resin in the chloride form. The decolorized juice produced is of high purity and low color, increasing sugar recovery and quality [29].

7.2.5 Color and Taste Correction in Alcoholic Beverages

Wines are sometimes treated with activated carbon for color and taste correction. Because of the variability of the grapes and the presence of complex organic compounds, it can be difficult to achieve a consistent color. Among the complex organic compounds are antho-cyanidins (polyphenolic compounds), which give a red coloration, and chlorophyll, which gives a yellow coloration. Other compounds such as carotenoids and tannins may also be present. PAC has been traditionally used for the complete or slight color modification of red, rosé, and white wine using batch techniques. In similar fashion, total decolorization is achieved in the production of vermouths. Quality, dosing rates, and treatment conditions are extensively described in various directives such as the "Codex Oenologique International."

An additional problem associated with the presence of phenolic compounds is the color darkening during storage resulting from chemical reactions involving these compounds. Browning is an important problem in white wines and also in beer. To reduce the concentration of brown compounds that shorten the commercial life of these beverages, the winemaking and beer industries have been using several adsorbents, mainly activated carbon and polyvinylpolypyrrolidone (PVPP) [30]. Alternatively, yeasts and their cell walls have been successfully tested [31].

7.2.6 Elimination of Color in Flavorings

Hydrolyzed vegetal proteins are extracted from corn, soy, or wheat and are decomposed into amino acids by acid or enzymatic hydrolysis. They are used as flavorings in cooking. The process starts with the acid hydrolysis of the proteins followed by neutralization. The mixture obtained is dark and has small photic particles in suspension. Activated carbon is added with the double objective of decolorizing the mixture and helping in the posterior filtration. The process concludes with the evaporation and drying of the final solid product [14].

7.2.7 Purification of Carbon Dioxide for Use in Carbonated Drinks

One source of CO_2 is the excess of production during the fermentation process in breweries. To enable its use in the beverage industry, the CO_2 must be purified by activated carbon to remove taste and odor-causing compounds such as H_2S, mercaptanes, and other organic compounds. For soft drink producers, the CO_2 can be produced via combustion of fossil fuels or via extraction from existing gas sources. It is common practice that this sourced CO_2 is treated by activated carbon in safety filters before it is used as an additive in order to assure that traces of taste and odor compounds as well as traces of aromatic hydrocarbons are completely eliminated [14]. Zeolites may also be used [32].

7.2.8 Decaffeination of Tea and Coffee

Caffeine is a natural substance that is present in the leaves (teas), seeds (coffee), and fruits of more than 60 plant species worldwide. The interest in its extraction lies in the commercialization of decaffeinated products. However, caffeine can be further used in the soft drink and pharmaceutical industries.

In the decaffeination of green coffee beans and tea by a water extraction process or by liquid [33] and supercritical [34] CO_2, caffeine may be removed by contact with substantially neutral active carbon. To remove the extracted caffeine from the activated carbon, an acid [35] or steam may be used.

7.2.9 Removal of Unwanted Odor or Color Compounds from Glycerin

Glycerin, also well known as glycerol, is a colorless, odorless, hygroscopic, and sweet-tasting viscous liquid. Refined glycerin serves as a humectant in candy, cakes, and casings for meats and cheeses; a solvent for flavors (such as vanilla); a sweetener; a food coloring; and a filler in low-fat food products (i.e., cookies) as well as a thickening agent in liqueurs. It is also used in the manufacture of mono- and diglycerides for use as emulsifiers and of polyglycerol esters used in shortenings and margarine.

Natural glycerin is the main by-product of biodiesel and soap production (by transesterification of edible oils and fats with acid, alkali, superheated steam, or an enzyme) or by fermentation of glucose. After the synthesis, the colored matter and odor-causing substances can be removed by activated carbon in the final stages ("bleaching") of purification prior to its use. Activated carbon can also be used in the primary stage of crude glycerin purification to reduce bulk color and fatty acids [14].

7.2.10 Purification of Fruit Juices

During processing of fruit juices and also during storage, development of undesirable odors and tastes and browning reactions can occur [36]. The problem of browning due to the presence of phenolic compounds is very important because changes in color and development of undesirable haze and turbidity seriously compromise acceptability of commercial juice. To prevent these problems and in many cases to optimize taste characteristics, a deliberate reduction of phenolics is necessary. Stabilization by means of activated carbon [37], gelatin, bentonite, silica gel, and PVPP is a widespread, conventional treatment in the juice industry, although the use of adsorbent resins has gained increasing importance as a final treatment after clarification [38].

Another application of adsorption in the juice industry is the removal of bitter flavanone glycosides, such as naringin and limonin in citrus products, particularly in grapefruit, because excessive bitterness is an important problem for its commercialization. Debittering units in commercial operation mainly use food-grade polystyrene divinylbenzene cross-linked polymeric resins previously acidified to prevent protein precipitation [39], although many other different adsorbents have been tested as cited in Singh et al. [40]. The process run in a continuous-use, fixed-bed column may be combined with a previous ultrafiltration to augment the efficiency of the whole process. The so-obtained debittered product is just slightly paler [41].

Finally, adsorption can also be used to remove traces of pesticides and fungicide residues such as the mycotoxin Patulin. This compound is highly undesirable because of carcinogenic and teratogenic characteristics and can be removed by the use of activated carbon or bentonite [42].

7.2.11 Purification of Starch-Based Sweeteners

Starch hydrolysates such as glucose, dextrose, maltose, fructose, and maltodextrins are produced using hydrolysis and isomer conversion techniques [43, 44]. These compounds are predominantly used as sweeteners in the food industry but also as intermediate materials in the production of sorbitol, citric acid, lactic acid, and MSG. During the process of hydrolysis of starch, color compounds are formed from the original starch and from the thermal decomposition of the sugars. In addition, hydroxymethylfurfural (HMF) is also formed, which must be removed to obtain color stability in the end product and to protect the immobilized enzyme system used to convert D-glucose to high-fructose syrup. To assist with processing, it is also necessary to remove foaming agents.

High-purity PAC is generally used to decolorize the glucose. The process is conducted batchwise. PAC is prepared as a slurry and is added to the mixing tank. Continuous agitation is applied for the required contact time at a temperature of 70°C–80°C. Subsequent filtering is used to remove the PAC after the treatment. GAC is also used in continuous flow using fixed-bed adsorbers for the final polishing of these products in order to comply with the critical sensory requirements of the soft drink industry or to meet the most highly stringent requirements when the finished product is used in intravenous fluids [14].

7.2.12 Decolorization of Citric Acid

Citric acid is predominantly produced by surface fermentation or the submerged fermentation of molasses using the mold *Aspergillus niger*. Citric acid is widely used in carbonated beverages and sweets to provide a fresh acidic taste or as a preservative in many food products. Refined sucrose, although expensive, is the substrate most commonly used for producing citric acid by fermentation [45]. To reduce production costs, sucrose from beet molasses may be also used.

There are several kinds of technologies currently used for the separation of citric acid from the fermentative broth, such as calcium salt precipitation and solvent extraction. These methods are complex and expensive, and they generate substantial amounts of waste for disposal. Adsorption is a simpler alternative for separation and purification. Therefore, several solid adsorbents have been considered for subsequent product recovery [46, 47] and purification [14]. Moreover, a packed column with an anion-exchange resin attached to a fermenter has proven to highly benefit the process in terms of conversion and reduction of the input water requirement [48].

7.2.13 Other Applications

Table 7.2 lists recent investigations on the use of adsorption for bulk separations of amino acids, saccharides, and lactoses. Additionally, the table contains improvements of the established adsorption processes by using better adsorbents, for example, in the decolorization of soy oil. It also explores new applications such as the elimination of cholesterol in different food products for its harmful effect on the health and concentration for encapsulation of flavors. New procedures where adsorption is coupled with a supercritical extraction to render a process of higher selectivity and purity are also described. Finally, the table includes applications where adsorption is used to recover highly valuable products such as proteins and enzymes from different sources including wastes.

TABLE 7.2
Examples of Other Applications of Adsorption in the Food Industry

Process	Adsorbent	Reference
Amino acids containing OH and SH groups from different types of amino acid	Titanium oxide or its hydrate	[49]
Isoamylase from impurities	Starch	[50]
Glucose isomerase from impurities	Weekly basic ion exchange material	[51]
Bulk lactulose from lactose	Zeolite molecular sieves	[52]
Polygalacturonase from recycled cucumber picle brinees	Pure-FLO B80 clay	[53]
Monosaccharides from oligosaccharides	Zeolitic molecular sieves	[54]
Cis/trans isomers of fatty acid compounds	Microporous zeolite	[55]
Polyhydric alcohols	Zeolitic molecular sieves	[56]
Proteins from aqueous food processing streams	Chitosan-alginate	[57]
Proteins from fermented aqueous food	Silica gels	[58]
β-Carotene from soy oil	Activated rice hull ash	[59]
Lutein from soy oil	Dispersed silicic acid	[60]
Lutein from soy oil	Rice hull ash	[61]
Cholesterol from egg yolk	Chitosan beads	[62]
Cholesterol from butter oil using supercritical CO_2 and adsorption	Alumina	[63]
Cholesterol from butter oil using supercritical ethange and adsorption	Alumina	[64]
Brines from green table olive processing	Activated carbon	[65]
Free fatty acids from used frying oils	Calcium silicate, magnesium silicate and a porous rhyolitic material and silicon dioxide	[66]
Flavors	Typical materials used in a box of tobacco	[67]
Flavors for encapsulation	Microporous pillared clay mineral	[68]
Flavors for encapsulation	Porous carbohydrates	[69]

Table 7.3 lists applications of adsorption in the selective recovery, purification, and concentration of bioactive compounds. These applications have been considered apart from the previously discussed ones for the growing interest in obtaining biologically active compounds from natural sources. Phytochemicals, especially polyphenols in plants, are the major bioactive compounds because of their antioxidative, antimicrobial, antiproliferative, antiviral, and anti-inflammatory properties, as cited in Bayçin et al. [89]. Vitamins and some amino acids may be considered nutraceuticals too. New opportunities are coming for these natural compounds in the growing segments of dietary supplements and functional food production and because of their possible utilization by the pharmaceutical and cosmetic industries.

The extraction of bioactive compounds from plants is usually done with organic solvents or hydroalcoholic mixtures; however, further purification is essential in order to obtain concentrated specific components because other compounds, such as sugars, proteins, and metals, may exist in the plant extracts. For the selective recovery of target plant metabolites from the crude solvent extracts, adsorption has been preferred for many researchers, because it is a low-cost separation. This aspect is especially important if the aim is to isolate the bioactive compounds from residues to balance the waste disposal costs.

On other occasions, adsorption has been coupled with novel processes, such as the use of an ultrafiltration membrane [93] or after a supercritical extraction [70] to achieve higher concentration and purity.

Different adsorbents have been used for the recovery of bioactive compounds. To a lesser extent, natural materials have been tested. For example, the biopolymer silk fibroin has been investigated in the recovery of oleuropein and rutin from olive leaf [89]. Rice hull ash has been used in the adsorption of antioxidants from rice bran oil [70]. On the other hand, activated carbon adsorption has been carried out in the recovery of phenolic compounds present in distilled grape pomace [90] and silica in the separation of vitamin E from palm fatty acid distillate [94, 95]. However, the

TABLE 7.3
Examples of Applications of Adsorption in the Recovering, Concentrating, and Purifying of Bioactive Compounds

Bioactive compound	Source	Adsorbent	Reference
Antioxidants	Rice bran oil	Rice hull ash	[70]
Anthocyanins and hydroxycinnamates	Pigmented pulp wash	Several commercial resins (EXA90, EXA118, EXA 31)	[71]
Anthocyanins	Pigmented pulp wash	Six commercial food-grade resins	[72]
Anthocyanins	Grape pomace extracts	Amberlite XAD 16 HP	[73]
Catechin thio conjugates	Pine bark	Resin XAD-16	[74]
Colorless L-carnitine extract	Aqueous meat or fish extract	Activated carbon	[75]

Bioactive compound	Source	Adsorbent	Reference
Cyanidin-3-glucoside	Aqueous solutions	Several resins	[76]
Deodorized garlic extract	Garlic	Several resins	[77]
EPA and DHA	Fish oil	Modified zeolite 13X	[78]
Flavonoid compounds	Leaf extract of *Ginkgo biloba*	Polycarboxyl ester resin XAD7	[79]
Flavonoid glycosides and terpene lactones	Leaf extract of *Ginkgo biloba*	Amberlite XAD-7HP	[80]
Flavonoid compounds	Leaf extract of *Ginkgo biloba*	Macroporous copolymer MA-DVB beads	[81]
Flavonol glycosides and terpene lactones	Leaf extract of *Ginkgo biloba*	Macroporous polymethacrylate beads	[82]
Hesperidin	Aqueous solutions	Styrene-divinylbenzene and acrylic resins	[83]
Hesperidin	Orange peel waste	Styrene-divinylbenzene resin	[84]
Hesperidin	Orange juice processing wastewater	Styrene-divinylbenzene resin	[85]
L-tryptophan	Organic aqueous systems with L-serine	Activated carbons and neutral polymeric resins (XAD-4 and XAD-7)	[86]
Narirutin	Water-extract of citrus unshiu peels	Amberlite XAD-7	[87]
Odorless garlic	Garlic	Cyclodextrin	[88]
Oleuropein and rutin antioxidants	Olive leaf	Silk fibroin	[89]
Phenolic compounds	Distilled grape pomace	Different shape activated carbons	[90]
Phenolic compounds	Inga edulis leaves	Several resins	[91]
Phenolic compounds	Apple juice	Polumethilmetracrylate resin	[92]
Tea polyphenol	Green tea leaves	Several resins	[93]
Vitamin B12 and cephalosporin-C	Fermentation products	Nonionic polymeric adsorbents (commercial)	[94]
Vitamin E (α-tocopherol)	Palm fatty acid distillate	Silica	[95]
Vitamin E (α-tocopherol)	Palm fatty acid distillate	Silica gel, aluminum oxide, synthetic brominated polyaromatic SP 207, and functionalized	[96]
Vitamin E (α-tocopherol)	Solutions with different polar and nonpolar solvents	Mesoporous carbons CMK-1, CMK-3	[97]
Vitexin and isovitexin	Pigeonpea extracts	Macroporous resins	[98]

most explored adsorbents have been commercial or specifically designed resins (see Table 7.3).

Apart from the selection of the best adsorbent, many of these works are focused on the optimization of the process. The variables tested are the composition, pH, and polarity of the hydroalcoholic extract solution as a previous step affecting the posterior recovery [76, 91]. Also, temperature, the presence of competing compounds in the solution, the agitation, the adsorbent mass [89, 95], and the compound's initial concentration are the parameters affecting the adsorption itself on the chosen adsorbent [89, 98].

The influence of all these variables was discussed from the given isotherms. The models of Freundlich and Langmuir were preferred to fit the experimental data obtained in batch experiments. To a lesser extent, dynamic systems such as fixed-bed processes were used to optimized the adsorption and desorption processes [98].

The main disadvantage with the use of adsorption as the method for the recovery of valuable compounds is the need of a further step in order to recuperate the adsorbate from the adsorbent. Little investigation has been conducted in this aspect and when done, it has been reduced to test the best adsorbent to facilitate desorption [94] and the selection of the most appropriate eluent among the conventional hydroalcoholic mixtures [83, 90] and organic solvents [86]. In this aspect, an interesting work has been conducted by Di Mauro et al. [84], who successfully used alkaline eluents in the desorption and immediate precipitation of hesperidin [84]. More recently, Cao et al. [78] compared the use of hexane containing ethanol and supercritical CO_2, discovering that this latter option was more beneficial in terms of selectivity and recovery [78].

7.3 NOMENCLATURE

Symbol	Definition	Units in SI system	Dimensions in M, N, L, T, and θ
a_p	Outer surface area of the particle	m^2	L^2
a_w	Water activity, p_w/p_w^o		
A	Total surface area	m^2	L^2
C	Concentration of the adsorbate in the fluid phase		
c_i	Concentration of the adsorbate i in the fluid phase	$kmol \cdot m^{-3}$	NL^{-3}
c_i^*	Concentration of the adsorbate i in equilibrium with the adsorbed phase concentration, q_i	$kmol \cdot m^{-3}$	NL^{-3}
c_{Bi}	Concentration of the adsorbate i in the bulk fluid	$kmol \cdot m^{-3}$	NL^{-3}
c_{oi}	Concentration of the adsorbate i in the feed	$kmol \cdot m^{-3}$	NL^{-3}
c_{Si}	Concentration of the adsorbate i in the surface of the particle	$kmol \cdot m^{-3}$	NL^{-3}
c_p	Specific heat capacity of the fluid	$J \cdot kg^{-1} \cdot K^{-1}$	$L^2T^{-2}\theta^{-1}$
d_p	Particle diameter	m	L
d'_p	Equivalent particle diameter, $d_p\psi$	m	L
D	Impeller diameter	m	L
D_e	Effective diffusivity	$m^2 \cdot s^{-1}$	L^2T^{-1}

Symbol	Definition	Units in SI system	Dimensions in M, N, L, T, and θ
D_K	Knudsen diffusivity	$m^2 \cdot s^{-1}$	L^2T^{-1}
D_m	Molecular diffusivity	$m^2 \cdot s^{-1}$	L^2T^{-1}
D_p	Average pore diameter	m	L
D_s	Surface diffusivity	$m^2 \cdot s^{-1}$	L^2T^{-1}
FBU	Fractional bed utilization	%	
F_{CO_2}	CO_2 flow rate	kgh^{-1}	MT^{-1}
h	Film heat transfer coefficient	$W \cdot m^{-2} \cdot K^{-1}$	$MT^{-3}\theta^{-1}$
H	Enthalpy per mole	J/kmol	$MN^{-1}L^2T^{-2}$
k	Thermal conductivity of the fluid	$W \cdot m^{-1} \cdot K^{-1}$	$MT^{-3}\theta^{-1}$
k_F	Parameter in the Freundlich equation		
k_g	Film mass transfer coefficient	$m \cdot s^{-1}$	LT^{-1}
K	Adsorption equilibrium constant		
K_i	Adsorption equilibrium constant for the component i in a multicomponent mixture		
K_j	Adsorption equilibrium constant for the component j in a multicomponent mixture		
K_o	Adsorption equilibrium constant at standard conditions		
k_L	Overall mass transfer coefficient in the liquid phase	$m \cdot s^{-1}$	LT^{-1}
M	Molecular weight	$g \cdot mol^{-1}$	
M	Moisture of a food	$kg \cdot kg^{-1}$	
M_l	Moisture corresponding of a monolayer	$kg \cdot kg^{-1}$	
n_F	Index in the Freundlich equation		
n_i	Molar rate of the adsorbate i due to external transport	$kmol \cdot s^{-1}$	NT^{-1}
N_i	Moles of the adsorbate i transferred due to external transport	kmol	N
N_{ik}	Flux of the adsorbate i due to Knudsen diffusion	$kmol \cdot m^{-2} \cdot s^{-1}$	$NL^{-2}T^{-1}$
N_{im}	Flux of the adsorbate i due to molecular diffusion	$kmol \cdot m^{-2} \cdot s^{-1}$	$NL^{-2}T^{-1}$
N_{is}	Flux of the adsorbate i due to surface migration diffusion	$kmol \cdot m^{-2} \cdot s^{-1}$	$NL^{-2}T^{-1}$
P	Partial pressure	$N \cdot m^{-2}$	$ML^{-1}T^{-2}$
i	Partial pressure of the adsorbate i	$N \cdot m^{-2}$	$ML^{-1}T^{-2}$
p_w	Partial pressure of the water vapor in a food	$N \cdot m^{-2}$	$ML^{-1}T^{-2}$
p_w^o	Partial pressure of water vapor	$N \cdot m^{-2}$	$ML^{-1}T^{-2}$
P	Adsorption pressure	MPa	$ML^{-1}T^{-2}$
P	Input power per unit of fluid volume	$W \cdot m^{-3}$	$L^{-1}T^{-3}$
q	Concentration of the adsorbate in the adsorbed phase	$kmol \cdot m^{-3}$	NL^{-3}
q_b	Adsorption capacity at breakthrough point	g_{SOLUTE}/g_{CARBON}	
q_s	Saturation adsorption capacity	g_{SOLUTE}/g_{CARBON}	
q_i	Concentration of the adsorbate i in the adsorbed phase		
q_j	Concentration of the adsorbate j in the adsorbed phase	$kmol \cdot m^{-3}$	NL^{-3}
q_{max}	Concentration of the adsorbate in the adsorbed phase in a monolayer	$kmol \cdot m^{-3}$	NL^{-3}
$q_{i,max}$	Concentration of the adsorbate i in the adsorbed phase in a monolayer	$kmol \cdot m^{-3}$	NL^{-3}
Q	Heat transferred due to external transport	J	M

Symbol	Definition	Units in SI system	Dimensions in M, N, L, T, and θ
q'	Rate of heat transferred due to external transport	W	MT^{-1}
r	Distance along the radius of the adsorbent particle	m	L
R	Gas constant	$J{\cdot}kmol^{-1}{\cdot}K^{-1}$	$MN^{-1}L^2T^{-2}\theta^{-1}$
R_o	External radius of the adsorbent particle	m	L
S	Surface area per mass of adsorbent	$m^2{\cdot}kg^{-1}$	L^2M^{-1}
t	Time	s	T
t_b	Breakthrough time	min	T
t_s	Saturation time	min	T
T	Absolute temperature	K	θ
T_B	Temperature of the bulk fluid	K	θ
T_S	Temperature of the solid surface	K	θ
V	Volume of the liquid	m^3	L^3
W	Mass of adsorbent	kg	M
Greek letter			
Δ	Change in property		
ε_p	Particle porosity		
ι	Distance in the pore	m	L
η	Rotation speed	s^{-1}	T^{-1}
μ	Viscosity	$N{\cdot}s{\cdot}m^{-2}$	$ML^{-1}T^{-1}$
υ	Fluid velocity	$m{\cdot}s^{-1}$	LT^{-1}
ρ	Fluid density	$kg{\cdot}m^{-3}$	ML^{-3}
ρ_p	Particle density	$kg{\cdot}m^{-3}$	ML^{-3}
Ψ	Sphericity		
Dimensionless number			
Bi	Biot number for mass transfer (k_gR_o/D_e)		
Bi	Biot number for heat transfer (hR_o/k)		
Nu	Nusselt number (hd_p/k)		
Po	Power number (PV/η^3D^5)		
Pr	Prandlt number ($c_p\mu/k$)		
Re	Reynolds number ($\rho vd_p/\mu$)		
Sc	Schmidt number ($\mu/\rho D_m$)		
Sh	Sherwood number (k_gd_p/D_m)		

7.4 FURTHER READING

The following books are recommended for further reading on the fundamentals of adsorption and adsorption processes.

1. Suzuki, M. 1990. *Adsorption engineering.* Tokyo: Elsevier.
2. Ruthven, D. M. 1984. *Principles of adsorption and adsorption processes.* New York: John Wiley & Sons.
3. Le Van, M. D. 1996. *Fundamentals of adsorption, V.* Norwell, MA: Kluwer Academic Publishers.

4. Thomas, W. J., and B. D. Crittenden. 1998. *Adsorption, technology and design*. Oxford: Butterworth-Heinemann.
5. Perry, R. H. M., D. W. Green, and J. O. Maloney. 1997. *Perry's chemical engineers' handbook*. 7th ed. New York: McGraw-Hill Book Company.
6. Seader, J. D., and E. J. Henley. 2006. *Separation process principles*. 2nd ed. New York: John Wiley & Sons.
7. Richardson, J. F., and J. H. Harker. 2006. *Coulson & Richardson's chemical engineering, particle technology & separation processes*, Vol. 2, 5th ed. Amsterdam: Butterworth-Heinemann.
8. Wankat, P. C. 1990. *Rate-controlled separations*. New York: Elsevier Applied Science.

7.5 REFERENCES

1. Bird, R. B., W. E. Stewart, and E. N. Lightfoot. 2000. *Transport phenomena*. 2nd ed. New York: John Wiley & Sons.
2. Wakao, N., and T. Funazkri. 1978. Effect of fluid dispersion coefficients on particle-to-mass transfer coefficients in packed beds. *Chemical Engineering Science* 33:1375–1384.
3. Kunii, D., and O. Levenspiel. 1991. *Fluidization engineering*, 2nd ed. Boston: Butterworth-Heinemann, Chapter 3.
4. Schneider, P., and J. M. Smith. 1968. Chromatographic study of surface diffusion. *AIChE Journal* 14:886–895.
5. Markham, E. C., and A. F. Benton. 1931. The adsorption of gas mixtures by silica. *Journal of the American Chemical Society* 53:497–507.
6. Yon, C. M., and P. H. Turnock. 1971. Multicomponent adsorption equilibria on molecular sieves. *AIChE Symposium Series* 67 (117): 75–83.
7. Kipling, J. J. 1995. *Adsorption from solutions of nonelectrolytes*. London: Academic Press.
8. Cheftel, J. C., and H. Cheftel. 1992. *Introduction a la biochimie et a la technologie des aliments*. Paris: Technique et Documentation.
9. Brunauer, S., P. H. Emmett, and E. Teller. 1938. Adsorption of gases in multimolecular layers. *Journal of the American Chemical Society* 60:309–319.
10. Suzuki, M. 1990. *Adsorption engineering*. Tokyo: Elsevier.
11. Ruthven, D. M. 1984. *Principles of adsorption and adsorption processes*. New York: John Wiley & Sons.
12. Seader, J. D., and E. J. Henley. 2006. *Separation process principles*. 2nd ed. New York: John Wiley & Sons.
13. Chemviron Homepage. http://www.chemvironcarbon.com/en/ (accessed September 2007).
14. Norit Homepage. http://www.norit.com/ (accessed September 2007).
15. Ruthven, D. M., S. Farooq, and K. S. Knaebel. 1994. *Pressure-swing adsorption*. New York: VCH.
16. Wankat, P. C. 1986. *Large scale adsorption and chromatography*. Boca Raton: CRC Press.
17. Broughton, D. B. 1968. Molex: Case history of a process. *Chemical Engineering Progress* 64:60–65.
18. Humphrey, J. L., and G. E. Keller II. 1997. *Separation process technology*. New York: McGraw-Hill.
19. Neuzil, R. W., and R. A. Jensen. 1978. *85th National Meeting of AIChE*, Philadelphia.
20. Neuzil, R. W., and J. W. Priegnitz. 1982. US Patent 4,349,668.
21. Hamm, W., and R. J. Hamilton. 2000. *Edible oil processing*. Boca Raton: CRC Press.

22. Christidis, G. E., and S. Kosiari. 2003. Decolorization of vegetable oils: A study of the mechanism of adsorption of ß-carotene by an acid activated bentonite from Cyprus. *Clays and Clay Minerals* 51:327–333.
23. Molina, C., and R. Guardeño. 2004. Spanish patent 2,200,690.
24. Bernardini, E. 1973. *The new oil and fat technology*. 2nd ed. Rome: Tecnologie SRL.
25. Willians, R. B., and G. L. Culp. 1986. *Handbook of public water systems*. New York: Van Nostrand Reinhold Company.
26. CDTI, Centro para el Desarrollo Industrial. 1993. *Tecnología de Alimentos*. Spain.
27. Chung, Ch. Ch. 2000. *Handbook of sugar refining: A manual for the design and operation of sugar refining facilities*. New York: John Wiley & Sons.
28. Rossiter G., C. Jensen, and W. Fechter. 2002. *Proceedings of the Sugar Processing Research Conference*, New Orleans, 162–177.
29. Broadhurst, H. A., and P. W. Rein. 2003. Modeling adsorption of cane-sugar solution colorant in packed-bed ion exchangers. *AIChE Journal* 49:2519–2532.
30. McMurrough, I., D. Madigan, and M. R. Smith. 1995. Adsorption by polyvinylpolypyrrolidone of catechins and proanthocyanidins from beer. *Journal of Agricultural and Food Chemistry* 43:2687–2691.
31. Razmkhab, S., A. López-Toledano, J. M. Ortega, M. Mayen, J. Mérida, and M. Medina. 2002. Adsorption of phenolic compounds and browning products in white wines by yeasts and their cell walls. *Journal of Agricultural and Food Chemistry* 50:7432–7437.
32. Lansbarkis, J. R., and J. S. Ginrich. 2000. US Patent 5,858,068.
33. Lack, E., and H. Seidlitz. 1992. In *Extraction of natural products using near-critical solvents*, ed. M. B. King and T. R. Bott, 101–139. London: Blacky Academic and Professional.
34. Zosel, K. 1980. CA Patent 1,089,699.
35. Pieter, J. N., R. Klamer, and L. Kaper. 1990. AU Patent 598,544B.
36. Ashurst, P. R. 2005. *Chemistry and technology of soft drinks and fruit juices*. Oxford: Blackwell Publishing.
37. Carabasa, M., A. Ibarz, S. Garza, and G. V. Barbosa-Cánovas. 1998. Removal of dark compounds from clarified fruit juices by adsorption processes. *Journal of Food Engineering* 37:25–41.
38. Gokmen, V., and A. Serpen. 2002. Equilibrium and kinetic studies on the adsorption of dark colored compounds from apple juice using adsorbent resin. *Journal of Food Engineering* 53:221–227.
39. Shaw, P. E. 1990. Citrus juice debittering—current status worldwide. *The Citrus Industry* 71 (6): 54–55.
40. Singh, S. V., A. K. Gupta, and R. K. Jain. 2008. Adsorption of naringin on nonionic (neutral) macroporus adsorbent resin from its aqueous solutions. *Journal of Food Engineering* 86:259–271.
41. Lee, H. S., and J. G. Kim. 2003. Effects of debittering on red grapefruit juice concentrate. *Food Chemistry* 82:177–180.
42. Bissessur, J., K. Permaul, and B. Odhav. 2001. Reduction of patulin during apple juice clarification. *Journal of Food Protection* 64:1216–1219.
43. Dziedzic S. Z., and M. W. Kearsley. 1984. *Glucose syrups: Science and technology*. New York: Elsevier Applied Science Publishers.
44. Kearsley, M. W., and S. Z. Dziedzic. 1995. *Handbook of starch hydrolysis and their derivatives*. London: Blacky Academic and Professional.
45. Berovic, M., and M. Legisa. 2007. Citric acid technology. *Biotechnology Annual Review* 13:303–343.
46. Bradley, K. J., M. K. Toledo, and R. T. Toledo. 1987. Physicochemical factors affecting ethanol adsorption by activated carbon. *Biotechnology & Bioengineering* 28:445–452.

47. Pitt, W. W., G. L. Hang, and K. K. Lee. 1983. Recovery of ethanol from fermentation broths using selective sorption-desorption. *Biotechnology & Bioengineering* 22:123–131.
48. Jianlong, W., W. Xiangua, and Z. Ding. 2000. Production of citric acid from molasses integrated with in-situ product separation by ion-exchange resin adsorption. *Bioresource Technology* 75:231–234.
49. Nakada, K., M. Takagi, and M. Hirai. 1979. Japanese Patent 54,046,714.
50. Fang, T. Y., L. L. Lin, and W. H. Hsu. 1998. Japanese Patent 63,134,055.
51. Johnson, R. A., R. L. Antrim, and N. E. Lloyd. 1986. US Patent 4,610,965.
52. Chao, C. C., and J. D. Sherman. 1983. US Patent 4,394,178.
53. Buescher, R., and C. Hamilton. 2002. Adsorption of polygalacturonase from recycled cucumber pickle brines by Pure-Flo B80 clay *Journal of Food Biochemistry* 26 (2): 153–165.
54. Neuzil, R. W., and J. W. Priegnitz. 1981. GB Patent 1,585,369.
55. Nakada, K., M. Takagi, and M. Hirai. 1979. Japanese Patent 54,046,714.
56. Sherman, J. D., and C. C. Chao. 1985. EU Patent 0,137,063.
57. Wibobo, S., G. Velázquez, V. Savant, and J. A. Torres. Surimi wash water treatment for protein recovery: Effect of chitosan-alginate complex concentration and treatment time on protein adsorption. *Bioresource Technology* 96 (6): 665–671.
58. Ito, M., K. Yamauchi, and K. Matsuzawa. 1993. The adsorption of proteins in fermented aqueous food by silica gels. *Colloids and Surfaces A: Physicochemical and Engineering Aspects* 74 (1): 107–113.
59. Chen, Ch-Y., Ch-I. Lin, and H-K Chen. 2003. Kinetics of adsorption of β-carotene from soy oil with activated rice hull ash. *Journal of Chemical Engineering of Japan* 36 (3): 265–270.
60. Proctor A., and H. E. Snyder. 1987. Adsorption of lutein from soybean oil on silicic acid I. Isotherms. *Journal of the American Oil Chemists' Society* 64 (8): 1163–1168.
61. Palaniapan, S., and A. Proctor. 1990. Competitive adsorption of lutein from soy oil onto rice hull ash. *Journal of the American Oil Chemists' Society* 67 (9): 572–577.
62. Chiu, S-H., T-W. Chung, R. Giridhar, and W-T. Wu. 2004. Immobilization of β-cyclodextrin in chitosan beads for separation of cholesterol from egg yolk. *Food Research International* 37:217–223.
63. Mohamed, R. S., G. B. M. Neves, and T. G. Kieckbusch. 1998. Reduction of cholesterol and fractionation of butter oil using supercritical CO_2 with adsorption. *International Journal of Food Science and Technology* 33 (5): 445–454.
64. Mohamed, R. S., M. D. A. Saldaña, F. H. Socantaype, and T. G. Kieckbusch. 2000. Reduction in the cholesterol content of butter oil using supercritical ethane extraction and adsorption on alumina. *Journal of Supercritical Fluids* 16:225–233.
65. Garrido, A., P. García, and M. Brenes. 1992. The recycling of table olive brine using ultrafiltration and activated carbon adsorption. *Journal of Food Engineering* 17 (4): 291–305.
66. Casimir, A., and A. E. Reynolds, Jr. 1999. WO Patent 9,964,545.
67. Miyauchi, M., M. Atsuko, Y. Nakanishi, and Y. Sagara. 1996. Binary adsorption equilibria of various flavours and water for materials contained in a box of tobacco product. *Food Science and Technology International (Tsukuba, Japan)* 2:65–71.
68. Ishii, R., Y. Imai, M. Wada, T. Ebina, T. Hanaoka, and F. Mizukami. 2006. Adsorption and desorption behaviors of flavor molecules into a microporous pillared clay mineral and the application to flavor capsule composites. *Applied Clay Science* 33 (2): 99–108.
69. Zeller, B. L., F. Z. Saleeb, and R. D. Ludescher. 1998. Trends in development of porous carbohydrate food ingredients for use in flavour encapsulation. *Trends in Food Science & Technology* 9:389–394.
70. Patel, B., and T. Walker. 2004. *ASAE Annual International Meeting* 7191–7199.

71. Scordino, M., A. Di Mauro, A. Passerini, and E. Maccarone. 2005. Selective recovery of anthocyanins and hydroxycinnamates from a byproduct of citrus processing. *Journal of Agricultural and Food Chemistry* 53:651–658.
72. Di Mauro, A., E. Arena, B. Fallico, A. Passerini, and E. Maccarone. 2002. Recovery of anthocyanins from pulp wash of pigmented oranges by concentration on resins. *Journal of Agricultural and Food Chemistry* 50 (21): 5968–5974.
73. Kammerer, D., J. G. Kljusuric, R. Carle, and A. Schieber. 2005. Recovery of anthocyanins from grape pomace extracts (Vitis vinifera L. cv. Cabernet mitos) using a polymeric adsrober resin. *European Food Research and Technology* 220 (3-4): 431–437.
74. Selga, A., and J. L. Torres. 2005. Efficient preparation of catechin thio conjugates by one step extraction/depolymerization of pine (*Pinus pinaster*) bark procyanidins. *Journal of Agricultural and Food Chemistry* 53 (20): 7760–7765.
75. Kamiyama, S., N. Tamura, K. Osada, and M. Watanuki. 1993. Japanese Patent 05,095,793.
76. Scordino, M., A. Dimauro, A. Passerini, and E. Maccarone. 2004. Adsorption of flavonoids on resins: cyanidin 3-glucoside. *Journal of Agricultural and Food Chemistry* 52 (7): 1965–1972.
77. Kimizuka, A., Y. Ueda, M. Sakaguchi, and R. Miyajima. 1988. US Patent 4,741,914.
78. Cao, X., and B. Hur. 2005. Separation of EPA and DHA from fish oil using modified zeolite 13X and supercritical CO_2. *Journal of Industrial & Engineering Chemistry* 11 (5): 762–768.
79. Yoon, S. Y., W. J. Choi, J. M. Park, and J. Yang. 1997. Selective adsorption of flavonoid compounds from the leaf extract of *Ginkgo biloba* L. *Biotechnology Techniques* 11 (8): 553–556.
80. Lai, S., R. Chen, and S. Suen. 2003. Adsorption separation for the extracts from *Ginkgo biloba* leaves using intermediate polarity resins. *Journal of Liquid Chromatography & Related Technologies* 26 (17): 2941–2960.
81. Xu, M., Z. Shi, R. Shi, J. Liu, Y. Lu, and B. He. 2000. Synthesis of the adsorbent based on macroporous copolymer MA-DVB beads and its application in purification for the extracts from *Ginkgo biloba* leaves. *Reactive and Functional Polymers* 43 (3): 297–304.
82. Ma, X., C. Jiang, Y. Yang, and L. Zhang. 1997. Adsorption of flavonoids in *Ginkgo biloba* L. leaves by macroporous adsorptive resins. *Zhongguo Zhongyao* 22 (9): 539–542.
83. Scordino, M., A. Di Mauro, A. Passerini, and E. Maccarone. 2003. Adsorption of flavonoids on resins: Hesperidin. *Journal of Agricultural and Food Chemistry* 51 (24): 6998–7004.
84. Di Mauro, A., B. Fallico, A. Passerini, P. Rapisarda, and E. Maccarone. 1999. Recovery of hesperidin from orange peel by concentration of extracts on styrene-divinylbenzene resin. *Journal of Agricultural and Food Chemistry* 47 (10): 4391–4397.
85. Di Mauro, A., B. Fallico, A. Passerini, and E. Maccarone. 2000. Waste water from citrus processing as a source of hesperidin by concentration on styrene-divinylbenzene resin. *Journal of Agricultural and Food Chemistry* 48 (6): 2291–2295.
86. Ribeiro, M. H. L., D. M. F. Prazeres, J. M. S. Cabral, and M. M. R. DaFonseca. 1995. Adsorption studies for the separation of L-tryptophan from L-serine and índole in a bioconversion médium. *Bioprocess Engineering* 12 (1–2): 95–102.
87. Kim, M. R., W. C. Kim, D. Y. Lee, and C. W. Kim. 2007. Recovery of narirutin by adsorption on a non-ionic polar resin from a water-extract of Citrus unshiu peels. *Journal of Food Engineering* 78 (1): 27–32.
88. Kawashima, Z. 1986. Japanese Patent 61,091,128.
89. Bayçin, D., E. Altiok, S. Ulkü, and O. Bayraktar. 2007. Adsorption of olive leaf (*Olea europaea* L.) antioxidants on silk fibroin. *Journal of Agricultural and Food Chemistry* 55 (4): 1227–1236.

90. Soto, M. L., A. Moure, H. Domínguez, and J. C. Parajó. 2008. Charcoal adsorption of phenolic compounds present in distilled grape pomace. *Journal of Food Engineering* 84 (1): 156–163.
91. Silva, E. M., D. R. Pompeu, Y. Larondelle, and H. Rogez. 2007. Optimisation of the adsorption of polyphenols from *Inga edulis* leaves on macroporous resins using an experimental design methodology. *Separation and Purification Technology* 53 (3): 274–280.
92. Kammerer, D. R., Z. S. Saleh, R. Carle, and R. A. Stanley. 2007. Adsorptive recovery of phenolic compounds from apple juice. *European Food Research and Technology* 224 (5): 605–631.
93. Li, P., Y. Wang, M. Runyu, and X. Zhang. 2005. Separation of tea polyphenol from green tea leaves by a combined CATUFM-adsorption resin process. *Journal of Food Engineering* 67 (3): 253–260.
94. Ramos, A. M., M. Otero, and A. E. Rodrigues. 2004. Recovery of vitamin B12 and cephalosporin-C from aqueous solutions by adsorption on non-ionic polymeric adsorbents. *Separation and Purification Technology* 38 (1): 85–98.
95. Chu, B. S., B. S. Baharin, Y. B. CheMan, and S. Y. Quek. 2004. Separation of vitamin E from palm fatty acid distillate using silica: I. Equilibrium of batch adsorption. *Journal of Food Engineering* 62 (1): 97–103.
96. Chu, B. S., B. S. Baharin, Y. B. CheMan, and S. Y. Quek. 2005. Comparison of selected adsorbents for adsorption and desorption of vitamin E from palm fatty acid distillate. *Journal of Food Lipids* 12:23–33.
97. Hartmann, M., G. Chandrasekar, and A. Vinu. 2005. Adsorption of vitamin E on mesoporous carbon molecular sieves. *Chemistry of Materials* 17 (4): 829–833.
98. Fu, Y., Y. Zu, W. Liu, C. Hou, L. Chen, S. Li, X. Shi, and M. Tong. 2007. Preparative separation of vitexin and isovitexin from pigeonpea extracts with macroporous resins. *Journal of Chromatography A* 1139 (2): 206–213.

Index

A

Absinthe, 97
Absorbents, 416
Absorption, microwave-assisted extraction, 151, 153
Accuracy ranges, cost estimation, 48–49
Acerola juices, 117
Acetaldehyde, 99
    continuous tray column distillation, 110, 113–114
    distillation concentration, 106–107
    hangover syndrome, 102
Acetaldehyde diethyl acetal, 99
Acetic acid
    distillate concentration, 106, 113
    equilibrium pressure, 90
    –ethanol relative volatility, 95–96
    hydrothermal generation, 306
    oxidation, 111
Acetone, 141, 152, 169
    Soxhlet percolation extraction process, 163
    vapor phase cooling, 344
Acetonitrile, 141
Acidification
    extraction yields, 313
    phenolic compounds, 194, 196
Acidity
    adsorbents, 416
    chemical refining, 221, 245
    coconut oil refining, 23–25
    edible oil glycerin esters, 13, 18–19
    liquid–liquid extraction, 256
    spirit quality, 100, 102, 110, 113
Activated carbons
    adsorbents, 414–415
    coffee aroma volatile recovery, 370–371
    ecologically refined oil, 422
    reactivation, 418
    starch-based sweetener purification, 428
    syrup decolorization, 424–425
    wine color/taste correction, 426
Adenosine triphosphate (ATP), 329–330
Adsorbates, 404, 411
Adsorbed phase, 404, 409
Adsorbents, 414–417
    citrus oil fractionation processing, 360
    regeneration, 417–418
    regeneration of aroma, 370–371
    supercritical carbon dioxide extraction, 376
Adsorption, 2, 271
    coffee aroma extraction, 375–376
    concentration technique, 403–422
    edible oil refining processing, 422–423
    flow rates, 381–382
    food processing applications, 422–432
    orange volatile oil countercurrent extraction, 360–362
    processes, 418–422
    solid matrix supercritical extraction, 379
Adsorption isotherms, 361
Affinity, 404–405
Agitated tanks, 183, 198, 200, 404
Agitated vessels, 418, 419
Agitation
    adsorbents, 414
    adsorption parameters, 432
    continuous application, 428
    extraction columns, 223
    solute concentration reduction, 419
    solvent extraction, 167
    speed for piperine, 161
Agitation power estimation, 200–201
Aglycons, 190
Agricultural waste, 289, 310
Agrochemical crops, 336
Alcohol gradation
    batch cachaça distillation, 104–105
    continuous cachaça tray column distillation, 110–111
    distillation profiles, 107–108
    ethanol concentration continuous distillation, 115
Alcoholic beverages, 76, 97–109, 426. *See also* Spirits
Alcoholic extracts, 244

Alcohols
chain length, 247
distillation equilibrium equations, 88
short chain liquid–liquid equilibrium, 224–225
steam distillation, 14
vegetable oil deacidification, 246
Aldehydes, 14
coffee aroma, 371
desorption, 360
distillate concentration, 99, 107, 120–123
distillation congeners, 88
essential oils, 243
Alembic distillation, 103–104
cachaça production, 102
copper/stainless steel effects, 99
simulation equations, 82–83
Alembics, 77, 102–109
Aliphatic waxes, 30–31
Alkali refining, 246
Alkyl chain length, 328
Almond shells, 311
Alumina, 360, 415
Aluminum oxide, 415
Amino acids, 429
Anacardic acids (AAs), 327–328, 347–348
bioactivity and uncoupling effects, 329–331
cashew processing, 332–334, 333–334
separation ratios, 345
supercritical extraction, 336–338
Anhydrous ethanol, 94
Anise seed, 11
cost estimation, 52–55
cost of manufacturing, 55–58
pressure and yield, 42–43
steam distillation, 43–45
volatile oil market pricing, 72
Anthocyanins, 189–191, 192–193, 426
Antimicrobial properties, 38
Antioxidants, 2, 249
aromatic/condimentary/medicinal plants, 38
chemical classes, 4, 7
condimentary plant, 139–140
GRAS solvent extraction, 185–189
hot water/pressurized extraction thermal degradation, 313
olive oil percolation extraction, 164–165
solvent modifiers, 303–304
supercritical fluid cosolvents, 302–306
supercritical fluid extraction, 288–315
Aqueous solution mass transfer, 157
Aqueous two-phase systems (ATPS), 241–242
Arnica, 177
Aroma industry, 38–39
Aromas
cashew juice distillation concentration/purification, 117–129
coffee, 2
coffee supercritical adsorption process, 370–385
distillation processes, 75–76
distillation process recovery, 101
mixture distillation simulation, 97–101
orange volatile oil fractionation, 352–366
removal with supercritical extraction, 301
volatile compound complexity, 371–372
volatile/essential oil distinction, 11
Aromatic compounds
activated carbon filtration, 427
coffee, 371–372
distillation, 75
olive oil percolation extraction, 165
Aromatic plants
antioxidants, 38
polyphenols, 4
pretreatment, 290–291
solid–liquid extraction, 138
steam distillation, 11
steam distillation oil release, 14–15
supercritical carbon dioxide extraction processing, 289
volatile steam distillation, 38–43
Arrhenius function, 278, 409–410
Artemisia, 39
Ascorbic acid, 140
ASOG (analytical solution of groups) model, 91, 236–238
Aspen wood, 312
Asphalathin, 188
Association for the Advancement of Cost Engineering International (AACEI), 47, 50
Autohydrolysis, 298, 309–311
Axial dispersion, 273, 274, 281
Azeotropic distillation, 94
Azeotropic mixtures, 246

B

Balance equations
microwave-assisted extraction, 154
solid–liquid low pressure, 142–144
supercritical fluid extraction, 273–281
ultrasound-assisted extraction, 156–158
Balm extracts, 189
Bamboo leaf extract (BLE), 159–160
Batch deodorizer modeling, 19–23
Batch distillation, 102–109
equations, 83–84
scheme, 77
Batch equipment
distillation columns, 117–129
slurry extraction, 162

Batch extraction
continuously stirred, 197
saponified/esterified soy deodorized distillate, 291
screw extractors, 165–167
solid–liquid processes, 165–167
Batch operations
adsorption, 404, 418–422
solid matrices, 162
Bath systems
air, 342–343
cooling, 313
thermostatic cooling, 53
ultrasound-assisted extraction, 171, 173, 175–178, 180, 182, 184
Bed density
clove bud costs, 395
cost of manufacturing, 53, 54, 393, 394
ginger essential oil cost, 398
Bed extraction, 290
chamomile steam distillation, 58–60
fixed beds, 419–420, 428
fluidized, 418
moving, 420–422
Bentonite, 415–416
Benzene, 141
Bergamot oil, 353, 359, 365
Berries, 190, 192–193
ß-carotene, 140, 297
Beverages
adsorption, purification of carbonated, 426–427
alcoholic, 426
spirit quality and distillation, 100
Bid estimates, cost estimation, 49
Binary interaction parameters, 238–239, 356–357
Binary mixtures
analytical methods, 342
carbon dioxide as pseudo, 340
two-liquid model, 90–91
Binodal curve, 233
equilibrium phase behavior, 242
temperature, 247
vegetable oil deacidification, 247–249
Bioactive compounds, 1–7
adsorption/desorption concentration, 403–404
adsorption recovery, 430–432
availability and pretreatment, 289
cashew separation scheme, 347–348
cashew trees and nuts, 327–328
liquid–liquid extraction vegetable oil deacidification, 249–258
microwave-assisted extraction, 172–174
solid–liquid extraction, 138
ultrasound-assisted extraction, 172–174
vegetal matrices steam distillation, 41
Bioactivity, uncoupling effects, 329–331
Biofuels, 93, 94, 100
Biomass refining, 306
Biopolymer hydrothermal processing, 307
Black pepper, 11, 53, 290
methanol solid–liquid extraction, 160
superheated steam, 42
volatile oil market pricing, 72
volatile oil steam distillation cost, 63–65
volatile steam distillation, 44
Black tea wastes, 186
Bleaching, 427
Boiling points
distillation process description, 82–83
terpenoids, 4
Boiling temperatures, 103
Boiling water
Roselle petal extracts, 190
steam distillation temperature, 11
Borneol, 181
Boundary conditions, ultrasound mass transfer, 157
Brandy, 103
Brunauer–Emmet–Teller (BET) equation, 412–413
Budget authorization, 49
Building costs, 388
Business plan cost estimation, 50
Butane, 288
Butylhydroxyanisol (BHA), 310
Butylhydroxytoluene (BHT), 310
Byproducts
cashew processing, 333
cashew volatile batch distillation, 129
continuous tray cachaça distillation, 115–116
nonsaccharides, 309
recovery and purification, 19
steam explosion reaction, 312

C

Cachaça, 2
distillation process, 83, 101–117
phase equilibrium equations, 88
Caffeine extraction, 159, 163–164
Cailletet apparatus, 339–340
Calorimetric methods, ultrasonic intensity, 156
Cane sugar
refining adsorption, 425–426
volatile component distillation, 99–100
Canola oil
binodal curve, 247–248, 249
isomerization and steam deacidification, 28–30
wax decomposition, 31

Capital costs
condimentary plant extract manufacturing, 388–390
estimation for solvent extraction, 202
pressure swing adsorption process, 363
Carbonated drinks, adsorption purification of carbon dioxide, 426–427
Carbon chains, 94
Carbon dioxide
adsorption purification, 426–427
antioxidant supercritical extraction, 288
aroma recovery, 374–375
capital costs, 388
cashew nut shell liquid fractionation, 340–346
dissolution, 344–346
distillation degassing, 114
environmental friendliness, 391
fennel extract, 4
flow rates, 381–382
phase behavior, 338–339
single-stage supercritical steam separation, 294–296
supercritical fluid properties, 272
supercritical fluid successive extraction, 299–301
thermophysical property, 3
Carbon tetrachloride, 141
Carboxylic acids, 90
cardanols, 328, 331, 333, 345
cardols, 328, 345
Carnisic compounds, 188, 189
Carnosic acid, 184, 189
Carnosol, 139
Carnosolic acid, 139
Carotenoids, 18, 426
edible oil refining processes, 256–258
GRAS solvent extraction, 191–192
palm oil, 253–254
single-stage supercritical steam extraction, 295
stagewise supercritical fluid extraction, 299
supercritical fluid extraction cosolvents, 305
Carvone, 180
Cashew nut shell liquid (CNSL), 327, 332–334, 335, 336
constituent properties, 346
liquid–liquid–vapor equilibrium, 340–342
supercritical CO_2 separation, 334–339
Cashews, 271
bioactive compound extraction, 327
bioactive compounds separation, 347–348
cultivation and production, 331–332
juice aroma batch distillation concentration/ purification, 117–129
phenolic lipids, 328–329
processing, 332–334
trees and processing, 332–334
Catechol, 312, 328
Cavitation, ultrasound-assisted extraction, 155–156, 180, 183
Cell structure, 140
essential oils, 293
low pressure solvent selection, 141
solvent extraction, 152–153
ultrasound-assisted extraction, 157
vegetal material pretreatment, 290
Cellulase, 187, 191–192
Cellulose, hydrothermal treatment, 308
Centrifugal extractors, liquid-liquid extraction equipment, 223
Centrifugation
cane sugar refining, 425
olive oil extraction, 164–165
slurry extraction, 162
Cetyltrimetylammonium bromide, 160–161
Chamomile, 11, 290
steam distillation, 40, 42
ultrasound-assisted extraction dry, 177
volatile oil distillation costs, 58–60
volatile oil market pricing, 72
volatile steam distillation, 44
Chemical classes
antioxidant/healthful bioactive compounds, 4, 7
phase equilibrium estimation, 3
volatile oils and terpenes, 2–3
Chemical reactions, distillation mixture, 83
Chemical refining, 221, 245–246
Chemisorption, 404–405
Chestnut burs, 311
Chilton method, cost estimation, 50
Chlorine, 424
Chloroform, 141, 163–164, 244
Chlorophyll, 18, 426
Chromatographic analysis, 118
Chromatographic separation, 244, 293
Ciclohexane, 94
Cineole, 181
Cis-isomers, steam deacidification, 19, 26–30
Citric acid
adsorption decolorization, 428–429
liquid–liquid extraction production, 242–243
Citrus oils
alcoholic extracts, 244
component phase equilibrium, 352–354
countercurrent extraction, 355
liquid–liquid extraction production, 243–244
pressure swing adsorption, 364
pressure swing adsorption process, 366
Clarification, cane sugar refining, 425
Clove basil, 290
Clove buds, 2, 70, 271, 281, 395–398
Coconut oil, 23–26, 103
Coextracts, antioxidant potential, 139–140

Coffee
adsorption for decaffeination, 427
beans, 377
oils, 377
optimal commercial processing conditions, 382–383
smell aroma volatile, 370–371
soluble aroma, 2
supercritical oil extraction–adsorption, 383–385
Coffee aroma, 271
component analysis, 376
high-pressure adsorption/desorption, 370–384
solid matrix supercritical extraction, 379–383
supercritical extraction–adsorption, 383–385
Color. *See also* Decolorization
adsorption correction, 426
adsorption removal, 422–425, 427
Compressibility, supercritical fluids, 283–285
Concentration
bioactive compound adsorption/desorption, 403–404
cashew aroma/flavoring distillate, 120–126
cashew volatile batch distillation, 124–126
coffee aroma conventional, 372–374
convective mass flux, 274
equilibrium of adsorption, 409
solvent-to-flow ratio factors, 357–359
Condensation
adsorbent regeneration, 418
aroma recovery, 373
capillary, 414
distillate, 17
flavor, 307
vapor sampling, 344
Condensed water, steam distillation process, 16
Condensers
balance equations, 84, 85
continuous tray column distillation, 110
distillation process efficiencies, 86
steam distillation process, 17
Condimentary plants, 138
antioxidant action, 139
polyphenols, 4
steam distillation, 11, 38–43
supercritical fluid extract, cost of manufacturing, 388–400
volatile oil steam distillation, cost of manufacturing, 47–48, 50–52, 52–70
Congeners, 88
Conical extractors, 165
Contaminants, distillation processes, 99–101
Continuous contact, liquid–liquid extraction equipment, 222–223
Continuous countercurrent extractors, 165
Continuous distillation
of cachaça in tray columns, 109–117
neutral spirits, 100–101
Continuous extraction, solid–liquid processes, 165–167
Continuous multistage countercurrent extractor, liquid–liquid mass balance equations, 232–234
Continuous processing, liquid feeds, 292–293
Continuous stirring, batch extraction, 197
Contract value, cost estimation, 49
Control baseline, cost estimation, 49
Convective flux, 274
Convective transport, 273–275
Copper, 99–100
Corn cobs, 310, 311
Corn syrup, 421
Cosolvents
aroma supercritical extraction, 375
supercritical fluid antioxidant, 302–306
Cost/capacity curves, 49
Cost estimation, 50
classes, 47, 48–49
condimentary plant steam distillation, 52–70
Cost of manufacturing (COM), 139, 197, 271
anise seed steam distillation, estimated, 55–58
black pepper steam distillation, 63–65
chamomile steam distillation, estimated, 58–60
condimentary plant supercritical fluid extracts, 388–400
costs classes, 48–49
estimation method, steam distillation, 47–54
estimation methods, 50
extraction techniques, 1
market price and volatile oil steam distillation, 70–72
rosemary steam distillation, estimated, 60–63
solvent extraction, 206
steam distillation, economic viability, 43
thyme steam distillation, estimated, 65–69
Cost of operational labor (C_{OL}), 205–206
Cost of process, 13
Cost of time (C_{TM}), 50
Cost of utilities (C_{UT}), 203–205, 392
Cost of waste treatment (Cwt), 206
Costs
adsorbents, 405
adsorption on solid matrices operating, 379
estimate weighting factors, 50
freeze drying, 290–291
steam distillation, 51–52
Cottonseed oil, 256–258
Countercurrent extraction
essential oil mutual solubility, 353
liquid materials, 354–359

moving bed adsorption operations, 420–421
solid–liquid low pressure, 148–150
supercritical processing, 297–298
Countercurrent extractors, 165
Crosscurrent extraction
solid–liquid low pressure, 147–148
solvent-to-feed ratio, 188–189
Crown Iron immersion extractor, 166–167
Cup horn, 178–180
Cylindrical mixing extractors, 165

D

Dalton's law, 16
Deacidification
bioactive compounds from liquid–liquid extraction vegetable oil, 249–258
edible fats/oil steam distillation, 13–14
fixed oil steam distillation, 10
free fatty acid liquid–liquid extraction processing, 220–221
free fatty acid removal oil purification, 245–247
liquid–liquid extraction from vegetable oils, 247–249
vegetable oil liquid–liquid extraction, 245, 258
vegetable oils, 1
vegetable oil solvent extract, 246
vegetable oil stripping, 18–32
Dearomatization, successive supercritical, 301
Debittering, 427–428
Decaffeination, 427
Decanters, 12
Decarboxylate, 333–334
Decoction, 140
Decolorization
adoption purification, 422–423
cane sugar refining, 426
citric acid, 428–429
hydrolyzed vegetal proteins, 426
syrups, 424–425
Deetherification, 308
Degassing, distillation, 109, 110, 113–114, 115
Density
bed, 53, 54, 393, 394, 395
solute, 275
supercritical fluids, 354
vapor phase molecular, 89–90
Deodorization
edible fats/oil steam distillation, 13–14
edible tocopherol content, 253
mass stripping with steam, 19
oil composition estimation, 22–23
successive supercritical carbon dioxide extraction, 301
vegetable oil steam distillation, 18–32
Deodorized distillates, 291
Depolymerization, 308, 310
Depreciation, cost of manufacturing, 393
Design quantities, cost estimation, 49
Desorption, 432
adsorbent regeneration, 417–418
concentration technique for bioactive compounds, 403–404
curves from silica gel, 363
essential oil supercritical carbon dioxide, 360
fixed bed operations, 419–420
overall extraction curve, 278
pressure swing process, 363–366
ultrasonic extraction, 157
Deterpenation, 244
Dextran, 241
Dextrose, 428
Dichloroethane, 141
Dicot woody tissue, 308
Dielectric properties, microwave-assisted extraction, 151–152
Diethylene glycol, 244, 245
Diffusion
internal transport adsorption, 408
particle size, 189
Diffusion coefficient
overall extraction curve modeling, 277
ultrasound intraparticle, 157
Diffusion rate, liquid solvent selection, 141
Dilution/distribution coefficients, 238–239
Dimethylsulfite, 99
Direct costs, condimentary plant extract manufacturing, 390–391
Discrepancy functions, 85–86
Displacement, adsorbent regeneration, 418
Dissolution
carbon dioxide pressure, 345
hydrophobic isolates, 40
phenolic compounds, 196
solvent solid–liquid extraction, 140, 142–143
Distillates
condensation, 17
continuous flows, 110–111
volatile compound gradations, 77–78
Distillation, 2, 12–13. *See also* Steam distillation
aroma and spirit processing, 75–76
cachaça, 101–117
cashew juice aroma concentration/purification, 117–129
coffee volatile compounds, 373
cycle cost estimation scaling up, 51
deterpenation, 244
double, 99
dry steam, 12
equipment design and evaluation, 86

rate continuous tray column, 111
simulation and design, 86
Distillation columns
cashew aroma concentration/purification, 117–129
continuous, 78–80
liquid–liquid extraction mass balance/ equilibrium, 239–240
packing, 81–82
tray, 80–82, 86, 222
Distillers, 12
Diterpene oils, 375
Dixon rings, 292
Downstream processing, enzyme/protein purification, 241
Drinking water
availability, 388
purification, 423–424
Dry ice, 344
Drying
cylindrical mixing extractors, 165
pretreatment, 290–291
Duplicate oils, 39

E

Ecologically refined oil 422
Economics, solvent extraction, 197–206
Edible fats, 221
Edible oils, 221
adsorption refining processes, 422–423
nutritive value categories, 252–253
steam distillation, 13–14
Electron-transport mechanisms, 329–330
Elution, 140
Emulsification, 221
Energy
adsorption activation, 405
costs, 37
mixture boiling point maintenance, 83
pressure swing adsorption process, 363
steam distillation, 13, 35–36
Engineering, cost estimation, 49
Enthalpy balance equations, 84, 85
Environmental friendliness, 2, 391
Enzymes
antioxidant compound extraction, 187
commercial production, 241
liquid–liquid vegetable oil extraction, 241
lycopene extraction, 191–192
mitochondrial, 329
starch-based sweetener purification, 428
Equations of state (EOS)
height equivalent to theoretical stage models, 356
Peng–Robinson, 283–285, 352
phase equilibrium, 3
Soave–Redilich–Kwong, 283–285
Equilibrium. *See also* Mass balance equations; Phase equilibrium; Vapor–liquid equilibrium
adsorption, 404
liquid–liquid extraction, 222, 224–225, 227–228
liquid–liquid extraction column simulation, 240
liquid–liquid–vapor, 327
mass balance equations, 84–85
separation processes by adsorption, 409–414
single stage solvent extraction, 146
steam distillation vapor–liquid, 21–22
supercritical thermodynamic, 281–285
ultrasound mass transfer equations, 157
vaporization process mass, 82–83
wine distillation curve, 94–95
Equipment
cost index, 201
liquid–liquid extraction, 221–223
purchase costs estimation, 201–202
recirculating static apparatus, 342–343
sizing/solvent extraction economics, 198
slurry extraction, 159, 162–163
solid–liquid solvent extraction, 159–167
steam distillation patents, 37–38
utilization economics, 197
Equipment costs, 389
Essential oils, 11. *See also* Volatile oils
cell structure, 293
glandular trichomes pressure, 338
liquid–liquid extraction solvent selection, 244–245
liquid stream extraction, 291
microwave-assisted extraction, 169
solubility and phase equilibria, 352–354
stagewise extraction, 298
steam distillation, 14, 36
supercritical carbon dioxide desorption, 360
Essential unsaturated fatty acids (EFAs), 249
Esters, cashew fruit juice, 118
Estimation
cost of manufacturing condimentary plant extracts, 392–395
solvent extraction process costs, 201–207
solvent extraction process economics, 197
Ethanoate esters, 99
Ethanol, 141
anthocyanin GRAS solvent extraction, 190, 191
antioxidant compound extraction, 185, 186–187, 187
antioxidant supercritical extraction, 302, 304
binodal curves of anhydrous, 247–248
cachaça production, 102
coffee aroma cosolvent, 375

concentration in distilled spirits, 93–97
continuous tray column distillation, 110
distillation, 76
distillation vapor-phase equilibrium, 88–97
hydrated, 93
limonene–linalool phase equilibria, 352
liquid–liquid extraction solvent selection, 244
multistage crosscurrent extraction, 228
phenolic compound extraction, 194
Soxhlet percolation extraction process, 163
ultrasound-assisted extraction process, 181, 184
utilities cost estimates for solvent extraction, 204–205
volatility values in spirits distillation, 89
water use with vegetable oil deacidification, 249

Ethyl acetate, 141
autohydrolysis liquor antioxidants production, 309–310
lignocellulosic material, autohydrolysis liquor extracts, 311
solid matrix supercritical extraction, 380–382

Ethyl carbamate, 100

Ethylene, 288

Ethylene glycol, 94

Ethylmethylketone, 141

Eucalyptus, 281, 309, 310, 311

Eugenol, 171

European Community, 422

Evaporation
aroma recovery, 101
cane sugar refining, 425–426
chemical alteration, 139
concentration process, 117, 118, 242
extraction vessel cooling, 181
solvent extraction step, 206
solvent recovery, 171
solvent removal, 165
solvent stripping, 246
syrup purification, 425–426
volatile oil extraction, 14

External transport, adsorption process steady-state film theory, 406–407

Extractable substances (ES), ultrasound-assisted, 183

Extraction
bioactive compounds, 1–7
emerging technologies, 312–315
method choice, 3

Extraction columns. *See also* Distillation columns; Packed columns
liquid–liquid, 222–223
liquid–liquid simulation, 239–241
mass balance concentration, 274
pulsed, 223
rectification systems, 99
supercritical, 272, 273, 390, 394
thermodynamic phase equilibrium, 281

Extraction curve
cost of manufacturing, 393–394
ginger, 281
ginger, cost of manufacturing, 398–399
Martínez mass transfer model, 280
mathematical model of overall, 276–281
supercritical mass balance, 275–276

Extraction cycles
chamomile distillation, 59
column systems, 390, 394
cost of manufacture, 62
rosemary distillation, 60
steam distillation, 51

Extraction efficiency
phenolic compounds, 193
supercritical/solid-phase methods, 375
ultrasound-assisted processes, 184

Extraction plant construction, 388–390

Extraction rates
supercritical fluid densities, 354
ultrasound-assisted, 180

Extraction tanks, 159

Extraction techniques. *See also* Bed extraction; Distillation; Solvent extraction; Steam distillation; Supercritical fluid extraction; Ultrasound-assisted extraction
cost of manufacturing estimates, 1
selection of solid–liquid, 167
solvent low pressure, 140

Extraction time
anise seed steam distillation, 55
microwave-assisted, 170–171
microwave-assisted process, 153

Extraction vessels, 168

Extraction yields
anise seed steam distillation, 42–43, 55–57
antioxidant thermal degradation, 313
cashew shell nut liquid, 337–338
chamomile steam distillation, 58–60
microwave power increments, 169–170
phenolic compound and pH, 196
solvent modifiers for antioxidants, 303–304
steam distillation, 40
steam distillation flow rate, 42
subcritical hot/pressurized water, 313
thyme steam distillation, 65–68
ultrasound-assisted, 192

Extractors
centrifugal, 223
conical, 165
continuous countercurrent, 165, 232–234
Crown Iron immersion, 166–167

immersion, 166–167
liquid–liquid mechanically agitated, 223
screw batch, 165–167
solid–liquid low pressure, 144
Extracts. *See also* Antioxidants; Aromas; Distillates; Pigments; Volatile oils
antioxidant supercritical fluid successive, 299–301
cost of manufacturing clove bud, 395–398
lignocellulosic autohydrolysis liquors, 311
manufacturing cost estimation, 392–395

F

Fats and fat-related substances
phase equilibrium estimation, 3
steam distillation and edible, 13–14
Fatty acids, 1, 18
countercurrent extraction, 355
liquid–liquid extraction, 220–221
oil composition, 23
oil deodorization distillate byproducts, 291
vegetable oil stripping, 103
wax steam deacidification degradation, 31–32
Fatty alcohols, 31
Fatty systems
binary interaction parameters, model components, 238–239
liquid–liquid equilibrium diagram, 224–225
Fedor's groups, 22
Feed mass
anise seed steam distillation, 56
countercurrent extraction, 355
single stage solvent extraction, 146–147
Feedstocks
hot water/high pressure technology, 313
hydrothermal treatment of xylan-containing, 308–310
residue phenolics and terpenoids, successive extraction, 302
single-stage supercritical steam extraction fraction separation, 294–296
supercritical carbon dioxide extraction cosolvents, 305
supercritical carbon dioxide extraction processing, 289
Fennel extract, 4
Fenske rings, 292
Fermentation, 102, 311–312
Fermented must, 93
Fick's law, 142, 274, 408
Filtration
hot water extraction, 307
slurry extraction, 162, 419
Fixed beds
adsorption processes, 419–420
fruit juice debittering, 428
Fixed capital investment (FCI), 392
Fixed costs
condimentary plant extract manufacturing, 391
steam distillation, 51
Fixed oils, 1
deacidification, 10–11
solid–liquid extraction, 138
Flavanone glycosides, 427–428
Flavanones, 297–298
Flavonoids, 2, 139
cane sugar refining, 425
liquid stream extraction, 291
solubility and supercritical extraction, 302
supercritical fluid extraction cosolvent, 305
Flavonols, 2, 313
Flavor compounds, cashew, 118–119, 129
Flavorings
adsorption removal of color from, 426
alcohol/cachaça production, 102
essential oils, 14
Florentine, 12
Flowers, 15
Flow rates
batch/continuous extraction, 166–167
countercurrent separation, 356, 359
fixed bed adsorption, 419–420
optimal commercial processing, 383
solid matrix supercritical extraction, 381–382
steam distillation, 42, 67
steam distillation cost, 51, 53–54
steam distillation, cost of manufacture, 53–54
supercritical solvents, 281
Fluidized beds, 418
Food industry
activated carbon, 414
adsorption processing, 404–405, 429
antioxidant use, 38, 185
carbon dioxide solvent adoption, 374–375
citric acid use, 242
distillation, 75–76, 88, 91, 97
extractor systems, 145
pigment use, 189
polymer–polymer systems, 241
solid–liquid extraction, 138, 140, 141
solvent extraction, 150
Sorbex/Sarex processes, 421
starch-based sweeteners, 428
steam distillation, 13
thyme volatile oils, 45
volatile oils use, 14, 39
water purification, 423

Food processing, 1, 2
  adsorption applications, 422–432
  antioxidant use, 185
  steam distillation fundamentals, 9–17
Food storage, sorption isotherm, 412–414
4(5)-methylimidazole, 375
Fractionation, 271
  alcohol distillation profiles, 107–109
  antioxidant compound extraction, 186
  cachaça distillation portions, 103
  essential oil distillation, 75–76
  lignocellulosic material, antioxidant extraction, 306–312
  multistage crosscurrent extraction, 228–229
  orange volatile oil, 352–367
  process objectives, 35
  single supercritical fluid separation, 293–298
  solvent modifiers for antioxidants, 303–304
  Sovavá supercritical mass transfer model, 279–280
  stagewise supercritical fluid extraction, 299
  supercritical carbon dioxide extraction, 289
  supercritical chromatography, 293
Fragrance
  citrus oil liquid–liquid extraction, 243–244
  distillation, 76
Fragrance industry, 39
Free fatty acids (FFA), 224–225
  binary interaction parameters, 238
  edible fat/oil deacidification, 13–14
  glycerol hydrolysis, 18
  liquid–liquid extraction, 220–221, 239
  liquid stream extraction, 291–292
  oil purification, 245–247
  refining processes, 255
  steam distillation, 14
Freeze drying, 290–291
Freundlich isotherm, 410–411, 412, 432
Fructose, 421, 424–425, 428
Fruit, steam distillation, 15
Fruit juices
  adsorption purification, 427–428
  debittering, 428
  distillation, 75
  evaporation concentration process, 117
  supercritical freeze drying extraction, 291
Fugacity, supercritical equilibrium, 282–283
Fugacity coefficients
  distillation vapor–liquid phase, 87–88
  vapor–liquid equilibrium, 90
  volatile oil extraction phase equilibrium, 3
Fuller's earth, adsorbents, 415–416
Fungicide removal, 428
Furans, 309, 311
Furfural, 309, 380–382

G

γ-oryzanol, 249, 254–258
Gardenia fruit, 157
Gas adsorption, equilibrium, 409–414
Gas chromatography
  coffee aroma, 372, 376
  equilibrium ratio measurement, 342
Gas-like fluid densities, 354
Gas-liquid systems, packed columns, 223
Gas mixtures, pressure swing adsorption, 362, 364
Gasoline, 94
Generally recognized as safe (GRAS)
  bioactive compound solvent extraction, 185–196
  solvents, 2
  supercritical carbon dioxide extraction, 288
General manufacturing expenses, 392
Gentian, 177
Gibbs free energy
  liquid–liquid mass balance equations, 234
  volatile oil extraction phase equilibrium, 3
Ginger, 281, 290
  cost of manufacturing extracts, 398–400
  microwave-assisted extraction, 169
Gingko, 328
Ginseng, 176–177
Glandular trichomes, 15
Glucose, 241, 421, 424–425, 426, 428
Glycerin, adsorption odor/color removal, 427
Glycerin esters, 1, 13–14, 18
Glycerol triesters, 220
Goodloe knitted-mesh packing, 292
Good manufacturing processes (GMP), 141
Goto model, 278–279
Grape
  pomace autohydrolysis liquors, 311
  seed phenolic compound extraction, 196
  seeds, 290
  skin anthocyanin solvent extraction, 190–191
Green solvents, 2
Green tea leaves, 186
Grinding, 290
Group contribution models, liquid–liquid mass balance, 236–239

H

Hangover syndrome, 102
Health products, 1
Heat
  adsorption processes, 382, 404, 406–407, 410, 412
  batch deodorization, 20
  cashew shell nut liquor processing, 334–335, 344

cell processes, 329–330
conduction, 277
cost estimation, 49
direct application, 103
distillation mass balance equations, 82–86, 97
energy-to-mass calculation, 204
evaporation, 205
fusion molar, 151
hydrodiffusion, 40
loss, 109
microwave-assisted extraction, 151–152
percolation extraction, 163–164
phenolic compound extraction, 187
sound and ultrasound production, 155
source intensity, 104
terpene processing, 243
transport, 409
vaporization, 14, 21, 204–205
wine distillation, 103
Heat duty, 13
Heat exchangers, 31, 53, 76, 272, 337, 389, 390, 391, 425
Heat transfer
adsorption, 407
liquid distillation separation, 76
microwave-assisted extraction, 154
ultrasound-assisted extraction, 156–158
vaporization rate, 83
Height equivalent to theoretical stage (HETS) models, 356–358
Hemicellulose, 306–309
Hemicellulose hydrolysis, 311
Henry's law, 285, 409–410
Herbal plants, pretreatment, 289–291
Hexane, 141, 180, 244, 375
Hibberts ketones, 312
High-pressure extraction, 312–315
antioxidants, 306–315
phase equilibrium, 3
Hops, 186
Hot water extraction (HWE)
antioxidants, 187
herbal antioxidants, 186
stages and experimental techniques, 314
supercritical fluid, 271
vegetal biomass technologies for lignocellulosic materials, 312–315
Humidity
microwave-assisted extraction, 169
solvent extraction and material, 142
sorption equilibrium, 412
Hydroalcoholic solvents, 189, 432
Hydrodiffusion, steam distillation, 39, 40
Hydrodistillation. *See* Water distillation
Hydrolysates, 428
Hydrolysis, 18, 291
Hydrolytic degradation, hemicellulose and antioxidant solubilization, 306
Hydrolyzed vegetal proteins, 426
Hydroquinone, 328
Hydrosol, 36, 37, 38, 39, 42, 44, 52, 53
Hydrothermal liquors, hemicellulose-derived oligosaccharides, 309
Hydrothermal treatment
lignocellulosic material antioxidant extraction, 306–312
zeolites, 415
Hydrotropic solvents, 160–161
Hydroxymethylfurfural, 428
Hysteresis loop, 413–414

I

Ideal behavior
activity coefficients, 89
gas, 16
gas vapor phase, 87
liquid phase, 90
mixtures, 87, 89
vapor–liquid equilibrium, 89–90
Ideal gas, 21, 89, 283
Ideal heat duty, 13
Ideal stages
cooling column, 83
distillation column, 80
experimental design, 119
reboiler/condenser, 86
Ideal temperature, 40
Immersion extractors, 166–167
Impellers, 222
Indirect sonication, 176
Industrial installations, capital costs, 388–390
Industrial location, 388
Industrial production
batch distillation process scale, 78
distillation degassing, 114
ultrasound-assisted extraction, 183
Industrial waste, supercritical extraction processing, 289
Inert matrix, 11–12
Inert solids
crosscurrent extraction, 147–148
single stage solvent extraction, 146
Inflation rate, 389
Initial conditions, ultrasound mass transfer, 157
Initialization procedure, 124
Instantaneous concentrations, 120, 124–126
Interaction parameters, UNIFAC, 92
Interfacial mass flux, 274
Interfacial mass transfer models, 278–281
Internal transport, adsorption processes, 408–409
Investment costs, 51

Ion-exchange resins, 415–416
Ion-exchange separation, 426
Ionic migration, microwave-assisted extraction, 151
Isoflavones, 170
Isomer formation, steam deacidification, 26–30
Isomerization, 19
Isopropanol, 194
Isothermal systems, supercritical equilibrium, 282
Isotherms
  Freundlich, 410–411, 412, 432
  Henry, 409–410
  Langmuir, 361, 410, 411, 432
  sorption, 412–414

J

Joback's technique, 22
Jojoba oil, 246
Juices
  adsorption purification, 427–428
  aroma evaporation, 101
  aroma/flavor distillation, 117–118
  batch distillation concentration/purification, 118–129
  boiling processes, 139
  concentration process, 117
  distillation, 75
  fixed-bed debittering, 428
  phase equilibrium equations, 88
  supercritical freeze drying extraction, 291

K

Ketones, 14, 312
Khüni columns, 223
Kinetic assays, batch/continuous extraction, 166
Kinetics
  Goto supercritical mass transfer, 278
  microwave-assisted extraction, 154
  overall extraction curve modeling, 277
  solid–liquid low pressure, 142–144
  supercritical extraction processing, 273
  supercritical fluid extraction, 273–281
  ultrasound-assisted extraction, 156–158
Knudson diffusion, 408

L

Labor costs, 390
  solvent extraction, 205–206
  steam distillation, 51–52
Lactose, 429
Lang factors, 50, 202
Langmuir equation, 158, 361, 410–411
Langmuir isotherm, 361, 410, 411, 432
Laurel essential oil, 169
Lavender, 39
Leaching, 140, 167
  ultrasound-assisted extraction, 171, 175–176
  ultrasound extraction, 157
Leaves, 15, 139, 290
  green tea, 186
  mate, 185–186
  olive, 182, 292, 430
Lemon
  oil, 301, 359
  peel, 360
  verbena, 290
Lever-arm rule, liquid-liquid mass balance equations, 225–227
Lignin, 306–312
Lignocellulosic materials (LCM), high-pressure water extraction, 306–312
Limonene
  black pepper, 44
  desorption, 360
  distillation, 101
  essential oil deterpenation/solvents, 244–245
  phase equilibria, 352–354
  solvent-to-feed ratio countercurrent extraction, 357–359
  ultrasound-assisted extraction, 180, 181
Linalool
  countercurrent extraction, 357–358
  desorption, 360
  essential oil deterpenation/solvents, 244–245
  phase equilibria, 352–354
Linear isotherm of Henry, 409–410
Linoleic acid, 26, 28, 249
Liposomal membranes, 330
Liquid adsorption, 411
Liquid carbon dioxide extraction, 373
Liquid chromatographic-electrospray mass spectrometric quantification, 375
Liquid chromatography, 342
Liquid film, 81–82
Liquid-like fluid density, 354
Liquid–liquid equilibrium, 220
  fatty and short-chain alcohol systems, 224–225
  vegetable oil deacidification, 247–249
Liquid–liquid extraction, 2
  alkali refining, 246
  antioxidant, 293
  equipment, 221–223
  optimization and aroma, 375
  ultrasound-assisted systems, 183
  vegetable oil processing, 219–221
  vegetable oil processing literature, 241–247

Liquid–liquid extraction columns, 239–241
Liquid–liquid–vapor equilibrium, 327, 340–347
Liquid materials
steam distillation, 14
supercritical fluid extraction, 354–366
Liquid mixtures
distillation separation, 76
vapor–liquid equilibrium data, 91
Liquid phase
fugacity, 87
separation process, 339
supercritical extraction sampling, 344
Liquid streams, antioxidant supercritical fluid extraction, 291–292
Liquid–vapor interface, 15–17
Lixiviation, 140
Low-pressure extraction, 139–140, 140
microwave-assisted, 151, 168–171
solid-liquid, 140–151, 158–167
ultrasound-assisted, 171–185
LRPEK curve, 224–225
Lycopene, 191–192, 297

M

Macela, 7
costs of utilities (C_{UT}), 205
raw materials cost estimation, 202–203
solvent extraction cost estimation, 202
solvent extraction economics, 203
Maceration, 156, 167
Maltodextrins
polymer + polymer systems, 241–242
purification, 428
Maltose, 428
Mango, 290, 328
Marigold, 177
Market prices, volatile oil cost, 70–72
Marshal & Swift Equipment Cost Index, 201
Martínez model, 280
Mass balance
countercurrent extraction, 149–150
crosscurrent extraction, 147–148
liquid–liquid extraction column simulation, 240
single stage solvent extraction, 146
Mass balance equations
distillation processes and heat, 82–86
liquid–liquid extraction, 225–234
single stage extraction, 144
Mass/energy balances, juice aroma/flavor distillation capture, 117–118
Mass transfer, 271
adsorption, 405
liquid–liquid extraction, 221–222, 225–234
solid–liquid low pressure extraction, 142–144
solid matrices operating pressure, 379
steam distillation, 16–17
stepwise distillation, 80
supercritical fluid extraction, 273–281
supercritical temperature/pressure data correlation, 346–347
ultrasound-assisted devices, 177–181
ultrasound-assisted extraction, 156–158
Mate leaves, 185–186
Materials
selection for microwave-assisted extraction, 151
solid–liquid extraction preparation, 141
Measurement
adsorbate-adsorbent affinity, 404, 410–411
cashew nut shell liquid separation process, 339–346
coffee aroma analysis, 372
gas chromatography, 342
internal reflux ratio, 356
solubility supercritical equilibrium, 281–282
surface diffusion, 408
temperature, 343
Measuring devices
Cailletet apparatus, 339–340
gas chromatograph, 376
gas flow meter, 337
recirculating static apparatus, 342–344
thermocouples, 53
Medicinal plants, 4, 7, 138
antioxidant extraction, 297
pretreatment, 290
steam distillation, 11
successive supercritical carbon dioxide extraction, 301
supercritical carbon dioxide extraction processing, 289
volatile steam distillation, 38–43
Melon fruit spirits, 99
Methanol, 94, 99, 141
binodal curves of anhydrous, 247–248
cachaça production, 102
continuous tray column distillation, 113
essential oil deterpenation/solvent selection, 245
Methanol extraction, 160–161
Methyl chloride, 373
Microwave-assisted extraction (MAE), 138, 140, 168–171
solid–liquid low pressure, 151–154
steam distillation, 43
Microwave extraction, 15
Microwave ovens, 168
Milling processes, 39
Mint, 177
Mitochondria, 327, 329–331
Mixtures

aromas, 97–101
azeotropic, 246
binary, 90–91, 340, 342
boiling point maintenance, 83
concentration and adsorbent use, 405
distillation chemical reactions, 83
gas, 362, 364
ideal behavior, 87, 89
liquid, 76, 91
optimization, 292
UNIFAC method and complex, 93
volatility and evaluation, 88–90
water–ethanol, 194
Moisture content, sorption equilibrium, 412–413
Molecular motion
adsorption mechanisms, 408
microwave-assisted extraction, 151–152
Molecular structure, UNIFAC model, 91–92
Monocots, lignified tissues, 308
Moving bed adsorption processes, 420–422
Multistage crosscurrent extraction, 293
continuous, 232–234
liquid–liquid mass balance equations, 228–232
Murphree efficiency, 17, 80, 86
Mushrooms
juice boiling processes, 139
shiitake, 160, 290

N

Natural products
costs and duplicates, 39
solvent characteristics, 158
Neutral oil
chemical refining, 221
free fatty acid removal, oil purification, 245
steam deacidification/deodorization loss, 23–26
Neutral spirits, 100–101
Newton–Raphson method, 240–241
Nonionic polymeric adsorbents, 415–416
Nonlinear programming model, countercurrent extraction, 359
Nonrandom two-liquid (NRTL) model, 90–91
binary interaction parameters, model components, 238–239
fermented must phase equilibrium, 93
liquid–liquid mass balance equations, 234–236
Nonsaccharide byproducts, 309
Nusselt number (Nu), 407
Nutraceuticals
edible oil refining, 256–258
liquid–liquid extraction, 246–247
palm oil refining, 253–255
steam deacidification/deodorization, 19
ultrasound-assisted extraction, 314
Nutrition categories, edible oils, 252

O

Oak wood, 99
Odor adsorption, 427
Oil deodorized distillates (ODD), 291
Oils. *See also* Edible oils; Volatile oils
acidity, 18
steam deodorization and deacidification, 22–23
steam distillation, 13–17
steam distillation release, 14–15
supercritical extraction from coffee matrices, 375
Oilseeds, liquid–liquid extraction refining, 245
Oleic acid, 247–248, 249
Oleoresin fractionating, 2, 4, 7
Oleuropein, 182–183, 430
Olfactometric data, cashew fruit juice, 118
Oligosaccharides, 309
Olive
biophenols (OBPs), ultrasound-assisted extraction, 182–183
leaves, 182, 292, 430
steam explosion reaction byproducts, 312
Olive oil
deacidification, 246
percolation extraction, 164–165
ultrasound-assisted extraction, 176, 177
1,1-dichloroethane, 141
1,1,1-tricholoroethane, 141
1-propanol, 141
1,2-dichloroethane, 141
1,2,3-trihydroxypropane, 220
Onions, 139–140
Operating conditions
adsorption on solid matrices, 379
cashew volatiles distillation, 126–127
clove bud extract manufacturing, 397–398
ethyl acetate-soluble phenolics antioxidant properties, 310
process/solvent cycle nonlinear modeling, 359
solid–liquid low pressure, 144
solid matrix supercritical carbon dioxide extraction, 377–379
steam distillation, cost of manufacture, 53, 54
steam distillation, volatile oil, 36
stepwise distillation, 81
supercritical fluid extraction cosolvent, 305
turmeric, steam distillation, 42
yield/volatile oil composition steam distillation, 43
Operational labor costs (C_{OL}), 51–52, 392

Operational methods
hot water emerging technologies, 312–315
supercritical fluid compound separation, 297
supercritical fluid extraction, 291, 293
Operational variables
antioxidant supercritical extraction, 292–293
solid–liquid solvent extraction, 159–167
Optimization
ginger essential oil manufacturing, 400
isolate properties, 288
manufacturing costs/market price, 70
mixture and temperature profiles, 292
practical size and extraction kinetics, 290
solvent flow rate, 28[illegible]
solvent-to-raw material ratio, 292
supercritical carbon dioxide extraction, 375
supercritical fluid extraction cosolvent selection, 305
Orange
juice aroma evaporation, 101
oil fractions, 2
oil supercritical fluid fractionation, 352–367
peel enzymatic extraction process, 191–192
volatile oil, 271
Oregano, 139, 163, 186
Organic products, 12
Organic solvents, 2, 4, 7
Oryzanol, 249, 254–258
overall extraction curve (OEC), 276–281
Oxygenated compounds, 244
concentration and solvent-to-flow ratio, 358
orange/lemon/bergamot oils, 359
orange oil fractionation, 352
silica gel adsorption/desorption, 361–362

P

Packed columns
citric acid decolorization, 428–429
heat sensitive purification, 82
liquid extraction selectivity, 292
liquid–liquid extraction, 223
overall extraction curve modeling, 276–277
raw material pretreatment, 290
Palm oil
carotenoid concentrations, 256
liquid stream extraction, 292
refining processes, 246
tocol composition, 252
Paprika, 169
Parametric cost factors, 49
Particles
adsorption mechanisms, 408–409
mass transfers, 157
phenolic compound extraction, 196
Particle size
antioxidant GRAS bioactive compound extraction, 189
phenolic compound GRAS solvent extraction, 196
Partition coefficients, edible oil liquid-liquid extraction, 256–257
Patents, steam distillation, 37–38
Patulin, 428
Pectinase, 191–192
Peng–Robinson equations of state, 3, 283–285, 352, 356
Pentane ether, 373
Peppers, 171
Percolation extraction, 159–162
coffee volatile compounds, 373
fixed bed adsorption, 419–420
olive oil, 165
temperature/pressure conditions, 163
water treatment, 424
Permissible daily exposures, 141
Pesticide removal, 428
Phase behavior
cashew nut shell liquid separation process fractionation, 327, 340–347
extraction columns, 281
supercritical extraction, 281
supercritical extraction data correlation, 346–347
supercritical fluids, 283–285
Phase equilibrium, 271. *See also* Mass balance equations
aqueous two-phase systems, 241–242
Cailletet apparatus measurement, 339–340
citrus oil components, 352–354
liquid–liquid mass balance equations, 234–236
multistage crosscurrent extraction, 228, 230–231, 233
solid–liquid low pressure extraction, 150–151
UNIFAC interaction parameters, 92–93
volatile oils compounds, 3–4
Phenolic antioxidants
boiling processes, 139
ethyl acetate-soluble, 310
extraction, 186
single-stage supercritical steam extraction fraction separation, 294–295
stagewise supercritical fluid stagewise extraction, 299
steam explosion, 311–312
successive extraction, 300, 302
Phenolic compounds, 430
autohydrolysis liquor antioxidants, 309
cane sugar refining, 425
extraction of high-quality, 188

GRAS bioactive compound solvent extraction, 193–196
reduction, 427
solvent-to-feed ratio, 189
Phenolic lipids, 328
anacardic acids, 327–328
carbon dioxide phase behavior, 338–339
cashews, 328–329
Phenols
olive oil percolation extraction, 165
Soxhlet percolation extraction process, 163
ultrasound-assisted extraction, 156
pH gradients, mitochondria, 329–330
Phosphoric acid, 242
pH yield effect, solvent extraction, 196
Phytochemical adsorption separation, 430
Piezoelectric materials, 178
Pigments, 18
adsorption removal, 422
condimentary plants, 139
extraction and temperature, 192
GRAS solvent bioactive compound solvent extraction, 189
Pine wood
ethyl acetate extraction, 309–310
lignocellulosic material autohydrolysis liquors, 311
steam explosion phenolics, 312
Piperine, 160
Plant extracts
bioactive compounds, 2
phenolic compound extraction, 193–194
Plant materials complexity, 139
Plant matrices
antioxidant compound extraction, 187
large molecule substances extraction, 4, 7
Plant metabolism, 2
Plant oil bags/cells, 39
Poison ivy, 328
Poison sumac, 328
Polyethylene glycol (PEG), aqueous two-phase systems, 241–242
Polyglycerol esters, 427
Polymeric resins, 428
Polymer + polymer systems, polyethylene glycol/dextran, 241
Polyphenols, 2, 4, 139
Polystyrene divinylbenzene, 428
Polyunsaturated fatty acids (PUFAs), 19, 249
Polyvinylpolypyrrolidone, 426
Poplar, 312
Potato extracts, 190
Power increments, microwave-assisted extraction, 169–170
Poynting factor, 88
Prandlt number (Pr), 407
Prebiotic food ingredients, 309
Preservative properties, 38
Pressure
adsorbent regeneration, 418
anise seed steam distillation, 42–43
essential oil glandular trichomes, 338
essential oil mutual solubility, 353
ethyl acetate and furfural solid matrix supercritical extraction, 380
microwave-assisted extraction, 153, 168–169
optimal commercial processing conditions, 382
percolation extraction, 163
supercritical equilibrium modeling, 282
supercritical extraction data correlation, 346–347
supercritical fluid densities, 354
supercritical fluid properties, 292–293
supercritical steam stagewise extraction, 298–299
vapor–liquid equilibria, 21–22
Pressure-swing, supercritical extraction steps, 337–338
Pressure-swing adsorption (PSA), 352, 362–366, 419, 420
Pressurized fluid extraction, 269–287
Pressurized solvent extraction, vegetal biomass technologies for lignocellulosic materials, 312
Pretreatment, solid raw materials, 289–291
Proanthocyanidins, 314, 315
Probe systems, ultrasound-assisted extraction, 171, 175–176
Process capacity, steam distillation, 13
Process design optimization, volatile oil extraction phase equilibrium, 3
Process efficiency
distillation, 86
steam distillation mass transfer, 16–17
Process flow
autohydrolysis, 298
hot water extraction, 307
single stage supercritical fluid extraction, 297
steam distillation, 12–13
supercritical extraction–adsorption pilot plant, 378
Processing plants, cost estimation, 50
Processing techniques
antioxidant supercritical fluid extraction, 293–302
cashews, 332–334
coffee aroma concentration, 372–374
Processing technology, supercritical carbon dioxide, 374–383
Process parameters, 139
microwave-assisted extraction, 152–153
solid liquid extraction, 140–142
ultrasound-assisted extraction, 184–185

Process scheme, countercurrent supercritical fluid extraction, 354
Product flows, heat and mass balance equations, 85–86
Production units, capital costs, 388–390
Propane, 288
Propolis tincture, 291
Propyl acetate, 141
Proteins
commercial production, 241
decolorization of hydrolyzed vegetal, 426
lignocellulosic hydrothermal treatment, 308
liquid–liquid vegetable oil extraction, 241–242
Pulp manufacture, 311
Pulsed columns, 223
Purge
adsorbent regeneration, 418
thermal-swing adsorption (TSA), 420
Purification, 2
adsorption and starch-based sweeteners, 428
adsorption for drinking water, 423–424
adsorption for fruit juices, 427–428
antioxidant extracts, 297
application dependency, 2
bioactive compounds, 1–7
cane sugar refining, 425
carbon dioxide by adsorption, 426–427
cashew aroma/flavoring distillate, 120
cashew volatile batch distillation, 121–122
cashew volatiles distillation, 126–128
enzyme/protein production, 241
hemicellulose-derived oligosaccharides, 309
successive supercritical carbon dioxide extraction, 301
supercritical carbon dioxide extraction, 289
vegetable oil deacidification, 18–19
zeolites, 415
Pyrolysis, 40

Q

Quality
alcohol distillation cuts, 107, 109
alcoholic beverages, 426
congeners and alcohol, 88–89
continuous tray column distillation, 113–115
GRAS solvents and bioactive compounds, 185–189
spirit distillation, 100
steam distillation, 36, 40
thyme steam distillation, 69
vegetable oil steam deacidification, 19
volatile compound distillation, 129
Quercetin, 7, 168
Quercetin glycosides, 314, 315

R

Radical scavenging, antioxidant compound extraction, 187
Raoult's law, 94
Rasching rings, 292
Raw material costs (C_{RM}), 390, 394–395
capital costs, 388
estimation cost of manufacturing, 392
estimation for solvent extraction, 202–203
rosemary, 61–62
scaling-up estimation, 51
steam distillation, volatile oil, 70
Raw materials
antioxidant conventional/supercritical fluid extraction, 299
antioxidant supercritical fluid extraction, 289–292
clove buds, 396
lignocellulosic material autohydrolysis liquors, 311
pretreatment in antioxidant supercritical fluid extraction, 289–291
solid–liquid extraction variables, 159–160
steam distillation, 11–12
steam distillation oil release, 14–15
variability and industrialization, 139
Reactive batch deodorizers, 18–32
Reboilers
distillation process efficiencies, 86
mass and enthalpy balance, and equilibrium equations, 85
Recirculating static apparatus, 342–343
Rectification column systems, 99
Recycling, solvent costs, 202–203
Red grape pomace, 310
Refined oils, 221
Refining processes
adsorption for edible oils, 422–423
adsorption in cane sugar, 425–426
edible oil nutritive value retention, 253–254
Reflux, 109, 355
aroma/flavor distillation capture, 117
batch distillation flow, 124
continuous cachaça tray column distillation, 110–111
distillation process, 80
drums, 84, 85
ratio, 126, 127
Residual activity coefficients, 92
Residues, 141
antioxidant supercritical fluid, successive solvent processes, 299–301

phenolics and terpenoids, successive extraction, 302
single stage solvent extraction stream, 145
supercritical fluid extraction, processing of solid, 301–302
ultrasound-assisted dry extraction, 177
Resonant tube, 178
Resorcinol, 328
Resveratrol, 186
Retention index
crosscurrent extraction, 148
single stage solvent extraction, 146
single stage solvent extraction processes, 144–145
Retinal, 253
Reynolds numbers, 407
Rice bran oils (RBO)
bioactive component, 254–256
tocol composition, 252
Ripeness, target compound, 139
Roots, 15
Roselle petal extracts, 190
Rosemarinic acid, 139, 184
Rosemary, 11, 53, 139, 290
antioxidant compound extracts, 186
hydrodistillation, 42
percolation extraction process, 161–162
ultrasound-assisted extraction, 177–178
volatile oil market pricing/cost of manufacturing, 72
volatile oil steam distillation, cost of manufacturing, 60–63
volatile steam distillation, 44
Rotating disk contractor (RDC) columns, 223

S

Sabine, 39–40
Saccharides, 429
Sage, 139
antioxidant compound extracts, 186
costs of utilities (C_{UT}), 204–205
raw material cost estimation, 202–203
solvent extraction process, economic evaluation, 200–201
solvent-to-feed ratio, 188–189
Soxhlet percolation extraction process, 163
ultrasound-assisted extraction, 177, 181, 183–184
Saponification, 221
Saponified/esterified soy deodorized distillate (SODD), 291
Saponins, 176–177
Scale of operations
cost estimation, 49
cost of manufacturing, 394
overall extraction curve modeling, 276
Sovavá supercritical mass transfer model, 280
steam distillation costs, 51, 53–54
Schmidt numbers (Sc), 407
Screw extractors, 165–167
Seasonings, 1
Seeds, 14–17
Selectivity
adsorbents, 414
adsorption separation applications, 429
citrus oil countercurrent extraction, 355–356
countercurrent extraction, 357
packing material separation, 292
solvent, 184, 245
supercritical fluid processes, 272
Separation
adsorption bulk applications, 429–432
adsorption technique, 404
batch distillation, 118
cashew nut shell liquid separation process fractionation, 339–346
cashew supercritical CO_2 extraction, 334–338
equilibrium of adsorption, 409
liquid–liquid extraction, 222
liquid mixture distillation, 76
mixture volatility values, 89
phase equilibrium, 360
phenolic compound extraction, 193
pressure swing adsorption process, 363–366
ratios estimation in vapor–liquid equilibrium, 345
Sarex moving bed adsorption, 421
scheme cashew compounds, 347
solid–liquid extraction process variables, 159–160
supercritical cashew processing scheme, 347
zeolites, 415
Separation tanks, performance, 197–198
Separation vessels, supercritical fluid extraction, 293–298
Shiitake mushrooms, 160, 290
Short-chain alcohols, 224–225
Side-stream cuts, 359
Sieve-plate columns, 222
Silica gel, 352
adsorbents, 415
limonene–linalool desorption, 360–361
Silicate adsorbents, 415–416
Silk fibroin, 430
Simulation
alembic distillation, 82–83
aroma and spirit distillation, 97–101
batch alembic distillation, 104
cashew aroma/flavoring fractionation/capture, 119–129

continuous cachaça tray column distillation, 110
continuous tray column distillation, 114
distillation processes, 100–101
distillation vapor-liquid phase equilibrium, 87–88
liquid–liquid vegetable oil extraction column, 239–241
solvent extraction, 197–200
steam deacidification, 23
volatile compound distillation, 129
Single-stage extraction
antioxidant supercritical fluid, 293–298
solid–liquid low pressure, 144–147
Single-state equilibrium extraction, liquid–liquid mass balance equations, 227–228
Slurry extraction
coffee aroma compounds, 374
solid–liquid equipment and process, 159, 162–163
Soave–Redilich–Kwong equations of state, 3, 283–285, 356
Sodium butyl monoglycol sulfate, 160–161
Sodium hypochlorite, 424
Sodium lauryl sulfate, 160–161
Software applications, 93, 100–101, 110, 197–200, 394
Solid adsorbates, 404
Solid feedstocks, stagewise supercritical fluid extraction, 299
Solid–fluid extraction, 272
Solid–liquid extraction, 158–167
raw material pretreatment, 289–290
ultrasound-assisted systems, 183
Solid–liquid low pressure extraction, 140–151
Solid matrices
antioxidant extraction processes, 293
extraction equipment, 162
microwave-assisted extraction, 153
solvent diffusion coefficient, 142–143
steam distillation, 35
supercritical carbon dioxide extraction process, 377–379
supercritical fluid extraction, 271–287
Solid-phase extraction, supercritical, 375
Solid preparation, 141
Solid raw materials
pretreatment for supercritical extraction, 289–291
steam distillation oil release, 15
Solid residues, supercritical fluid extraction processing, 301–302
Solid-to-liquid ratio, percolation extraction, 159
Solid-to-solid ratio, phenolic compound extraction, 193
Solid-to-solvent ratio
percolation extraction, 164
ultrasound-assisted extraction, 181
vegetable material extraction process, 198
Solubility
antioxidant supercritical extraction, 302
equilibrium conditions and mutual, 353–354
limonene essential oil, 352
supercritical equilibrium measurement, 282
Solute density, 275
Solute diffusion coefficient, 142–143
Solute solubility, 150–151
Solvent extraction, 171–185, 199
coffee aromas, 375–376
coffee volatile compounds, 373
deterpenation, 244
economics, 197–206
GRAS solvent bioactive compound, 185–196
hemicellulose-derived oligosaccharides, 309
liquid stream supercritical carbon dioxide refining, 289
mass transfer, 142–144
microwave-assisted, 152, 168–171
sage process economic evaluation, 200–201
single stage processes, 144–147
solid–liquid, 137–140, 158–167
thermodynamic phase equilibrium, 150–151
ultrasound-assisted, 156, 171–185
volatile oils, 2
Solvent feed, percolation extraction, 163–164
Solvent-free microwave-assisted extraction, 171
Solvent movement, interfacial, 274
Solvent properties, 141
Solvent recovery, 414
Solvent recycling, 165
Solvents
antioxidant GRAS solvent extraction, 185–187
carbon dioxide, 272
cost estimation for vegetable extraction processes, 202–203
high-temperature microwave-assisted extraction, 168
methanol solid–liquid extraction, 160
moving bed adsorption operations, 421
phenolic compound GRAS solvent extraction, 194–195
pigment GRAS solvent extraction, 189–192
power in supercritical extraction, 292–293
regulatory classification, 141
solid–fluid extraction, 272
solid–liquid extraction, 142, 158–159
solid residue supercritical fluid extraction processing, 301–302
supercritical carbon dioxide extraction modifying, 303–306
supercritical concentration, 273

supercritical extraction, environmental friendliness, 391
supercritical fluid successive extraction, 299–301
supercritical thermodynamic equilibrium, 281–285
ultrasound-assisted extraction, 183
volatile oil extraction phase equilibrium, 3–4
water in liquid–liquid extraction vegetable oil deacidification, 249
Solvent selection
criteria for low-pressure processes, 140–142
liquid–liquid citrus oil extraction processes, 243–244
microwave-assisted extraction, 152, 153
natural product extraction, 157
Solvent selectivity
alcohol chain length, 245
ultrasound-assisted extraction process, 184
Solvent-to-feed (S/F) ratio
anise seed steam distillation, 55, 56–67
antioxidant GRAS solvent extraction, 188–189
black pepper steam distillation, 63–65
chamomile steam distillation, 59–60
citrus oil countercurrent separation selectivity, 356
height equivalent to theoretical stage (HETS) models, 356–357
liquid extraction efficiency, 292
orange oil countercurrent processing, 357–359
phenolic compound GRAS solvent extraction, 195
pigment GRAS solvent extraction, 192
rosemary costs and steam distillation, 62–63
steam distillation cost estimation scaling-up, 51
supercritical countercurrent extraction, 297–298
thyme steam distillation, 65
Solvent-to-raw material ratio, 161
Solvent usage, microwave-assisted extraction, 151
Solvent velocity, 273–274
Sonication, ultrasound-assisted extraction, 176
Sonochemistry, 154, 155
Sonotubes, 178, 179
Sorbex process, 421
Sorption isotherm, 412–414
Sovavá model, 279–280
Soxhlet extraction, 163, 167
antioxidant compounds, 186–187
diterpene oil extraction, 375
Soy
deodorized distillate, 291
isoflavone microwave-assisted extraction, 170
oil deacidification, 246
Spice plants
antioxidant extraction, 139
polyphenols, 4
Soxhlet percolation extraction process, 163
Spirits
characteristics, 98–99
distillation, 75–78, 97
ethanol concentration, 76, 93–94
mixture distillation simulation, 97–101
vapor-phase equilibrium, 88–97
Spray columns, 222
Squalene, 291, 355
Stage efficiency parameters, 16–17
Stagewise extraction, 298
liquid–liquid, 222, 239–241
supercritical steam, 298–299
Stainless steel alembics, 99
Standard-state fugacity, 87–88
Starch-based sweeteners, purification, 428
Steady-state film theory, external transport adsorption, 406–407
Steam
batch deodorization, 20
continuous tray column distillation, 109–110
superheated temperature, 42
Steam deacidification
alembic batch distillation, 103
cis–trans isomer formation, 26–30
neutral oil loss, 23–26
oil composition estimation, 22–23
simulation, 23
Steam deodorization, 23–26
Steam distillation (SD), 1, 9–11, 40. *See also* Distillation
coffee aroma compounds, 374
condimentary plant volatile oil, cost of manufacturing, 52–70
costs, 51–52
costs of volatile oil manufacturing, 47–72
edible oil tocopherol content, 253
equipment, 53
fundamentals, 11–17
vegetable oil stripping, 18–32
volatile oil extraction, 2
volatile oils, 35–45
waxes degradation, 30–32
Steam explosion, 311–312
Steam mass costs, 51
Stepwise mode
distillation process, 80
single stage supercritical fluid extraction, 297
Sterols, 18
Stills. *See* Distillers
Stochastic cost estimation, 48, 49
Strategic decisions, cost estimations, 50
Stripping

batch deodorization, 20
coffee volatile compounds, 373–374
continuous tray column distillation, 109–110, 114–115
countercurrent supercritical fluid extraction, 354–355, 359
distillation processes, 79
edible tocopherol content, 253
steam deacidification/deodorization, 19
vapor–liquid equilibria, 21–22
vegetable oil deacidification, 10
vegetable oils fatty acids, 103
Subcritical water extraction, 162, 313
Successive extraction
phenolics and terpenoids from residues, 302
solvents and antioxidant supercritical fluid extract/residue, 299–301
Sucrose, 424–425
Sugar cane
cachaça, 2, 101–102
juice, 88
spirits, 76, 103
steam explosion, 311
Sugar refining, 425–426
Sugars, 309
Sulfate, 99
Sulfur compounds, 99
Sulfur olive oil miscella, 246
Sulzer rings, 292
Summer savory, 163
Supercritical CO_2 extraction, 15
cashew separation, 327, 334–338
coffee aroma recovery, 374–383
orange volatile oil aroma, 352–366
pressure swing adsorption, 364
Supercritical equilibrium modeling, 282
Supercritical fluid extraction (SFE)
adsorption and phase equilibrium separation, 360–362
adsorption separation applications, 429
antioxidants, 288–315
condimentary plant extracts, cost of manufacturing, 388–400
densities and separation rates, 354
deterpenation, 244
liquid material processing, 354–366
orange volatile oil fractions, 352–367
phase equilibrium separation, 360
solid matrices, 269–287
Supercritical fluids, 272
phase and fugacity, 283
solvating power, 292–293
Supercritical freeze drying extraction, 291
Supercritical technology, economics, 197–198
Superheated steam, 42
Superheated water extraction, 161–162
Superior alcohols, 102, 106–109
Surfactants, 160–161
Sweeteners, purification, 428
Sweet grass, 186–187
Sweet potatoes, 191–192
Synthetic duplicate oils, 39
Syrups, 424–425

T

Tanks
adsorption in agitated, 404
solid–liquid extraction, 159
solved extraction agitated, 197, 198, 200
supercritical extraction separation, 197–198
Tannins, 426
Target compounds, 39, 139–140
Taste, adsorption correction in alcoholic beverages, 426
Tea
adsorption for decaffeination, 427
leaf antioxidant compound extracts, 186
tree steam distillation, 39, 40
Technological know-how, steam distillation, 13
Temperature
adsorption, 405
antioxidant GRAS solvent extraction, 187–188
Arrhenius function, 409–410
batch cachaça distillation, 105
black pepper steam distillation, 65
cashew volatile batch distillation, 123–124
continuous feed liquid extraction, 292
deacidification of vegetable oils, 19
essential steam distillation, 36
ethyl acetate and furfural solid matrix supercritical extraction, 380–381
high-quality extracts, 188
liquid–liquid extraction vegetable oil deacidification, 247
low pressure solvent selection, 141
microwave-assisted extraction, 153, 168–170, 171
optimal commercial processing conditions, 382–83
percolation extraction, 163
phenolic compound GRAS solvent extraction, 195–196
pigment GRAS solvent extraction, 192–193
solute solubility, 151
sonochemical effects, 155
steam distillation, 38, 40
supercritical equilibrium modeling, 282
supercritical extraction data correlation, 346–347
supercritical fluid densities, 354
supercritical fluid solvating properties, 292–293

terpenoid boiling point, 4
ultrasound-assisted extraction, 156, 184
vegetal biomass technologies for lignocellulosic materials, 313
Temperature-sensitive materials, steam distillation, 11
Terpenes, 244
citrus oil countercurrent continuous extraction, 356
countercurrent extraction, 355
mutual solubility conditions, 354
orange juice aroma evaporation, 101
orange/lemon/bergamot oils, 359
orange volatile oil fractionation, 352
Terpenoids, 11
single-stage supercritical steam extraction fraction separation, 294–295
stagewise supercritical fluid extraction, 299
steam distillation, 40
successive extraction, 300, 302
supercritical fluid extraction cosolvent, 305
thermophysical properties, 4
volatile oils, 2–3
Thermal conductivity, 407
Thermal degradation
antioxidant extraction yield/selectivity, 313
cashew processing, 334–335
steam distillation, 13
vegetable oil steam deacidification, 19
Thermal reactivation, adsorbents, 417–418
Thermal-swing adsorption (TSA), 420
Thermodynamic equilibrium, distillation vapor-liquid phase, 87
Thermodynamics
essential oil deterpenation/solvent selection, 245
liquid–liquid mass balance equations, 234–236
solid–liquid low pressure extraction, 150–151
supercritical fluid extraction equilibrium, 281–285
utilities cost estimates for solvent extraction, 204
Thermophysical properties
phase equilibrium, 3–4
volatile oil components, 4
volatile oil compounds list, 7–8
Thujones, 181
Thyme, 11, 53
antioxidant compound extracts, 186
microwave-assisted extraction processing, 171
superheated steam, 42
volatile oil, cost of manufacturing, 65–70
volatile oil, market pricing, 72
volatile oil, steam distillation, 44–45
Thymol, 171
Time
antioxidant GRAS solvent extraction, 187–188
phenolic compound GRAS solvent extraction, 195–196
pigment GRAS solvent extraction, 192–193
thyme steam distillation, 66
Tocols, 18
Tocopherols, 140, 249
countercurrent extraction, 355
olive oil percolation extraction, 165
refining methods, 256–258
separation selectivity, 292
single-stage supercritical steam extraction fraction separation, 295–296
value and retention, 252–253
Toluene, 141
Tomato skins, 290
Toxicity, solvent regulatory classification, 141
Transducers, 178
Trans-isomers, steam deacidification, 19, 26–30
Trans-2-hexenal, 164
Tray columns
balance equations, 84, 85
cachaça batch continuous distillation, 109–117
distillation, 80–82, 86, 222
Triacylglycerols (TAG), 18, 23, 220–221
Trichloroacetic acid, 186
Trichomes, 15
Triglycerides, 13–14
Trihalomethanes (THM), 424
Tropical juices, 117
Turbidity, 30–31
Turmeric oil, 2, 42
2-acetyl-4(5)-(1,2,3,4)-tetrahydroxybutyl-imidazole, 375
2-methylbutanoic acid, cashew distillate flow profiles, 125–126, 127
2-methylcardols, 328
2-propanol, 141

U

Ultrasonic intensity (UI), 156
Ultrasound-assisted extraction, 138, 140, 154–158, 171–185, 192, 314
Ultrasound devices, 177–181
Ultrasound probe systems, 178–180, 181
UNIFAC (UNIQUAC functional-group activity coefficient) model, 234, 285
Universal quasi-chemical (UNIQUAC) model, 90
binary interaction parameters, model components, 238–239

liquid–liquid mass balance equations, 234–336
liquid–liquid vegetable oil extraction group contribution, 237
Utilities, 390–391
costs estimation, 203–205
steam distillation costs, 52

V

Vacuum operations, 11
Vacuums, 16
Valerian, 181
Vaporization
efficiency steam distillation vapor–liquid equilibrium, 21–22
liquid distillation separation, 76
oil acidity, 25
plant milling process, 39
steam distillation process, 15–16
utilities cost estimates for solvent extraction, 204
vegetable oil purification, 19
Vaporization rate
alembic distillation simulation, 82–83
cachaça batch distillation, 104
steam stripping processes, 21
Vapor–liquid contact distillation, 76
Vapor–liquid equilibrium, 16
cashew nut shell liquid separation process fractionation, 344–347
distillation processes and heat, 86–97
juice aroma/flavor distillation capture, 117–118
orange peel oil countercurrent extraction, 359
recirculating static apparatus measurement, 342–344
steam distillation vaporization efficiency, 21–22
wax decomposition, 31
Vapor phase
cachaça distillation, 116–117
density and molecular interactions, 89–90
fugacity, 87
separation process, 339
supercritical extraction sampling, 344
Vegetable materials
antioxidant extraction, 299–300
continuously stirred batch extraction, 197
solvent extraction economics, 198
Vegetable matrices
anthocyanin GRAS solvent extraction, 190
ultrasound-assisted extraction, 180
volatile oil steam distillation, 36
Vegetable oils
deacidification, 1, 246–247
deacidification by stripping, 18–32
fatty acids stripping, 103
liquid–liquid extraction, 219–220, 220
liquid stream extraction technologies, 291
solid–liquid extraction, 138
stripping and deacidification, 10
Vegetal biomass
hydrothermal treatments, 307–308
lignocellulosic material hot water extraction technologies, 312–315
Vegetal compounds, 288
Viral equations, 90
Vitamin A, 253
Vitamin E, 251–252, 430
Vitamins, 18, 430
Void–particle interface, 273–276
Volatile compounds
cashew distillate flow profiles, 125–126
cashew fruit juice, 118–119
coffee aroma, 371–372
supercritical extraction from coffee matrices, 375–376
Volatile liquid mixture distillation, 75
Volatile oils (VO), 1, 2–4
bioactive compounds, 2
cost of manufacture estimates, 50
manufacturing costs/market prices, 70–72
phase equilibrium, 3–4
steam distillation, 10–11, 14, 35–45, 39–40
Volatile terpenoids, 2
Volatility values
distillation separation, 89
ethanol concentration, 93–94
wine alcoholic components, 96–97
Volume, microwave-assisted extraction, 153

W

Waste treatment
adsorption processes, 422
adsorption techniques, 405
Waste treatment costs (C_{WT})
estimation, 206
estimation, cost of manufacturing, 392
steam distillation, 52
Wastewater
steam distillation hydrosol, 37
treatment, 156
Water
anthocyanin GRAS solvent extraction, 190, 191
antioxidant compound extraction, 187
carotenoid extraction, 191
cost estimates for solvent extraction, 204–205

ethanol vegetable oil deacidification, 249
liquid–liquid extraction solvent selection, 244
steam distillation hydrosol, 37
supercritical carbon dioxide extraction cosolvent, 304–305
Water adsorption isotherms, 412–413
Water distillation, 11
microwave-assisted extraction, 169
rosemary, 42
Water–ethanol mixtures, phenolic compounds, 194, 195
Water-extracted products, 306, 314
Water extraction, high-pressure, 306
Waxes, 18, 30–32
Wheat bran, 187, 188
Whisky, 103
Willow wood, 312
Wilson equations, 90, 91
Wine, 104
color/taste correction, 426
component/concentration ranges, 88
distillation, 94–95, 106

X

Xylans, 306–308
Xylose, 308–309

Y

Yellow bell papers, 139–140
York–Scheible columns, 223

Z

Zeolites
adsorbents, 415
coffee volatile compounds, 373
fructose moving bed adsorption, 421